A Textbook of

Engineering Mathematics

Empowering Budding Engineers with Sound Mathematical Skills

"We want that education by which character is formed, strength of mind is increased, the intellect is expanded, and by which one can stand on one's own feet..."

. . . Swami Vivekananda

"The meaning of education is to encourage in people a sense of self-introspection, the ability to grasp and to develop the capacity for constructive works..."

. . . Dr. S. Radhakrishnan

"Dream, Dream, Dream,
Dream translates into thoughts, and
Thoughts result into actions"

. . . Dr. A. P. J. Kalam

"Always surround yourself with the kind of people who can make you believe that you can do it..."

. . . R. S. Salaria

"Treat others the way you wish to be treated..."

. . . R. S. Salaria

A Textbook of Engineering Mathematics

Empowering Budding Engineers with Sound Mathematical Skills

Revised & Enlarged Edition

for

B. Tech. (All Branches), Semester II

A Textbook Designed Strictly According to the Syllabi of

Punjab Technical University

R.S. Salaria

Rayat-Bahra Institute of Engineering &

Nano-Technology, Hoshiarpur.

Formerly with

Thapar University, Patiala.

Amanpreet Singh

SGTB Khalsa College, Anandpur sahib.

Formerly with

Rayat-Bahra Institute of Engineering &

Nano-Technology, Hoshiarpur.

Assisted by

Anita Salaria

KHANNA BOOK PUBLISHING CO. (P) LTD.

Publisher of Science, Technology and Engineering Books

4C/4344, Ansari Road, Darya Ganj, New Delhi-110002

Phone: 011-23244447-48 **Mobile:** +91-99109 09320

E-mail: contact@khannabooks.com

Website: www.khannabooks.com

Price : ₹ 350/-

A Textbook of Engineering Mathematics
R.S. Salaria, Amanpreet Singh

ISBN: 978-93-82609-91-9

Edition: 2016

Published by:

KHANNA BOOK PUBLISHING CO. (P) LTD.
4C/4344, Ansari Road, Darya Ganj, New Delhi-110 002
Phone: 011-23244447-48 **Mobile:** +91-9910909320
E-mail: contact@khannabooks.com

Laser Typeset by:
Book One Graphics, New Delhi

Printed in India by:
India Book Printers & Binders, Delhi

Dedication

Our Dearest Parents,
Daughters Surbhi & Mehak,
Beloved Readers,

AND

Our Country Men/Women at Large.

My Slogan for My Fellow Citizens

India . . .

. . . is a Great Country

AND

Indians . . .

. . . are Outstanding.

Preface to the Revised Edition

Continuing with the tradition of providing the rich content on the subject, the present edition lays more emphasis on the overall layout of the subject matter so as to make it more appealing and attractive. We are certain that the present edition will help the students in inculcating the habit of reading, self-learning and problem solving.

Most of the topics are thoroughly revised, presented in a kind of manner so that the students can grasp them with quite each.

Number of illustrative examples is increased so as to cover all the aspects of topics and kind of problems which can be asked in university examination

Plenty of questions are added in almost all the exercises that will provide an intellectual challenge to the students, and the students are advised to convert this challenge to an opportunity and at the end of the day, you will be a very satisfied person.

Recent university question papers, after the revision of the syllabus, and their solutions are incorporated, which will help the students to know the university pattern of examinations and how the solutions need to be presented.

We are hopeful that this edition will be going beyond your expectations. Do keep on writing, as your valuable feedback and suggestions will help us in tuning the book to your expectations.

. . . R. S. Salaria

. . . Amanpreet Singh

Preface to the First Edition

Dear readers, we take pride in sharing with you our passion about the lovely and beautiful subject of Mathematics which we love from the bottom of our heart. At the same time we would like share our concern about the state of teaching and learning of Mathematics. A lot has to be done on both the fronts, and we invite you all to join our mission.

The book in your hand is a first step of our commitment to uplift the standard of teaching and learning of Mathematics. Since we started our journey in writing the book little late, you may find few gaps and omissions in the book, but we assure you that in our continuous journey in this endeavour and with your whole hearted support, we hope to make significant contribution in the teaching and learning of Mathematics. We hope to see that day when we hear from our students that Mathematics is a wonderful subject and you all are relishing it.

This book is designed to serve as a textbook for the students of B. Tech. (All Branches) Sem-II, of Punjab Technical University, Jalandhar.

While writing the book, special emphasis was given to keep the language very simple and lucid; level of presentation is kept simple and illustrative so that even an average reader can grasp the subject matter with quite ease.

We have made best efforts to make the book increasingly useful for the students and fellow teachers. Efforts have also been put to keep the errors to minimum but still if certain errors creep in, we would like to apologize to our readers.

We are certain that you will enjoy reading this book. We do need a little favour from your side in the form of your response about the book. Your free and frank comments, suggestions, and positive criticism will help us to improve the book further to meet your expectations.

Hope this book will be a valuable source for conceptual clarity of the subject and will also help the students to achieve handsome score in the university examination.

. . . R. S. Salaria

Email: rssalaria@yahoo.com

. . . Amanpreet Singh

Email: aman.rbient@yahoo.com

Acknowledgments

It is our pleasure to take this opportunity to sincerely thank all the learned teachers who have adopted our book and introduced it to their students. We are also thankful for their appreciations that we have been receiving from them in the form of emails and telephone calls.

We would be failing in our duty if we go without appreciating the response and interest shown by our beloved students in our book. They are always a source of inspiration for us and we feel honoured because of their faith in us.

We would like to place on record, the significant contributions to the revised edition by

1. Ms. Sonia Bansal, Rayat-Bahra Institute of Engineering & Nano-Technology, Hoshiarpur.
2. Ms. Amanjot Kaur, Rayat-Bahra Institute of Engineering & Nano-Technology, Hoshiarpur.
3. Ms. Sarabjit Kaur Chawla, Sant Baba Bhag Singh University (Formerly Sant Baba Bhag Singh Institute of Engineering & Technology), Khiala, Jalanddhar.
4. Mr. Hardeep Singh, IITT, Pojewal.

We thank all these wonderful people, from the core of our heart, for sparing precious time from their busy schedule, and reviewing the entire subject matter of the book and providing valuable suggestions.

We would also like to acknowledge valuable inputs from our brilliant former student, Er. Jaipuneet Singh, who completed his BTech (CSE) from Rayat-Bahra Institute of Engineering & Technology, Hoshiarpur.

Finally, we thank Mr. Puneet Khanna and Mr. Mukul Seth, Khanna Book Publishing Co (P) Ltd., for their aggressive marketing for establishing the book as one of the best textbooks on the subject of "Engineering Mathematics".

. . . R. S. Salaria

. . . Amanpreet Singh

Word of Advice for Students

Our beloved students, we would like to remind you that we can categorize people as

People who watch the things happening,

People who look into how things are happening, and

People who make the things happen.

Now you are to find out where you fit-in and where you want to be. We firmly believe that youth of our country is very talented, and they can do wonders if guided and handled the right way.

Take some time to discover yourself and try to assess how much talent you possess. Utilize your period of study to harness that talent. We can assure you that you will have a great future.

As rightly said, we are living in an era where we have guided missiles but unguided youth.

We would like to take this opportunity to guide our budding engineers the right way to handle the subject of Mathematics. Read the next few lines very carefully and just follow them, we assure you of results that will be amazing.

"Mathematics is mother of all sciences."

"Mathematics is a science of sciences and art of arts."

These quotes are sufficient to highlight the importance of the subject of Mathematics.

When we say science, it means there is proper methodology that needs to be comprehensively understood. When we say art, it means skill that needs to be practiced. As rightly said – "Practice makes a man perfect". Therefore, once the concepts are well understood and rigorously practiced, surely you will excel.

Since you have joined B. Tech. programme and therefore, after a minimum of four years time, you will be qualified to be called an engineer. But we would like to tell you that if you really wish to be known as a great engineer, it will not be possible without strong foundation in Mathematics. Just look at the history, you will find that all great engineers, past and present, had strong Mathematical skills.

Therefore, our sincere advice to you all is that involve yourself in the learning of Mathematics and with no time you will find that you are enjoying and just having fun with Mathematics. That day you will feel that Mathematics is one of the easiest subjects and has applications in all fields of study as well in our daily life.

The final word, you may set any goal for yourself and that may be of becoming a successful engineer, an administrator or a businessman, but you must also try to be a nice human being. You know a nice human being is like a fruit of all seasons, and is welcome every where, anytime. We believe, the basic mantra to become a nice human being is too simple and can be stated as

"Treat the others the way you wish to be treated."

Enjoy your student life and make best use of it . . .

Wish you good luck and a nice time ahead.

. . . R. S. Salaria

. . . Amanpreet Singh

Syllabus

BTAM-102: Engineering Mathematics-II

PART-A

1. Ordinary Differential Equations of First Order: Exact Differential equations, Equations reducible to exact form by integrating factors; Equations of the first order and higher degree. Clairaut's equation. Leibniz's linear and Bernoulli's equation.
2. Linear Ordinary Differential Equations of Second & Higher Order: Solution of linear Ordinary Differential Equations of second and higher order; methods of finding complementary functions and particular integrals. Special methods for finding particular integrals: Method of variation of parameters, Operator method. Cauchy's homogeneous and Legendre's linear equation, Simultaneous linear equations with constant coefficients.
3. Applications of Ordinary Differential Equations: Applications to electric R-L-C circuits, Deflection of beams, Simple harmonic motion, Simple population model.

PART-B

4. Linear Algebra: Rank of a matrix, Elementary transformations, Linear independence and dependence of vectors, Gauss-Jordan method to find inverse of a matrix, reduction to normal form, Consistency and solution of linear algebraic equations, Linear transformations, Orthogonal transformations, Eigen values, Eigen vectors, Cayley-Hamilton Theorem, Reduction to diagonal form, orthogonal, unitary, Hermitian and similar matrices.
5. Infinite Series: Convergence and divergence of series, Tests of convergence (without proofs): Comparison test, Integral test, Ratio test, Rabee's test, Logarithmic test, Cauchy's root test and Gauss test. Convergence and absolute convergence of alternating series.
6. Complex Numbers and Elementary Functions of Complex Variable: De-Moivre's theorem and its applications. Real and Imaginary parts of exponential, logarithmic, circular, inverse circular, hyperbolic, inverse hyperbolic functions of complex variables. Summation of trigonometric.

Contents at a Glance

Chapter 0

Useful Formulae, Theorems and Results

Learning Objectives

After reading this chapter, students will be able to

- grasp various formulae, theorems and results
- make use of them to solve various problems

TRIGONOMETRIC RATIOS

$\sin(-x) = -\sin x$ $\quad\cos(-x) = \cos x$ $\quad\sin\left(\frac{\pi}{2} - x\right) = \cos x$

$\cos\left(\frac{\pi}{2} - x\right) = \sin x$ $\quad\sin\left(\frac{\pi}{2} + x\right) = \cos x$ $\quad\cos\left(\frac{\pi}{2} + x\right) = -\sin x$

$\sin(\pi - x) = \sin x$ $\quad\cos(\pi - x) = -\cos x$ $\quad\sin(\pi + x) = -\sin x$

$\cos(\pi + x) = -\cos x$ $\quad\sin(2\pi - x) = -\sin x$ $\quad\cos(2\pi - x) = \cos x$

$\sin(2n\pi + x) = \sin x$ $\quad\cos(2n\pi + x) = \cos x$

TRIGONOMETRIC IDENTITIES

$\cos^2 x + \sin^2 x = 1$ $\quad\sec^2 x = 1 + \tan^2 x$

$\operatorname{cosec}^2 x = 1 + \cot^2 x$ $\quad\sin(x + y) = \sin x \cos y + \cos x \sin y$

$\sin(x - y) = \sin x \cos y - \cos x \sin y$ $\quad\cos(x + y) = \cos x \cos y - \sin x \sin y$

$\cos(x - y) = \cos x \cos y + \sin x \sin y$

$$\tan(x + y) = \frac{\tan x + \tan y}{1 - \tan x \tan y} \qquad \tan(x - y) = \frac{\tan x - \tan y}{1 + \tan x \tan y}$$

$$\cot(x + y) = \frac{\cot x \cot y - 1}{\cot y + \cot x} \qquad \cot(x - y) = \frac{\cot x \cot y + 1}{\cot y - \cot x}$$

$$\cos 2x = \cos^2 x - \sin^2 x = 2\cos^2 x - 1 = 1 - 2\sin^2 x - 1 = \frac{1 - \tan^2 x}{1 + \tan^2 x}$$

$$\sin 2x = 2\sin x \cos x = \frac{2\tan x}{1 + \tan^2 x} \qquad \tan 2x = \frac{2\tan x}{1 - \tan^2 x}$$

$$\sin 3x = 3\sin x - 4\sin^3 x \qquad \cos 3x = 4\cos^3 x - 3\cos x$$

$$\tan 3x = \frac{3\tan x - \tan^3 x}{1 - 3\tan^2 x}$$

$$\cos x + \cos y = 2\cos\frac{x + y}{2}\cos\frac{x - y}{2} \qquad \cos x - \cos y = -2\sin\frac{x + y}{2}\sin\frac{x - y}{2}$$

$$\sin x + \sin y = 2\sin\frac{x + y}{2}\cos\frac{x - y}{2} \qquad \sin x - \sin y = 2\cos\frac{x + y}{2}\sin\frac{x - y}{2}$$

$2\cos x \cos y = \cos(x + y) + \cos(x - y)$ $\quad 2\sin x \sin y = -\cos(x + y) + \cos(x - y)$

$2\sin x \cos y = \sin(x + y) + \sin(x - y)$ $\quad 2\cos x \sin y = \sin(x + y) - \sin(x - y)$

HYPERBOLIC FUNCTIONS

$\sinh x=\dfrac{e^x-e^{-x}}{2}$ $\cosh x=\dfrac{e^x+e^{-x}}{2}$ $\tanh x=\dfrac{e^x-e^{-x}}{e^x+e^{-x}}$

$\coth x=\dfrac{e^x+e^{-x}}{e^x-e^{-x}}$ $\text{sech}\, x=\dfrac{2}{e^x+e^{-x}}$ $\text{cosech}\, x=\dfrac{2}{e^x-e^{-x}}$

RELATIONS BETWEEN HYPERBOLIC AND CIRCULAR FUNCTIONS

$\cos(i\theta)=\cosh\theta$

$\cot(i\theta)=\dfrac{\cos(i\theta)}{\sin(i\theta)}=\dfrac{\cosh\theta}{i\sinh\theta}=-i\coth\theta$

$\sin(i\theta)=i\sinh\theta$

$\sec(i\theta)=\dfrac{1}{\cos(i\theta)}=\dfrac{1}{\cosh\theta}=\text{sech}\,\theta$

$\tan(i\theta)=\dfrac{\sin(i\theta)}{\cos(i\theta)}=\dfrac{i\sinh\theta}{\cosh\theta}=i\tanh\theta$

$\text{cosec}(i\theta)=\dfrac{1}{\sin(i\theta)}=\dfrac{1}{i\sinh}=-i\,\text{cosech}\,\theta$

$\left[\because \dfrac{1}{i}=-i\right]$

HYPERBOLIC IDENTITIES

$\cosh^2 x-\sinh^2 x=1$

$\text{sech}^2 x+\tanh^2 x=1$

$\coth^2 x-\text{cosec}\,\text{h}^2 x=1$

$\sinh 2x=2\sinh x\cosh x=\dfrac{2\tanh x}{1-\tanh^2 x}$

$\cosh 2x=\cosh^2 x+\sinh^2 x=2\cosh^2 x-1=1+\sinh^2 x=\dfrac{1+\tanh^2 x}{1-\tanh^2 x}$

$\sinh 3x=3\sinh x-4\sinh^3 x$

$\cosh 3x=4\cosh^3 x-3\cosh x$

$\tanh 3x=\dfrac{3\tanh x+\tanh^3 x}{1-3\tanh^2 x}$

DEFINITION OF A DERIVATIVE

If $y=f(x)$, the derivative of y or $f(x)$ with respect to x is defined as

$$\frac{dy}{dx}=\underset{h\to 0}{\text{Lt}}\frac{f(x+h)-f(x)}{h}=\underset{\Delta x\to 0}{\text{Lt}}\frac{f(x+\Delta x)-f(x)}{\Delta x}$$

where $h=\Delta x$. The derivative is also denoted by y', $\dfrac{df}{dx}$, $f'(x)$. The process of taking (finding) a derivative is called *differentiation*.

GENERAL RULES OF DIFFERENTIATION

In the following u, v, w are functions of x; a, b, c are constants.

$$\frac{d}{dx}(c)=0$$

$$\frac{d}{dx}(x)=1$$

$$\frac{d}{dx}(x^n)=nx^{n-1}$$

$$\frac{d}{dx}(u\pm v\pm w\pm\ldots)=\frac{du}{dx}\pm\frac{dv}{dx}\pm\frac{dw}{dx}\pm\ldots$$

$$\frac{d}{dx}(cu)=c\frac{du}{dx}$$

$$\frac{d}{dx}(uv)=u\frac{dv}{dx}+v\frac{du}{dx}$$

$$\frac{d}{dx}(uvw)=uv\frac{dw}{dx}+uw\frac{dv}{dx}+vw\frac{du}{dx}$$

$$\frac{d}{dx}\left(\frac{u}{v}\right)=\frac{v\frac{du}{dx}-u\frac{dv}{dx}}{v^2}$$

$$\frac{dy}{dx}=\frac{dy}{du}\cdot\frac{du}{dx} \quad \text{(Chain Rule)}$$

$$\frac{d}{dx}(u^n)=nu^{n-1}\frac{du}{dx}$$

DERIVATIVES OF TRIGNOMETIC AND INVERSE TRIGNOMETRIC FUNCTIONS

$$\frac{d}{dx}(\sin x)=\cos x$$

$$\frac{d}{dx}(\cos x)=-\sin x$$

$$\frac{d}{dx}(\tan x)=\sec^2 x$$

$$\frac{d}{dx}(\cot x)=-\operatorname{cosec}^2 x$$

$$\frac{d}{dx}(\sec x)=\sec x\tan x$$

$$\frac{d}{dx}(\operatorname{cosec} x)=-\operatorname{cosec} x\cot x$$

$$\frac{d}{dx}(\sin^{-1}x)=\frac{1}{\sqrt{1-x^2}}$$

$$\frac{d}{dx}(\cos^{-1}x)=-\frac{1}{\sqrt{1-x^2}}$$

$$\frac{d}{dx}(\tan^{-1}x)=\frac{1}{1+x^2}$$

$$\frac{d}{dx}(\cot^{-1}x)=-\frac{1}{1+x^2}$$

$$\frac{d}{dx}(\sec^{-1}x)=\frac{1}{x\sqrt{x^2-1}}$$

$$\frac{d}{dx}(\operatorname{cosec}^{-1}x)=-\frac{1}{x\sqrt{x^2-1}}$$

DERIVATIVES OF EXPONENTIAL AND LOGARITHMIC FUNCTIONS

$$\frac{d}{dx}(e^x)=e^x$$

$$\frac{d}{dx}(\log x)=\frac{1}{x}$$

$$\frac{d}{dx}(a^x)=a^x\log a$$

DERIVATIVES OF HYPERBOLIC AND INVERSE HYPERBOLIC FUNCTIONS

$$\frac{d}{dx}(\sinh x)=\cosh x$$

$$\frac{d}{dx}(\cosh x)=\sinh x$$

$$\frac{d}{dx}(\tanh x)=\operatorname{sech}^2 x$$

$$\frac{d}{dx}(\coth x)=-\operatorname{cosech}^2 x$$

$$\frac{d}{dx}(\operatorname{sech} x)=-\sec x\tan x$$

$$\frac{d}{dx}(\operatorname{cosech} x)=-\operatorname{cosech} x\coth x$$

$$\frac{d}{dx}(\sinh^{-1}x)=\frac{1}{\sqrt{x^2+1}}$$

$$\frac{d}{dx}(\cosh^{-1}x)=\frac{1}{\sqrt{x^2-1}}$$

$$\frac{d}{dx}(\tanh^{-1}x)=\frac{1}{1-x^2}$$

$$\frac{d}{dx}(\coth^{-1}x) = -\frac{1}{1-x^2} \qquad \frac{d}{dx}(\text{sech}^{-1}x) = -\frac{1}{x\sqrt{1-x^2}} \qquad \frac{d}{dx}(\text{cosech}^{-1}x) = -\frac{1}{x\sqrt{1+x^2}}$$

DIFFERENTIALS

Let $y = f(x)$ and $\Delta y = f(x+\Delta x) - f(x)$, then

$$\frac{\Delta y}{\Delta x} = \frac{f(x) - f(x+\Delta x)}{\Delta x} = f'(x) + \varepsilon = \frac{dy}{dx} + \varepsilon$$

where $\varepsilon \to 0$ as $\Delta x \to 0$, thus $\Delta y = f'(x)\Delta x + \varepsilon \Delta x$

If we call $\Delta x = dx$, the differential of x, then we define the differential of y to be

$$dy = f'(x)dx$$

RULES FOR DIFFERENTIALS

Rules for differential are exactly analogous to those for derivatives. As examples, we observe that

$$d(u \pm v \pm w \pm ...) = du \pm dv \pm dw \pm ... \qquad d(uv) = udv + vdu \qquad d\left(\frac{u}{v}\right) = \frac{vdu - udv}{v^2}$$

$$d(u^n) = nu^{n-1}du \qquad d(\sin u) = \cos u du \qquad d(\cos u) = -\sin u du$$

DEFINITION OF AN INDEFINITE INTEGRAL

If $\frac{dy}{dx} = f(x)$, then y is the function whose derivative is $f(x)$ and is called the anti-derivative of $f(x)$ or the *indefinite integral* of $f(x)$, denoted by $\int f(x)dx$. Similarly, $y = \int f(u)du$, then if $\frac{dy}{du} = f(u)$. Since, the derivative of a constant is zero, all indefinite integrals differ by an arbitrary constant. The process of taking (finding) an integral is called *integration*.

GENERAL RULES FOR INTEGRATION

In the following *u*, *v*, *w* are functions of *x*; *a*, *b*, *c* are constants.

$$\int c dx = cx \qquad \int (u \pm v \pm w \pm ...)dx = \int udx \pm \int vdx \pm \int wdx \pm ...$$

$$\int udv = uv - \int vdu \quad \text{[Integration by parts]} \qquad \int f(ax)dx = \frac{1}{a}\int f(u)du$$

$$\int x^n dx = \frac{x^{n+1}}{n+1}$$

INTEGRALS OF EXPONENTIAL FUNCTIONS

$\int \frac{1}{x} dx = \log x$

$\int e^x dx = e^x$

$\int a^x dx = \frac{a^x}{\log a}$

INTEGRALS OF TRIGONOMETIC FUNCTIONS

$\int \cos x dx = \sin x + c$

$\int \sin x dx = -\cos x + c$

$\int \tan x dx = \log|\sec x| + c$

$\int \cot x dx = \log|\sin x| + c$

$\int \sec x dx = \log|\sec x + \tan x| + c$

$\int \operatorname{cosec} x dx = \log|\operatorname{cosec} x - \cot x| + c$

$\int \sec^2 x dx = \tan x + c$

$\int \operatorname{cosec}^2 x dx = -\cot x + c$

$\int \sec x \tan x dx = \sec x + c$

$\int \operatorname{cosec} x \cot x dx = -\operatorname{cosec} x + c$

INTEGRALS OF HYPERBOLIC FUNCTIONS

$\int \cosh x dx = \sinh x$

$\int \sinh x dx = \cosh x$

$\int \tanh x dx = \log \cosh x$

$\int \coth x dx = \log \sinh x$

$\int \operatorname{sech} x dx = \sin^{-1}(\tanh x)$

$\int \operatorname{cosech} x dx = \log \tanh \frac{x}{2}$ or $-\coth^{-1} e^x$

$\int \operatorname{sec} h^2 x dx = \tanh x$

$\int \operatorname{cosech}^2 x dx = -\coth u$

$\int \tanh^2 x dx = x - \tanh x$

$\int \coth^2 x dx = x - \coth x$

INTEGRALS OF OTHER IMPORTANT FUNCTIONS

$$\int \frac{dx}{x^2 - a^2} = \frac{1}{2a} \log\left|\frac{x-a}{x+a}\right|$$

$$\int \frac{dx}{a^2 - x^2} = \frac{1}{2a} \log\left|\frac{a+x}{a-x}\right|$$

$$\int \frac{dx}{\sqrt{x^2 - a^2}} = \log\left|x + \sqrt{x^2 - a^2}\right|$$

$$\int \frac{dx}{\sqrt{x^2 + a^2}} = \log\left|x + \sqrt{x^2 + a^2}\right|$$

$$\int \frac{dx}{x^2 + a^2} = \frac{1}{a} \tan^{-1} \frac{x}{a}$$

$$\int \frac{dx}{\sqrt{a^2 - x^2}} = \sin^{-1} \frac{x}{a}$$

$$\int \sqrt{x^2 - a^2}\, dx = \frac{x\sqrt{x^2 - a^2}}{2} - \frac{a^2}{2}\left(\log\left|x + \sqrt{x^2 - a^2}\right|\right)$$

$$\int \sqrt{a^2 - x^2}\, dx = \frac{x\sqrt{a^2 - x^2}}{2} + \frac{a^2}{2} \sin^{-1} \frac{x}{a}$$

$$\int \sqrt{x^2 + a^2}\, dx = \frac{x\sqrt{x^2 + a^2}}{2} + \frac{a^2}{2}\left(\log\left|x + \sqrt{x^2 + a^2}\right|\right)$$

$$\int [f(x)]^n f'(x)\, dx = \frac{[f(x)]^{n+1}}{n+1}$$

$$\int \frac{f'(x)}{f(x)} dx = \log f(x)$$

DEFINITION OF A DEFINITE INTEGRAL

Let $f(x)$ be defined in interval $a \le x \le b$. Divide the intervals into n equal parts of length Δx as $\frac{b-a}{n}$. Then the definite integral of $f(x)$ between $x = a$ and $x = b$ is defined as

$$\int_a^b f(x)dx = \underset{n\to\infty}{\text{Lt}} \left[f(a)\Delta x + f(a+\Delta x)\Delta x + f(a+2\Delta x)\Delta x + ... + f(a+(n-1)\Delta x)\right]$$

The limit will certainly exists if $f(x)$ is piecewise continuous.

If $f(x) = \frac{d}{dx}g(x)$, then by fundamental theorem of integral calculus, the above definite integrals can be evaluated by using the result

$$\int_a^b f(x)dx = \int_a^b \frac{d}{dx}g(x)dx = [g(x)]_a^b = g(b) - g(a)$$

GENERAL FORMULAS INVOLVING DEFINITE INTEGRALS

$$\int_a^b [f(x) \pm g(x) \pm h(x) \pm ...]dx = \int_a^b f(x)dx + \int_a^b g(x)dx + \int_a^b h(x)dx + ...$$

$$\int_a^b f(x)dx = -\int_b^a f(x)dx \qquad \int_a^b f(x)dx = \int_a^c f(x)dx + \int_c^b f(x)dx$$

$$\int_0^a f(x)dx = \int_0^a f(a-x)dx \quad \text{[trigonometrical property]}$$

$$\int_{-a}^a f(x)dx = \begin{cases} 0 & \text{if } f(x) \text{ is odd function} \\ 2\int_0^a f(x)dx & \text{if } f(x) \text{ is even function} \end{cases}$$

$\int_a^b f(x)dx = (b-a)f(c)$ where c is between a and b. This is called *mean value theorem* for definite integrals and is valid if $f(x)$ is continuous in $a \le x \le b$.

$$\int_c^{c+2\pi} \sin nx\, dx = \left[\frac{-\cos nx}{n}\right]_c^{c+2\pi} = 0,\ n \ne 0; \qquad \int_c^{c+2\pi} \cos nx\, dx = \left[\frac{\sin nx}{n}\right]_c^{c+2\pi} = 0,\ n \ne 0$$

$$\int_c^{c+2\pi} \sin mx \cos nx\, dx = 0,\ m \ne n \qquad \int_c^{c+2\pi} \cos mx \cos nx\, dx = 0,\ m \ne n$$

$$\int_c^{c+2\pi} \sin mx \sin nx\, dx = 0,\ m \ne n$$

$$\int_c^{c+2\pi} \cos^2 nx\, dx = \pi,\ n \neq 0; \quad \int_c^{c+2\pi} \sin^2 nx\, dx = \pi,\ n \neq 0$$

$$\int_c^{c+2\pi} \sin nx \cos nx\, dx = \frac{1}{2}\int_c^{c+2\pi} \sin 2nx\, dx = 0,\ n \neq 0$$

$$\int e^{ax} \sin bx\, dx = \frac{e^{ax}}{a^2+b^2}(a \sin bx - b\cos bx) + c$$

$$\int e^{ax} \cos bx\, dx = \frac{e^{ax}}{a^2+b^2}(a \cos bx + b\sin bx) + c$$

$\int uv dx = uv_1 - u'v_2 + u''v_3 - u'''u_4 + \ldots,$ where u and v are functions of x, i.e.,

Integral of the product of two functions
= 1st function × integral of 2nd — go on differentiating 1st, integrating 2nd, signs alternately +ve and –ve.

[**Hint:** Simplification should be done only when the integration is over].

For example, $\int x^2 \cos nx\, dx = x^2\left(\frac{\sin nx}{n}\right) - 2x\left(\frac{-\cos nx}{n^2}\right) + 2\left(\frac{-\sin nx}{n^3}\right)$

$$= \frac{x^2}{n}\sin nx + \frac{2x}{n^2}\cos nx - \frac{2}{n^3}\sin nx$$

BASIC DIFFERENTIAL EQUATIONS AND SOLUTIONS

SNo.	Differential Equation	Solution
1.	Separation of Variables $f(x_1)g(y_1)dx + f(x_2)g(y_2)dy = 0$	$\int \frac{f(x_1)}{f(x_2)}dx + \int \frac{g(y_2)}{g(y_1)}dy = c$
2.	Exact Equation $M(x, y)dx + N(x, y)dy = 0$	$\int_y Mdx + \int(\text{terms of } N \text{ not containing } x)dy = c$
3.	Clairaut's equation $y = px + f(p)$	$p = cx + f(c)$
4.	Leibnitz's Equation $\frac{dy}{dx} + P(x)y = Q(x)$	$ye^{\int Pdx} = \int Qe^{\int Pdx} + c$
5.	Legendre's Equation $(1-x^2)\frac{d^2y}{dx^2} - 2x\frac{dy}{dx} + n(n+1)y = 0$	$y = c_1P_n(x) + c_2Q_n(x)$

LINEAR ALGEBRA

Matrix: A set of $m \cdot n$ numbers (real or complex) arranged in a rectangular form having m rows (horizontal lines) and n columns (vertical lines), the numbers being enclosed by bracket [] or (), is called an $m \times n$ matrix (read as m by n). Elements of the matrix are denoted by a_{ij} where i represent the row and j represent the column. In terms of its elements, and the matrix is written as $\left[a_{ij}\right]_{(m,n)}$ or $\left[a_{ij}\right]_{m\times n}$.

Rank of Matrix - The number r is called the rank of matrix A if

- every minor of order $(r+1)$ of A is zero.
- there exists at least one minor of order r which is non-zero.

Normal Form of a Matrix - Any non-zero matrix A_{mxn} can be reduced to any one of the following forms by performing elementary (row, column or both) transformations:

(i) I_r (ii) $(I_r, 0)$ (iii) $\begin{pmatrix} I_r \\ 0 \end{pmatrix}$ (iv) $\begin{pmatrix} I_r & 0 \\ 0 & 0 \end{pmatrix}$

All those form are known as *normal forms* of the matrix.

Inverse of a matrix - Let A be any square non-singular matrix, then

$$A = IA$$

Apply suitable E-row transformation to A on the left hand side so that A is reduced to I. Simultaneously, apply the same E-row transformations to I on the right hand side. Let I reduce to B, so that

$$I = BA$$

Post-multiplying by A^{-1}, we get

$$IA^{-1} = BAA^{-1} \quad \Rightarrow \quad B = A^{-1}$$

Consistency and solution of linear algebraic equation

In a matrix form, the system of linear algebraic equations can be written as

$$AX = B$$

which need not always have a solution. It may have no solution or a unique solution or an infinite number of solutions.

- **Consistent System of Equation -** A system of linear equation having one or more solution is called a consistent system of equations.
- **Inconsistent System of Equations -** If a system of equation has no *solution*, it is said to be inconsistent.

Criteria for Consistency

In terms of rank of the coefficient matrix and rank of the augmented matrix, the criteria for *a unique solution, no solution, infinite number of solutions* of a system of linear equations is as follows:

- For a system of non-homogeneous of linear equations $AX = B$:
  - If $\rho[A:B] \neq \rho(A)$ the system is inconsistent.
  - If $\rho[A:B] = \rho(A) =$ number of the unknowns, the system has a unique solution .
  - If $\rho[A:B] = \rho(A) <$ number of unknowns, the system has a infinite number of solutions.

Eigen Values and Eigen Vectors - If A is the square matrix of order n, we can form the matrix $A - \lambda I$, where λ is the scalar and I is the unit matrix of order n, is called the *characteristics equation* of A.

On expanding the determinant $|A - \lambda I|$, and equating it to zero, the characteristics equation can be written as a polynomial equation of degree n in λ of the form

$$(-1)^n \lambda^n + k_1 \lambda^{n-1} + k_2 \lambda^{n-1} + \ldots + k_n = 0$$

The roots of this equation are called the *characteristics roots* or *latent roots* or *eigen value* of A.

Cayley-Hamilton theorem - Every square matrix satisfies its characteristics equation, *i.e.*, if

$$|A - \lambda I| = (-1)^n \lambda^n + p_1 \lambda^{n-1} + p_2 \lambda^{n-2} + \ldots + p_{n-2} \lambda^2 + p_{n-1} \lambda + p_n = 0$$

is the characteristics equation of the nth order square matrix A ,then

$$(-1)^n A^n + p_1 A^{n-1} + p_2 A^{n-2} + \ldots + p_{n-2} A^2 + p_{n-1} A + p_n I = 0$$

Diagonalization of a matrix - If a square matrix A of order n is said to be diagonalizable if there exist an invertible matrix P of order n such that $P^{-1}AP = D$, where D is a diagonal matrix and the elements of D are eigen values of A.

Similar Matrices - Let A and B be the square matrices of the same order. The matrix A is said to be similar to B if their exist an invertible matrix P such that $A = P^{-1}BP$ or $PA = BP$.

Orthogonal transformation - A linear transformation $Y = AX$, where

$$Y = \begin{pmatrix} y_1 \\ y_2 \\ \vdots \\ y_n \end{pmatrix}, A = \begin{pmatrix} a_{11} & a_{12} & \dots & a_{1n} \\ a_{21} & a_{22} & \dots & a_{2n} \\ \vdots & \vdots & \vdots & \vdots \\ a_{n1} & a_{n2} & \dots & a_{nn} \end{pmatrix}, X = \begin{pmatrix} x_1 \\ x_2 \\ \vdots \\ x_n \end{pmatrix}$$

is said to be orthogonal if it transforms $y_1^2 + y_2^2 + \dots + y_n^2$ into $x_1^2 + x_2^2 + \dots + x_n^2$.

INFINITE SERIES

Convergent Sequence – A sequence $\{a_n\}$ is said to be convergent if $\underset{n\to\infty}{\text{Lt}}\, a_n$ is finite.

Divergent Sequence – A sequence $\{a_n\}$ is said to be divergent if $\underset{n\to\infty}{\text{Lt}}\, a_n$ is not finite, i.e., if $\underset{n\to\infty}{\text{Lt}}\, a_n = +\infty$ or $-\infty$

Oscillatory Sequence – If a sequence $\{a_n\}$ neither converges to a finite number nor diverges to $+\infty$ or $-\infty$, it is called an *oscillatory sequence.* An oscillatory sequence can oscillate finitely or infinitely.

Limit of a Sequence – A sequence $\{a_n\}$ is said to approach to the limit l (say), when $n \to \infty$, if for each $\varepsilon > 0$, $\exists$ a positive integer *m* (depending upon ε) such that $|a_n - l| < \varepsilon \quad \forall\, n \geq m$.

Infinite Series – If $\{u_n\}$ is a sequence of real numbers, then the expression $u_1 + u_2 + u_3 + \dots + u_n + \dots$, i.e., the sum of the terms of the sequence, which are infinite in number, is called an *infinite series.* The above series is usually denoted by $\sum_{n=1}^{\infty} u_n$ or simply as Σu_n.

Nature of Geometric Series – A geometric series in *x* is of the form

$$1 + x + x^2 + x^3 + \dots + x^n + \dots$$

(*a*) It converges if $-1 < x < 1$, i.e., $|x| < 1$

(*b*) It diverges if $x \geq 1$

(*c*) It oscillates finitely if $x = -1$

(*d*) It oscillates infinitely if $x < -1$

Art. : Necessary condition for convergence of a positive term series is that if a positive term series Σu_n is convergent, then $\underset{n\to\infty}{\text{Lt}}\, u_n = 0$.

Art. : A positive term series either converges or diverges to $+\infty$.

Art. : The necessary and sufficient condition for the convergence of a positive term series $\sum u_n$ is that the sequence $\{S_n\}$ of its partial sums is bounded above.

Cauchy's General Principle of Convergence of Series: The necessary and sufficient condition for the infinite series $\sum_{n=1}^{\infty} u_n$ to converge is that given $\varepsilon > 0$, however small, $\exists$ a positive integer p such that $|S_{n+p} - S_n| < \varepsilon,\ \forall\ n \geq m;\ m$ and $p \in N$, i.e.,

$$|u_{n+1} + u_{n+2} + ... + u_{n+p}| < \varepsilon \text{ for } n \geq m,\ m \text{ and } p \in N.$$

Art. : If m is a given positive integer, then the two series

$$u_1 + u_2 + u_3 + ... + u_m + u_{m+1} + ... + u_n \text{ and } u_{m+1} + u_{m+2} + u_{m+3} + ... + u_n$$

converge or diverge together.

Comparison Tests for Series of Positive Terms

Test 1: Let $\sum a_n$ and $\sum b_n$ be positive terms series. Let k be a positive constant independent of n and m be a fixed positive integers.

(1) If $a_n \leq kb_n\ \forall\ n \geq m$ and $\sum b_n$ is convergent then $\sum a_n$ is convergent.

(2) If $a_n > kb_n\ \forall\ n \geq m$ and $\sum b_n$ is divergent then $\sum a_n$ is divergent.

Test 2: Let $\sum a_n$ and $\sum b_n$ be two positive terms series.

(1) If $\underset{n\to\infty}{\text{Lt}} \frac{a_n}{b_n} = l$, l being a finite non-zero constant, the $\sum a_n$ and $\sum b_n$ both converges or diverges together.

(2) If $\underset{n\to\infty}{\text{Lt}} \frac{a_n}{b_n} = 0$ and $\sum b_n$ converges, then $\sum a_n$ is also converges.

(3) If $\underset{n\to\infty}{\text{Lt}} \frac{a_n}{b_n} = \infty$ and $\sum b_n$ diverges, then $\sum a_n$ also diverges.

(4) If $\underset{n\to\infty}{\text{Lt}} \frac{a_n}{b_n} = \infty$ and $\sum a_n$ converges, then $\sum b_n$ also converges.

Test 3: Let $a_n > 0$, $b_n > 0$ and $c_n > 0$. Let $\sum b_n$ be convergent and $\sum c_n$ be divergent.

(1) $\sum a_n$ converges if $\overline{\text{lt}}\left(\frac{a_n}{b_n}\right) = \lambda \neq \infty$

(2) $\sum a_n$ converges if $\overline{\text{lt}}\left(\frac{a_n}{c_n}\right) = \mu \neq \infty$

Test 4: Let Σa_n and Σb_n be two positive term series.

(1) If $\frac{a_{n+1}}{a_n} < \frac{b_{n+1}}{b_n}$ $\forall\, n \geq m, m \in N$ and Σb_n is convergent, then Σa_n is also convergent.

(2) If $\frac{a_{n+1}}{a_n} > \frac{b_{n+1}}{b_n}$ $\forall\, n \geq m,\; m \in n$ and Σb_n is divergent, then Σa_n is also divergent.

P–Test for the series of type $\sum \frac{1}{n^p}$: A series of type

$$\sum \frac{1}{n^p} = \frac{1}{1^p} + \frac{1}{2^p} + \frac{1}{3^p} + \ldots + \frac{1}{n^p} + \ldots, \quad p > 0$$

converges if $p > 1$ and divergent if $p \leq 1$.

Cauchy's Integral Test: If for $x \geq 1$, *f(x)* is a non negative monotonic decreasing and integral function of *x* such that $f(n) = u_n$ $\forall\, n \in N$, then the series $\sum_{n=1}^{\infty} u_n$ and the integral $\int_1^{\infty} f(x)dx$, either both converge or both diverge.

D' Alembert's Ratio Test: If Σu_n is a positive term series and $\underset{n\to\infty}{\text{Lt}} \left[\frac{u_n}{u_{n+1}}\right] = l$, then

(*i*) Σu_n is convergent if $l > 1$ (*ii*) Σu_n is divergent if $l < 1$

Raabe's Test: If Σu_n is a series of positive terms and $\underset{n\to\infty}{\text{Lt}} \left(n\left(\frac{u_n}{u_{n+1}} - 1 \right) \right) = l$, then the series is convergent if $l > 1$ and divergent if $l < 1$.

Logarithmic Test: A positive term series Σu_n converges or diverges according as

$$\underset{n\to\infty}{\text{Lt}} \left(n \log \frac{u_n}{u_{n+1}} \right) > 1 \text{ or } < 1$$

Gauss Test: If $a_n > 0$ and $\frac{a_n}{a_{n+1}} = 1 + \frac{\mu}{n} + O\left(\frac{1}{n^2}\right)$, then Σa_n converges for $\mu > 1$ and diverges for $\mu \leq 1$.

Cauchy's Root Test: If $\sum u_n$ is a real series of positive terms and $\underset{n\to\infty}{\text{Lt}} (u_n)^{1/n} = r$, then the series $\sum u_n$ is convergent if $r < 1$ and divergent if $r > 1$.

Alternating Series: A series whose terms are alternatively positive and negative is called an *alternating series.*

Leibinitz (or Alternating series) Test: If $\{a_n\}$ is monotone decreasing sequence of positive terms, i.e., $a_{n+1} > a_n$, and converges to zero, i.e., $\underset{n\to\infty}{\text{Lt}}\, a_n = 0$, then $\sum_{n=1}^{\infty}(-1)^{n-1}\, a_n$ is convergent.

Power Series: A *power series* is a series of the form

$$\sum_{n=0}^{\infty} a_n x^n = a_0 + a_1 x + a_2 x^2 + \ldots + a_n x^n + \ldots$$

where a's are constants and x is a variable. We can also denote power series by $\sum a_n x^n$. This is called *power series* about $x = 0$.

Uniform Convergence: A sequence $\{(u_n(x))\}$, $n \in N$, $\forall n \in I$ is said to converge uniformly to $u(x)$ in I if for every $\varepsilon > 0$, however small $\exists$ a positive integer m (independent of $x\ \forall\ x \in I$) such that $|u_n(x) - u(x)| < \varepsilon \quad \forall n \geq m$ and $x \in I$.

Weierstrass's M-Test: A series $\sum_{n=1}^{\infty} u_n(x)$ of functions converges uniformly and absolutely in I if $\exists$ a convergent series $\sum_{n=1}^{\infty} M_n$ of positive constants such that $|u_n(x)| \leq M_n\ \forall\ n \in N$ and $\forall x \in I$.

COMPLEX NUMBERS

A complex number is generally written as $a + ib$ where a and b are real numbers and i, called the *imaginary unit*, has the property that $i^2 = -1$. The real number a and b are called the *real* and *imaginary* parts of $a + ib$, respectively.

The complex numbers $a + ib$ and $a - ib$ are called *complex conjugate* of each other.

Arithmetic Operations on Complex Numbers	
Equality of Complex Numbers	$a + ib = c + id$ iff $a = c$ and $b = d$
Addition of Complex Numbers	$(a + ib) + (c + id) = (a + c) + i(b + d)$
Subtraction of Complex Numbers	$(a + ib) - (c + id) = (a - c) + i(b - d)$
Multiplication of Complex Numbers	$(a + ib) \times (c + id) = (ac - bd) + i(ad - bc)$
Division of Complex Numbers	$\frac{(a + ib)}{(c + id)} = \frac{(a + ib)}{(c + id)} \cdot \frac{(c - id)}{(c - id)} = \left(\frac{ac + bd}{c^2 + d^2}\right) + i\left(\frac{bc - ad}{c^2 + d^2}\right)$
Note that the above operations are obtained by using the ordinary rules of algebra and replacing i^2 by -1 wherever it occurs.	

Polar form of a Complex Number

$$a+ib=r(\cos\theta+i\sin\theta)=r\text{ cis }\theta$$

where $r=\sqrt{a^2+b^2}$ is called modulus and θ the amplitude of $a+ib$.

Multiplication of Complex Numbers in Polar Form

$$[r_1(\cos\theta_1+i\sin\theta_1)]\times[r_2(\cos\theta_2+i\sin\theta_2)]=r_1r_2[\cos(\theta_1+\theta_2)+i\sin(\theta_1+\theta_2)]$$

In short $(r_1\text{cis}\,\theta_1)(r_2\text{cis}\,\theta_2)=r_1r_2\text{cis}(\theta_1+\theta_2)$

Division of Complex Numbers in Polar Form

$$\frac{r_1(\cos\theta_1+i\sin\theta_1)}{r_2(\cos\theta_2+i\sin\theta_2)}=\frac{r_1}{r_2}[\cos(\theta_1-\theta_2)+i\sin(\theta_1-\theta_2)]$$

In short $\dfrac{r_1\text{cis}\,\theta_1}{r_2\text{cis}\,\theta_2}=\dfrac{r_1}{r_2}\text{cis}(\theta_1-\theta_2)$

De Movire's Theorem

If n is any integer, positive or negative, then

$$[r(\cos\theta+i\sin\theta)]^n=r^n[\cos n\theta+i\sin n\theta]$$

Polar form of a Complex Number Expressed as an Exponential

$$a+ib=r(\cos\theta+i\sin\theta)=re^{i\theta}$$

Operations with Complex Number in Polar Form

$$(r_1e^{i\theta_1})\times(r_2e^{i\theta_2})=r_1r_2e^{i(\theta_1+\theta_2)}$$

$$\frac{r_1e^{i\theta_1}}{r_2e^{i\theta_2}}=\frac{r_1}{r_2}e^{i(\theta_1-\theta_2)}$$

$$(re^{i\theta})^n=r^ne^{in\theta}$$

$$(re^{i\theta})^{\frac{1}{n}}=(re^{i(\theta+2k\pi)})^{\frac{1}{n}}=r^{\frac{1}{n}}e^{i(\theta+2k\pi)}$$

Logarithm of a Complex Number

$$\text{Log}(re^{i\theta})=\log r+i\theta+2n\pi i$$

❑ ❑ ❑

Chapter 1

Ordinary Differential Equations of First Order

Learning Objectives

After reading this chapter, students will be able to

- understand how differential equations are being used in modeling various physical process
- understand the basic terminology related to differential equations
- define an exact differential equation
- prove required necessary and sufficient condition for an differential equation bto be an exact differential equation
- solve different problems related to exact differential equations
- reduce the differential equations which are not exact to exact differential equation and then solve them
- explain the integration factor and various methods to find integration factors
- explain various methods to solve differential equation of the first degree and higher order
- describe Clairaut's equation and solve different problem related to it
- describe Leibniz's linear equation and solve different problem related to it
- describe Bernoulli's equation and solve different problem related to it

1.1 INTRODUCTION

The subject of ordinary differential equations (ODEs) is not only a fascinating part of mathematics but also an essential tool for modelling many physical processes. In addition, most of the scientific laws are expressed in terms of differential equations. This is illustrated below:

1. In Thermodynamics, the differential equation

$$\frac{dT}{dt} = -0.27(u-60)^{\frac{5}{4}}$$

describe the rate of change of temperature T of a body losing heat to its surroundings.

2. In Probability theory, the differential equation

$$\frac{dp_r}{dt} = (r+1)(n-r)p_{r+1} - r(n-r+1)p_r$$

is used to study the spread of an epidemic. Here p_r is the probability that in a sample of size n, r people are still uninfected at a time t.

3. In Mechanics, the differential equation

$$m\frac{dv}{dt} = mg - kv^2$$

describe the motion of a body falling freely under gravity through a medium which offers resistance proportional to the square of the velocity.

4. In Physics, the solution of the differential equation

$$m\frac{d^2y}{dx^2} = a - y + by^2$$

was used by Einstein in his investigation of the orbital motion of planets under the assumption of general relativity.

5. Another very important differential equation encountered in the study of physics is Vander Pol's equation

$$m\frac{d^2y}{dt^2} - \varepsilon(1-y^2)\frac{dy}{dt} + ay = 0$$

which has also been used to model the working of the heart.

6. In economics, the differential equation

$$\frac{dx}{dt} = sf(x) - g(x)$$

provides a simple model of economic growth, where x equals ratio of supply of labour, $f(x)$ is the productivity function, $g(x)$ represents the growth in the supply of labour, and s is a constant.

The above examples emphasize the importance of differential equations in mathematical investigations.

1.2 BASIC TERMINOLOGY OF DIFFERENTIAL EQUATIONS

Let us make ourselves familiar with the basic terms used related to the subject of differential equations.

Ordinary Derivative - If y is a function of x only, i.e., $y = f(x)$, then $\frac{dy}{dx}$ is called the ordinary derivative of y w. r. t. x. Physically it means the rate of change of the dependent variable y with respect to the independent variable x.

Partial Derivative - If u is a function of x and y, i.e. $u = f(x, y)$, then $\frac{\partial u}{\partial x}$ is called the partial derivative of u with respect to x keeping y constant, and $\frac{\partial u}{\partial y}$ is called the partial derivative of u with respect to y keeping x constant. Physically it means the rate of change of the dependent variable with respect to one of the independent variable keeping others fixed.

And the derivatives of higher order are interpreted as rates of rates, and hence differential equations are equations involving rates.

Differential Equation - A *differential equation* is an equation that in addition to independent and dependent variables also contains one or more derivatives of the dependent variables. These derivatives can be ordinary or partial.

Ordinary Differential Equation - An *ordinary differential equation* is an equation involving only ordinary derivatives of one or more functions with respect to a single independent variable. Following are some examples of ordinary differential equations.

$$\frac{dy}{dx} = -\frac{x}{y} \qquad \frac{dy}{dx} = x^2 + y^2 \qquad \frac{d^2y}{dx^2} + y = 0$$

Partial Differential Equation - A *partial differential equation* is an equation involving partial derivatives of one or more functions with respect to more than one independent variable. Following are some examples of partial differential equations.

$$\frac{\partial u}{\partial t} = k\frac{\partial^2 u}{\partial x^2} \qquad \frac{\partial^2 u}{\partial x^2} + k\frac{\partial^2 u}{\partial y^2} = 0$$

Order of a Differential Equation - The *order of a differential equation* is the order of the highest order derivative in the differential equation. For example

$$m\frac{d^2y}{dx^2} + c\frac{dy}{dx} + ky = 0$$

is a second order ordinary differential equation.

Degree of a Differential Equation - The *degree of a differential equation* is the power of the highest order derivative in the differential equation, provided it is free from radicals and fractional powers of the derivatives. For example

$$\left[\frac{d^2y}{dx^2}\right]^2 + c\frac{dy}{dx} + ky = 0$$

is a second degree ordinary differential equation.

Linear and Non-linear Differential Equations - An *ordinary differential equation* is said to be linear if dependent variable and all of its derivatives occurring in the equation occur only in the first degree and are not multiplied together, in any term. Following are some examples of linear differential equations.

$$\frac{dy}{dx} = \sin x \qquad \frac{d^2y}{dx^2} + y = 0$$

A differential equation which is not linear is called non-linear differential equation.

Following are some examples of non-linear differential equations.

$$\frac{d^2y}{dx^2}+\left(\frac{dy}{dx}\right)^2+y^2=0 \qquad \left(\frac{d^2y}{dx^2}\right)^2+x^2\left(\frac{dy}{dx}\right)^3=0$$

Note: A linear differential equation is always of the first degree but every first degree differential equation need not be linear. For example, the differential equation

$$\frac{d^2y}{dx^2}+\left(\frac{dy}{dx}\right)^2+y^2=0$$

is not linear, though its degree is 1.

Initial Value and Boundary Value Problems - An ordinary differential equation of nth order is of the form

$$f\left(x,y,\frac{dy}{dx},\frac{d^2y}{dx^2},\ldots,\frac{d^ny}{dx^n}\right)=0$$

Its general solution contains n arbitrary constants and is of the form

$$\phi(x,y,c_1,c_2,c_3,\ldots,c_n)=0$$

To obtain its solution, n conditions must be given so that the constants c_1, c_2, c_3,…, c_n can be determined.

If these conditions are prescribed at *one point only*, say x_1, the differential equations together with the conditions constitute an *initial value problem* (IVP) of the nth order.

However, if the conditions are prescribed at *two or more points*, then differential equations together with the conditions constitute a *boundary value problem* (BVP) of the nth order.

Solution of a Differential Equation - A solution of a differential equation is a relation, free from derivatives, between the variables which satisfies the given differential equation.

A solution of a differential equation in which the number of arbitrary constants is equal to the order of the differential equation is called the *general or complete solution* of the given differential equation.

A solution of a differential equation obtained from the general solution by giving particular values to arbitrary constants is called a *particular solution* of the given differential equation.

1.3 SOLUTION OF DIFFERENTIAL EQUATIONS OF FIRST ORDER AND FIRST DEGREE

Every first order and first degree differential equation is not solvable. Only those differential equations which belong to (or can be reduced to) one of the following categories can be solved by the standard methods:

- Equations in which variables are separable.
- Equations of the form $\frac{dy}{dx} = f(ax+by+c)$.
- Homogeneous equations.
- Equations reducible to homogeneous equations.
- Linear equations.
- Exact equations.

The first four types of differential equations and the methods to find their solution have already been covered in lower classes. The topic of linear equations and to find their solutions will be covered in next chapter (*Chapter 3*).

This chapter will focus on exact differential equations as well as equations reducible to exact.

1.4 EXACT DIFFERENTIAL EQUATIONS

Definition 1: A differential equation obtained from its primitive directly by differentiation, without any operation of multiplication, elimination or reduction, etc., is said to be an exact differential equation.

Thus, a differential equation of the form

$$M(x,y)\,dx + N(x,y)\,dy = 0$$

is an exact differential equation if it can be obtained by differentiating the equation

$$u(x, y) = C$$

which is its primitive, i.e., if

$$du = Mdx + Ndy$$

Definition 2: A differential equation of the form $Mdx + Ndy = 0$ is said to be exact if there exists another function u such that $Mdx + Ndy = du$ or $Mdx + Ndy$ is obtained from its primitive directly by differentiation, without any operation of multiplication, elimination or reduction, etc., is said to be an exact differential equation.

Art. The necessary and sufficient condition for the differential equation $Mdx + Ndy = 0$ to be exact is $\dfrac{\partial M}{\partial y} = \dfrac{\partial N}{\partial x}$.

Proof: Condition is Necessary:

The equation $Mdx + Ndy = 0$ will be exact, if $du = Mdx + Ndy$

But $$du = \frac{\partial u}{\partial x}dx + \frac{\partial u}{\partial y}dy$$

By total derivative theorem,

$$dz = \frac{\partial z}{\partial x}dx + \frac{\partial z}{\partial y}dy$$

$$\therefore \quad Mdx + Ndy = \frac{\partial u}{\partial x}dx + \frac{\partial u}{\partial y}dy$$

Equating coefficients of dx and dy, we get

$$M = \frac{\partial u}{\partial x} \quad \text{and} \quad N = \frac{\partial u}{\partial y}$$

$$\therefore \quad \frac{\partial M}{\partial y} = \frac{\partial^2 u}{\partial y \partial x} \quad \text{and} \quad \frac{\partial N}{\partial x} = \frac{\partial^2 u}{\partial x \partial y}$$

Also $$\frac{\partial^2 u}{\partial y \partial x} = \frac{\partial^2 u}{\partial x \partial y}$$

$$\therefore \quad \frac{\partial M}{\partial y} = \frac{\partial N}{\partial x}$$

which is the necessary condition of exactness.

Condition is Sufficient: Let $u = \int_y M dx$

$$\therefore \quad \frac{\partial u}{\partial x} = M \quad \text{and} \quad \frac{\partial^2 u}{\partial y \partial x} = \frac{\partial M}{\partial y}$$

But $$\frac{\partial^2 u}{\partial y \partial x} = \frac{\partial^2 u}{\partial x \partial y} \quad \text{and} \quad \frac{\partial^2 u}{\partial y \partial x} = \frac{\partial M}{\partial y} = \frac{\partial N}{\partial x}$$

$$\therefore \quad \frac{\partial N}{\partial x} = \frac{\partial^2 u}{\partial x \partial y} = \frac{\partial}{\partial x}\left(\frac{\partial u}{\partial y}\right)$$

Integrating both sides w.r.t x treating y as constant, we get

$$N = \frac{\partial u}{\partial y} + f(y) \qquad \text{where } f(y) \text{ is a function of } y \text{ free from } x.$$

$$\therefore \quad Mdx + Ndy = \frac{\partial u}{\partial x}dx + \left(\frac{\partial u}{\partial y} + f(y)\right)dy \qquad \left[\because M = \frac{\partial u}{\partial x}, N = \frac{\partial u}{\partial y} + f(y)\right]$$

$$= \left(\frac{\partial u}{\partial x}dx + \frac{\partial u}{\partial y}dy\right) + f(y)dy = du + f(y)dy$$

$$= d\left[u + \int f(y)dy\right]$$

which shows that $Mdx + Ndy$ is an exact differential and hence $Mdx + Ndy = 0$ is an exact differential equation.

Remarks:

Since $Mdx + Ndy = d\left[u + \int f(y)dy\right]$

$\therefore \quad Mdx + Ndy = 0 \quad \Rightarrow \quad d\left[u + \int f(y)dy\right] = 0$

Integrating, we get $u + \int f(y)dy = c$

But $u = \int_y M dx$ and $f(y)$ = terms of N not containing x.

Hence the solution of $Mdx + Ndy = 0$ is

$\int_y M dx + \int(\text{terms of } N \text{ not containing } x)\,dy = c.$

ILLUSTRATIVE EXAMPLES 1.1

Example 1.1: Solve differential equation $(x^2 - ay)dx = (ax - y^2)dy.$ [PTU May 2005]

Solution: Given differential equation is

$$(x^2 - ay)dx - (ax - y^2)dy = 0$$

$$\Rightarrow \quad (x^2 - ay)dx + (y^2 - ax)dy = 0 \qquad \ldots (1)$$

Compare it with $Mdx + Ndy = 0$, we get

$$M = x^2 - ay, \; N = y^2 - ax$$

$$\frac{\partial M}{\partial y} = -a, \quad \frac{\partial N}{\partial x} = -a$$

Since $\dfrac{\partial M}{\partial y} = \dfrac{\partial N}{\partial x}$

∴ Eq. (1) is exact differential equation.

∴ It's solution is

$$\int_y Mdx + \int (\text{terms of } N \text{ not containing } x)dy = c$$

$$\Rightarrow \quad \int_y (x^2 - ay)dx + \int y^2 dy = c$$

$$\Rightarrow \quad \frac{x^3}{3} - ayx + \frac{y^3}{3} = c$$

$$\Rightarrow \quad x^3 + y^3 - 3axy = c' \quad \text{where } c' = 3c$$

Example 1.2: Solve differential equation

$$(\sec x \tan x \tan y - e^x)dx + \sec x \sec^2 y \, dy = 0.$$ [PTU Dec 2005]

Solution: Given differential equation is

$$(\sec x \tan x \tan y - e^x)dx + \sec x \sec^2 y \, dy = 0 \qquad \ldots (1)$$

Compare it with $Mdx + Ndy = 0$, we get

$$M = \sec x \tan x \tan y - e^x, \quad N = \sec x \sec^2 y$$

$$\frac{\partial M}{\partial y} = \sec x \tan x \sec^2 y, \quad \frac{\partial N}{\partial x} = \sec x \tan x \sec^2 y$$

Since $\dfrac{\partial M}{\partial y} = \dfrac{\partial N}{\partial x}$

$\therefore$ Eq. (1) is exact differential equation.

$\therefore$ It's solution is

$$\int_y (\sec x \tan x \tan y - e^x) dx + \int (\text{term of } N \text{ not containing} x) dy = c$$

$$\Rightarrow \quad \tan y \int_y \sec x \tan x dx - \int_y e^x dx + \int 0 \, dy = c$$

$$\Rightarrow \quad \tan y \sec x - e^x = c$$

Example 1.3: Solve differential equation

$$(5x^4 + 3x^2 y^2 - 2xy^3) dx + (2x^3 y - 3x^2 y^2 - 5y^4) dy = 0.$$

Solution: Given differential equation is

$$(5x^4 + 3x^2 y^2 - 2xy^3) dx + (2x^3 y - 3x^2 y^2 - 5y^4) dy = 0 \quad \ldots (1)$$

Compare it with $Mdx + Ndy = 0$, we get

$$M = 5x^4 + 3x^2 y^2 - 2xy^3 \text{ and } N = 2x^3 y - 3x^2 y^2 - 5y^4$$

$$\therefore \quad \frac{\partial M}{\partial y} = 6x^2 y - 6xy^2 \quad \text{and} \quad \frac{\partial N}{\partial x} = 6x^2 y - 6xy^2$$

Since $\dfrac{\partial M}{\partial y} = \dfrac{\partial N}{\partial x}$

$\therefore$ Eq. (1) is exact differential equation.

$\therefore$ It's solution is

$$\int_y M dx + \int (\text{terms of } N \text{ not containing } x)\, dy = c$$

$$\Rightarrow \quad \int_y \left(5x^4 + 3x^2y^2 - 2xy^3\right) dx + \int -5y^4 dy = c$$

$$\Rightarrow \quad x^5 + x^3y^2 - x^2y^3 - y^5 = c$$

Example 1.4: Solve differential equation

$$\left(\cos x \ \tan y + \cos(x+y)\right) dx + \left(\sin x \ \sec^2 y + \cos(x+y)\right) dy = 0.$$

Solution: Given differential equation is

$$\left(\cos x \ \tan y + \cos(x+y)\right) dx + \left(\sin x \ \sec^2 y + \cos(x+y)\right) dy = 0 \qquad \ldots (1)$$

Compare it with $Mdx + Ndy = 0$, we get

$$M = \cos x \tan y + \cos(x+y) \quad \text{and} \quad N = \sin x \sec^2 y + \cos(x+y)$$

$$\therefore \quad \frac{\partial M}{\partial y} = \cos x \sec^2 y - \sin(x+y) \quad \text{and} \quad \frac{\partial N}{\partial x} = \cos x \sec^2 y - \sin(x+y)$$

Since $\dfrac{\partial M}{\partial y} = \dfrac{\partial N}{\partial x}$

$\therefore$ Eq. (1) is exact differential equation.

$\therefore$ It's solution is

$$\int_y M dx + \int (\text{terms of } N \text{ not containing } x)\, dy = c$$

$$\Rightarrow \quad \int_y \left(\cos x \tan y + \cos(x+y)\right) dx + \int 0.dy = c$$

$$\Rightarrow \quad \tan y \int_y \cos x dx + \int \cos(x+y)\, dx = c$$

$$\Rightarrow \quad \sin x \tan y + \sin(x+y) = c$$

Example 1.5: Consider the differential equation

$$\left(1+e^{k\frac{x}{y}}\right)dx+e^{\frac{x}{y}}\left(1-\frac{x}{y}\right)dy=0$$

For what value of k the above equation will be exact? [PTU May 2010]

Solution: Given differential equation is

$$\left(1+e^{k\frac{x}{y}}\right)dx+e^{\frac{x}{y}}\left(1-\frac{x}{y}\right)dy=0 \qquad \ldots(1)$$

Compare it with $Mdx+Ndy=0$, we get

$$M=1+e^{k\frac{x}{y}} \quad \text{and} \quad N=e^{\frac{x}{y}}\left(1-\frac{x}{y}\right)$$

$$\therefore \quad \frac{\partial M}{\partial y}=e^{k\frac{x}{y}}\left(-k\frac{x}{y^2}\right)$$

$$\frac{\partial N}{\partial x}=e^{\frac{x}{y}}\frac{\partial}{\partial x}\left(1-\frac{x}{y}\right)+\left(1-\frac{x}{y}\right)\frac{\partial}{\partial x}\left(e^{\frac{x}{y}}\right) \qquad \left[\because \frac{\partial}{\partial x}(uv)=u\frac{\partial v}{\partial x}+v\frac{\partial u}{\partial x}\right]$$

$$=e^{\frac{x}{y}}\left(-\frac{1}{y}\right)+\left(1-\frac{x}{y}\right)e^{\frac{x}{y}}\frac{1}{y}=e^{\frac{x}{y}}\left(-\frac{1}{y}\right)+\left(\frac{1}{y}-\frac{x}{y^2}\right)e^{\frac{x}{y}}$$

$$=e^{\frac{x}{y}}\left(-\frac{1}{y}+\frac{1}{y}-\frac{x}{y^2}\right)=e^{\frac{x}{y}}\left(-\frac{x}{y^2}\right)$$

Eq. (1) will be exact differential equation if

$$\frac{\partial M}{\partial y}=\frac{\partial N}{\partial x}$$

$$\therefore \quad e^{k\frac{x}{y}}\left(-k\frac{x}{y^2}\right)=e^{\frac{x}{y}}\left(-\frac{x}{y^2}\right) \Rightarrow ke^{k\frac{x}{y}}=e^{\frac{x}{y}}$$

which holds only if $k=1$

$\therefore$ for $k=1$, Eq. (1) is an exact differential equation.

Example 1.6: Solve differential equation $\left(y^2e^{xy^2}+4x^3\right)dx+\left(2xy\, e^{xy^2}-3y^2\right)dy=0$.

Solution: Given differential equation is

$$\left(y^2e^{xy^2}+4x^3\right)dx+\left(2xy\, e^{xy^2}-3y^2\right)dy=0 \qquad \ldots (1)$$

Compare it with $Mdx+Ndy=0$, we get

$$M=y^2e^{xy^2}+4x^3,\ N=2xy\, e^{xy^2}-3y^2$$

$$\therefore \quad \frac{\partial M}{\partial y}=\frac{\partial}{\partial y}\left(y^2e^{xy^2}+4x^3\right)=y^2.\, e^{xy^2}\ .2xy+e^{xy^2}\ .2y+0$$

$$=e^{xy^2}\left(2xy^3+2y\right)$$

$$\frac{\partial N}{\partial x}=\frac{\partial}{\partial x}\left(2xy\, e^{xy^2}-3y^2\right)=2y\frac{\partial}{\partial x}\left(xe^{xy^2}\right)-\frac{\partial}{\partial x}\left(3y^2\right)$$

$$=2y\left(x.e^{xy^2}.y^2+e^{xy^2}.1\right)-0=e^{xy^2}\left(2xy^3+2y\right)$$

Since $\dfrac{\partial M}{\partial y}=\dfrac{\partial N}{\partial x}$

$\therefore$ Eq. (1) is exact differential equation.

$\therefore$ It's solution is

$$\int_y Mdx+\int(\text{terms of } N \text{ not containing } x)\,dy=c$$

$$\Rightarrow \quad \int_y\left(y^2e^{xy^2}+4x^3\right)dx+\int\left(-3y^2\right)dy=c$$

$$\Rightarrow \quad y^2\int e^{xy^2}dx+x^4-3\int y^2dy=c$$

$$\Rightarrow \quad y^2\frac{e^{xy^2}}{y^2}+x^4-y^3=c$$

$$\Rightarrow \quad e^{xy^2}+x^4-y^3=c$$

which is the required solution.

Example 1.7: Solve differential equation $xdy + ydx + \frac{xdy - ydx}{x^2 + y^2} = 0$.

Solution. Given differential equation is

$$xdy + ydx + \frac{xdy - ydx}{x^2 + y^2} = 0 \qquad \ldots (1)$$

which can be written as

$$\left(y - \frac{y}{x^2 + y^2}\right)dx + \left(x + \frac{x}{x^2 + y^2}\right)dy = 0 \qquad \ldots (2)$$

Comparing it with $Mdx + Ndy = 0$, we get

$$M = y - \frac{y}{x^2 + y^2} \qquad N = x + \frac{x}{x^2 + y^2}$$

$$\therefore \qquad \frac{\partial M}{\partial y} = 1 - \frac{(x^2 + y^2).1 - y.2y}{(x^2 + y^2)^2} = 1 - \frac{(x^2 - y^2)}{(x^2 + y^2)^2}$$

$$\frac{\partial N}{\partial x} = 1 + \frac{(x^2 + y^2).1 - x.2x}{(x^2 + y^2)^2} = 1 - \frac{x^2 - y^2}{(x^2 + y^2)^2}$$

Since $\frac{\partial M}{\partial y} = \frac{\partial N}{\partial x}$

$\therefore$ Eq. (1) is exact differential equation.

$\therefore$ It's solution is

$$\int_y Mdx + \int (\text{terms of } N \text{ not containing } x)\, dy = c$$

$$\Rightarrow \quad \int_y \left(y - \frac{y}{x^2 + y^2}\right)dx + \int 0 dy = c$$

$$\Rightarrow \quad y\int dx - y.\int \frac{1}{x^2 + y^2}dx = c \qquad \Rightarrow \qquad xy - y.\frac{1}{y}\tan^{-1}\frac{x}{y} = c$$

$$\Rightarrow \quad xy - \tan^{-1}\frac{x}{y} = c$$ which is the required solution.

Example 1.8: Solve differential equation $\frac{dy}{dx}=-\frac{ax+hy+g}{hx+by+f}$. Show that this differential equation represents a family of conics.

Solution. Given differential equation is

$$\frac{dy}{dx}=-\frac{ax+hy+g}{hx+by+f}. \qquad \ldots (1)$$

which can be written as

$$(ax+hy+g)\,dx+(hx+by+f)\,dy=0 \qquad \ldots (2)$$

Comparing it with $Mdx+Ndy=0$, we get

$$M=ax+hy+g, \quad N=hx+by+f$$

$$\therefore \quad \frac{\partial M}{\partial y}=h, \quad \frac{\partial N}{\partial x}=h$$

Since $\frac{\partial M}{\partial y}=\frac{\partial N}{\partial x}$

$\therefore$ Eq. (2) is exact differential equation.

$\therefore$ It's solution is

$$\int_y Mdx+\int(\text{terms of } N \text{ not containing } x)\,dy=c$$

$$\Rightarrow \quad \int_y (ax+hy+g)\,dx+\int(by+f)\,dy=c$$

$$\Rightarrow \quad \frac{ax^2}{2}+hxy+gx+\frac{by^2}{2}+fy=c$$

$$\Rightarrow \quad ax^2+2hxy+by^2+2gx+2fy+c'=0 \qquad \text{where } c'=-2c$$

which is the required solution, and this equation also represents a family of conics.

EXERCISE 1.1

Solve the following differential equations:

1. $\left(x^2-4xy-2y^2\right)dx+\left(y^2-4xy-2x^2\right)dy=0$
2. $\left(3x^2+6xy^2\right)dx+\left(6x^2y+4y^3\right)dy=0$ [PTU May 2009]
3. $\left(xy\,e^{\frac{x}{y}}+y^2\right)dx-x^2\,e^{\frac{x}{y}}dy=0$.
4. $\left(x^2+y^2-a^2\right)xdx+\left(x^2-y^2-b^2\right)y\,dy=0$
5. $\left(1+e^{\frac{x}{y}}\right)dx+\left(1-\frac{x}{y}\right)e^{\frac{x}{y}}dy=0$
6. $\left(y\cos x+1\right)dx+\sin x\,dy=0$
7. $\left(\sec x\tan x\tan y-e^x\right)dx+\sec x\sec^2 y\,dy=0$
8. $\left(y+\frac{y^3}{3}+\frac{x^2}{2}\right)dx+\frac{1}{4}\left(x+xy^2\right)dy=0$
9. $ye^{xy}dx+\left(xe^{xy}+2y\right)dy=0$
10. $\left(2xy\cos x^2-2xy+1\right)dx+\left(\sin x^2-x^2\right)dy=0$
11. $\left(\sin x\cos y+e^{2x}\right)dx+\left(\cos x\sin y+\tan y\right)dy=0$
12. $\left(y\left(1+\frac{1}{x}\right)+\cos y\right)dx+\left(x+\log x-x\sin y\right)dy=0$
13. $\frac{dy}{dx}+\frac{y\cos+\sin y+y}{\sin x+x\cos y+x}$
14. $\left(\cos x\tan y+\cos(x+y)\right)dx+\left(\sin x\sec^2 y+\cos(x+y)\right)dy=0$
15. Under what conditions on '*a*, *b*, *c* and *d*', the differential equation
 $(a\sinh x\cos y+b\cosh x\sin y)dx+\left(c\sinh x\cos y+d\cosh x\sin y\right)dy=0$
 is exact? [PTU May 2012]

ANSWERS

1. $x^3-6x^2y-6xy^2+y^3=c$
2. $x^3+3x^2y^2+y^4=c$
3. $e^{\frac{x}{y}}+\log x=c$
4. $x^4+2x^2y^2-2a^2x^2-y^4-2b^2y^2=c$

5. $x + ye^{\frac{x}{y}} = c$

6. $y\sin x + x = c$

7. $\sec x \tan y - e^x = c$

8. $\frac{1}{4}x^4 y + \frac{x^4 y^3}{12} + \frac{x^6}{12} = c$

9. $e^{xy} + y^2 = c$

10. $y\sin x^2 - x^2 y + x = c$

11. $-\cos x \cos y + \frac{1}{2}e^{2x} + \log\sec y = c$

12. $y(x + \log x) + x\cos y = c$

13. $y\sin x + (\sin y + y)x = c$

14. $\sin x \tan y + \sin(x + y) = c$

15. $a = -d, b = c$

1.5 EQUATIONS REDUCIBLE TO EXACT DIFFERENTIAL EQUATION

Differential equations which are not exact can sometimes be made exact after multiplying by a suitable factor (a function of x or y or both) called *integrating factor (I.F.)*.

For example, consider the equation

$$ydx - xdy = 0 \qquad \ldots (1)$$

Comparing it with $Mdx + Ndy = 0$, we get

Here $M = y$ and $N = -x$

$$\frac{\partial M}{\partial y} = 1, \quad \frac{\partial N}{\partial x} = -1$$

Since $\frac{\partial M}{\partial y} \neq \frac{\partial N}{\partial x}$, therefore, the equation is not exact.

- Multiplying the Eq. (1) by the factor $\frac{1}{y^2}$, it becomes

$$\frac{ydx - xdy}{y^2} = 0 \quad \text{or} \quad d\left(\frac{x}{y}\right) = 0$$

which is exact.

- Multiplying the Eq. (1) by the factor $\frac{1}{x^2}$, it becomes

$$\frac{xdy - ydx}{x^2} = 0 \quad \text{or} \quad d\left(\frac{y}{x}\right) = 0$$

which is exact.

- Multiplying the Eq. (1) by the factor $\frac{1}{xy}$, it becomes

$$\frac{dx}{x}-\frac{dy}{y}=0 \quad \text{or} \quad d(\log x-\log y)=0$$

which is exact.

$\therefore$ $\frac{1}{y^2}$, $\frac{1}{x^2}$ and $\frac{1}{xy}$ are integrating factors of Eq. (1).

Remark: If a differential equation has a solution, then it has an infinite number of integrating factors. We can prove it as shown in next article.

Art. Prove that integrating factors of differential equation $Mdx+Ndy=0$ are infinite.

Proof: Given differential equation is

$$Mdx+Ndy=0 \quad \ldots (1)$$

Let μ be its I.F.

Multiplying both sides of (1) by μ, we get

$$\mu Mdx+\mu Ndy=0$$

which is an exact differential equation.

$$\therefore \quad \mu Mdx+\mu Ndy=du \quad \ldots (2)$$

where u is a function of x and y.

Multiplying both sides of (2) by $f(u)$, where $f(u)$ is any function of u, we get

$$\therefore \quad \mu f(u)Mdx+\mu f(u)Ndy=f(u)du \quad \ldots (3)$$

which is also an exact differential equation.

$\therefore$ $\mu f(u)$ is also an I.F. of Eq. (1).

But $f(u)$ is any function of u, which can be chosen in infinite ways.

$\therefore$ the number of integrating factors are infinite.

1.6 FINDING INTEGRATING FACTORS BY INSPECTION

In certain situations, a careful observation can help to find the integrating factor. Following differential equations can be helpful in finding a suitable integrating factor.

- $ydx + xdy = d(xy)$
- $\dfrac{xdy - ydx}{x^2} = d\left(\dfrac{y}{x}\right)$
- $\dfrac{ydx - xdy}{y^2} = d\left(\dfrac{x}{y}\right)$
- $\dfrac{xdy - ydx}{x^2 + y^2} = d\left(\tan^{-1}\dfrac{x}{y}\right)$
- $\dfrac{xdy - ydx}{xy} = d\left(\log\left(\dfrac{x}{y}\right)\right)$
- $\dfrac{ydx + xdy}{xy} = d\left(\log(xy)\right)$
- $\dfrac{xdy + ydx}{x^2 + y^2} = d\left(\dfrac{1}{2}\log\left(x^2 + y^2\right)\right)$
- $\dfrac{xdy - ydx}{x^2 + y^2} = d\left(\dfrac{1}{2}\log\left(\dfrac{x+y}{x-y}\right)\right)$

ILLUSTRATIVE EXAMPLES 1.2

Example 1.9: Given the following differential equation

$$ydx - xdy = -3x^2y^2e^{x^3}dx$$

Find its integration factor by inspection and hence solve it.

Solution: Given differential equation is

$$ydx - xdy = -3x^2y^2e^{x^3}dx \qquad \ldots (1)$$

Rewriting it as

$$ydx - xdy + 3x^2y^2e^{x^3}dx = 0 \qquad \ldots (2)$$

Since $3x^2e^{x^3} = d\left(e^{x^3}\right)$, the term $3x^2y^2e^{x^3}dx$ should not involve y^2.

This suggests that $\dfrac{1}{y^2}$ may be an integrating factor.

Multiplying both sides of Eq. (2) by $\dfrac{1}{y^2}$, we get

$$\frac{ydx - xdy}{y^2} + 3x^2e^{x^3}dx = 0$$

$$\Rightarrow \quad d\left(\frac{x}{y}\right)+d\left(e^{x^3}\right)=0 \qquad \ldots(3)$$

which is an exact differential equation.

Integrating Eq. (3),we get

$$\frac{x}{y}+e^{x^3}=c$$

which is the required solution.

Example 1.10: Given the following differential equation

$$(y-1)\,dx = xdy$$

Find its integration factor by inspection and hence solve it.

Solution: Given differential equation is

$$(y-1)\,dx = xdy \qquad \Rightarrow \qquad (y-1)\,dx - xdy = 0 \qquad \ldots(1)$$

Comparing it with $Mdx + Ndy = 0$, we get

$$M = y-1 \quad \text{and} \quad N = -x$$

$$\therefore \quad \frac{\partial M}{\partial y}=1 \quad \text{and} \quad \frac{\partial N}{\partial x}=-1$$

Since $\frac{\partial M}{\partial y} \neq \frac{\partial N}{\partial x}$, therefore, Eq. (1) is not exact.

Rewriting Eq. (1) as

$$ydx - xdy = dx$$

Multiplying it by $\frac{1}{x^2}$, we get

$$\frac{ydx - xdy}{x^2} = \frac{1}{x^2}dx$$

$$\Rightarrow \quad d\left(-\frac{y}{x}\right)=d\left(-\frac{1}{x}\right) \qquad \ldots(2)$$

$\therefore$ factor $\frac{1}{x^2}$ makes the Eq. (1) exact differential equation.

$\therefore$ $\frac{1}{x^2}$ is an integration factor.

Integrating Eq. (2), we get

$$-\frac{y}{x} = -\frac{1}{x} + c \quad \text{or} \quad y = 1 - cx$$

which is the required solution.

Example 1.11: Solve $xdy - y\,dy = x\sqrt{x^2 - y^2}\,dx$.

Solution: Given differential equation is

$$xdy - y\,dy = x\sqrt{x^2 - y^2}\,dx \qquad \ldots (1)$$

$$\Rightarrow \quad xdy - y\,dy = x^2\sqrt{1 - \frac{y^2}{x^2}}\,dx$$

$$\Rightarrow \quad \frac{\dfrac{xdy - y\,dy}{x^2}}{\sqrt{1 - \dfrac{y^2}{x^2}}} = dx$$

$$\Rightarrow \quad d\left(\sin^{-1}\left(\frac{y}{x}\right)\right) = dx$$

$$\left[\begin{array}{l} \because \dfrac{d}{dx}\sin^{-1} x = \dfrac{1}{\sqrt{1-x^2}}\dfrac{d(x)}{dx} = \dfrac{1}{\sqrt{1-x^2}} \\ \therefore \dfrac{d}{dx}\sin^{-1}\left(\dfrac{y}{x}\right) = \dfrac{1}{\sqrt{1-\left(\dfrac{y}{x}\right)^2}}\dfrac{d}{dx}\left(\dfrac{y}{x}\right) = \dfrac{\dfrac{xdy - y\,dy}{x^2}}{\sqrt{1-\left(\dfrac{y}{x}\right)^2}} \end{array}\right]$$

Integrating, we get

$$\sin^{-1}\left(\frac{y}{x}\right) = x + c \quad \Rightarrow \quad \frac{y}{x} = \sin(x+c)$$

$$\Rightarrow \quad y = x\sin(x+c)$$

which is the required solution.

1.7 FIVE RULES FOR FINDING INTEGRATING FACTORS

It is not easy to find integration factors by inspection, therefore, we have five rules for finding integrating factors. We shall discuss each of them in this section.

1.7.1 First Rule

If the equation $Mdx+Ndy=0$ is homogeneous in x and y, *i.e.*, if M and N are homogeneous functions of the same degree in x and y, then $\frac{1}{Mx+Ny}$ is an I.F., provided $Mx+Ny\neq 0$.

Proof: The given differential equation is

$$Mdx+Ndy=0 \qquad \ldots (1)$$

Now $Mdx+Ndy=\frac{1}{2}\left[(Mx+Ny)\left(\frac{dy}{x}+\frac{dy}{y}\right)+(Mx-Ny)\left(\frac{dx}{x}-\frac{dy}{y}\right)\right]$

(Note this step)

$$\therefore \frac{1}{Mx+Ny}[Mdx+Ndy]=\frac{1}{2}\left[\frac{dx}{x}+\frac{dy}{y}+\frac{Mx-Ny}{Mx+Ny}\left(\frac{dx}{x}-\frac{dy}{y}\right)\right] \qquad \ldots (2)$$

$\because$ M, N are homogeneous functions of the same degree in x and y

Let $M=x^n\phi_1\left(\frac{y}{x}\right)$, $N=x^n\phi_2\left(\frac{y}{x}\right)$

$$\therefore \frac{Mx-Ny}{Mx+Ny}=\frac{x^n\phi_1\left(\frac{y}{x}\right).x-x^n\phi_2\left(\frac{y}{x}\right).y}{x^n\phi_1\left(\frac{y}{x}\right).x+x^n\phi_2\left(\frac{y}{x}\right).y}=\frac{x^{n+1}\phi_1\left(\frac{y}{x}\right)-x^n y\phi_2\left(\frac{y}{x}\right)}{x^{n+1}\phi_1\left(\frac{y}{x}\right)+xx^n y.\phi_2\left(\frac{y}{x}\right)}$$

$$=\frac{\phi_1\left(\frac{y}{x}\right)-\frac{y}{x}\phi_2\left(\frac{y}{x}\right)}{\phi_1\left(\frac{y}{x}\right)+\frac{y}{x}\phi_2\left(\frac{y}{x}\right)} \qquad \text{(Divide numerator \& denominator by } x^{n+1}\text{)}$$

$$\therefore \frac{Mx-Ny}{Mx+Ny}=f\left(\frac{y}{x}\right), \text{ say}$$

$\therefore$ from Eq. (2), we get

$$\frac{1}{Mx+Ny}[Mdx+Ndy]=\frac{1}{2}\left[d(logx)+d(\log y)+f\left(\frac{y}{x}\right)\{d(\log x)-d(\log y)\}\right]$$

$$=\frac{1}{2}\left[d\{\log(xy)\}-f\left(\frac{y}{x}\right)d\left(\log\frac{y}{x}\right)\right] \qquad \left[\because \log\frac{x}{y}=-\log\frac{y}{x}\right]$$

$$=\frac{1}{2}d\left[\log(xy)-\psi\left(\log\frac{y}{x}\right)\right] \qquad \left[\because f\left(\frac{y}{x}\right)=f\left(e^{\log\frac{y}{x}}\right)=\psi\left(\log\frac{y}{x}\right), say\right]$$

which is an exact differential.

$\therefore$ $\dfrac{1}{Mx+Ny}$ is an I.F. of Eq. (1).

Note 1: This method is suitable when $Mx+Ny$ consists of only one term. Otherwise it is better to use $y=vx$, separate the variables and then integrate.

Note 2: If $Mx+Ny=0$, then this method fails.

In this case, $Mx+Ny=0 \quad \Rightarrow \quad \dfrac{M}{N}=\dfrac{-y}{x}$

$\therefore \quad Mdx+Ndy=0 \quad \Rightarrow \quad -\dfrac{y}{x}dx+dy=0 \quad \Rightarrow \quad \dfrac{dy}{y}-\dfrac{dx}{x}=0$

Integrating, we get

$$\log|y|-\log|x|=c \;\Rightarrow\; \log\left|\frac{y}{x}\right|=c \;\Rightarrow\; |y|=e^c|x| \text{ or } y=Ax \text{ where } A=\pm e^c$$

which is required solution.

ILLUSTRATIVE EXAMPLES 1.3

Example 1.12: Solve differential equation $x^2ydx-\left(x^3+y^3\right)dy=0$.

Solution: Given differential equation is

$$x^2ydx-\left(x^3+y^3\right)dy=0 \qquad \ldots (1)$$

which is homogeneous in x and y, as degree of each term is same, i.e., degree = 3.

Comparing (1) with $Mdx+Ndy=0$, we get

$$M=x^2y, N=-\left(x^3+y^3\right)$$

$$\therefore \quad \text{I.F.}=\frac{1}{Mx+Ny}=\frac{1}{x^3y-y\left(x^3+y^3\right)}=\frac{-1}{y^4}$$

$\therefore$ Multiplying both sides of (1) by $\left(\frac{-1}{y^4}\right)$, we get

$$\frac{-1}{y^4}x^2ydx+\frac{1}{y^4}\left(x^2+y^3\right)dy=0$$

$$\Rightarrow \quad \frac{-x^2}{y^3}dx+\left[\frac{x^2}{y^4}+\frac{1}{y}\right]dy=0 \qquad \text{which is exact.}$$

$$\because M=\frac{-x^2}{y^3}, N=\frac{x^2}{y^4}+\frac{1}{y}$$

$$\Rightarrow \frac{\partial M}{\partial y}=\frac{\partial N}{\partial x}$$

Its solution is

Using $\int\limits_{y} Mdx+\int(\text{terms of } N \text{ not containging } x)dy=c$

$$\int\frac{-x^2}{y^3}dx+\int\frac{1}{y}dy=c$$

$$\Rightarrow \quad \frac{-1}{y^3}\int x^2dx+\int\frac{1}{y}dy=c \qquad [\because y \text{ is constant}]$$

$$\Rightarrow \quad \frac{-x^3}{3y^3}+\log|y|=c$$

$$\Rightarrow \quad \log|y|=c+\frac{x}{3y^2}$$

$$\Rightarrow \quad |y|=e^{c+\frac{x^3}{2y^2}}=e^ce^{\frac{x^3}{2y^2}}$$

$$\Rightarrow \quad y=Ae^{\frac{x^3}{2y^2}} \qquad \text{where} \quad A=\pm e^c$$

Example 1.13: Solve differential equation is $\left(x^2y-2xy^2\right)dx-\left(x^3-3x^2y\right)dy=0.$

[PTU Dec 2003]

Solution: Given differential equation is

$$\left(x^2y-2xy\right)dx-\left(x^3-3x^2y\right)dy=0 \qquad \ldots (1)$$

which is homogeneous in x and y

Comparing Eq. (1) with $Mdx+Ndy=0$, we get

$$M = x^2y - 2xy^2, N = -x^3 + 3x^2y$$

$\therefore$ $\text{I.F.} = \dfrac{1}{Mx+Ny} = \dfrac{1}{x^3y - 2x^2y^2 - x^3y + 3x^2y^2} = \dfrac{1}{x^2y^2}$

$\therefore$ Multiplying both sides of (1) by $\dfrac{1}{x^2y^2}$ we get

$$\left(\frac{1}{y} - \frac{2}{x}\right)dx - \left(\frac{x}{y^2} - \frac{3}{y}\right)dy = 0, \text{ which is exact}$$

$\because M = \dfrac{1}{y} - \dfrac{2}{x}, N = -\left(\dfrac{x}{y^2} - \dfrac{3}{y}\right)$

$\Rightarrow \dfrac{\partial M}{\partial y} = \dfrac{\partial N}{\partial x}$

Its solution is

$$\int\left(\frac{1}{y} - \frac{2}{x}\right)dx - \int\left(\frac{-3}{y}\right)dy = c$$

$\Rightarrow$ $\dfrac{x}{y} - 2\log|x| + 3\log|y| = c$ which is the required solution.

1.7.2 Second Rule

If the equation $Mdx + Ndy = 0$ is of the form

$$f_1(xy)\, y\, dx + f_2(xy)\, x\, dy = 0$$

then $\dfrac{1}{Mx - Ny}$ is an I.F., provided $Mx - Ny \neq 0$.

Proof: The given equation is

$$Mdx + Ndy = 0 \qquad \ldots (1)$$

Now $Mdx + Ndy = \dfrac{1}{2}\left[(Mx+Ny)\left(\dfrac{dx}{x} + \dfrac{dy}{y}\right) + (Mx - Ny)\left(\dfrac{dx}{x} - \dfrac{dy}{y}\right)\right]$

$$\therefore \quad \frac{1}{(Mx-Ny)}[Mdx + Ndy] = \frac{1}{2}\left[\frac{Mx+Ny}{Mx-Ny}\left(\frac{dx}{x} + \frac{dy}{y} + \frac{dx}{x} - \frac{dy}{y}\right)\right] \qquad \ldots (2)$$

$\because$ $Mdx + Ndy = 0$ is of form $f_1(xy)\, y\, dx + f_2(xy)\, x\, dy = 0$

$\therefore$ $M = f_1(xy).y, \quad N = f_2(xy).x$

$\therefore$ $\dfrac{Mx+Ny}{Mx-Ny} = \dfrac{f_1(xy).y\,x + f_2(xy).xy}{f_1(xy).yx - f_2(xy).xy} = \dfrac{f_1(xy) + f_2(xy)}{f_1(xy) - f_2(xy)} = \phi(xy)$, say.

$\therefore$ From Eq. (2), we get

$$\frac{1}{Mx-Ny}[Mdx+Ndy]=\frac{1}{2}\left[\phi(xy)\{d\log(x)+d(\log y)\}+d(\log x)-d(\log y)\right]$$

$$=\frac{1}{2}\left[\phi(xy)\,d\left\{\log(xy)\,d\{\log(xy)\}+d\left(\log\frac{x}{y}\right)\right\}\right]$$

$$=\frac{1}{2}\left[\psi\{\log(xy)\}d\{\log(xy)\}+d\left(\log\frac{x}{y}\right)\right]$$

$$\left[\because \phi(xy)=\phi\left(e^{\log xy}\right)=\psi\{\log(xy)\}\right]$$

$\therefore \dfrac{1}{Mx-Ny}$ is an I.F. of Eq. (1)

Note: This rule fails if $Mx-Ny=0$.

In that case, $Mx-Ny=0 \Rightarrow \dfrac{M}{N}=\dfrac{y}{x}$

Also $Mdx+Ndy=0 \Rightarrow \dfrac{M}{N}dx+dy=0$

$\Rightarrow \dfrac{y}{x}dx+dy=0 \Rightarrow \dfrac{M}{N}dx+dy=0$

$\Rightarrow \dfrac{y}{x}dx+dy=0 \Rightarrow \dfrac{dx}{x}+\dfrac{dy}{y}=0$

Integrating both sides, we get

$\log|x|+\log|y|=\log c$

$\therefore$ $|xy|=c$ which is the required solution.

ILLUSTRATIVE EXAMPLES 1.4

Example 1.14: Solve differential equation $y(1-xy)\,dx-x(1+xy)\,dy=0$.

Solution: Given differential equation is

$$y(1+xy)\,dx-x(1+xy)\,dy=0 \qquad \ldots(1)$$

which is of the form $f_1(xy)y\;dx+f_2(xy)x\;dy=0$

Comparing Eq. (1) with $Mdx + Ndy = 0$, we get

$$M = y(1-xy), N = -x(1+xy)$$

$$\therefore \quad \text{I.F.} = \frac{1}{Mx - Ny} = \frac{1}{xy(1-xy) + xy(1+xy)} = \frac{1}{2xy}$$

$\therefore$ Multiplying both sides of Eq. (1) by $\frac{1}{2xy}$, we get

$$\frac{1-xy}{2x}dx - \frac{1+xy}{2y}dy = 0$$

which is exact and its solution is

$$\frac{1}{2}\int\frac{1}{x}dx - \frac{y}{2}\int 1dx - \frac{1}{2}\int\frac{1}{y}dy = c$$

$$\Rightarrow \quad \frac{1}{2}\log|x| - \frac{1}{2}xy - \frac{1}{2}\log|y| = c$$

$$\Rightarrow \quad \log\left|\frac{x}{y}\right| - xy = C, \text{ where } C = 2c.$$

Example 1.15: Solve differential equation $y(2xy+1)dx + x(1+2xy-x^3y^3)dy = 0$,

Solution: Given differential equation is

$$y(2xy+1)dx + x(1+2xy-x^3y^3)dy = 0 \qquad \ldots (1)$$

which is of form $f_1(xy)y\,dx + f_2(xy)x\,dy = 0$

Comparing (1) with $Mdx + Ndy = 0$

$$M = y(2xy+1), N = x(1+2xy-x^3y^3)$$

$$\therefore \quad \text{I.F.} = \frac{1}{Mx - Ny} = \frac{1}{xy(2xy+1) - xy(1+2xy-x^3y^3)} = \frac{1}{x^4y^4}$$

Multiplying both sides of (1) by $\frac{1}{x^4y^4}$, we get

$$\frac{2xy+1}{x^4y^3}dx + \frac{1+2xy-x^3y^3}{x^3y^4}dy = 0$$

which is exact.

Its solution is

$$\frac{2}{y^2}\int\frac{1}{x^3}dx+\frac{1}{y^3}\int\frac{1}{x^4}dx=c \qquad \Rightarrow \qquad \frac{2}{y^2}\left(\frac{x^{-2}}{-2}\right)+\frac{1}{y^3}\left(\frac{x^{-3}}{-3}\right)=c$$

$$\Rightarrow \qquad \frac{-1}{x^2y^2}-\frac{1}{3x^3y^3}=c$$

1.7.3 Third Rule

If in equation $Mdx+Ndy=0$, $\dfrac{\dfrac{\partial M}{\partial y}-\dfrac{\partial N}{\partial x}}{N}$ is a function of x only $= f(x)$, say,

then $e^{\int f(x)dx}$ is an I.F.

Proof: The given equation is

$$Mdx+Ndy=0 \qquad \ldots(1)$$

Now $e^{\int f(x)dx}$ will be an I.F. of Eq. (1), if

$$Me^{\int f(x)dx}dx+Ne^{\int f(x)dx}dy=0 \qquad \ldots(2)$$

is an exact differential equation.

i.e., if $\dfrac{\partial}{\partial y}\left[Me^{\int f(a)dx}\right]=\dfrac{\partial}{\partial x}\left[Ne^{\int f(x)dx}\right]$

$\Rightarrow$ if $e^{\int f(x)dx}\cdot\dfrac{\partial M}{\partial y}=\dfrac{\partial N}{\partial x}e^{\int f(x)dx}+Ne^{\int f(x)dx}.f(x)$

$\Rightarrow$ if $\dfrac{\partial M}{\partial y}=\dfrac{\partial N}{\partial x}+Nf(x)$ $\qquad \left[\text{Divide by } e^{\int f(x)dx}\right]$

$\Rightarrow$ if $\dfrac{\partial M}{\partial y}-\dfrac{\partial N}{\partial x}=Nf(x)$

$\Rightarrow$ if $\dfrac{\dfrac{\partial M}{\partial y}-\dfrac{\partial N}{\partial x}}{N}=f(x)$, which is true

$\therefore$ $e^{\int f(x)dx}$ is an I.F. of Eq. (1).

ILLUSTRATIVE EXAMPLES 1.5

Example 1.16: Find I.F. for $\left(e^y + xe^y\right)dx + xe^y dy = 0$.

Solution: Given differential equation is

$$\left(e^y + xe^y\right)dx + xe^y dy = 0 \qquad \ldots (1)$$

Comparing Eq. (1) with $Mdx + Ndy = 0$, we get

$$M = e^y + xe^y, N = xe^y$$

$$\frac{\partial M}{\partial y} = e^y + xe^y, \frac{\partial N}{\partial x} = e^y$$

$$\therefore \quad \frac{\frac{\partial M}{\partial y} - \frac{\partial N}{\partial x}}{N} = \frac{e^y + xe^y - e^y}{xe^y} = \frac{xe^y}{xe^y} = 1 = f(x)$$

$$\therefore \quad \text{I.F.} = e^{\int f(x)dx} = e^{\int dx} = e^x$$

Example 1.17: Solve differential equation $\left(x^2 + y^2 + 2x\right)dx + 2ydy = 0$.

Solution: Given differential equation is

$$\left(x^2 + y^2 + 2x\right)dx + 2ydy = 0 \qquad \ldots (1)$$

Comparing Eq. (1) with $Mdx + Ndy = 0$, we get

$$M = x^2 + y^2 + 2x, N = 2y$$

$$\frac{\partial M}{\partial y} = 2y, \frac{\partial N}{\partial x} = 0$$

$$\frac{\frac{\partial M}{\partial y} - \frac{\partial N}{\partial x}}{N} = \frac{2y}{2y} = 1$$

$$\text{I.F.} = e^{\int dx} = e^x$$

$\therefore$ Multiplying Eq. (1) by e^x, we get

$$e^x\left(x^2 + y^2 + 2x\right)dx + 2ye^x dy = 0$$

which is exact and its solution is

$$\int_{y} e^x(x^2+y^2+2x)dy + \int 0dx = c$$

$$\Rightarrow \quad \int x^2e^x dx + y^2\int e^x dx + 2\int xe^x dx = c$$

$$\Rightarrow \quad x^2e^x - \int 2xe^x dx + y^2e^x + \int 2xe^x dx = c$$

$$\Rightarrow \quad \left(x^2+y^2\right)e^x = c \qquad \text{which is the required solution.}$$

1.7.4 Fourth Rule

If in the equation $Mdx + Ndy = 0, \dfrac{\dfrac{\partial N}{\partial x} - \dfrac{\partial M}{\partial y}}{M}$ is a function of y only $= f(y)$, say, then $e^{\int f(y)dy}$ is an I.F.

Proof: The given equation is $Mdx + Ndy = 0$ . . . (1)

Now $e^{\int f(y)dy}$ will be an I.F. of Eq. (1), if

$$Me^{\int f(y)dy}dx + Ne^{\int f(y)dy}dy = 0 \qquad \text{. . . (2)}$$

is an exact differential equation.

i.e., if $\dfrac{\partial}{\partial y}\left[Me^{\int f(y)dy}\right] = \dfrac{\partial}{\partial x}\left[Ne^{\int f(y)dy}\right]$

$$\Rightarrow \text{ if } \frac{\partial M}{\partial y}e^{\int f(y)dy} + Me^{\int f(y)dy}f(y) = \frac{\partial N}{\partial x}.e^{\int f(y)dy}$$

$$\Rightarrow \text{ if } \frac{\partial M}{\partial y} + Mf(y) = \frac{\partial N}{\partial x} \quad \Rightarrow \quad \text{if } \frac{\partial N}{\partial x} - \frac{\partial M}{\partial y} = Mf(y)$$

$$\Rightarrow \text{ if } \frac{\dfrac{\partial N}{\partial x} - \dfrac{\partial M}{\partial y}}{M} = f(y), \text{ which is true.}$$

$\therefore \quad e^{\int f(y)dy}$ is an I.F. of Eq. (1).

ILLUSTRATIVE EXAMPLES 1.6

Example 1.18: Solve differential equation $y(x+y+1)dx + x(x+3y+2)dy = 0,\ y > 0.$

Solution: Given differential equation is

$$y(x+y+1)dx + x(x+3y+2)dy = 0 \qquad \ldots (1)$$

Compare Eq. (1) with $Mdx + Ndy = 0$, we get

$$M = y(x+y+1), \quad N = x(x+3y+2)$$

$$\frac{\partial M}{\partial y} = x+2y+1, \quad \frac{\partial N}{\partial x} = 2x+3y+2$$

$$\therefore \quad \frac{\dfrac{\partial N}{\partial x} - \dfrac{\partial M}{\partial y}}{M} = \frac{2x+3y+2-x-2y-1}{y(x+y+1)} = \frac{x+y+1}{y(x+y+1)} = \frac{1}{y} = f(y)$$

$$\text{I.F.} = e^{\int f(y)dy} = e^{\int \frac{1}{y}dy} = e^{\log y} = y$$

Multiplying Eq. (1) by y, we get

$$y^2(x+y+1)dx + xy(x+3y+2)dy = 0$$

which is exact and its solution is

$$\int y^2(x+y+1)dx = c \qquad [\because y \text{ is constant}]$$

$$\Rightarrow \quad y^2 \int x dx + y^2(y+1)\int 1 dx = c \quad \Rightarrow \quad \frac{x^2y^2}{2} + xy^2(y+1) = c$$

which is the required solution.

1.7.5 Fifth Rule

If the equation is of the form $x^a y^b(mydx + nxdy) + x^{a'} y^{b'}(m'ydx + n'xdy) = 0$, then $x^h y^k$ is an I.F., where $\dfrac{a+h+1}{m} = \dfrac{b+k+1}{n}$ and $\dfrac{a'+h+1}{m'} = \dfrac{b'+k+1}{n'}$.

Proof: Given equation is

$$x^a y^b(mydx + nxdy) + x^{a'} y^{b'}(m'ydx + n'xdy) = 0 \qquad \ldots (1)$$

Multiplying both sides by $x^h y^k$, we get

$$x^{a+h} y^{b+k} (mydx + nxdy) + x^{a'+h} y^{b'+k} (m'ydx + n'xdy) = 0 \qquad \ldots (2)$$

If $x^h y^k$ is an I.F. of Eq. (2), then

$$\therefore \quad \frac{\partial}{\partial y}\left[mx^{a+h} y^{b+k+1} + m'x^{a'+h} y^{b'+k+1}\right] = \frac{\partial}{\partial x}\left[nx^{a+h+1} y^{b+k} + n'x^{a'+h+1} y^{b'+k}\right]$$

$$\Rightarrow \quad m(b+k+1)x^{a+h} y^{b+k} + m'(b'+k+1)x^{a'+h} y^{b'+k}$$

$$= n(a+h+1)x^{a+h} y^{b+k} + n'(a'+h+1)x^{a'+h} y^{b'+k}$$

$$\Rightarrow \quad m(b+k+1) = n(a+h+1), \; m'(b'+k+1) = n'(a'+h+1)$$

$$\Rightarrow \quad \frac{a+h+1}{m} = \frac{b+k+1}{n}, \frac{a'+h+1}{m'} = \frac{b'+k+1}{n'}.$$

which is true.

Hence $x^h y^k$ is an I.F. of the given Eq. (1).

ILLUSTRATIVE EXAMPLES 1.7

Example 1.19: Solve differential equation $(3x^2y^4 + 2xy)dx + (2x^3y^3 - x^2)dy = 0$.

Solution: Given differential equation is

$$(3x^2y^4 + 2xy)dx + (2x^3y^3 - x^2)dy = 0 \qquad \ldots (1)$$

$$\Rightarrow \quad x^2y^3(3ydx + 2xdy) + x(2ydx - xdy) = 0$$

$\therefore$ Comparing Eq. (1) with the equation

$$x^a y^b (mydx + nxdy) + x^{a'} y^{b'} (m'ydx + n'xdy) = 0$$

we get

$$a = 2, b = 3, m = 3, n = 2, a' = 1, b' = 0, m' = 2, n' = -1$$

$$\text{I.F.} = x^h y^k \text{ where } \frac{a+h+1}{m} = \frac{b+k+1}{n}, \frac{a'+h+1}{m'} = \frac{b'+k+1}{n'}.$$

$$i.e., \quad \frac{2+h+1}{3} = \frac{3+k+1}{2}, \quad \frac{1+h+1}{2} = \frac{0+k+1}{-1}$$

$\Rightarrow \quad 6 + 2h = 12 + 3k, \; -2 - h = 2k + 2$

$\Rightarrow \quad 2h - 3k - 6 = 0, \; h + 2k + 4 = 0$

Solving these equations, we get $h = 0, k = -2$ as its solution.

$\therefore$ I.F. $= x^h y^k = x^0 y^{-2} = \dfrac{1}{y^2}$

$\therefore$ Multiplying both sides of (1) by $\dfrac{1}{y^2}$, we get

$$\left(3x^2y^2 + \frac{2x}{y}\right)dx + \left(2x^3y - \frac{x^2}{y^2}\right)dy = 0$$

which is exact and its solution is

$$3y^2\int x^2dx + \frac{2}{y}\int xdx + 0 = c$$

$\Rightarrow \quad x^3y^2 + \dfrac{x^2}{y} = c$ which is the required solution.

EXERCISE 1.2

Solve the following differential equations:

1. $\left(xy^2 - e^{\frac{1}{x^3}}\right)dx - x^2ydy = 0$
2. $(y^3 - 2yx^2)dx + (2xy^2 - x^3)dy = 0$
3. $y(xy + 2x^2y^2)dx + x(xy - x^2y^2)dy = 0$
4. $\left(xy^2 - e^{\frac{1}{x^3}}\right)dx - x^2ydy = 0$ [PTU Dec 2003, Dec 2012]
5. $(x^3 + xy^4)dx + 2y^3dy = 0$
6. $ydx - xdy + \log|x| = 0$
7. $(y^4 + 2y)dx + (xy^3 + 2y^4 - 4x)dy = 0$
8. $\left(xy\, e^{\frac{x}{y}} + y^2\right)dx - x^2\, e^{\frac{x}{y}}dy = 0$
9. $y^3dx + (x^2y^2 + 2)dy = 0$

10. $\left(x^4+y^4\right)dx-xy^3dy=0$
11. $\left(3xy^2-y^3\right)dx-\left(2x^2y-xy^2\right)dy=0$
12. $y\left(x^2y^2+2\right)dx+x\left(2-x^2y^2\right)dy=0$
13. $(1+xy)\,y\,dx+(1-xy)\,xdy=0$
14. $\left(x^3y^3+x^2y^2+xy+1\right)ydx+\left(x^3y^3-x^2y^2-xy+1\right)xdy=0$
15. $\left(xy^2+2x^2y^3\right)dx+\left(yx^2-x^3y^2\right)dy=0$
16. $\left(y^2+2x^2y\right)dx+\left(2x^3-xy\right)dy=0$
17. $\left(2x^2y-3y^4\right)dx+\left(3x^3+2xy^3\right)dy=0$
18. $\left(y+\frac{1}{3}y^3+\frac{1}{2}x^2\right)dx+\frac{1}{4}\left(x+xy^2\right)dy=0$
19. $\left(3x^2y^4+2xy\right)dx+\left(2x^3y^2-x^2\right)dy=0$
20. $(xy\sin xy+\cos xy)\,ydx+(xy\sin xy-\cos xy)\,xdy=0$
21. $(3x^2y^3e^y+y^3+y^2)dx+(x^3y^3e^y-xy)dy=0$ [PTU May 2012]

ANSWERS

1. $\frac{-y^2}{2x^2}+\frac{1}{3}e^{\frac{1}{x^3}}=c$
2. $x^2y^2\left(y^2-x^2\right)=c$
3. $\frac{-1}{xy}+2\log|x|=c$
4. $\frac{-y}{2x^2}+\frac{1}{3}e^{\frac{1}{x^3}}=c$
5. $\left(x^2-1+y^4\right)e^{x^3}=3$
6. $cx+1+y+\log|x|=0, x\neq 0$
7. $\left(y+\frac{2}{y^2}\right)x+y^2=c$
8. $e^{\frac{x}{y}}+\log x=c$
9. $\frac{1}{2}y^2x^2+\log|y|=c$
10. $y^4=4x^4\log|x|+cx^4$
11. $3\log|x|+\frac{y}{x}-2\log|y|=c$
12. $x=cye^{\frac{1}{x^2y^2}}$
13. $\frac{-1}{xy}+\log\left|\frac{x}{y}\right|=c$
14. $xy-\frac{1}{xy}-2\log|y|=c$
15. $\log x^2=\log|y|+c+\frac{1}{xy}$
16. $-x^{\frac{-3}{2}}y^{\frac{3}{2}}+6x^{\frac{1}{2}}y^{\frac{1}{2}}=c$

17. $5x^{\frac{-36}{13}}y^{\frac{24}{13}}-12x^{\frac{-10}{13}}y^{\frac{-15}{13}}=c$

18. $\frac{1}{4}x^4y+\frac{x^4y^3}{12}+\frac{x^6}{12}=c$

19. $x^3y^2+x^2y^{-1}=c$

20. $x\sec xy=cy$

21. $e^yx^3+x+\frac{x}{y}=c$

1.8 DIFFERENTIAL EQUATIONS OF THE FIRST ORDER AND HIGHER DEGREE

An equation of the form

$$P_0p^n+P_1p^{n-1}+P_2p^{n-2}+...+P_n=0 \qquad ...(1)$$

where $P_0\neq 0,\ P_1,\ P_2,...,P_n$ are functions of x and y and $p=\frac{dy}{dx}$; is a differential equation of first order and n^{th} degree. If $n>1$, then Eq. (1) is differential equation of first order but not of first degree.

These equations are generally divided into following types:

- Equations solvable for p.
- Equations solvable for y
- Equations solvable for x
- Clairaut's equation
- Equations of the type $f(y,p)=0; f(x,p)=0$.
- Equations reducible to Clairaut's form.

1.8.1 Equations Solvable for p

Suppose the equation

$$P_0p^n+P_1p^{n-1}+P_2p^{n-2}+...+P_n=0 \qquad ...(1)$$

be solvable for p so that it can be put in the form

$$[p-f_1(x,y)]\times[p-f_2(x,y)]\times...\times[p-f_n(x,y)]=0$$

$$\Rightarrow \quad p-f_1(x,y)=0,\ p-f_2(x,y)=0,\ ...,\ p-f_n(x,y)=0$$

$$\Rightarrow \quad \frac{dy}{dx}=f_r(x,y) \qquad \text{for } r=1,\ 2,\ 3,\ ...,n\,.$$

Let $\phi_r(x, y, c_r) = 0$ for $r = 1, 2, 3, ..., n$

be the solutions of above linear equations each of which is of first order, then

$$\phi_1(x, y, c_1)\times\phi_2(x, y, c_2)\times...\times\phi_n(x, y, c_n) = 0$$

is the most general solution of the given equation.

Since the given equation is of first order, it cannot have more than one arbitrary constant.

$\therefore$ we take $c_1 = c_2 = ... = c_n = c$

$\Rightarrow$ the general solution of Eq. (1) becomes

$$\phi_1(x, y, c)\times\phi_2(x, y, c)\times...\times\phi_n(x, y, c) = 0$$

ILLUSTRATIVE EXAMPLES 1.8

Example 1.19: Solve differential equation $p^3 + 2xp^2 - y^2p^2 - 2xy^2p = 0.$

Solution: Given differential equation can be written as

$$p\left(p^2 + 2xp - y^2p - 2xy^2\right) = 0 \qquad \ldots (1)$$

$$\Rightarrow \quad p\left[p(p+2x) - y^2(p+2x)\right] = 0$$

$$\Rightarrow \quad p(p+2x)\left(p - y^2\right) = 0$$

$$\Rightarrow \quad \frac{dy}{dx} = 0 \quad \text{or} \quad \frac{dy}{dx} + 2x = 0 \quad \text{or} \quad \frac{dy}{dx} - y^2 = 0 \qquad \left[\because p = \frac{dy}{dx}\right]$$

$$\Rightarrow \quad \frac{dy}{dx} = 0 \quad \text{or} \quad dy = -2x\,dx \quad \text{or} \quad \frac{dy}{y^2} = dx$$

$$\Rightarrow \quad y = c \quad \text{or} \quad y = -x^2 + c \quad \text{or} \quad -\frac{1}{y} = x + c$$

$$\Rightarrow \quad y - c = 0 \quad \text{or} \quad y + x^2 - c = 0 \quad \text{or} \quad xy + cy + 1 = 0$$

$\therefore$ the general solution of Eq. (1) is

$$(y-c)\left(y+x^2-c\right)(xy+cy+1) = 0$$

Example 1.20: Solve differential equation $p^3 = ax^4$.

Solution: The given differential equation is

$$p^3 = ax^4 \quad \Rightarrow \quad p = a^{\frac{1}{3}} x^{\frac{4}{3}} \qquad \ldots (1)$$

is the only real value of p

$$\Rightarrow \quad \frac{dy}{dx} = a^{\frac{1}{3}} x^{\frac{4}{3}}$$

Integrating w. r. t. x, we get

$$y = a^{\frac{1}{3}} \frac{3}{7} x^{\frac{7}{3}} + c \quad \text{or} \quad 7(y-c) = 3a^{\frac{1}{3}} x^{\frac{7}{3}}$$

Cubing (i.e., taking cube on both sides), we get

$$343(y-c)^3 = 27ax^7$$

which is the required general solution of Eq. (1).

Example 1.21: differential equation $xyp^2 + (x^2 + xy + y^2)p + x^2 + xy = 0.$

Solution: The given differential equation is

$$xyp^2 + (x^2 + xy + y^2)p + x^2 + xy = 0 \qquad \ldots (1)$$

which can be written as

$$px(yp + x) + x(yp + x) + y(yp + x) = 0$$

$$\Rightarrow \quad (yp + x)(px + x + y) = 0$$

$$\Rightarrow \quad y\frac{dy}{dx} + x = 0 \quad \text{or} \quad x\frac{dy}{dx} + x + y = 0$$

$$\Rightarrow \quad ydy + xdx = 0 \quad \text{or} \quad xdy + (x + y)dx = 0$$

$$\Rightarrow \quad \frac{y^2}{2} + \frac{x^2}{2} = c \quad \text{or} \quad xdy + ydx + xdx = 0$$

$$\Rightarrow \quad \frac{y^2}{2} + \frac{x^2}{2} = c \quad \text{or} \quad x.y + y.x + \frac{x^2}{2} = 0$$

$$\Rightarrow \quad x^2 + y^2 = 2c \quad \text{or} \quad xy + \frac{x^2}{2} = c$$

$$\Rightarrow \quad x^2 + y^2 - 2c = 0 \quad \text{or} \quad x^2 + 2xy = 2c$$

$\therefore$ the general solution of Eq. (1) is

$$\left(x^2 + y^2 - 2c\right)\left(x^2 + 2xy - 2c\right) = 0$$

Example 1.22: Solve differential equation $p^2 + 2py\cot x = y^2$. [PTU Jan 2009]

Solution: Given differential equation is

$$p^2 + 2py\cot x = y^2 \quad \Rightarrow \quad p^2 + 2py\cot x - y^2 = 0 \qquad \ldots (1)$$

which is quadratic in p, with $a = 1,\ b = 2y\cot x,\ c = -y^2$, therefore

$$p = \frac{-2y\cot x \pm \sqrt{(2y\cot x)^2 - 4\times 1\times(-y^2)}}{2} \qquad \left[\begin{array}{l} ax^2 + bx + c = 0 \\ \Rightarrow x = \dfrac{-b \pm \sqrt{b^2 - 4ac}}{2a} \end{array}\right]$$

$$= \frac{-2y\cot x \pm \sqrt{4y^2\cot^2 x + 4y^2}}{2} = \frac{-2y\cot x \pm 2y\sqrt{\cot^2 x + 1}}{2}$$

$$= \frac{2y(-\cot x \pm \sqrt{\text{cosec}^2 x})}{2} \qquad \left[\because\ \text{cosec}^2 x = \cot^2 x + 1\right]$$

$$= y(-\cot x \pm \text{cosec}x)$$

Thus,

$$p = y\left(-\cot x + \text{cosec } x\right) \qquad \ldots (2)$$

or

$$p = y\left(-\cot x - \text{cosec } x\right) \qquad \ldots (3)$$

From Eq. (2), we have

$$\frac{dy}{dx} = y\left(-\cot x + \text{cosec } x\right) \quad \text{or} \quad \frac{dy}{y} = \left(-\cot x + \text{cosec } x\right)dx$$

Integrating, we get

$$\log y = -\log\sin x + \log\tan\frac{x}{2} + \log c = \log\frac{c\tan\frac{x}{2}}{\sin x}$$

$$\Rightarrow \quad y = \frac{c\tan\frac{x}{2}}{2\sin\frac{x}{2}\cos\frac{x}{2}} = \frac{c\frac{\sin\frac{x}{2}}{\cos\frac{x}{2}}}{2\sin\frac{x}{2}\cos\frac{x}{2}} = \frac{c}{2\cos^2\frac{x}{2}} \qquad \left[\because \sin x = 2\sin\frac{x}{2}\cos\frac{x}{2}\right]$$

$$\Rightarrow \quad y\cos^2\frac{x}{2} = c', \quad \text{where } c' = \frac{c}{2}$$

$$\Rightarrow \quad y\cos^2\frac{x}{2} - c' = 0$$

From Eq. (3), we have

$$\frac{dy}{dx} = y(\cot x - \operatorname{cosec} x) \quad \text{or} \quad \frac{dy}{y} = (\cot x - \operatorname{cosec} x)\,dx$$

Integrating, we get

$$\log y = -\log\sin x - \log\tan\frac{x}{2} + \log c$$

$$= \log\frac{c}{\sin x\tan\frac{x}{2}}$$

$$= \log\frac{c}{\left(2\sin\frac{x}{2}\cos\frac{x}{2}\right)\left(\frac{\sin\frac{x}{2}}{\cos\frac{x}{2}}\right)} = \log\frac{c}{2\sin^2\frac{x}{2}} \qquad \left[\because \sin x = 2\sin\frac{x}{2}\cos\frac{x}{2}\right]$$

$$\Rightarrow \quad y = \frac{c}{2\sin^2\frac{x}{2}}$$

$$\Rightarrow \quad y\sin^2\frac{x}{2} = c'' \quad \text{where } c'' = \frac{c}{2}$$

$$\Rightarrow \quad y\sin^2\frac{x}{2} - c'' = 0$$

$\therefore$ The general solution of the given equation is

$$\left(y\cos^2\frac{x}{2} - c'\right)\left(y\sin^2\frac{x}{2} - c''\right) = 0$$

EXERCISE 1.3

Solve the following equations:

1. (*a*) $p^2-5p+6=0$

   (*b*) $p^2-7p+12=0$ [PTU Dec 2006]

   (*c*) $p^2-9p+18=0$

   (*d*) $p^2-ax=0$

2. $xy\left(\frac{dy}{dx}\right)^2-(x^2+y^2)\frac{dy}{dx}+xy=0$

3. $yp^2+(x-y)p-x=0$

4. $x^2\left(\frac{dy}{dx}\right)^2+3xy\frac{dy}{dx}+2y^2=0$

5. $\frac{dy}{dx}-\frac{dx}{dy}=\frac{x}{y}-\frac{y}{x}$

6. $p^2-2p\sinh x-1=0$

7. $p(p+y)=x(x+y)$ [PTU May 2007]

8. $4y^2p^2+2pxy(3x+1)+3x^3=0$

ANSWERS

1. (*a*) $(y-2x-c)(y-3x-c)=0$ (*b*) $(y-4x+c)(y-3x+c)=0$

   (*c*) $(y-6x+c)(y-3x+c)=0$ (*c*) $25(y+c)^2=4ax^5$

2. $(y^2-x^2-c)(y-cx)=0$

3. $(y-x-c)(x^2+y^2-c)=0$

4. $(xy-c)(x^2y-c)=0$

5. $(xy-c)(x^2-y^2-c)=0$

6. $(y-e^x-c)(y-e^{-x}-c)=0$

7. $\left(y-\frac{1}{2}x^2+c\right)(y+c+ce^{-x}-1)=0$

8. $(y^2+x^3-c)\left(y^2+\frac{1}{2}x^2-c\right)=0$

1.8.2 Equations Solvable for y

An equation $F(x, y, p)=0$ is said to be solvable in y, if it can be expressed as

$$y=\phi(x, p) \qquad \ldots(1)$$

Differentiating w.r.t. x, we get

$$\frac{dy}{dx}=p=f\left(x, p, \frac{dp}{dx}\right) \qquad \ldots(2)$$

which is a differential equation in two variables p and x.

Suppose

$$\psi(p, x, c) = 0 \qquad \ldots (3)$$

be the solution of the Eq. (2).

Eliminating p from Eq. (1) and Eq. (3), we get the solution of the given equation.

Note I: If the elimination of p from Eq. (1) and Eq. (3) is not possible, then express Eq. (1) and Eq. (3) as

$$x = \psi_1(p, c),\ y = \psi_2(p, c)$$

These two relations taken together constitute the general solution of Eq. (1).

Note II: If the equation $p - f\left(x, p, \frac{dp}{dx}\right) = 0$, *i.e.*, Eq. (2) can be written as

$$f_1(x, p) \times f_2\left(x, p, \frac{dp}{dx}\right) = 0$$

Here we neglect the factor $f_1(x, p)$ which does not contain $\frac{dp}{dx}$ and then proceed as discussed above to get the required solution.

Remark: We neglect the factor $f_1(x, p)$ because if $f_1(x, p) = 0$, then eliminating p from this equation and Eq. (1), we don't get a general solution.

ILLUSTRATIVE EXAMPLES 1.9

Example 1.23: Solve differential equation $y + px = x^4 p^2$.

Solution: Given differential equation is

$$y = -px + x^4 p^2 \qquad \ldots (1)$$

Differentiating both sides w. r. t. x, we get

$$\frac{dy}{dx} = p = -p - x\frac{dp}{dx} + 4x^3 p^2 + 2x^4 p\frac{dp}{dx}$$

$$\Rightarrow \quad 2p + x\frac{dp}{dx} - 2px^3\left(2p + x\frac{dp}{dx}\right) = 0 \quad \text{or} \quad \left(2p + x\frac{dp}{dx}\right)(1 - 2px^3) = 0$$

$$\Rightarrow \quad \left(2p + x\frac{dp}{dx}\right)\left(1 - 2px^3\right) = 0$$

Discarding the factor $\left(1 - 2px^3\right)$, we have

$$2p + x\frac{dp}{dx} = 0 \quad \text{or} \quad \frac{dp}{p} + 2\frac{dx}{x} = 0$$

Integrating, we get

$$\log|p| + 2\log|x| = \log c, \quad c > 0$$

$$\Rightarrow \quad \log\left|px^2\right| = \log c \quad \Rightarrow \quad px^2 = c \quad \Rightarrow \quad p = \frac{c}{x^2}$$

Putting this value of p in Eq. (1), we have

$$y = \frac{-c}{x} + c^2$$

which is the required solution.

Example 1.24: Solve differential equation $y = 2px - p^2$

Solution: The given differential equation is

$$y = 2px - p^2 \qquad \ldots (1)$$

Differentiating both sides w. r. t. x, we get

$$\frac{dy}{dx} = p = 2p + 2x\frac{dp}{dx} - 2p\frac{dp}{dx}$$

$$\Rightarrow \quad p + (2x - 2p)\frac{dp}{dx} = 0$$

$$\Rightarrow \quad p\frac{dx}{dp} + 2x - 2p = 0$$

$$\Rightarrow \quad \frac{dx}{dp} + \frac{2}{p}x = 2 \qquad \ldots (2)$$

which is a linear equation

$$\therefore \quad \text{I.F.} = e^{\int \frac{2}{p}dp} = e^{2\log p} = p^2$$

∴ The solution of Eq. (2) is

$$x(I.F.) = \int 2(I.F.)\,dp + c$$

$$\Rightarrow \quad xp^2 = \int 2p^2\,dp + c$$

$$\Rightarrow \quad x = \frac{2}{3}p + cp^{-2} \qquad \ldots (3)$$

Putting this value of x in Eq. (1), we have

$$y = 2p\left(\frac{2}{3}p + cp^{-2}\right) - p^2$$

$$\Rightarrow \quad y = \frac{1}{3}p^2 + 2cp^{-1} \qquad \ldots (4)$$

Eqs. (3) and (4) together constitute the general solution of Eq. (1).

EXERCISE 1.4

Solve the following equations:

1. $xp^2 - 2yp + ax = 0$
2. $y - 2px = \tan^{-1}\left(xp^2\right)$ [PTU May 2010]
3. $16x^2 + 2p^2y - p^3x = 0$
4. $y = x + 2\tan^{-1} p$
5. $y = 3x + \log p$
6. $x - yp = ap^2$
7. $x^2\left(\frac{dy}{dx}\right)^4 + 2x\frac{dy}{dx} - y = 0$

ANSWERS

1. $2y = cx^2 + \frac{a}{c}$
2. $y = \sqrt{cx} + \tan^{-1} c$
3. $16 + 2c^2y - c^3x^2 = 0$
4. $x + c = \log\frac{p-1}{\sqrt{p^2+1}} - \tan^{-1} p$
5. $y = 3x + \log\frac{3}{1-ce^{3x}}$
6. $x = \frac{p}{\sqrt{1-p^2}}\left(c + a\sin^{-1} p\right)$, $y = \frac{1}{\sqrt{1-p^2}}\left(c + a\sin^{-1} p\right) - ap$
7. $y = c^2 + 2\sqrt{cx}$

1.8.3 Equations Solvable for x

An equation $F(x, y, p)$ is solvable for x if it can be put in the form

$$x=\phi(y, p) \qquad \ldots (1)$$

Differentiating Eq. (1) w.r.t. y, we get

$$\frac{dx}{dy}=\frac{1}{p}=f\left(y, p, \frac{dp}{dy}\right) \qquad \ldots (2)$$

which is an equation in two variables p and y.

Suppose

$$\psi(p, y, c)=0 \qquad \ldots (3)$$

be the solution of Eq. (2). Eliminating p from Eq. (1) and Eq. (3), we get the solution of Eq. (1).

Note I: If the elimination of p from Eq. (1) and Eq. (3) is not possible, then express Eq. (1) and Eq. (3) as

$$x=\psi_1(p, c),\ y=\psi_2(p, c)$$

These two relations taken together constitute the general solution of Eq. (1).

Note II: If the equation $\frac{1}{p}-f\left(y, p, \frac{dp}{dy}\right)=0$, *i.e.*, Eq. (2) can be written as

$$f_1(y, p)\times f_2\left(y, p, \frac{dp}{dy}\right)=0$$

Here we neglect the factor $f_1(y, p)$ which does not contain $\frac{dp}{dy}$ and then proceed as discussed above to get the required solution.

Remark: We neglect the factor $f_1(y, p)$ because if $f_1(y, p)=0$, then eliminating p from this equation and Eq. (1), we don't get a general solution.

ILLUSTRATIVE EXAMPLES 1.3

Example 1.25: Solve differential equation $y+px=x^4 p^2$.

Solution: Given differential equation is $y = -px + x^4 p^2$ . . . (1)

Differentiating both sides w. r. t. x, we get

$$\frac{dy}{dx} = p = -p - x\frac{dp}{dx} + 4x^3 p^2 + 2x^4 p\frac{dp}{dx}$$

$$\Rightarrow \quad 2p + x\frac{dp}{dx} - 2px^3\left(2p + x\frac{dp}{dx}\right) = 0$$

$$\Rightarrow \quad \left(2p + x\frac{dp}{dx}\right)\left(1 - 2px^3\right) = 0$$

Discarding the factor $\left(1 - 2px^3\right)$, we have

$$2p + x\frac{dp}{dx} = 0 \quad \Rightarrow \quad \frac{dp}{p} + 2\frac{dx}{dp} = 0$$

Integrating, we get

$$\log p + 2\log x = \log c \quad \Rightarrow \quad \log px^2 = \log c$$

$$\Rightarrow \quad px^2 = c \quad \Rightarrow \quad p = \frac{c}{x^2}$$

Putting this value of p in Eq. (1), we have

$$y = \frac{-c}{x} + c^{-2}$$

which is the required solution.

Example 1.26: Solve differential equation $ayp^2 + (2x - b)p - y = 0$.

Solution: The given differential equation can be written as

$$2x = \frac{y}{p} - ayp + b \qquad \ldots (1)$$

Differentiating Eq. (1) w.r.t y, we have

$$2\frac{dx}{dy} = \frac{1}{p} - \frac{y}{p^2}\frac{dp}{dy} - ap - ay\frac{dp}{dy}$$

$$\Rightarrow \quad \frac{2}{p} = \frac{1}{p} - \frac{y}{p^2}\frac{dp}{dy} - ap - ay\frac{dp}{dy}$$

$$\Rightarrow \qquad \frac{1}{p}+\frac{y}{p^2}\frac{dp}{dy}+ap+ay\frac{dp}{dy}=0$$

$$\Rightarrow \qquad y\frac{dp}{dy}\left(\frac{1}{p^2}+a\right)+p\left(\frac{1}{p^2}+a\right)=0$$

$$\Rightarrow \qquad \left(\frac{1}{p^2}+a\right)\left(y\frac{dp}{dy}+p\right)=0$$

Neglecting $\left(\frac{1}{p^2}+a\right)$ term, we get

$$p+y\frac{dp}{dy}=0 \qquad \Rightarrow \qquad pdy+ydp=0 \qquad \Rightarrow \qquad d(py)=0$$

Integrating, we get

$$py=c \qquad \Rightarrow \qquad p=\frac{c}{y}$$

Putting in Eq. (1), we get

$$2x=\frac{y^2}{c}-ac+b$$

which is the required solution.

Example 1.27: Solve differential equation $p=\tan\left(x-\frac{p}{1+p^2}\right)$

Solution: Given differential equation is

$$p=\tan\left(x-\frac{p}{1+p^2}\right) \qquad \Rightarrow \qquad \tan^{-1}p=x-\frac{p}{1+p^2}$$

Solving for *x*, we have

$$x=\tan^{-1}p+\frac{p}{1+p^2} \qquad \ldots (1)$$

Differentiating both sides w. r. t *y*, we get

$$\frac{dx}{dy}=\frac{1}{p}=\frac{1}{1+p^2}\frac{dp}{dy}+\frac{(1+p^2)(2p^2)}{(1+p^2)^2}\frac{dp}{dy}$$

$$\Rightarrow \quad \frac{1}{p}=\frac{2\left(1+p^2\right)-2p^2}{\left(1+p^2\right)^2}\frac{dp}{dy} \qquad \Rightarrow \qquad dy=\frac{2p}{\left(1+p^2\right)}dp$$

Integrating, we get

$$y=c-\frac{1}{1+p^2} \qquad \ldots (2)$$

Eqs. (1) and (2) together constitute the general solution.

EXERCISE 1.5

Solve the following equations:

1. (*a*) $x=4p+4p^3$

   (*b*) $x^2=a^2\left(1+p^2\right)$

2. $y=3px+6p^2y^2$

3. $y=2px+p^2y$ [PTU Dec 2012]

4. $p^3-4xyp+8y^2=0$

5. $y^2\log y=xyp+p^2$

6. $x=y+a\log p$

ANSWERS

1. (*a*) $x=4p+4p^3$ and $y=c-p^2-2p-2\log|p-1|$

   (*b*) $x=a\sqrt{1+p^2}$ and $2y+c=ap\sqrt{1+p^2}-\log\left|p+\sqrt{1+p^2}\right|$

2. $y^3=3cx+6c^2$

3. $y=\frac{2cx}{y}+\frac{c}{y^2}$

4. $64y=c\left(c-4x\right)^2$

5. $\log y=cx+c^2$

6. $y=c-a\log|1-p|$

1.9 CLAIRAUT'S EQUATION

[PTU May 2007, Jan 2009]

An equation of the form

$$y = px + f(p) \qquad \dots (1)$$

is known as Clairaut's equation

Differentiating Eq. (1) w.r.t x, we get

$$p = p + x\frac{dp}{dx} + f(p)\frac{dp}{dx} \quad \text{or} \quad [x + f'(p)]\frac{dp}{dx} = 0$$

Discarding the factor $[x + f'(p)]$, we have

$$\frac{dp}{dx} = 0$$

Integrating it, we get

$$p = c$$

Putting $p = c$ in (1), we get

$$p = cx + f(c)$$

as the required solution.

Thus, the solution of Clariaut's equation is obtained by writing c for p.

Note: Sometimes, the equation is not in Clairaut's form, but can be reduced to Clairaut's form by suitable substitutions or transformations

Following are some suggestive useful transformations:

- If equation contains y^2; $2py$; Put $y^2 = Y$.
- If equation contains x^2; $\frac{p}{2x}$; Put $x^2 = X$.
- If equation contains x^2, y^2, then put $x^2 = X,\ y^2 = Y$ so that

$$\frac{ydy}{xdx} = \frac{dY}{dX} \quad \Rightarrow \quad \frac{y}{x}p = P, \text{ where } \boxed{\frac{dY}{dX} = P}$$

- If equation contains e^x, e^y then put $e^x = X$, $e^y = Y$ so that

$$\frac{e^y dy}{e^x dx} = \frac{dY}{dX} \quad \text{or} \quad \boxed{p = \frac{X}{Y} P}$$

Sometimes it may be more useful to put $e^{ax} = X$ and $e^{by} = Y$ and then proceed as above.

ILLUSTRATIVE EXAMPLES 1.4

Example 1.28: Solve differential equation $y = p \sin p + \cos p$.

Solution: The given differential equation is

$$y = p \sin p + \cos p \qquad \ldots (1)$$

Differentiating Eq. (1) w. r. t. x, we get

$$\frac{dy}{dx} = (p \cos p + \sin p - \sin p)\frac{dp}{dx}$$

$$\Rightarrow \quad p = p \cos p \frac{dp}{dx} \quad \text{or} \quad dx = \cos p dp$$

Integrating, we get

$$x = \sin p + c$$

which is the required solution.

Example 1.29: Solve differential equation $x + \dfrac{p}{\sqrt{1+p^2}} = a$.

Solution: The given differential equation is

$$x + \frac{p}{\sqrt{1+p^2}} = a \qquad \ldots (1)$$

Differentiating Eq. (1) w. r. t. y, we get

$$\frac{dx}{dy} + \frac{\sqrt{1+p^2}.1 - p\dfrac{2p}{2\sqrt{1+p^2}}}{(1+p^2)} \frac{dp}{dy} = 0$$

1.10 LEIBNIZ'S LINEAR EQUATION

[PTU June 2003, May 2005, May 2007]

A differential equation is said to be linear if the dependent variable and its derivative occur only in the first degree and are not multiplied together.

The general form of linear differential equation of first order is

$$\frac{dy}{dx} + Py = Q \qquad \ldots (1)$$

where P and Q are either constants or functions of x only.

Sometimes a differential equation takes linear form if we regard x as a dependent variable and y as independent variable. The equation then can be put as

$$\frac{dx}{dy} + Px = Q \qquad \ldots (2)$$

where P and Q are either constants or functions of y only.

Equations (1) and (2) are also known as *Leibnitz's linear equations.*

Solving Leibnitz's Linear Equations

[PTU Dec 2006]

Let us first take linear equation of the type

$$\frac{dy}{dx} + Py = Q$$

Multiply both sides by its integrating factor $e^{\int P dx}$, to get

$$\frac{dy}{dx} e^{\int P dx} + Pye^{\int P dx} = Qe^{\int P dx}$$

$$\Rightarrow \quad \frac{dy}{dx} e^{\int P dx} + y\left(Pe^{\int P dx}\right) = Qe^{\int P dx}$$

$$\Rightarrow \quad \frac{d}{dx}\left(ye^{\int P dx}\right) = Qe^{\int P dx}$$

Integrating both sides, we get

$$ye^{\int P dx} = \int Q e^{\int P dx} dx + c$$

It can be written as

$$y(I.F.) = \int Q.(I.F.) dx + c$$

which is the required solution.

Note: For linear equation of the type

$$\frac{dx}{dy} + Px = Q$$

The integrating factor for this linear equation is $e^{\int P dy}$.

Proceeding on the similar lines, we get

$$xe^{\int P dy} = \int Q e^{\int P dy} dy + c$$

It can be written as

$$x(I.F.) = \int Q.(I.F.) dy + c$$

which is the required solution.

ILLUSTRATIVE EXAMPLES 1.5

Example 1.31: Solve differential equation $\frac{dy}{dx} + \frac{y}{x} = e^x, \; x > 0.$

Solution: Given differential equation is

$$\frac{dy}{dx} + \frac{y}{x} = e^x \qquad \ldots (1)$$

Comparing Eq. (1) with

$$\frac{dy}{dx} + Py = Q$$

we have $P = \frac{1}{x}$ and $Q = e^x$

It's I.F. $= e^{\int P dx} = e^{\int \frac{1}{x} dx} = e^{\log x} = x$

$\therefore$ the solution is

$$y(I.F.) = \int Q.(I.F.)dx + c$$

$$\Rightarrow \quad y.x = \int e^x.x dx + c$$

$$xy = xe^x - \int e^x dx + c \qquad \text{integrating by parts.}$$

$$= xe^x - e^x + c = (x-1)e^x + c$$

$$\Rightarrow \quad y = \frac{(x-1)}{x}e^x + \frac{1}{x}c \qquad \text{which is the required solution.}$$

Example 1.32: Solve differential equation $x\log x\frac{dy}{dx} + y = 2\log x.$

Solution: Given differential equation is

$$x\log x\frac{dy}{dx} + y = 2\log x \qquad \text{. . . (1)}$$

Dividing both sides by $x\log x$, we get

$$\frac{dy}{dx} + \frac{y}{x\log x} = \frac{2}{x}$$

Comparing Eq. (1) with $\frac{dy}{dx} + Py = Q$, we have

$$P = \frac{1}{x\log x} \quad \text{and} \quad Q = \frac{2}{x}$$

It's I.F. $= e^{\int P dx} = e^{\int \frac{1}{x\log x} dx} = e^{\int \frac{\frac{1}{x}}{\log x} dx} = e^{\log\log x} = \log x$

$\therefore$ the solution is

$$y(I.F.) = \int Q.(I.F.)dx + c$$

$$\Rightarrow \quad y.\log x = \int \frac{2}{x}.\log x dx + c$$

$$\Rightarrow \quad y.\log x = 2\int \log x.\left(\frac{1}{x}\right)dx + c \qquad \left[\because \int [f(x)]^n f'(x)dx = \frac{[f(x)]^{n+1}}{n+1}\right]$$

$$= 2\frac{(\log x)^2}{2} + c = (\log x)^2 + c$$

$$\Rightarrow \quad y = \log x + \frac{1}{\log x}c$$ which is the required solution.

Example 1.33: Find the particular solution of the differential equation

$$\frac{dy}{dx} + y\cot x = 4x \operatorname{cosec} x \qquad (x \neq 0)$$

given that $y = 0$ when $x = \frac{\pi}{2}$.

Solution: Given differential equation is

$$\frac{dy}{dx} + y\cot x = 4x \operatorname{cosec} x \qquad \ldots(1)$$

Comparing Eq. (1) with $\frac{dy}{dx} + Py = Q$, we get

$$P = \cot x \quad \text{and} \quad Q = 4x \operatorname{cosec} x$$

It's I.F. $= e^{\int P dx} = e^{\int \cot x dx} = e^{\log \sin x} = \sin x$

$\therefore$ the solution is

$$y(I.F.) = \int Q.(I.F.)dx + c$$

$$\Rightarrow \quad y.\sin x = \int 4x \operatorname{cosec} x \sin x dx + c = \int 4x dx + c$$

$$\Rightarrow \quad y \sin x = 4\frac{x^2}{2} + c = 2x^2 + c$$

Given that $y = 0$ when $x = \dfrac{\pi}{2}$

$$\therefore \quad 0.\sin\frac{\pi}{2} = 2\left(\frac{\pi}{2}\right)^2 + c \Rightarrow 0 = \frac{\pi^2}{2} + c \qquad \Rightarrow \quad c = -\frac{\pi^2}{2}$$

$$\therefore \quad y\sin x = 2x^2 - \frac{\pi^2}{2}$$ which is the required solution.

Example 1.34: Solve differential equation $x\left(1-x^2\right)\dfrac{dy}{dx} + \left(2x^2 - 1\right)y = x^3$.

Solution: Given differential equation is

$$x\left(1-x^2\right)\frac{dy}{dx} + \left(2x^2-1\right)y = x^3 \qquad \ldots (1)$$

Dividing by $x\left(1-x^2\right)$, we get

$$\frac{dy}{dx} + \frac{2x^2-1}{x\left(1-x^2\right)}y = \frac{x^2}{1-x^2} \qquad \ldots (2)$$

Comparing Eq. (2) with $\dfrac{dy}{dx} + Py = Q$, we get

$$P = \frac{2x^2-1}{x\left(1-x^2\right)}, \; Q = \frac{x^2}{1-x^2}$$

Now, $P = \dfrac{2x^2-1}{x(1-x)(1+x)} = -\dfrac{1}{x} + \dfrac{1}{2(1-x)} - \dfrac{1}{2(1+x)}$ [By partial fractions]

$$\int P dx = -\log x - \frac{1}{2}\log(1-x) - \frac{1}{2}\log(1+x) = -\log\left[x(1-x)^{\frac{1}{2}}(1+x)^{\frac{1}{2}}\right]$$

$$= -\log\left[x\left(1-x^2\right)^{\frac{1}{2}}\right] = \log\frac{1}{x\sqrt{1-x^2}}$$

$$\therefore \quad I.F. = e^{\int P dx} = e^{\log\frac{1}{x\sqrt{1-x^2}}} = \frac{1}{x\sqrt{1-x^2}}$$

Thus, the solution is

$$y(I.F.)=\int Q(I.F.)dx+c$$

$$\Rightarrow \quad y.\frac{1}{x\sqrt{1-x^2}}=\int\frac{x^2}{1-x^2}\frac{1}{x\sqrt{1-x^2}}dx+c \quad \left[\because \int[f(x)]^n f(x)dx=\frac{[f(x)]^{n+1}}{n+1}, n\neq -1\right]$$

$$=-\frac{1}{2}\int(1-x^2)^{-\frac{3}{2}}(-2x)dx+c=(1-x^2)^{-\frac{1}{2}}+c$$

$$\Rightarrow \quad y=x+cx\sqrt{1-x^2}$$

which us the required solution.

Example 1.35: Solve differential equation $\left(\frac{e^{-2\sqrt{x}}}{\sqrt{x}}-\frac{y}{\sqrt{x}}\right)\frac{dx}{dy}=1$

Solution: Given differential equation can be written as

$$\frac{dy}{dx}+\frac{y}{\sqrt{x}}=\frac{e^{-2\sqrt{x}}}{\sqrt{x}} \qquad \ldots (1)$$

Comparing it with $\frac{dy}{dx}+Py=Q$, we have

$$P=\frac{1}{\sqrt{x}}, \; Q=\frac{e^{-2\sqrt{x}}}{\sqrt{x}}$$

$$\therefore \quad I.F.=e^{\int Pdx}=e^{\int\frac{1}{\sqrt{x}}dx}=e^{\frac{x^{1/2}}{1/2}}=e^{2\sqrt{x}}$$

$\therefore$ The solution is

$$y(I.F.)=\int Q(I.F.)dx+c$$

$$\Rightarrow \quad ye^{2\sqrt{x}}=\int\frac{e^{-2\sqrt{x}}}{\sqrt{x}}.e^{2\sqrt{x}}dx+c$$

$$\Rightarrow \quad ye^{2\sqrt{x}}=\int x^{-\frac{1}{2}}dx+c=2\sqrt{x}+c$$

which is the required solution.

Example 1.36: Solve differential equation $\left(1+y^2\right)dx=\left(\tan^{-1}y-x\right)dy$.

Solution: Given differential equation can be written as

$$\left(1+y^2\right)\frac{dx}{dy}+x=\tan^{-1}y \qquad \ldots (1)$$

Dividing it by $\left(1+y^2\right)$, we get

$$\frac{dx}{dy}+\frac{x}{1+y^2}=\frac{\tan^{-1}y}{1+y^2}$$

Comparing Eq. (1) with $\frac{dx}{dy}+Px=Q$, we get

$$P=\frac{1}{1+y^2},\ Q=\frac{\tan^{-1}y}{1+y^2}$$

It's $I.F.=e^{\int Pdx}=e^{\int\frac{1}{1+y^2}dy}=e^{\tan^{-1}y}$

$\therefore$ The solution is

$$x(I.F.)=\int Q(I.F.)dy+c=\int\left(\frac{\tan^{-1}y}{1+y^2}\right).e^{\tan^{-1}y}dy+c$$

$$\Rightarrow \qquad xe^{\tan^{-1}y}=\int\left(\frac{\tan^{-1}y}{1+y^2}\right).e^{\tan^{-1}y}dy+c$$

Put $t=\tan^{-1}y,\ dt=\frac{1}{1+y^2}dy$

$$\therefore \qquad xe^t=\int te^t\,dt+c=te^t-\int 1.e^t\,dt+c=te^t-\int e^t\,dt+c=te^t-e^t+c$$

$$\Rightarrow \qquad xe^{\tan^{-1}y}=\tan^{-1}ye^{\tan^{-1}y}-e^{\tan^{-1}y}+c$$

$$\Rightarrow \qquad x=\tan^{-1}y-1+ce^{-\tan^{-1}y}$$

EXERCISE 1.7

Solve the following differential equations:

1. $\frac{dy}{dx}+\frac{y}{x}=x^3-3$
2. $x\log x\frac{dy}{dx}+y=2\log x$
3. $(x+1)\frac{dy}{dx}-y=e^x(x+1)^2$
4. $(x^2+1)\frac{dy}{dx}+2xy=x^2$
5. $\cos^2 x\frac{dy}{dx}+y=\tan x$
6. $(1+x^3)\frac{dy}{dx}+6x^2y=1+x^2$
7. $\frac{dy}{dx}+y\cot x=\cos x$
8. $(1+x^2)\frac{dy}{dx}+y=e\tan^{-1}x$
9. $\frac{dy}{dx}+y\cot x=4x\operatorname{cosec} x$, if $y=0$ where $x=\frac{\pi}{2}$
10. $\frac{dy}{dx}+y\cot x=5e^{\cos x}$, if $y=-4$ where $x=\frac{\pi}{2}$
11. $\frac{dy}{dx}=y\tan x=3e^{-\sin x}$, if $y=4$ where $x=0$
12. $(1-x^2)\frac{dy}{dx}+2xy=x\sqrt{1-x^2}$
13. $x\frac{dy}{dx}+y=e^x-xy$
14. $\left(1+y^2\right)+\left(x-e-\tan^{-1}y\right)\frac{dy}{dx}=0$
15. $e^{-y}\sec^2 y\,dy=dx+x\,dy$
16. $\left(x+2y^3\right)\frac{dy}{dx}=y$

17. $ye^{g}\,dx = \left(g^{2} + 2xe^{g}\right)dy$

18. $\dfrac{dy}{dx} = \dfrac{y}{2y\log y + y - x}$

19. $x^{2}\dfrac{dy}{dx} = 3x^{2} - 2xy + 1$

20. $\sqrt{1-y^{2}}\,dx = \left(\sin^{-1} y - x\right)dy$

21. $x\left(\dfrac{dy}{dx} + y\right) = 1 - y$ [PTU May 2012]

ANSWERS

1. $10xy = 2x^{5} - 15x^{2} + c$

2. $y\log x = (\log x)^{2} + c$

3. $y = (x+1)\left(e^{x} + c\right)$

4. $y\left(x^{2}+1\right) = \dfrac{x^{3}}{3} + c$

5. $y = \tan x - 1 + ce^{-\tan x}$

6. $y\left(1+x^{3}\right)^{2} = y + \dfrac{x^{3}}{3} + \dfrac{x^{4}}{4} + \dfrac{x^{6}}{6} + c$

7. $y\sin x = \dfrac{1}{2}\sin^{2} x + c$

8. $2ye^{\tan^{-1} x} = e^{2\tan^{-1} x} + c$

9. $y\sin x = 2x^{2} - \dfrac{\pi^{2}}{2}$

10. $y\sin x = 5e^{\cos x} - 9$

11. $y\cos x = 7 - 3e^{-\sin x}$

12. $y = \sqrt{1-x^{2}} + c\left(1-x^{2}\right)$

13. $xy = \dfrac{1}{2}e^{x} + ce^{-x}$

14. $x = e^{-\tan^{-1} y}\left(\tan^{-1} y + c\right)$

15. $xe^{y} = \tan y + c$

16. $x + y^{3} + cy$

17. $x = y^{2}\left(1 - e^{-y}\right)$

18. $x = \dfrac{c}{y} + y\log y$

19. $y = x + x^{-1} + cx^{-2}$

20. $x = \sin^{-1} y - 1 + ce^{-\sin^{-1} y}$

21. $y = \dfrac{1}{x} + \dfrac{c}{xe^{x}}$

1.11 BERNOULLI'S EQUATION

[PTU May 2007, Jan 2009]

A differential equation of the form

$$\frac{dy}{dx}+Py=Qy^n \qquad \ldots (1)$$

where P and Q are either constants or functions of x only, is known as *Bernoulli's equation*.

It can be reduced to linear form.

Dividing both sides of (1) by y^n, we get

$$y^{-n}\frac{dy}{dx}+Py^{1-n}=Q \qquad \ldots (2)$$

Put $y^{1-n}=t$, $(1-n)y^{-n}\dfrac{dy}{dx}=\dfrac{dt}{dx}$

$$\Rightarrow \qquad y^{-n}\frac{dy}{dx}=\frac{1}{1-n}\frac{dt}{dx}$$

$\therefore$ Eq. (2) becomes

$$\frac{1}{1-n}\frac{dt}{dx}+Pt=Q \quad \Rightarrow \quad \frac{dt}{dx}+P(1-n)t=Q(1-n) \qquad \ldots (3)$$

which is a linear differential equation with t as dependent variable and x as independent variable.

ILLUSTRATIVE EXAMPLES 1.6

Example 1.37: Solve differential equation $2\dfrac{dy}{dx}=\dfrac{y}{x}+\dfrac{y^2}{x^2}$.

Solution: Given differential equation is

$$2\frac{dy}{dx}=\frac{y}{x}+\frac{y^2}{x^2}$$

$$\Rightarrow \quad \frac{dy}{dx} - \frac{1}{2x} y = \frac{1}{x^2} y^2 \qquad \ldots (1)$$

Comparing it with

$$\frac{dy}{dx} + Py = Qy^n$$

we get

$$n = 2$$

Dividing Eq. (1) by y^2, we get

$$y^{-2} \frac{dy}{dx} - \frac{1}{2x} y^{-1} = \frac{1}{x^2} \qquad \ldots (2)$$

Put $y^{-1} = t$, $\quad (-1) y^{-2} \frac{dy}{dx} = \frac{dt}{dx} \quad \Rightarrow \quad y^{-2} \frac{dy}{dx} = -\frac{dt}{dx}$

Substituting these values, Eq. (2) becomes

$$-\frac{dt}{dx} - \frac{1}{2x} t = \frac{1}{x^2}$$

$$\Rightarrow \quad \frac{dt}{dx} + \frac{1}{2x} t = -\frac{1}{x^2} \qquad \ldots (3)$$

It's I.F. $= e^{\int P dx} = e^{\int \frac{1}{2x} dx} = e^{\frac{1}{2} \int \frac{1}{x} dx} = e^{\frac{1}{2} \log x} = e^{\log \sqrt{x}} = \sqrt{x}$

Multiplying both sides of (3) by $\sqrt{x}$, we get

$$\sqrt{x} \frac{dt}{dx} + \frac{1}{2\sqrt{x}} t = -\frac{1}{x^{3/2}}$$

$$\Rightarrow \quad \sqrt{x} \frac{dt}{dx} + \frac{1}{2\sqrt{x}} t = -x^{-3/2} \quad \Rightarrow \quad \frac{d}{dx}\left(\sqrt{x}.t\right) = -x^{-3/2}$$

Integrating both sides, we get

$$\sqrt{x}.t = \int -x^{-3/2} dx + c = -\frac{x^{\frac{-1}{2}}}{-\frac{1}{2}} + c = \frac{2}{\sqrt{x}} + c$$

$\Rightarrow \quad \sqrt{x}.t = \frac{2}{\sqrt{x}} + c$

$\Rightarrow \quad t = \frac{2}{x} + \frac{c}{\sqrt{x}}$

Replacing t by y^{-1}, we get

$\frac{1}{y} = \frac{2}{x} + \frac{c}{\sqrt{x}}$ which is the required equation.

Example 1.38: Solve differential equation $\frac{dy}{dx} + \frac{x}{1-x^2} y = x\sqrt{y}$.

Solution: Given differential equation is

$$\frac{dy}{dx} + \frac{x}{1-x^2} y = x\sqrt{y}$$

Dividing by $\sqrt{y}$, we get

$$y^{-1/2}\frac{dy}{dx} + \frac{x}{1-x^2} y^{1/2} = x \qquad \ldots (1)$$

Put $y^{1/2} = z$ so that

$$\frac{1}{2} y^{-1/2}\frac{dy}{dx} = \frac{dz}{dx} \quad \Rightarrow \quad y^{-1/2}\frac{dy}{dx} = 2\frac{dz}{dx}$$

Eq. (1) becomes

$$2\frac{dz}{dx} + \frac{x}{1-x^2} z = x \quad \text{or} \quad \frac{dz}{dx} + \frac{x}{2(1-x^2)} z = \frac{x}{2}$$

which is linear in z.

It's $I.F. = e^{\int \frac{x}{2(1-x^2)} dx} = e^{-\frac{1}{4}\int \frac{-2x}{1-x^2} dx} = e^{-\frac{1}{4}\log(1-x^2)} = (1-x^2)^{-\frac{1}{4}}$

$\therefore$ The solution is

$$z(I.F.) = \int \frac{x}{2}(I.F.)\,dx + c$$

$$\Rightarrow \quad z\left(1-x^2\right)^{-1/4} = -\frac{1}{4}\int\left(1-x^2\right)^{-1/4}(-2x)dx + c = -\frac{1}{4}\cdot\frac{\left(1-x^2\right)^{3/4}}{3/4} + c$$

$$\Rightarrow \quad z = -\frac{1}{3}\left(1-x^2\right) + c\left(1-x^2\right)^{1/4}$$

$$\Rightarrow \quad \sqrt{y} = -\frac{1}{3}\left(1-x^2\right) + c\left(1-x^2\right)^{1/4}$$

Example 1.39: Solve differential equation $xy\left(1+xy^2\right)\dfrac{dy}{dx} = 1$

Solution: Given equation can be written as

$$xy + x^2y^3 = \frac{dx}{dy} \qquad \Rightarrow \qquad \frac{dx}{dy} - yx = y^3x^2$$

Dividing by x^2, we get

$$x^{-2}\frac{dx}{dy} - yx^{-1} = y^3 \qquad \text{. . . (1)}$$

Putting $x^{-1} = z$ so that $-x^{-2}\dfrac{dx}{dy} = \dfrac{dz}{dy}$ or $x^{-2}\dfrac{dx}{dy} = -\dfrac{dz}{dy}$

Equation (1) becomes

$$-\frac{dz}{dy} - yz = y^3 \quad \text{or} \quad \frac{dz}{dy} + yz = -y^3$$

which is linear in z.

It's $I.F. = e^{\int ydx} = e^{\frac{1}{2}y^2}$

$\therefore$ The solution is

$$z(I.F.) = \int\left(-y^3\right)(I.F.)dy + c$$

$$\Rightarrow \quad ze^{\frac{1}{2}y^2} = \int -y^3e^{\frac{1}{2}y^2}dy + c \quad \Rightarrow \quad ze^{\frac{1}{2}y^2} = -\int y^2e^{\frac{1}{2}y^2}ydy + c = -\int 2te^t dt + c$$

where $t = \dfrac{1}{2}y^2$

$$\Rightarrow \quad ze^{\frac{1}{2}y^2} = -2\left[te^t - \int 1 - e^t dt\right] + c = -2\left(te^t - e^t\right) + c = -2e^{\frac{1}{2}y^2}\left(\frac{1}{2}y^2 - 1\right) + c$$

$$\Rightarrow \quad z = -2\left(\frac{1}{2}y^2 - 1\right) + ce^{\frac{1}{2}y^2}$$

$$\Rightarrow \quad \frac{1}{x} = 2 - y^2 + ce^{\frac{1}{2}y^2} \qquad \left[\because z = \frac{1}{x}\right]$$

Example 1.40: Solve differential equation $\dfrac{dy}{dx} + x\sin 2y = x^3 \cos^2 y$.

Solution: Given differential equation is

$$\frac{dy}{dx} + x\sin 2y = x^3 \cos^2 y$$

Dividing by $\cos^2 y$, we get

$$\sec^2 y\frac{dy}{dx} + \frac{x(2\sin y\cos y)}{\cos^2 y} = x^3 \qquad [\because \sin 2y = 2\sin y\cos y]$$

$$\Rightarrow \quad \sec^2 y\frac{dy}{dx} + 2x\tan y = x^3 \qquad \ldots(1)$$

Putting $\tan y = z$ so that $\sec^2 y\dfrac{dy}{dx} = \dfrac{dz}{dx}$

Eq. (1) becomes

$$\frac{dz}{dx} + 2xz = x^3 \qquad \text{which is linear in } z.$$

It's $I.F. = e^{\int 2x dx} = e^{x^2}$

$\therefore$ The solution is

$$z.e^{x^2} = \int x^3 e^{x^2} dx + c$$

$$= \int x^2 e^{x^2} .x\ dx + c = \frac{1}{2}\int x^2 e^{x^2} .2x\ dx + c$$

$$= \frac{1}{2}\int te^t dt + c \quad \text{where } t = x^2,\ dt = 2xdx$$

$$=\frac{1}{2}(t-1)e^t+c=\frac{1}{2}(x^2-1)e^t+c$$

$$\Rightarrow \quad z=\frac{1}{2}(x^2-1)+ce^{-x^2}$$

$$\Rightarrow \quad \tan y=\frac{1}{2}(x^2-1)+ce^{-x^2} \qquad [\because z=\tan y]$$

Example 1.41: Sclve differential equation $\frac{dy}{dx}-(\tan x)y=-y^2\sec^2 x$.

Solution: Given differential equation is

$$\frac{dy}{dx}-(\tan x)y=-y^2\sec^2 x$$

Dividing by y^2, we get

$$\frac{1}{y^2}\frac{dy}{dx}-\tan x\frac{1}{y}=-\sec^2 x$$

Put $\frac{1}{y}=z$

$$\therefore \quad -\frac{1}{y^2}\frac{dy}{dx}=\frac{dz}{dx} \Rightarrow -\frac{1}{y^2}\frac{dy}{dx}=-\frac{dz}{dx}$$

$$-\frac{dz}{dx}-\tan x.z=-\sec^2 x$$

$$\Rightarrow \quad \frac{dz}{dx}+\tan x.z=\sec^2 x$$

which is linear differential equation in z.

$$I.F.=e^{\int \tan x dx}=e^{-\log\cos x}=e^{\log\sec x}=\sec x$$

$\therefore$ It's solution is

$$z\sec x=\int \sec^2 x.\sec x dx+c$$

$$=\int \sec^3 x\,dx+c \qquad \ldots(1)$$

Let $I = \int \sec^3 x\, dx = \int \sec x . \sec^2 x\, dx$

Integrating by parts, we get

$$I = (\sec x)(\tan x) - \int \sec x \tan x \tan x dx$$

$$= \sec x \tan x - \int \sec x \left(sex^2 x - 1 \right) dx$$

$$= \sec x \tan x - I + \int \sec x\, dx$$

$\therefore \quad 2I = \sec x \tan x + \log(\sec x + \tan x)$

$\therefore \quad I = \frac{1}{2}\left[\sec x \tan x + \log(\sec x + \tan x) \right]$

Substituting in equation (1), we get

$$z \sec x = \frac{1}{2}\left[\sec x \tan x + \log(\sec x + \tan x) \right] + c$$

Substituting the value of z, we get

$$\frac{1}{y} \sec x = \frac{1}{2}\left[\sec x \tan x + \log(\sec x + \tan x) \right] + c$$

Example 1.42: Solve differential equation $e^y \left(\frac{dy}{dx} + 1 \right) = e^x$.

Solution: Given differential equation is

$$e^y \left(\frac{dy}{dx} + 1 \right) = e^x \quad \Rightarrow \quad e^y \frac{dy}{dx} + e^y = e^x \qquad \ldots (1)$$

Put $e^y = z$

$$\therefore \quad e^y \frac{dy}{dx} = \frac{dz}{dx}$$

$\therefore$ Eq. (1) becomes

$$\frac{dz}{dx} + z = e^x \qquad \ldots (2)$$

Comparing it with

$$\frac{dy}{dx}+Py=Q$$

we have $P=1,\ Q=e^x$

It's I.F. $=e^{\int P dx}=e^{\int 1.dx}=e^x$

Multiplying Eq. (2) by e^x, we get

$$e^x\frac{dz}{dx}+ze^x=e^{2x} \quad\Rightarrow\quad \frac{d}{dx}\left(e^x z\right)=e^{2x}$$

Integrating both sides, we get

$$e^x z=\int e^{2x}dx+c \quad\Rightarrow\quad e^x z=\frac{e^{2x}}{2}+c$$

Replace z by e^x, we get

$$e^x e^y=\frac{e^{2x}}{2}+c$$

$$e^{x+y}=\frac{1}{2}e^{2x}+c$$ which is the required solution.

EXERCISE 1.8

Solve the following differential equations:

1. $x\frac{dy}{dx}+y=x^3y^6$ [PTU May 2007]
2. $2\frac{dy}{dx}=\frac{y}{x}+\frac{y^2}{x^2}$
3. $\frac{dy}{dx}-x^2y=y^2e^{-\frac{1}{3}x^3}$
4. $\left(x^3y^2+xy\right)dx=dy$
5. $(x+1)\frac{dy}{dx}+1=2e^{-y}$
6. $\frac{dy}{dx}-\frac{\tan y}{1+x}=(1+x)e^x\sec y$

7. $\left(xy^2 - e^{\frac{1}{x^3}}\right)dx - x^2 y dy = 0$

8. $\frac{dy}{dx} + y \tan x = y^3 \cos x$

9. $\frac{dy}{dx} + \frac{y \log y}{x} = \frac{y(\log y)^2}{x^2}$

10. $y - \cos x \frac{dy}{dx} = y^2 (1 - \sin x) \cos x$, given that $y = 2$ when $x = 0$.

11. $y\left(2xy + e^x\right)dx = e^x dy$

12. $\tan y \frac{dy}{dx} + \tan x = \cos y \sec^2 x$

ANSWERS

1. $\frac{1}{y^5} = \frac{5}{2}x^3 + cx^5$

2. $\frac{x}{y} = 1 + c\sqrt{x}$

3. $y(c - x) = e^{\frac{1}{3}x^3}$

4. $\frac{1}{y} = -x^2 + 2 + ce^{-\frac{1}{2}x^2}$

5. $(x + y)e^y = 2x + c$

6. $\sin y = (1 + x)\left(e^x + c\right)$

7. $3y^2 = x^2\left(2e^{x\frac{1}{3}} + c\right)$

8. $\cos^2 x = y^2\left(c - 2\sin x + \frac{2}{3}\sin^3 x\right)$

9. $\frac{1}{x \log y} = \frac{1}{2x^2} + c$

10. $2(\tan x + \sec x) = y(2\sin x + 1)$

11. $e^x = y\left(c - x^2\right)$

12. $\sec y = (c + \sin x)\cos x$

❑ ❑ ❑

Chapter 2

Linear Ordinary Differential Equations of Second and Higher Order

Learning Objectives

After reading this chapter, students will be able to

- understand linear differential equations with constant coefficients
- explain the concept of auxiliary equation
- describe various rules to find complementary functions
- explain the concept of particular integral
- describe various rules to find particular integrals
- explain the method variation of parameters to find particular integral
- describe Cauchy's homogeneous linear equation and how it can be reduced to linear differential equation with constant coefficients, and then solve it
- describe Legendre's linear equation and how it can be solved
- describe simultaneous linear equations with constant coefficients and how they can be solved
- solve various problems related to linear differential equations

2.1 INTRODUCTION

A *linear differential equation* is that in which the dependent variable and its derivatives occur only in the first degree and are not multiplied together. The general form of an nth order linear differential equation is

$$\frac{d^n y}{dx^n} + P_1 \frac{d^{n-1} y}{dx^{n-1}} + P_2 \frac{d^{n-2} y}{dx^{n-2}} + \ldots + P_n y = X \qquad \ldots (1)$$

where $P_1, P_2, \ldots, P_n$ and X are functions of x only.

The above linear differential reduces to a *linear differential equation with constant coefficients* if $P_1, P_2, \ldots, P_n$ all are constants, say $a_1, a_2, \ldots, a_n$, and X is a function of x only, i.e.,

$$\frac{d^n y}{dx^n} + a_1 \frac{d^{n-1} y}{dx^{n-1}} + a_2 \frac{d^{n-2} y}{dx^{n-2}} + \ldots + a_n y = X \qquad \ldots (2)$$

In this chapter, our discussion will be focused on linear differential equations with constant coefficients and other type of differential equations that can be reduced to linear differential equations with constant coefficients.

2.2 THE OPERATOR D

The part $\frac{d}{dx}$ of the $\frac{dy}{dx}$ can be regarded as an operator such that when it operate on y, the result is the derivative of y.

In the similar way, $\frac{d^2}{dx^2}, \frac{d^3}{dx^3}, \ldots, \frac{d^n}{dx^n}$ can be regarded as $D^2, D^3, \ldots, D^n$ operators such that when they operate on y, the result is the second order, third order, . . ., nth order derivatives of y.

Thus, the symbol D can be regarded as a *differential operator* or simply an operator.

Using operator D, linear differential equation with constant coefficients can be written as

$$D^n y + a_1 D^{n-1} y + a_2 D^{n-2} y + \ldots + a_n y = X \qquad \ldots (1)$$

which further can be re-written as

$$\left(D^n + a_1 D^{n-1} + a_2 D^{n-2} + ... + a_n\right) y = X$$

$$\Rightarrow \qquad f(D)y = X \qquad \text{... (2)}$$

Here $f(D) = D^n + a_1 D^{n-1} + a_2 D^{n-2} + ... + a_n$ is called the *characteristic polynomial* of Eq. (2).

This characteristic polynomial equated to zero gives us the auxiliary equation (*A.E.*) of $f(D)y = X$. We will be discussing about *A.E.* in next section.

Note: The operator D can be regarded as an algebraic quantity, i.e.,

$$D(u+v) = Du + Dv$$

$$D(\lambda u) = \lambda Du$$

$$D^p D^q u = D^q D^p u = D^{p+q} u$$

The characteristic polynomial $f(D)$ can be factorized by ordinary rules of algebra and the factors can be written in any order.

Art. If $\phi(D)$ be a factor of the characteristic polynomial $f(D)$ and y_1 is a solution of $\phi(D)y = 0$, then show that y_1 is also a solution of the linear differential equation $f(D)y_1 = 0$.

Proof: The given linear differential equation is

$$f(D)y = 0 \qquad \text{... (1)}$$

Let $\quad f(D) = \phi(D)\psi(D)$

$$f(D)y_1 = \phi(D)\psi(D)y_1$$

$$= \psi(D)\phi(D)y_1 \qquad [\because \text{ order of factor can be changed}]$$

$$= \psi(D)0 \qquad [\because y_1 \text{ is a solution of } \phi(D)y = 0]$$

$$= 0$$

Hence, the result $y = y_1$ is a solution of Eq. (1).

Art. If $y_1, y_2, y_3, \ldots, y_n$ are linearly independent solutions of the linear differential equation

$$D^n y + a_1 D^{n-1} y + a_2 D^{n-2} y + \ldots + a_n y = 0 \qquad \ldots (1)$$

then $u = c_1 y_1 + c_2 y_2 + c_3 y_3 + \ldots + c_n y_n$ is also its solution, where $c_1, c_2, c_3, \ldots, c_n$ are arbitrary constants.

Proof: Since $y_1, y_2, y_3, \ldots, y_n$ are solutions of Eq. (1), therefore

$$D^n y_1 + a_1 D^{n-1} y_1 + a_2 D^{n-2} y_1 + \ldots + a_n y_1 = 0$$

$$D^n y_2 + a_1 D^{n-1} y_2 + a_2 D^{n-2} y_2 + \ldots + a_n y_2 = 0$$

$$D^n y_3 + a_1 D^{n-1} y_3 + a_2 D^{n-2} y_3 + \ldots + a_n y_3 = 0$$

$$\vdots$$

$$D^n y_n + a_1 D^{n-1} y_n + a_2 D^{n-2} y_n + \ldots + a_n y_n = 0$$

Now,

$$\begin{aligned} D^n u + a_1 D^{n-1} u + a_2 D^{n-2} u + \ldots + a_n u &= D^n \left(c_1 y_1 + c_2 y_2 + c_3 y_3 + \ldots + c_n y_n \right) \\ &\quad + a_1 D^{n-1} \left(c_1 y_1 + c_2 y_2 + c_3 y_3 + \ldots + c_n y_n \right) \\ &\quad + a_2 D^{n-2} \left(c_1 y_1 + c_2 y_2 + c_3 y_3 + \ldots + c_n y_n \right) + \ldots \\ &\quad + a_n \left(c_1 y_1 + c_2 y_2 + c_3 y_3 + \ldots + c_n y_n \right) \\ &= c_1 \left(D^n y_1 + a_1 D^{n-1} y_1 + a_2 D^{n-2} y_1 + \ldots + a_n y_1 \right) \\ &\quad + c_2 \left(D^n y_2 + a_1 D^{n-1} y_2 + a_2 D^{n-2} y_2 + \ldots + a_n y_2 \right) \\ &\quad + c_3 \left(D^n y_3 + a_1 D^{n-1} y_3 + a_2 D^{n-2} y_3 + \ldots + a_n y_3 \right) + \ldots \\ &\quad + c_n \left(D^n y_n + a_1 D^{n-1} y_n + a_2 D^{n-2} y_n + \ldots + a_n y_n \right) \\ &= c_1(0) + c_2(0) + c_3(0) + \ldots. + c_n(0) \\ &= 0 \end{aligned}$$

which shows that $u = c_1 y_1 + c_2 y_2 + c_3 y_3 + \ldots + c_n y_n$ is also the solution of Eq. (1). Since it contains n arbitrary constants, therefore, it is the general solution of Eq. (1).

Art. If u is the complete solution of equation $f(D)y=0$ and v is the particular solution (containing no arbitrary constants) of the equation $f(D)y=X$, then $(u+v)$ is the complete solution of the equation $f(D)y=X$.

Proof: Since u is the complete solution of equation $f(D)y=0$

$$\therefore \quad f(D)u=0 \qquad \ldots (1)$$

Since v is the particular solution of equation $f(D)y=X$

$$\therefore \quad f(D)v=X \qquad \ldots (2)$$

Adding Eq. (1) and Eq. (2), we get

$$f(D)(u+v)=X \qquad \ldots (3)$$

Thus, $(u+v)$ satisfies the Eq. (3), hence it is the complete solution (*C.S.*) since it contains *n* arbitrary constants.

The first part (i.e., *u*) is called the complementary function (*C.F.*) and the second part (i.e., *v*) is called the particular integral (*P.I.*) of the Eq. (3).

$\therefore$ the complete solution of Eq. (3) is expressed as

$$y=C.F.+P.I.$$

Thus, in order to solve Eq. (3), we first find *C.F.*, i.e., the complete solution of Eq. (1) and then *P.I.*, i.e., the particular solution of Eq. (3) and finally we add them.

2.3 AUXILIARY EQUATION

Consider the differential equation

$$\left(D^n+a_1D^{n-1}+a_2D^{n-2}+\ldots+a_n\right)y=0 \qquad \ldots (1)$$

Let $y=e^{mx}$ be a solution of Eq. (1), then

$$Dy=me^{mx} \qquad \left[\because D=\frac{d}{dx},\ D^2=\frac{d^2}{dx^2},\ \ldots,\ D^n=\frac{d^n}{dx^n}\right]$$

$$D^2y=m^2e^{mx}$$

$$\vdots$$

$$D^{n-2}y = m^{n-2}e^{mx}$$

$$D^{n-1}y = m^{n-1}e^{mx}$$

$$D^{n}y = m^{n}e^{mx}$$

Substituting the value of $y, Dy, D^2y, ..., D^n y$ in Eq. (1) we get

$$\left(m^n + a_1 m^{n-1} + a_2 m^{n-2} + ... + a_n\right)e^{mx} = 0$$

$$\Rightarrow \quad m^n + a_1 m^{n-1} + a_2 m^{n-2} + ... + a_n = 0, \text{ since } e^{mx} \neq 0 \qquad ...(2)$$

Thus, $y = e^{mx}$ will be a solution of Eq. (1) if m satisfies Eq. (2).

Eq. (2) is called the *auxiliary equation* (A.E.) for the differential Eq. (1).

Replacing m by D in Eq. (2) we get

$$D^n + a_1 D^{n-1} + a_2 D^{n-2} + ... + a_n = 0 \qquad ...(3)$$

Eq. (2) gives the same value of m as Eq. (3) gives of D. In practice, we take Eq. (3) as the auxiliary equation, which is obtained by equating to zero the symbolic co-efficient of y in Eq. (1).

Note: The equation obtained by equating to zero the symbolic coefficient of y is called *auxiliary equation*, briefly written as *A.E.*

2.4 RULES FOR FINDING THE COMPLIMENTARY FUNCTION

Consider the differential equation

$$\left(D^n + a_1 D^{n-1} + a_2 D^{n-2} + ... + a_n\right)y = 0 \qquad ...(1)$$

It's A.E. is

$$D^n + a_1 D^{n-1} + a_2 D^{n-2} + ... + a_n = 0 \qquad ...(2)$$

Let $m_1, m_2, ..., m_n$ be the roots of the *A.E.* The solution of Eq. (1) depends on the nature roots of the A.E.

The following cases arise:

Case I: When all the roots of the auxiliary equation are real and different.

Let $m_1, m_2, ..., m_n$ be the real and different roots of the A.E.

$\therefore \quad f(D) = \phi(D)(D - m_1)$

where $\phi(D)$ is a polynomial of degree $(n-1)$.

Now the solution of $(D - m_1)y = 0$ is a solution of Eq. (1), *i.e.*,

$$\frac{dy}{dx} - m_1 y = 0 \quad \Rightarrow \quad \frac{dy}{y} = m_1 dx \quad \Rightarrow \quad \log y = m_1 x$$

$$\Rightarrow \quad y = e^{m_1 x} \qquad \text{[on integration]}$$

Omitting the constant of integration, it is a solution of Eq. (1).

Similarly $e^{m_2 x}, e^{m_3 x}, ..., e^{m_n x}$ are the other independent solutions of Eq. (1).

Hence

$$y = c_1 e^{m_1 x} + c_2 e^{m_2 x} + ... + c_n e^{m_n x} \qquad \text{where } c\text{'s are arbitrary constants,}$$

is a general solution of Eq. (1).

Case II: When two roots of the auxiliary equation are real and equal, while all other roots are real and different.

Let the roots of eq. (3) be $m_1, m_1, m_3, ..., m_n$, then

$$f(D) = (D - m_1)^2 (D - m_3)(D - m_4)...(D - m_n)$$

$\therefore$ eq. (1) becomes

$$(D - m_1)^2 (D - m_3)(D - m_4)...(D - m_n)\, y = 0$$

Now the part of complete solution of eq. (1) corresponding to different roots is given by

$$c_3 e^{m_3 x} + c_4 e^{m_4 x} + ... + c_n e^{m_n x}$$

and we have to find the part of the complete solution corresponding to the equal roots.

Consider $(D-m_1)^2 y=0$ *i.e.*, $(D-m_1)(D-m_1)y=0$

Take $(D-m_1)y=z$

$\therefore$ we get $(D-m_1)z=0$

$$\Rightarrow \quad \frac{dz}{dx}-m_1 z=0 \qquad \ldots (4)$$

This is a linear differential equation of the first order.

$\therefore$ It's integration factor, I.F. $= e^{\int -m_1 dx} = e^{-m_1 x}$

Then the solution of Eq. (4) is

$$z(I.F.)=\int 0(I.E.)dx+c_1$$

$$\therefore \quad ze^{-m_1 x}=c_1 \quad \Rightarrow \quad Z=c_1 e^{m_1 x}$$

Substituting the value of z in $(D-m_1)y=z$, we get

$$(D-m_1)y=c_1 e^{m_1 x}$$

$$\Rightarrow \quad \frac{dy}{dx}-m_1 y=c_1 e^{-m_1 x} \qquad \ldots (5)$$

$$I.F.=e^{\int -m_1 dx}=e^{-m_1 x}$$

Then the solution of Eq. (5) is

$$y(I.F.)=\int c_1 e^{m_1 x}(I.F.)dx+c_2$$

$$\Rightarrow \quad ye^{m_1 x}=\int c_1 e^{m_1 x}(e^{-m_1 x})dx+c_2$$

$$\Rightarrow \quad ye^{m_1 x}=c_1\int dx+c_2$$

$$\Rightarrow \quad ye^{-m_1 x}=c_1 x+c_2 \quad \Rightarrow \quad y=(c_1 x+c_2)e^{m_1 x}$$

Hence the complete solution of Eq. (1), is given by

$$y=(c_1 x+c_2)e^{m_1 x}+c_3 e^{m_3 x}+\ldots+c_n e^{m_n x}$$

If however, three roots of the *A.E.* are real and equal, while all other roots are real and different.

Let the roots of eq. (3) be $m_1, m_1, m_1, m_4, ..., m_n$, then

$$y = \left(c_1 x^2 + c_2 x + c_3\right) e^{m_1 x} + c_3 e^{m_4 x} + ... + c_n e^{m_n x}$$

Case III: When two roots of the auxiliary equation are imaginary and different, while the others are all different.

Let $m_1 = \alpha + i\beta$, $m_2 = \alpha - i\beta$, $m_3, m_4, ..., m_n$ be the two imaginary and different roots. Note that complex roots occur in congregate pairs, while the others are all different.

The complete solution of eq. (1) is given by

$$y = c_1' e^{(\alpha - i\beta)x} + c_2' e^{(\alpha - i\beta)x} + c_3 e^{m_3 x} + c_4 e^{m_4 x} + ... + c_n e^{m_n x}$$

$$= e^{\alpha x}\left[c_1' e^{i\beta x} + c_2' e^{-i\beta x}\right] + c_3 e^{m_3 x} + c_4 e^{m_4 x} + ... + c_n e^{m_n x}$$

$$= e^{\alpha x}\left[c_1'(\cos\beta x + i\sin\beta x) + c_2'(\cos\beta x - i\sin\beta x)\right]$$

$$+ c_3 e^{m_3 x} + c_4 e^{m_4 x} + ... + c_n e^{m_n x}$$

$$= e^{\alpha x}\left[(c_1' + c_2')\cos\beta x + i(c_1' - c_2')\sin\beta x\right] + c_3 e^{m_3 x} + c_4 e^{m_4 x} + ... + c_n e^{m_n x}$$

$$= e^{\alpha x}\left[c_1 \cos\beta x + i c_2 \sin\beta x\right] + c_3 e^{m_3 x} + c_4 e^{m_4 x} + ... + c_n e^{m_n x}$$

where $c_1' + c_2' = c_1$ and $(c_1' - c_2')i = c_2$

Hence $y = e^{\alpha c}\left[c_1 \cos\beta x + c_2 \sin\beta x\right] + c_3 e^{m_3 x} + c_4 e^{m_4 x} + ... + c_n e^{m_n x}$

is the complete solution of Eq. (1).

Case IV: When the two pairs of imaginary roots of the characteristic polynomial equation are equal, while all other are different.

Let $m_1 = m_2 = \alpha + i\beta$, $m_3 = m_4 = \alpha - i\beta$ and $m_5, m_6, ..., m_n$ be the different roots.

Then the complete solution of eq. (1) is given by

$$y = e^{\alpha x}\left[(c_1 + c_2 x)\cos\beta x + (c_3 + c_4 x)\sin\beta x)\right] + c_5 e^{m_5 x} + c_6 e^{m_5 x} + ... + c_n e^{m_n x}$$

Note: The corresponding part of the general solution, i.e., complementary function from the roots of the auxiliary equation may be written as under:

Roots of A.E.	General Solution
Case I: All roots $m_1, m_2, ..., m_n$ are distinct.	$y = c_1 e^{m_1 x} + c_2 e^{m_2 x} + ... + c_n e^{m_n x}$
Case II: (*i*) $m_1 = m_2$ and all other roots are real and distinct. (*ii*) $m_1 = m_2 = m_3$ and all other roots are real and distinct.	$y = (c_1 + c_2 x) e^{m_1 x} + c_3 e^{m_2 x} + ... + c_n e^{m_n x}$ $y = (c_1 + c_2 x + c_3 x^2) e^{m_1 x} + c_4 e^{m_4 x} + ... + c_n e^{m_n x}$
Case III: Here $a \pm ib$ is a pair of complex root.	Corresponding part of general solution is $e^{ax}[A \cos bx + B \sin bx]$
Case IV: Here $a \pm ib$ is a pair of complex roots occurring twice.	Corresponding part of general solution is $e^{ax}\left[(c_1 + c_2 x)\cos bx + (c_3 + c_4 x)\sin bx\right]$

ILLUSTRATIVE EXAMPLES 2.1

Example 2.1: Solve the differential equation $\frac{d^2 y}{dx^2} + 5\frac{dy}{dx} + 6y = 0.$

Solution: The given differential equation is

$$\frac{d^2 y}{dx^2} + 5\frac{dy}{dx} + 6y = 0 \qquad ...(1)$$

It can be written as

$$\left(D^2 + 5D + 6\right) y = 0 \qquad ...(2)$$

The characteristic polynomial of Eq. (2) is

$$f(D) = D^2 + 5D + 6$$

∴ Roots of the characteristic polynomial are given by

$$D^2 + 5D + 6 = 0 \quad \Rightarrow \quad (D+2)(D+3) = 0 \quad \Rightarrow \quad D = -2, -3$$

The roots are real and different.

$\therefore$ The complete solution of Eq. (1) is

$$y = c_1 e^{-2x} + c_2 e^{-3x}$$

Example 2.2: Solve the differential equation $\dfrac{d^4 y}{dx^4} + 2\dfrac{d^2 y}{dx^2} + y = 0.$

Solution: The given differential equation is

$$\frac{d^4 y}{dx^4} + 2\frac{d^2 y}{dx^2} + y = 0 \qquad \ldots (1)$$

It can be written as

$$\left(D^4 + 2D^2 + 1\right) y = 0 \qquad \ldots (2)$$

The characteristic polynomial of Eq. (2) is

$$f(D) = D^4 + 2D^2 + 1$$

$\therefore$ Roots of the characteristic polynomial are given by

$$D^4 + 2D^2 + 1 = 0 \quad \Rightarrow \quad \left(D^2 + 1\right)^2 = 0$$

$$\Rightarrow \quad D^2 = -1, -1 \quad \Rightarrow \quad D = \pm i, \pm i$$

$\therefore$ The complete solution of eq. (1) is

$$y = e^{0x}\left[(c_1 + c_2 x)\cos x + (c_3 + c_4 x)\sin x\right]$$

$$\Rightarrow \quad y = (c_1 + c_2 x)\cos x + (c_3 + c_4 x)\sin x$$

Example 2.3: If given differential equation is $\dfrac{d^4 y}{dx^4} - a^4 y = 0$, show that

$$y = c_1 \cos ax + c_2 \sin ax + c_3 \cosh ax + c_4 \sinh ax.$$

Solution: The given differential equation is

$$\frac{d^4 y}{dx^4} - a^4 y = 0 \quad \Rightarrow \quad \left(D^4 - a^4\right) y = 0$$

The characteristic polynomial is

$$f(D) = D^4 - a^4$$

∴ Zero of the characteristic polynomial is given by

$$D^4 - a^4 = 0 \Rightarrow (D^2 - a^2)(D^2 + a^2) = 0 \quad \Rightarrow \quad D = \pm a, \ \pm ia$$

∴ the complete solution of (1) is given by

$$y = A_1 e^{ax} + A_2 e^{-ax} + e^{0x}[c_1 \cos ax + c_2 \sin ax]$$

$$\Rightarrow \quad y = A_1\left[\frac{(e^{ax} + e^{-ax}) + (e^{ax} - e^{-ax})}{2}\right] + A_2\left[\frac{(e^{ax} + e^{-ax}) - (e^{ax} - e^{-ax})}{2}\right]$$

$$+ c_1 \cos ax + c_2 \sin ax$$

$$\Rightarrow \quad y = A_1\left[\frac{(e^{ax} + e^{-ax}) + (e^{ax} - e^{-ax})}{2}\right] + A_2\left[\frac{(e^{ax} + e^{-ax}) - (e^{ax} - e^{-ax})}{2}\right] +$$

$$+ c_1 \cos ax + c_2 \sin ax$$

$$\Rightarrow \quad y = A_1[\cosh ax + \sinh ax] + A_2[\cosh ax - \sinh ax] \qquad \left[\because \cosh\theta = \frac{e^{i\theta} + e^{-i\theta}}{2}, \text{ and } \sinh\theta = \frac{e^{i\theta} - e^{-i\theta}}{2}\right]$$

$$+ c_1 \cos ax + c_2 \sin ax$$

$$= (A_1 + A_2)\cosh ax + (A_1 - A_2)\sinh ax + c_1 \cos ax + c_2 \sin ax$$

where $(A_1 + A_2) = c_3$ and $(A_1 - A_2) = c_4$

$$\therefore \quad y = c_1 \cos ax + c_2 \sin ax + c_3 \cosh ax + c_4 \sinh ax.$$

Example 2.4: Solve the differential equation $\dfrac{d^4 y}{dx^4} + 4y = 0.$

Solution: The given differential equation is

$$\frac{d^4 y}{dx^4} + 4y = 0 \quad \Rightarrow \quad (D^4 + 4)y = 0$$

The characteristic polynomial is

$$f(D) = D^4 + 4$$

∴ Roots of the characteristic polynomial are given by

$$\Rightarrow \quad D^4 + 4 = 0 \quad \Rightarrow \quad D^4 = (2i)^2 \quad \Rightarrow \quad D^2 = \pm 2i$$

Either $D^2 = 2i = (1+i)^2$ or $D^2 = -2i = (1-i)^2$

$\Rightarrow \quad D = \pm(1+i)$ or $D = \pm(1-i)$

$\Rightarrow \quad D = 1 \pm i, -1 \pm i$

$\therefore$ The complete solution of (1) is given by

$$y = e^x[c_1 \cos x + c_2 \sin x] + e^{-x}[c_3 \cos x + c_4 \sin x]$$

Example 2.5: Solve the differential equation $\dfrac{d^3y}{dx^3} - 7\dfrac{dy}{dx} - 6y = 0.$

Solution: The symbolic form (S. F.) of the given differential equation is

$$(D^3 - 7D - 6)y = 0$$

It's A.E. is

$$D^3 - 7D - 6 = 0$$

$\Rightarrow \quad (D+1)(D+2)(D-3) = 0$

$\Rightarrow \quad D = -1, -2, 3$

Hence, the complete solution is

$$y = c_1e^{-x} + c_2e^{-2x} + c_3e^{3x}$$

Example 2.6: Solve the differential equation $(D^3 - 4D^2 + 4D)y = 0.$

Solution: The A.E. for the given differential equation is

$$D^3 - 4D^2 + 4D = 0$$

$\Rightarrow \quad D(D^2 - 4D + 4)y = 0$

$\Rightarrow \quad D(D-2)^2 = 0$ which gives $D = 0, 2, 2$

Hence, the complete solution is

$$y = c_1e^{0x} + (c_2 + c_3x)e^{2x} \quad \Rightarrow \quad y = c_1 + (c_2x + c_3)e^{2x}$$

Example 2.7: Solve the differential equation $\frac{d^4y}{dx^4}+13\frac{d^2y}{dx^2}+36y=0.$

Solution: The A.E. for given differential equation is

$$D^4+13D^2+36=0$$

$$\Rightarrow \quad (D^2+4)(D^2+9)=0$$

which gives

$$D=\pm 2i, \pm 3i$$

Hence, the complete solution is

$$x=e^{0x}(c_1\cos 2x+c_2\sin 2x)+e^{0x}(c_3\cos 3x+c_4\sin 3x)$$

Example 2.8: If $\frac{d^2x}{dt^2}+\frac{g}{b}(x-a)=0;(a>0,\ b>0,\ g>0)$ and $x=\alpha,\ \frac{dx}{dt}=0$ when t = 0,

show that $x=a+(\alpha-a)\cos\sqrt{\frac{g}{b}}t.$

Solution: Given $\frac{d^2x}{dt^2}+\frac{g}{b}(x-a)=0$

Put $\quad x-a=y$

$$\therefore \quad \frac{d^2x}{dt^2}=\frac{d^2y}{dt^2}$$

$$\frac{d^2y}{dt^2}+\frac{g}{b}y=0$$

The A.E. is $m^2+\frac{g}{b}=0,$

$$\therefore \quad m^2=-\frac{g}{b} \quad \Rightarrow \quad m=\pm i\sqrt{\frac{g}{b}}$$

$$\therefore \quad y=c_1\cos\sqrt{\frac{g}{b}}t+c_2\sin\sqrt{\frac{g}{b}}t$$

$$\Rightarrow \qquad x-a=c_1\cos\sqrt{\frac{g}{b}}+c_2\sin\sqrt{\frac{g}{b}}t$$

When $x=\alpha,\ t=0,\quad \alpha-a=c_1$

$$x-a=(\alpha-a)\cos\sqrt{\frac{g}{b}}t+c_2\sin\sqrt{\frac{g}{b}}t$$

$$\frac{dx}{dt}=-(\alpha-a)\sqrt{\frac{g}{b}}\sin\sqrt{\frac{g}{b}}t+c_2\sqrt{\frac{g}{b}}\cos\sqrt{\frac{g}{b}}t$$

$$t=0,\ \frac{dx}{dt}=0$$

$$\therefore\quad 0=c_2\sqrt{\frac{g}{b}}\quad\Rightarrow\quad c_2=0$$

$$x-a=(\alpha-a)\cos\sqrt{\frac{g}{b}}t$$

$$\Rightarrow\qquad \text{Hence}\quad x=a+(\alpha-a)\cos\sqrt{\frac{g}{b}}t$$

EXERCISE 2.1

Solve the following linear differential equations:

1. $\frac{d^2y}{dx^2}-3\frac{dy}{dx}-4y=0$
2. $\frac{d^2y}{dx^2}+(a+b)\frac{dy}{dx}+aby=0$
3. $\frac{d^2y}{dx^2}-4\frac{dy}{dx}+y=0$
4. $\frac{d^2y}{dx^2}+6\frac{dy}{dx}+9x=0$
5. $\frac{d^3y}{dx^3}-3\frac{d^2y}{dx^2}+3\frac{dy}{dx}-y=0$
6. $\frac{d^3y}{dx^3}+6\frac{d^2y}{dx^2}+11\frac{dy}{dx}+6y=0$

7. $\dfrac{d^4y}{dx^4}-5\dfrac{d^2y}{dx^2}+11y=0$

8. $\dfrac{d^4y}{dx^4}+8\dfrac{d^2y}{dx^2}+16y=0$

9. $\left(D^2+1\right)^3\left(D^2+D+1\right)y=0$

10. $\dfrac{d^3y}{dx^3}+y=0$ [PTU May 2012]

11. $\dfrac{d^2y}{dt^2}-3\dfrac{dx}{dt}+29y=0$, when $t=0,\ x=0,\ \dfrac{dy}{dt}=0$

12. $\dfrac{d^2y}{dx^2}+4\dfrac{dy}{dx}+29y=0$ given $t=0,\ x=0,\ \dfrac{dy}{dt}=15$

13. If $\dfrac{d^4x}{dt^4}=m^4x$, show that $x=c_1\cos mt+c_2\sin mt+c_3\cosh mt+c_4\sinh mt$

ANSWERS

1. $y=c_1e^{4x}+c_2e^{-x}$
2. $y=c_1e^{-ax}+c_2e^{-bx}$
3. $y=c_1e^{(2+\sqrt{3})x}+c_2e^{(2-\sqrt{3})x}$
4. $y=(c_1+c_2x)e^{-3x}$
5. $y=\left(c_1+c_2x+c_3x^2\right)e^x$
6. $y=c_1e^{-x}+c_2e^{-2x}+c_3e^{-3x}$
7. $y=c_1e^x+c_2e^{-x}+c_3e^{2x}+c_4e^{-2x}$
8. $y=(c_1+c_2x)\cos 2x+(c_3+c_4x)\sin 2x$
9. $y=\left(c_1+c_2x+c_3x^2\right)\cos x+\left(c_4+c_5x+c_6x^2\right)\sin x$
$$+e^{-\frac{1}{2}x}\left[(c_7+c_8x)\cos\frac{\sqrt{3}}{2}x+(c_9+c_{10}x)\sin\frac{\sqrt{3}}{2}x\right]$$
10. $y=c_1e^{-x}+e^{\frac{1}{2}x}\left(c_2\cos\dfrac{\sqrt{3}}{2}x+c_3\sin\dfrac{\sqrt{3}}{2}x\right)$
11. $x=0$
12. $y=3e^{-2x}\sin 5x$

2.5 THE INVERSE OPERATOR $\dfrac{1}{f(D)}$

Definition: $\dfrac{1}{f(D)}X$ is that function of x, free from arbitrary constants, which when operated on $f(D)$ gives X.

Thus $\quad f(D)\left[\dfrac{1}{f(D)}X\right]=X$

$\therefore \quad f(D)$ and $\dfrac{1}{f(D)}$ are inverse operators.

Theorem 1: $\dfrac{1}{f(D)}X$ is the particular integral of $f(D)y=X$

Proof: The given equation is $f(D)y=X$ . . . (1)

Putting $y=\dfrac{1}{f(D)}X$ in (1), we have $f(D)\left[\dfrac{1}{f(D)}X\right]=X$ or $X=X$

which is true.

$\therefore \quad y=\dfrac{1}{f(D)}X$ is a solution of (1).

Since it contains no arbitrary constants,

$\therefore$ it is the particular integral of $f(D)y=X$.

Theorem 2: $\dfrac{1}{D}X=\int Xdx$

Proof: Let $\quad \dfrac{1}{D}X=y$

Operating both sides by D, we have $D\left(\dfrac{1}{D}X\right)=Dy$ or $X=\dfrac{dy}{dx}$

Integrating both sides w.r.t. x

$$y=\int Xdx$$

No arbitrary constant being added since $y=\dfrac{1}{D}X$ contains no arbitrary constant.

$$\therefore \quad \frac{1}{D}X=\int Xdx$$

Theorem 3: $\dfrac{1}{D-a}X = e^{ax}\int Xe^{-ax}dx$

Proof: Let $\dfrac{1}{D-a}X = y$

Operating both sides by $(D-a)$, we get

$$(D-a)\left(\frac{1}{D-a}X\right) = (D-a)\,y$$

or $X = \dfrac{dy}{dx} - ay$ i.e., $\dfrac{dy}{dx} - ay = X$

which is a linear equation and I.F. $= e^{\int -adx} = e^{-ax}$

$\therefore$ It's solution is $ye^{-ax} = \int Xe^{-ax}dx$, no constant being added.

or $y = e^{ax}\int Xe^{-ax}dx$

Hence, $\dfrac{1}{D-a}X = e^{ax}\int e^{-ax}Xdx$

2.6 RULES FOR FINDING THE PARTICULAR INTEGRAL

Consider the differential equation

$$\left(D^n + a_1D^{n-1} + a_2D^{n-2} + \ldots + a_{n-1}D + a_n\right)y = X$$

$\therefore$ It can be written as $f(D)y = X$

where $f(D) = D^n + a_1D^{n-1} + a_2D^{n-2} + \ldots + a_{n-1}D + a_n$

Case I: When $X = e^{ax}$

Since $De^{ax} = ae^{ax}$

$$D^2e^{ax} = a^2e^{ax}$$

$$D^3e^{ax} = a^3e^{ax}$$

$$\vdots$$

$$D^{n-1}e^{ax} = a^{n-1}e^{ax}$$

$$D^{n}e^{ax} = a^{n}e^{ax}$$

Now, $f(D)e^{ax} = \left(D^n + a_1 D^{n-1} + a_2 D^{n-2} + \ldots + a_{n-1}D + a_n\right)e^{ax}$

$$= \left(a^n + a_1 a^{n-1} + a_2 a^{n-2} + \ldots + a_{n-1}a + a_n\right)e^{ax}$$

$$= f(a)e^{ax}$$

Operating on both sides by $\frac{1}{f(D)}$, we get

$$\frac{1}{f(D)}\left(f(D)e^{ax}\right) = \frac{1}{f(D)}\left(f(a)e^{ax}\right)$$

$$e^{ax} = f(a)\frac{1}{f(D)}e^{ax}$$

Dividing both sides by $f(a)$, provided $f(a) \neq 0$, we get

$$\frac{1}{f(a)}e^{ax} = \frac{1}{f(D)}e^{ax}$$

Hence, $\frac{1}{f(D)}e^{ax} = \frac{1}{f(a)}e^{ax}$, provided $f(a) \neq 0$

However, if $f(a) = 0$, the above method fails.

Since $f(a) = 0$, $D = a$ is a root of A.E. $f(D) = 0$

$\therefore$ $(D-a)$ is a factor of $f(D)$

Let $f(D) = (D-a)\phi(D)$, where $\phi(a) \neq 0$ . . . (1)

Then $\frac{1}{f(D)}e^{ax} = \frac{1}{(D-a)\phi(D)}e^{ax} = \frac{1}{(D-a)}.\frac{1}{\phi(D)}e^{ax}$

$$= \frac{1}{(D-a)}.\frac{1}{\phi(a)}e^{ax} \qquad \text{[By above case]}$$

$$=\frac{1}{\phi(a)}\cdot\frac{1}{(D-a)}e^{ax}$$

$$=\frac{1}{\phi(a)}e^{ax}\int e^{ax}.e^{-ax}dx \qquad \left[\because \frac{1}{D-a}X=e^{ax}\int Xe^{-ax}dx\right]$$

$$=\frac{1}{\phi(a)}e^{ax}\int 1dx=x.\frac{1}{\phi(a)}e^{ax}$$

$$=x.\frac{1}{\phi(a)}e^{ax} \qquad \ldots (2)$$

Differentiating both sides of Eq. (1) w. r. t. D, we get

$$f'(D)=(D-a)\phi'(D)+\phi(D) \quad \Rightarrow \quad f'(a)=\phi(a)$$

$\therefore$ from (2), we have

$$\frac{1}{f(D)}e^{ax}=x.\frac{1}{f'(a)}e^{ax} \qquad \text{provided } f'(a)\neq 0$$

If $f'(a)=0$, then $\frac{1}{f(D)}e^{ax}=x^2.\frac{1}{f''(a)}e^{ax}$ provided $f''(a)\neq 0$

and so on .

ILLUSTRATIVE EXAMPLES 2.2

Example 2.9: Find the P.I. of $\left(4D^2+4D-3\right)y=e^{2x}$.

Solution: P.I. $=\frac{1}{4D^2+4D-3}e^{2x}=\frac{e^{2x}}{4(2)^2+4(2)-3}=\frac{e^{2x}}{16+8-3}=\frac{e^{2x}}{21}$ $[\because D=2]$

Example 2.10: Find the P.I. of $\left(D^2+3D+2\right)y=5$

Solution: P.I. $=\frac{1}{D^2+3D+2}\left(5e^{0x}\right)$ $\left[\because e^{0x}=1\right]$

$$=5.\frac{e^{0x}}{0+0+2}=\frac{5}{2} \qquad \text{[Replace } D \text{ by 0]}$$

Example 2.11: Find the P.I. of $\left(D^3-3D^2+4\right)y=e^{2x}$.

Solution: $\text{P.I.}=\dfrac{1}{D^3-3D^2+4}e^{2x}$

The denominator vanishes when D is replaced by 2. It is case of failure.

We multiply the numerator by x and differentiate the denominator w. r. t. D, we get

$$\therefore \quad \text{P.I.}=x.\frac{1}{3D^2-6D}e^{2x}$$

It is again a case of failure. (when $D=2$)

We multiply the numerator by x and differentiate the denominator w. r. t. D to get

$$\therefore \quad \text{P.I.}=x^2.\frac{1}{6D-6}e^{2x}$$

$$=x^2.\frac{1}{6(2)-6}e^{2x}$$

$$=\frac{x^2}{6}e^{2x}$$

Case II: When $X=\sin(ax+b)$ or $\cos(ax+b)$

$$D\sin(ax+b)=a\cos(ax+b)$$

$$D^2\sin(ax+b)=\left(-a^2\right)\sin(ax+b)$$

$$D^3\sin(ax+b)=-a^3\cos(ax+b)$$

$$D^4\sin(ax+b)=a^4\sin(ax+b)$$

or $\left(D^2\right)^2\sin(ax+b)=\left(-a^2\right)^2\sin(ax+b)$

$$\vdots$$

In general, $\left(D^2\right)^n\sin(ax+b)=\left(-a^2\right)^n\sin(ax+b)$

$$\therefore \quad f\left(D^2\right)\sin(ax+b)=\left[\left(D^2\right)^n+a_1\left(D^2\right)^{n-1}+a_2\left(D^2\right)^{n-2}+\ldots+a_{n-1}\left(D^2\right)+a_n\right]\sin(ax+b)$$

$$=\left[\left(-a^2\right)^n+a_1\left(-a^2\right)^{n-1}+a_2\left(-a^2\right)^{n-2}+\ldots+a_{n-1}\left(-a^2\right)+a_n\right]\sin(ax+b)$$

$$=f\left(-a^2\right)\sin(ax+b)$$

Operating on both sides by $\dfrac{1}{f\left(D^2\right)}$, we get

$$\frac{1}{f\left(D^2\right)}\left[f\left(D^2\right)\sin(ax+b)\right]=\frac{1}{f\left(D^2\right)}\left[f\left(-a^2\right)\sin(ax+b)\right]$$

$$\Rightarrow \qquad \sin(ax+b)=f\left(-a^2\right)\frac{1}{f\left(D^2\right)}\sin(ax+b)$$

Dividing both sides by $f\left(-a^2\right)$, provided $f\left(-a^2\right)\neq 0$, we get

$$\frac{\sin(ax+b)}{f\left(-a^2\right)}=\frac{1}{f(D)^2}\sin(ax+b)$$

Hence $\dfrac{1}{f(D)^2}\sin(ax+b)=\dfrac{\sin(ax+b)}{f\left(-a^2\right)}$ provided $f\left(-a^2\right)\neq 0$

Similarly, $\dfrac{1}{f(D)^2}\cos(ax+b)=\dfrac{\cos(ax+b)}{f\left(-a^2\right)}$ provided $f\left(-a^2\right)\neq 0$

However, if $f\left(-a^2\right)=0$, the above method fails.

Since $\cos(ax+b)+i\sin(ax+b)=e^{i(ax+b)}$

$$\therefore \qquad \frac{1}{f\left(D^2\right)}\left[\cos(ax+b)+i\sin(ax+b)\right]=\frac{1}{f\left(D^2\right)}\,e^{i(ax+b)}$$

[If we replace D by $ia, f\left(D^2\right)=f\left(-a^2\right)=0,$ so that it is a case of failure]

$$=x.\frac{1}{f'\left(D^2\right)}e^{i(ax+b)}=x.\frac{1}{f'\left(D^2\right)}\left[\cos(ax+b)+i\sin(ax+b)\right]$$

Equating real parts

$$\frac{1}{f\left(D^2\right)}\cos(ax+b)=x.\frac{1}{f'\left(-a^2\right)}\cos(ax+b) \qquad \text{provided } f'\left(-a^2\right)\neq 0$$

Equating imaginary parts

$$\frac{1}{f(D^2)}\sin(ax+b)=x.\frac{1}{f'(-a^2)}\sin(ax+b) \qquad \text{provided } f'(-a^2)\neq 0$$

If $f'(-a^2)=0$, then

$$\frac{1}{f(D^2)}\sin(ax+b)=x^2.\frac{1}{f''(-a^2)}\sin(ax+b) \qquad \text{provided } f''(-a^2)\neq 0$$

$$\frac{1}{f(D^2)}\cos(ax+b)=x^2.\frac{1}{f''(-a^2)}\cos(ax+b) \qquad \text{provided } f''(-a^2)\neq 0$$

and so on.

ILLUSTRATIVE EXAMPLES 2.3

Example 2.12: Find the P.I. of $(D^3+1)y=\sin(2x+3)$.

Solution: P.I. $=\dfrac{1}{D^3+1}\sin(2x+3)$

$$=\frac{1}{D(-2^2)+1}\sin(2x+3) \qquad \left[\begin{array}{l}\because a=2\\ D^2=-a^2\end{array}\right]$$

$$=\frac{1}{1-4D}\sin(2x+3)$$

Multiplying and dividing RHS by $(1+4D)$, we get

$$=\frac{1+4D}{(1-4D)(1+4D)}\sin(2x+3)=\frac{1+4D}{1-16D^2}\sin(2x+3)$$

$$=\frac{1+4D}{1-16(-2^2)}\sin(2x+3) \qquad \left[\text{Put } D^2=-2^2\right]$$

$$=\frac{1}{65}\left[\sin(2x+3)+4D\sin(2x+3)\right]=\frac{1}{65}\left[\sin(2x+3)+8\cos(2x+3)\right]$$

Example 2.13: Find the P.I. of $(D^2+4)y=\cos 2x.$

Solution: $\quad \text{P.I.}=\dfrac{1}{D^2+4}\cos 2x$

Here, the denominator vanishes when D^2 is replaced by $-2^2=-4$. It is a case of failure. Multiply the numerator by x and differentiate the denominator w.r.t. D to get

$$\text{P.I.}=x.\frac{1}{2D}\cos 2x=\frac{x}{2}\int\cos 2x\,dx=\frac{x}{4}\sin 2x$$

Case III. When $X=x^m, m$ being a positive integer.

Here $\quad P.I.=\dfrac{1}{f(D)}x^m$

Take out the lowest degree term from $f(D)$ to make the first term unity (so that Binomial Theorem for a negative index is applicable).

The remaining factor will be of the form $[1+\phi(D)]$ or $[1-\phi(D)]$.

Take this factor in the numerator. It is of the form $[1+\phi(D)]^{-1}$ or $[1-\phi(D)]^{-1}$

Expand it in ascending powers of D as for as the term containing x^m, since

$$D^{m+1}(x^m)=0, D^{m+2}(x^m)=0 \quad \text{so on.}$$

Operate on x^m term by term.

ILLUSTRATIVE EXAMPLES 2.4

Example 2.14: Find the P.I. of $(D^2+5D+4)y=x^2+7x+9.$

Solution: $\quad P.I.=\dfrac{1}{D^2+5D+4}(x^2+7x+9)=\dfrac{1}{4\left(1+\dfrac{5D}{4}+\dfrac{D^2}{4}\right)}(x^2+7x+9).$

$$=\frac{1}{4}\left[1+\left(\frac{5D}{4}+\frac{D^2}{4}\right)\right]^{-1}(x^2+7x+9)$$

$$=\left[\frac{1}{4}1-\left(\frac{5D}{4}+\frac{D^2}{4}\right)+\left(\frac{5D}{4}+\frac{D^2}{4}\right)^2+\ldots\right](x^2+7x+9)$$

$$=\frac{1}{4}\left[1-\frac{5D}{4}-\frac{D^2}{4}+\frac{25D^2}{16}\ldots\right]\left(x^2+7x+9\right) \qquad \left[\because (1+A)^{-1}=1-A+A^2-A^2+\ldots\right]$$

$$=\frac{1}{4}\left[1-\frac{5D}{4}+\frac{21D^2}{16}\ldots\right]\left(x^2+7x+9\right)$$

$$=\frac{1}{4}\left[\left(x^2+7x+9\right)-\frac{5D}{4}\left(x^2+7x+9\right)+\frac{21}{16}D^2\left(x^2+7x+9\right)\right]$$

$$=\frac{1}{4}\left[\left(x^2+7x+9\right)-\frac{5}{4}(2x+7)+\frac{21}{16}(2)\right]=\frac{1}{4}\left(x^2+\frac{9}{2}x+\frac{23}{8}\right)$$

Case IV. When $X=e^{ax}V$, where V is a function of x.

Let u be a function of x, then by successive differentiation, we have

$$D\left(e^{ax}u\right)=e^{ax}Du+ax^{ax}u=e^{ax}\left(D+a\right)u$$

$$D^2\left(e^{ax}u\right)=D\left[e^{ax}\left(D+a\right)u\right]=e^{ax}\left(D^2+aD\right)u+ae^{ax}\left(D+a\right)u$$

$$=e^{ax}\left(D^2+2aD+a^2\right)u=e^{ax}\left(D+a\right)^2u$$

$$\vdots$$

$$D^n\left(e^{ax}u\right)=e^{ax}\left(D+a\right)^n u$$

$$f\left(D\right)\left(e^{ax}u\right)=e^{ax}f\left(D+a\right)u$$

Operating on both sides by $\frac{1}{f(D)}$, we get

$$\frac{1}{f(D)}\left[f(D)\left(e^{ax}u\right)\right]=\frac{1}{f(D)}\left[e^{ax}\left(D+a\right)u\right] \quad\Rightarrow\quad e^{ax}u=\frac{1}{f(D)}\left[e^{ax}f\left(D+a\right)u\right]$$

Now, let $f\left(D+a\right)u=V$, *ie.*, $\quad u=\frac{1}{f\left(D+a\right)}V$

$\therefore$ We have $e^{ax}\frac{1}{f\left(D+a\right)}V=\frac{1}{f(D)}\left(e^{ax}V\right) \quad\Rightarrow\quad \frac{1}{f(D)}\left(e^{ax}V\right)=e^{ax}\frac{1}{f\left(D+a\right)}V$

Thus, e^{ax} which is on the right of $\frac{1}{f(D)}$ may be taken out to the left provided D is replaced by $D+a$.

ILLUSTRATIVE EXAMPLES 2.5

Example 2.15: Find the P.I. of $(D^2-4D+3)y=e^x\cos 2x$.

Solution: P.I. $=\left(\dfrac{1}{D^2-4D+3}\right)e^x\cos 2x = e^x\dfrac{1}{(D+1)^2-4(D+1)+3}\cos 2x$

$$=e^x\frac{1}{D^2-2D}\cos 2x = e^x\frac{1}{-2^2-2D}\cos 2x \qquad \left[\text{putting } D^2=-2^2\right]$$

$$=\frac{-1}{2}e^x\frac{1}{2+D}\cos 2x = \frac{-1}{2}e^x\frac{2-D}{(2+D)(2-D)}\cos 2x \qquad [\text{By rationalizing}]$$

$$=\frac{-1}{2}e^x\frac{2-D}{4-D^2}\cos 2x = \frac{-1}{2}e^x\frac{2-D}{4-(-2^2)}\cos 2x$$

$$=\frac{-1}{16}e^x(2\cos 2x - D\cos 2x) = \frac{-1}{16}e^x(2\cos 2x+2\sin 2x) = \frac{-1}{8}e^x(\cos 2x+\sin 2x)$$

Case V. When X is any other function of x.

Resolve $f(D)$ into linear factors.

Let $\quad f(D)=(D-m_1)(D-m_2)...(D-m_n)X$

Then $\quad P.I.=\dfrac{1}{f(D)}X=\dfrac{1}{(D-m_1)(D-m_2)...(D-m_n)}X$

$$=\left(\frac{A_1}{D-m_1}+\frac{A_2}{D-m_2}+...+\frac{A_n}{D-m_2}\right)X$$

$$=A_1\frac{1}{D-m_1}X+A_2\frac{1}{D-m_2}X+...+A_n\frac{1}{D-m_n}X$$

$$=A_1e^{m_1x}\int Xe^{-m_1x}dx+A_2e^{m_2x}\int Xe^{-m_2x}dx+...+A_ne^{m_nx}\int Xe^{-m_nx}dx$$

$$\left[\because \frac{1}{D-a}X=e^{ax}\int Xe^{-ax}dx\right]$$

Example 2.16: Solve the differential equation $\dfrac{d^2y}{dx^2}+a^2y=\text{cosec } ax$.

Solution: Given differential equation is

$$\frac{d^2y}{dx^2}+a^2y=\text{cosec } ax \quad \Rightarrow \quad (D^2+a^2)y=\text{cosec } ax$$

It's A. E. is

$$D^2 + a^2 = 0$$

which on solving gives $D = \pm\, ai$

$\therefore \quad C.F. = c_1 \cos ax + c_1 \sin ax$

Now, $P.I. = \dfrac{1}{D^2 + a^2} \operatorname{cosec} ax = \dfrac{1}{(D - ai)(D + ai)} \operatorname{cosec} ax$

$$= \frac{1}{2ai}\left[\frac{1}{D - ai} - \frac{1}{D + ai}\right] \operatorname{cosec} ax \qquad \ldots \text{(A)}$$

Here, $\dfrac{1}{D - ai} \operatorname{cosec} ax = e^{iax} \int \operatorname{cosec} ax \cdot e^{-iax} dx = e^{iax} \int \dfrac{\cos ax - i \sin ax}{\sin ax} dx$

$\Rightarrow \quad \dfrac{1}{D - ai} \operatorname{cosec} ax = e^{iax} \int (\cot ax - i) dx \quad \Rightarrow \quad \dfrac{1}{D - ai} \operatorname{cosec} ax = e^{iax}\left[\dfrac{\log|\sin ax|}{a} - ix\right]$

Changing i to $-i$, we get

$$\frac{1}{D + ai} \operatorname{cosec} ax = e^{-iax}\left[\frac{\log|\sin ax|}{a} + ix\right]$$

$\therefore$ from (A), we have

$$P.I. = \frac{1}{2ai}\left[e^{iax}\left(\frac{\log|\sin ax|}{a} - ix\right) - e^{-iax}\left(\frac{\log|\sin ax|}{a} + ix\right)\right]$$

$$= \frac{1}{a}\left[\frac{\log|\sin ax|}{a} \cdot \frac{e^{iax} - e^{-iax}}{2i} e^{iax} - x \cdot \frac{e^{iax} - e^{-iax}}{2}\right]$$

$$= \frac{1}{a}\left[\frac{\log|\sin ax|}{a} \sin ax - x \cos ax\right] \qquad \left[\because \sin\theta = \frac{e^{i\theta} - e^{-i\theta}}{2i}, \cos\theta = \frac{e^{i\theta} + e^{-i\theta}}{2}\right]$$

$\therefore \quad C.S. = C.F. + P.I.$

$$\Rightarrow \quad y = c_1 \cos ax + c_1 \sin ax + \frac{1}{a}\left[\frac{\log|\sin ax|}{a} \sin ax - x \cos ax\right]$$

Working Rule to Solve the Equation

$$\frac{d^n y}{dx^n}+a_1\frac{d^{n-1}y}{dx^{n-1}}+a_2\frac{d^{n-2}y}{dx^{n-2}}+...a_{n-1}\frac{dy}{dx}+a_n y=X$$

Step 1. Write the equation in symbolic form

$$\left(D^n+a_1D^{n-1}+a_2D^{n-2}+...+a_{n-1}D+a_n\right)y=X$$

Step 2. Solve the auxiliary equation.

$$D^n+a_1D^{n-1}+a_2D^{n-2}+...+a_{n-1}D+a_n=0$$

Step 3. Write the complementary function with the help of following table.

Roots of the A.E.	C.F.
If roots are real and distinct say m_1, m_2, m_3	$c_1e^{m_1x}+c_2e^{m_2x}+c_3e^{m_3x}+...$
If two real roots are equal, say $m_1=m_2=m$	$(c_1+c_2x)e^{mx}+c_3e^{m_3x}+...$
If three roots are equal, say $m_1=m_2=m_3=m$	$\left(c_1+c_2x+c_3x^2\right)e^{mx}+c_4e^{m_4x}+...$
If roots are a pair of imaginary number $\alpha\pm i\beta$	$e^{\alpha x}(c_1\cos\beta x+c_2\sin\beta x)$
If pair of imaginary roots of repeated twice, i.e., $\alpha\pm i\beta$, $\alpha\pm i\beta$	$e^{\alpha x}\left[(c_1+c_2x)\cos\beta x+(c_3+c_4x)\sin\beta x\right]$

Step 4. Find the particular integral of, i.e., $P.I.=\dfrac{1}{D^n+a_1D^{n-1}+a_2D^{n-2}...a_n}X$ with the help of the following rules:

Functions	Particular Integral
$X=e^{ax}$	$P.I.=\dfrac{1}{f(D)}e^{ax}=\dfrac{e^{ax}}{f(a)}$ provided $f(a)\neq 0$ If case $f(a)=0$ then multiply numerator by x and differentiate the denominator w. r. t. D and continue this process until denominator ceases to be zero on putting $D=a$.
$X=\sin(ax+b)$ or $X=\cos(ax+b)$	$P.I.=\dfrac{1}{\phi\left(D^2\right)}\sin(ax+b)=\dfrac{\sin(ax+b)}{\phi\left(-a^2\right)}$, provided $\phi\left(-a^2\right)\neq 0$

Functions	Particular Integral
$X = x^m$	$P.I. = \left[f(D)\right]^{-1} x^m$ expand $f(D)$ by binomial. Thus upto D^m and then operate on x^m.
$X = e^{ax}V$	$P.I. = e^{ax} \dfrac{1}{f(D+a)} V$
If X is any other function x	$P.I. = \dfrac{1}{f(D)} X$. Resolve $\dfrac{1}{f(D)}$ into partial fractions and operate each partial fraction on X.

Step 5. Write the complete solution as $C.S. = C.F. + P.I.$

ILLUSTRATIVE EXAMPLES 2.6

Example 2.17: Solve the differential equation $\dfrac{d^3y}{dx^3} - 3\dfrac{dy}{dx} + 2y = e^{2x}$.

Solution: The symbolic form of given equation is

$$(D^3 - 3D + 2)\, y = e^{2x} \qquad \ldots (1)$$

It's auxiliary equation is

$$D^3 - 3D + 2 = 0$$

By inspection $D = 1$ satisfies this equation

$$\therefore \quad (D-1)(D^2 + D - 2) = 0 \quad \Rightarrow \quad (D-1)(D-1)(D+2) = 0$$

$$\Rightarrow \quad D = 1, 1, -2 \qquad \text{(two roots are equal)}$$

$\therefore$ Complementary function of (1) is

$$y_c = (c_1 + c_2 x)e^x + c_3 e^{-2x}$$

The particular integral of (1) is

$$y_p = \frac{1}{D^3 - 3D + 2} e^{2x} = \frac{e^{2x}}{2^3 - 2(3) + 2} = \frac{e^{2x}}{4} \qquad \left[\because \frac{1}{F(D)} e^{ax} = \frac{e^{ax}}{F(a)}, \text{here } a = 2\right]$$

Hence, the general solution (1) is

$$y = y_c + y_p = (c_1 + c_2 x)e^x + c_3 e^{-2x} + \frac{e^{2x}}{4}$$

Example 2.18: Find the P.I. of $(D-2)^2 y = e^{2x}$.

Solution: Here P.I. is $y_p = \dfrac{1}{(D-2)^2} e^{2x}$ ... (1)

Put $D = 2$, denominator $F(D) = (D-2)^2$ of (1) becomes zero
i.e., $F(2) = 0$ and hence it is a case of failure.

$$\therefore \quad y_p = \frac{1}{(D-2)^2} e^{2x}.1 = e^{2x} \frac{1}{(D+2-2)^2} \quad \left[\because \frac{1}{F(D)}\left(e^{ax}V\right) = e^{ax}\frac{1}{F(D)}V\right]$$

$$= e^{2x}.\frac{1}{D^2}.1 = e^{2x}\frac{x^2}{2}$$

Example 2.19: Solve the differential equation $\dfrac{d^2y}{dx^2} - 2\dfrac{dy}{dx} + y = xe^x \sin x$.

[PTU Dec 2003, Jan 2010]

Solution: The symbolic form of given equation is

$$\left(D^2 - 2D + 1\right) y = xe^x \sin x \qquad \ldots (1)$$

It's auxiliary equation is

$$D^2 - 2D + 1 = 0 \quad \Rightarrow \quad (D-1)^2 = 0 \quad \Rightarrow \quad D = 1, 1$$

$$\therefore \quad C.F. = (c_1 + c_2 x) e^x$$

Now $P.I. = \dfrac{1}{(D-1)^2} xe^x \sin x = e^x \cdot \dfrac{1}{(D+1-1)^2} x \sin x$

$$= e^x \cdot \frac{1}{D^2} x \sin x = e^x \cdot \frac{1}{D} \int x \sin x \, dx \qquad \text{[Integrating by parts]}$$

$$= e^x \cdot \frac{1}{D}\left[x(-\cos x) - \int 1.(-\cos x)\, dx\right] = e^x \cdot \frac{1}{D}\left[-x\cos x + \int \cos x \, dx\right]$$

$$= e^x \cdot \frac{1}{D}[-x\cos x + \sin x] = e^x \int (-x\cos x + \sin x) dx$$

$$= e^x \left[-\int x\cos x \, dx + \int \sin x \, dx\right]$$

$$= e^x \left[-\left\{x(\sin x) - \int 1 \cdot (\sin x)\, dx\right\} - \cos x\right]$$

$$= e^x \left[-\left\{x \sin x - \int \sin x \, dx\right\} - \cos x\right]$$

$$= e^x \left[-\left\{x \sin x + \cos x\right\} - \cos x\right] = e^x \left[-x\sin x - \cos x - \cos x\right]$$

$$= e^x\left[-x\sin x - 2\cos x\right] = -e^x\left[x\sin x + 2\cos x\right]$$

Hence, the complete solution (1) is

$$y = C.F. + P.I. = (c_1 + c_2 x)e^x - e^x\left[x\sin x + 2\cos x\right]$$

Example 2.20: Solve the differential equation $(D^2 - 6D + 13)y = 8e^{3x}\sin 4x + 2^x$.

[PTU Dec 2005]

Solution: Given differential equation is

$$(D^2 - 6D + 13)y = 8e^{3x}\sin 4x + 2^x \qquad \ldots (1)$$

It's A.E. is

$$D^2 - 6D + 13 = 0 \quad \text{which is quadratic in } D.$$

$$\Rightarrow \quad D = \frac{6 \pm \sqrt{36 - 52}}{2} = \frac{6 \pm \sqrt{-16}}{2} = \frac{6 \pm 4i}{2} = 3 \pm 2i$$

$$\therefore \quad C.F. = e^{3x}(c_1\cos 2x + c_2\sin 2x)$$

Now $P.I. = \dfrac{1}{D^2 - 6D + 13}(8e^{3x}\sin 4x + 2^x)$

$$= 8\frac{1}{D^2 - 6D + 13}e^{3x}\sin 4x + \frac{1}{D^2 - 6D + 13}2^x$$

$$= 8e^{3x}\frac{1}{(D+3)^2 - 6(D+3) + 13}\sin 4x + \frac{1}{D^2 - 6D + 13}e^{\log 2^x}$$

$$= 8e^{3x}\frac{1}{D^2 + 4}\sin 4x + \frac{1}{D^2 - 6D + 13}e^{x\log 2}$$

$$= 8e^{3x}\frac{1}{D^2 + 4}\sin 4x + \frac{1}{D^2 - 6D + 13}e^{x\log 2}$$

Put $D^2 = -16$ in first term and $D = \log 2$ in second term, we get

$$P.I. = 8e^{3x}\frac{1}{-16 + 4}\sin 4x + \frac{1}{(\log 2)^2 - 6\log 2 + 13}e^{x\log 2}$$

$$= -\frac{3}{2}e^{3x}\sin 4x + \frac{1}{(\log 2)^2 - 6\log 2 + 13}e^{x\log 2}$$

Hence, the complete solution (1) is

$$y = C.F. + P.I. = e^{3x}(c_1\cos 2x + c_2\sin 2x) - \frac{3}{2}e^{3x}\sin 4x + \frac{1}{(\log 2)^2 - 6\log 2 + 13}e^{x\log 2}$$

Example 2.21: Solve the differential equation $(D^2+2D+2)y=e^{-x}\sec x$.

[PTU Dec 2002]

Solution: Given differential equation is

$$(D^2+2D+2)y=e^{-x}\sec x \quad \ldots (1)$$

It's A.E. is

$$D^2+2D+2=0 \quad \text{which is quadratic in } D.$$

$$\Rightarrow \quad D=\frac{-2\pm\sqrt{4-8}}{2}=\frac{-2\pm\sqrt{-4}}{2}=\frac{-2\pm 2i}{2}=-1\pm i$$

$$\therefore \quad C.F.=e^{-x}\left(c_1\cos x+c_2\sin x\right)$$

Now $P.I.=\dfrac{1}{D^2+2D+2}e^{-x}\sec x$

$$=e^{-x}\frac{1}{(D-1)^2+2(D-1)+2}\sec x$$

$$=e^{-x}\frac{1}{D^2+1}\sec x=e^{-x}\frac{1}{(D+i)(D-i)}\sec x$$

$$=\frac{e^{-x}}{2i}\left(\frac{1}{D-i}-\frac{1}{D+i}\right)\sec x \quad \text{[By partial fractions]} \quad \ldots (A)$$

Here, $\dfrac{1}{D-i}\sec ax=e^{ix}\int \sec x\cdot e^{-ix}dx$ $\quad\left[\because \dfrac{1}{D-a}X=e^{ax}\int Xe^{-ax}dx\right]$

$$=e^{ix}\int\frac{\cos x-i\sin x}{\cos x}dx=e^{ix}\int(1-i\tan x)dx$$

$$=e^{ix}\left(x+i\log\cos x\right)$$

Changing i to $-i$, we get

$$\frac{1}{D+i}\sec x=e^{-ix}\left(x-i\log\cos x\right)$$

$\therefore$ from (A), we have

$$P.I.=\frac{e^{-x}}{2i}\left[e^{ix}\left(x+i\log\cos x\right)-e^{-ix}\left(x-i\log\cos x\right)\right]$$

$$= \frac{e^{-x}}{2i}\left[x(e^{ix}-e^{-ix})+i\log\cos x(e^{ix}+e^{-ix})\right]$$

$$= e^{-x}\left(x\frac{(e^{ix}-e^{-ix})}{2i}+\frac{1}{a}\log\cos x\frac{(e^{ix}+e^{-ix})}{2}\right)$$

$$= e^{-x}\left(x\sin x+\cos x\log\cos x\right) \qquad \left[\cos x=\frac{e^{ix}+e^{-ix}}{2},\ \sin x=\frac{e^{ix}-e^{-ix}}{2i}\right]$$

Hence, the complete solution (1) is

$$y = C.F.+P.I. = e^{-x}\left(c_1\cos x+c_2\sin x\right)+e^{-x}\left(x\sin x+\cos x\log\cos x\right)$$

$$\Rightarrow \quad y = e^{-x}\left(c_1\cos x+c_2\sin x+x\sin x+\cos x\log\cos x\right)$$

Example 2.22: Solve the differential equation $\frac{d^2y}{dx^2}+3\frac{dy}{dx}+2y=e^{e^x}$.

[PTU Dec 2003]

Solution: Given differential equation is

$$\frac{d^2y}{dx^2}+3\frac{dy}{dx}+2y=e^{e^x} \qquad \ldots (1)$$

It's S.F. is

$$(D^2+3D+2)y=e^{e^x} \qquad \ldots (2)$$

It's A.E. is

$$D^2+3D+2=0$$

$$\Rightarrow \quad (D+1)(D+2)=0 \quad \Rightarrow \quad D=-1,-2$$

$$\therefore \quad C.F.=c_1e^{-x}+c_2e^{-2x}$$

Now $P.I. = \frac{1}{D^2+3D+2}e^{e^x}$

$$= \frac{1}{(D+1)(D+2)}e^{e^x}$$

$$= \left(\frac{1}{D+1}-\frac{1}{D+2}\right)e^{e^x} \qquad \text{[By partial fractions]} \qquad \ldots (3)$$

$$=\frac{1}{D+1}e^{e^x}-\frac{1}{D+2}e^{e^x}=\frac{1}{D-(-1)}e^{e^x}-\frac{1}{D-(-2)}e^{e^x}$$

$$=e^{-x}\int e^x\cdot e^{e^x}dx-e^{-2x}\int e^{2x}\cdot e^{e^x}dx \qquad \left[\because \frac{1}{D-a}X=e^{ax}\int Xe^{-ax}dx\right]$$

Put $e^x=t$, $e^x dx=dt$

$$\therefore \quad P.I.=e^{-x}\int e^t dt-e^{-2x}\int te^t dt$$

$$=e^{-x}e^t-e^{-2x}\left[te^t-\int 1\cdot e^t dt\right]=e^{-x}e^t-e^{-2x}\left[te^t-\int e^t dt\right]$$

$$=e^{-x}e^t-e^{-2x}\left(te^t-e^t\right)=e^{-x}e^t-e^{-2x}(t-1)e^t=\left[e^{-x}-e^{-2x}(t-1)\right]e^t$$

$$=\left[e^{-x}-e^{-2x}\left(e^x-1\right)\right]e^{e^x}=\left[e^{-x}-e^{-x}+e^{-2x}\right]e^{e^x}=e^{-2x}e^{e^x}$$

Hence, the complete solution (1) is

$$y=C.F.+P.I.=c_1e^{-x}+c_2e^{-2x}+e^{-2x}e^{e^x}$$

EXERCISE 2.2

Solve the following differential equations:

1. $\frac{d^3y}{dx^3}+y=3+5e^x$

2. $\frac{d^2y}{dx^2}-4y=\left(1+e^x\right)^2$

3. $\frac{d^2y}{dx^2}-2\frac{dy}{dx}+5y=\sin 3x$

4. $\left(D^2-4D+3\right)y=\sin 3x\cos 2x$ [PTU Jan 2009]

5. $\frac{d^2y}{dx^2}+4y=x\sin 2x$

6. $\left(D^2-3D+2\right)y=2e^x\cos\frac{x}{2}$

7. $\frac{d^2y}{dx^2}+a^2y=\sec ax$ [PTU May 2012]

8. $\left(D^2+2D+2\right)y=e^{-x}\sec x$.

9. $\left(D^3+2D^2+D\right)y=x^2e^{2x}+\sin^2 x$

10. $\frac{d^2y}{dx^2}-4\frac{dy}{dx}+4y=8(x^2+e^{2x}+\sin 2x)$

11. $(D^2+a^2)y=\sin ax$ [PTU May 2009]

12. $\frac{d^2y}{dx^2}-4\frac{dy}{dx}+4y=x^2+e^{2x}+\cos 2x$

13. $\frac{d^2y}{dx^2}-2\frac{dy}{dx}+4y=e^x\cos x+x$

14. $\frac{d^3y}{dx^3}+4\frac{dy}{dx}=\sin 2x$ [PTU Dec 2012]

ANSWERS

1. $y = c_1 e^{-x} + e^{\frac{1}{2}x}\left(c_2 \cos\frac{\sqrt{3}}{2} + c_3 \sin\frac{\sqrt{3}}{2}\right) + 3 + \frac{5}{2}e^x$

2. $y = c_1 e^{2x} + c_2 e^{-2x} - \frac{1}{4} - \frac{2}{3}e^x + \frac{1}{4}xe^{2x}$

3. $y = e^x(c_1 \cos 2x + c_2 \sin 2x) + \frac{1}{26}(3\cos 3x - 2\sin 3x)$

4. $y = c_1 e^x + c_2 e^{3x} + \frac{1}{884}(10\cos 5x - 11\sin 5x) + \frac{1}{20}(\sin x + 2\cos x)$

5. $y = c_1 \cos 2x + c_2 \sin 2x - \frac{x^2}{8}\cos 2x + \frac{x}{16}\sin 2x$

6. $y = c_1 e^x + c_2 e^{2x} - \frac{8}{5}e^x\left(2\sin\frac{x}{2} + \cos\frac{x}{2}\right)$

7. $y = c_1 \cos ax + c_2 \sin ax + \frac{1}{a}\left(x\sin ax + x\cos ax \frac{\log\cos ax}{a}\right)$

8. $y = e^{-x}(c_1 \cos x + c_2 \sin x + x\sin x + \cos x \log\cos x)$.

9. $y = c_1 + (c_2 + c_3 x)e^{-x} + \frac{e^{2x}}{18}\left(x^2 - \frac{7}{8}x + \frac{11}{6}\right) + \frac{1}{100}(3\sin 2x + 4\cos 2x)$

10. $y = (c_1 + c_2 x)e^{2x} + 2x^2 + 4x + 3 + 4x^2 e^{2x} + \cos 2x$

11. $y = c_1 \cos ax + c_2 \sin ax - \frac{x\cos ax}{2a}$

12. $y = (c_1 + c_2 x)e^{2x} + \frac{1}{4}\left(x^2 + 2x + \frac{3}{2}\right) + e^x - \frac{1}{8}\sin 2x$

13. $y = e^x(c_1 \cos\sqrt{3}x + c_2 \sin\sqrt{3}x) + \frac{1}{2}e^x \cos x + \frac{1}{4}x + \frac{1}{8}$

14. $y = c_1 + (c_1 \cos 2x + c_2 \sin 2x) - \frac{1}{4}\cos 2x$

2.7 METHOD OF VARIATION OF PARAMETERS

We shall now explain another method to find the particular integral of non-homogeneous linear equation whose complementary function is known. In this method the constants of complementary function are taken as function of the independent variable. In order to explain this method, we consider a linear second order differential equation.

$$\frac{d^2y}{dx^2} + P\frac{dy}{dx} + Qy = f(x) \quad \text{where } P \text{ and } Q \text{ are constants} \qquad \ldots (1)$$

Let the complementary function of (1) be

$$y = c_1 u + c_2 v \qquad \ldots (2)$$

where c_1, c_2 are constants.

Clearly u, v are the integral solutions of the corresponding homogenous Eq. (1), i.e., u, v are the integral solutions of

$$\frac{d^2y}{dx^2} + P\frac{dy}{dx} + Qy = 0 \qquad \ldots (3)$$

$$\Rightarrow \quad u'' + Pu' + Qu = 0 \quad \text{and} \quad v'' + Pv' + Qv = 0 \qquad \ldots (4)$$

Let $\quad y_p = y_1(x)u + y_2(x)v \qquad \ldots (5)$

be a solution of Eq. (1).

Differentiating Eq. (5), we get

$$\frac{dy}{dx} = y_1'(x)u + y_1(x)u' + y_2'(x)v + y_2(x)v'$$

$$= [y_1(x)u' + y_2(x)v'] + [y_1'(x)u + y_2'(x)v] \qquad \ldots (6)$$

Let the function $y_1(x), y_2(x)$ Satisfy the additional condition

$$y_1'(x)u + y_2'(x)v = 0 \qquad \ldots (7)$$

So that $\dfrac{dy}{dx} = y_1(x)u' + y_2(x)v' \qquad \ldots (8)$

Differentiating Eq. (8), we get

$$\frac{d^2y}{dx^2} = y_1(x)u'' + y_1'(x)u' + y_2(x)v'' + y_2'(x)v'$$

Substituting these values in (1), we get

$$\frac{d^2y}{dx^2} + P\frac{dy}{dx} + Qy = [y_1(x)u'' + y_1'(x)u' + y_2(x)v'' + y_2'(x)v']$$

$$+P[y_1(x)u' + y_2(x)v'] + Q[y_1(x)u + y_2(x)v] = f(x)$$

$$\Rightarrow \quad y_1(x)[u'' + Pu' + Qu] + y_2(x)[v'' + Pv' + Qv] + y_1'(x)u' + y_2'(x)v' = f(x)$$

$$\Rightarrow \quad y_1'(x)u' + y_2'(x)v' = f(x) \qquad \ldots (9) \qquad \text{[Using 4]}$$

Eqs. (7) and (9) are

$$y_1'(x)u + y_2'(x)v = 0 \quad \text{and} \quad y_1'(x)u' + y_2'(x)v' = f(x)$$

Solving these eq. for $y_1'(x), y_2'(x)$ we get

$$\left.\begin{aligned} y_1'(x) &= \frac{vf(x)}{\begin{vmatrix} u & v \\ u' & v' \end{vmatrix}} = -\frac{vf(x)}{w(u,v)} \\ \text{and} \quad y_2'(x) &= \frac{uf(x)}{\begin{vmatrix} u & v \\ u' & v' \end{vmatrix}} = \frac{uf(x)}{w(u,v)} \end{aligned}\right\} \quad \ldots (10)$$

Integration Eq. (10), we get

$$y_1(x) = -\int \frac{vf(x)}{w(u,v)} dx$$

$$\text{and} \quad y_2(x) = \int \frac{uf(x)}{w(u,v)} dx$$

where $w(u,v) = \begin{vmatrix} u & v \\ u' & v' \end{vmatrix}$ is known as Wronskian of *u*, *v*.

Substituting these values in Eq. (5), we get

$$y_p = -u \int \frac{vf(x)}{w(x,v)} dx + v \int \frac{uf(x)}{w(x,v)} dx$$

which is particular solution of (1).

Note: The above method is applicable even if the co-efficient *P*, *Q* are function of *x*.

ILLUSTRATIVE EXAMPLES 2.7

Example 2.23: Apply the method of variation of parameter to solve

$$\frac{d^2y}{dx^2} + 4y = 4\sec^2 2x$$

Solution: Given equation is symbolic form is

$$(D^2+4)y = 4\sec^2 2x$$

It's A.E. is

$$D^2+4=0 \text{ So that } D=\pm 2i$$

$\therefore$ It's C.F. is

$$y = c_1\cos 2x + c_2\sin 2x$$

Here $u=\cos 2x,\ v=\sin 2x,\ f(x)=4\sec^2 2x$

$$\therefore \quad w=\begin{vmatrix} u & v \\ u' & v' \end{vmatrix} = \begin{vmatrix} \cos 2x & \sin 2x \\ -2\sin 2x & 2\cos 2x \end{vmatrix} = 2$$

$$\text{P.I.} = uy_1 + vy_2\ , \quad \text{where } y_1(x) = -\int \frac{vf(x)}{w(u,v)}dx \quad \text{and} \quad y_2(x) = \int \frac{uf(x)}{w(u,v)}dx$$

$$= -\cos 2x \int \frac{\sin 2x\ 4\sec^2 2x}{2}dx + \sin 2x \int \frac{\cos 2x\, 4\sec^2 2x}{2}dx$$

$$= -2\cos 2x \int \sec 2x \tan 2x\, dx + 2\sin 2x \int \sec 2x\, dx$$

$$= -2\cos 2x \frac{\sec 2x}{2} + 2\sin 2x.\frac{1}{2}\log(\sec 2x + \tan 2x)$$

$$= -1 + \sin 2x \log(\sec 2x + \tan 2x)$$

General Solution is

$$y = c_1\cos 2x + c_2\sin 2x - 1 + \sin 2x\log(\sec 2x + \tan 2x)$$

Example 2.24: Solve by method of variation of parameter the differential equation $\dfrac{d^2y}{dx^2} + y = \sec x.$

Solution: Symbolic form of the equation is

$$(D^2+1)y = \sec x.$$

It's A.E. is

$$D^2+1=0, D=\pm i$$

It's C.F. = $c_1 \cos x + c_2 \sin x = c_1 u + c_2 v$

where $u = \cos x,\quad v = \sin x,\quad f(x) = \sec x$

$$w = \begin{vmatrix} u & v \\ u' & v' \end{vmatrix} = \begin{vmatrix} \cos x & \sin x \\ -\sin x & \cos x \end{vmatrix} = 1$$

$$P.I. = uy_1 + vy_2,\ \text{where } y_1 = -\int \frac{vf(x)}{w} dx,\quad y_2 = \int \frac{uf(x)}{w} dx$$

$$P.I. = -\cos x \int \frac{\sin x \sec x}{1} dx + \sin x \int \frac{\cos x \sec x}{1} dx$$

$$= -\cos x \int \tan x\, dx + \sin x \int 1.dx \ = \cos x \log \cos x + x \sin x$$

General solution is

$$y = c_1 \cos x + c_2 \sin^x + \cos x \log \cos x + x \sin x.$$

EXERCISE 2.3

1. $\dfrac{d^2 y}{dx^2} + y = \operatorname{cosec} x$
2. $\dfrac{d^2 y}{dx^2} + 4y = \tan 2x$
3. $\dfrac{d^2 y}{dx^2} + 4y = \sec 2x$
4. $\dfrac{d^2 y}{dx^2} + y = x \sin x$
5. $y'' - 2y' + 2y = e^x \tan x$
6. $y'' + 3y' + 2y = \sin e^x$ [PTU May 2012]
7. $y'' - 6y' + 9y = e^{3x} / x^2$ [PTU Dec 2012]

ANSWERS

1. $y = c_1 \cos x + c_2 \sin x \ - x \cos x + \sin x \log \sin x.$
2. $y = c_1 \cos 2x + c_2 \sin 2x - \dfrac{1}{4} \cos 2x \log(\sec 2x + \tan 2x).$
3. $y = c_1 \cos 2x + c_2 \sin 2x + \dfrac{1}{4} \cos 2x \log \cos 2x + \dfrac{x}{2} \sin 2x$
4. $y = c_1 \cos x + c_2 \sin x + \dfrac{x}{2} \sin x - \dfrac{x^2}{4} \cos x$
5. $y = e^n (c_1 \cos + c_2 \sin x) - e^n \cos x \log(\sec x + \tan x).$
6. $y = (\cos e^x + c_1) e^{-x} + (-e^x \cos e^x + \sin e^x + c_2) e^{-2x}$
7. $y = (-\log x) e^{3x} - e^{3x} + c_1 e^{3x} + x c_2 e^{3x}$

2.8 CAUCHY'S HOMOGENEOUS LINEAR EQUATION

An equation of the type

$$x^n \frac{d^n y}{dx^n} + a_1 x^{n-1} \frac{d^{n-1} y}{dx^{n-1}} + a_2 x^{n-2} \frac{d^{n-2} y}{dx^{n-2}} + \ldots + a_n y = X \qquad \ldots (1)$$

where a_i's are constants and X is a function of x, is called Cauchy's linear homogeneous equation of the nth order.

Such an equation can be reduced to linear differential equation with constant coefficients as follows.

Put $\quad x = e^z$ i.e., $z = \log x$

$$\therefore \quad \frac{dz}{dx} = \frac{1}{x} \qquad \therefore \quad \frac{dy}{dx} = \frac{dy}{dz} \cdot \frac{dz}{dx} = \frac{1}{x} \cdot \frac{dy}{dz}$$

$$\Rightarrow \quad x \cdot \frac{dy}{dx} = \frac{dy}{dz} \qquad \ldots (2)$$

Also $$\frac{d^2 y}{dx^2} = \frac{d}{dx} \cdot \left(\frac{dy}{dx} \right) = \frac{d}{dx} \left(\frac{1}{x} \frac{dy}{dz} \right) = -\frac{1}{x^2} \cdot \frac{dy}{dz} + \frac{1}{x} \frac{d^2 y}{dz^2} \cdot \frac{dy}{dx}$$

$$= -\frac{1}{x^2} \cdot \frac{dy}{dz} + \frac{1}{x} \frac{d^2 y}{dz^2} \cdot \frac{1}{x} = \frac{1}{x^2} \left(\frac{d^2 y}{dz^2} - \frac{dy}{dz} \right)$$

$$\Rightarrow \quad x^2 \frac{d^2 y}{dx^2} = \frac{d^2 y}{dz^2} - \frac{dy}{dz} \qquad \ldots (3)$$

Also $$\frac{d^3 y}{dx^3} = \frac{d}{dx} \left(\frac{d^2 y}{dx^2} \right) = \frac{d}{dx} \left[\frac{1}{x^2} \left(\frac{d^2 y}{dz^2} - \frac{dy}{dz} \right) \right]$$

$$= -\frac{2}{x^3} \left(\frac{d^2 y}{dz^2} - \frac{dy}{dz} \right) + \frac{1}{x^2} \frac{d}{dx} \left(\frac{d^2 y}{dz^2} - \frac{dy}{dz} \right)$$

$$= -\frac{2}{x^3} \left(\frac{d^2 y}{dz^2} - \frac{dy}{dz} \right) + \frac{1}{x^2} \left(\frac{d^3 y}{dz^3} \cdot \frac{dz}{dx} - \frac{d^2 y}{dz^2} \cdot \frac{dz}{dx} \right)$$

$$= -\frac{2}{x^3}\left(\frac{d^2y}{dz^2} - \frac{dy}{dz}\right) + \frac{1}{x^2}\left(\frac{d^3y}{dz^3} - \frac{d^2y}{dz^2}\right)\frac{1}{x}$$

$$= \frac{1}{x^3}\left(\frac{d^3y}{dz^3} - 3\frac{d^2y}{dz^2} + 2\frac{dy}{dz}\right)$$

$$\Rightarrow \quad x^3\frac{d^3y}{dx^3} = \frac{d^3y}{dz^3} - 3\frac{d^2y}{dz^2} + 2\frac{dy}{dz} \qquad \ldots (4)$$

and so on

Putting $\frac{d}{dz} = D$ in (2), (3) and (4), we get

$$x\frac{dy}{dx} = Dy$$

$$x^2\frac{d^2y}{dx^2} = D^2y - Dy = D(D-1)y$$

$$x^3\frac{d^3y}{dx^3} = D^3y - 3D^2y + 2Dy = \left(D^3 - 3D^2 + 2D\right)y$$

$$= D(D-1)(D-2)y$$

and so on

In general, $x^n\frac{d^ny}{dx^n} = \left[D(D-1)(D-2)\ldots(D-(n-1))\right]y$

$$\Rightarrow \quad x^n\frac{d^ny}{dx^n} = \left[D(D-1)(D-2)\ldots(D-n+1)\right]y$$

Substituting the values of

$$x\frac{dy}{dx},\ x^2\frac{d^2y}{dx^2},\ x^3\frac{d^3y}{dx^3},\ \ldots,\ x^n\frac{d^ny}{dx^n}$$

in (1), we get

$$\left[D(D-1)(D-2)\ldots(D-n+1)\right]y + \left[D(D-1)(D-2)\ldots(D-n+2)\right]y$$

$$+\left[D(D-1)(D-2)\ldots(D-n+3)\right]y + \ldots + Dy = X$$

$$\Rightarrow \quad f(D)\,y = X \qquad \ldots(5)$$

where $f(D)$ stands for a polynomial in D.

Now (5) is a linear differential equation with constant coefficients which can be solved by the methods already discussed in this chapter.

ILLUSTRATIVE EXAMPLES 2.8

Example 2.25: Solve differential equation

$$x^2y'' + 4xy' + 2y = 0 \quad \text{or} \quad x^2\frac{d^2y}{dx^2} + 4x\frac{dy}{dx} + 2y = 0.$$

Solution: Given differential equation is

$$x^2\frac{d^2y}{dx^2} + 4x\frac{dy}{dx} + 2y = 0$$

which is Cauchy's homogeneous linear equation.

Put $\quad x = e^z$ i.e., $z = \log x$

Substituting $x\dfrac{dy}{dx} = Dy,\ x^2\dfrac{d^2y}{dx^2} = D(D-1)\,y$, we get

$$D(D-1)\,y + 4Dy + 2y = 0 \quad \Rightarrow \quad (D^2 + 3D + 2)\,y = 0$$

It's A.E. is

$$D^2 + 3D + 2 = 0$$

which is quadratic equation in D, and on solving gives

$$D = -1,\ -2$$

Solution is $\quad y = c_1e^{-z} + c_2e^{-2z} = c_1\dfrac{1}{x} + c_2\dfrac{1}{x^2}$

Example 2.26: Solve the differential equation $x^3\dfrac{d^3y}{dx^3} + 2x^2\dfrac{d^2y}{dx^2} + 2y = 10\left(x + \dfrac{1}{x}\right)$.

Solution: Given differential equation is

$$x^3\frac{d^3y}{dx^3} + 2x^2\frac{d^2y}{dx^2} + 2y = 10\left(x + \frac{1}{x}\right)$$

which is Cauchy's homogeneous linear equation.

Put $x=e^z$ i.e., $z=\log x$

Substituting $x^2\dfrac{d^2y}{dx^2}=D(D-1)y$, $x^3\dfrac{d^3y}{dx^3}=D(D-1)(D-2)y$, we get

Substituting these values in the given equation, it reduces to

$$D(D-1)(D-2)y+2D(D-1)y+2y=10\left(e^z-e^{-z}\right)$$

$$\Rightarrow \quad \left[D(D-1)(D-2)+2D(D-1)+2\right]y=10\left(e^z-e^{-z}\right)$$

$$\Rightarrow \quad \left(D^3-D^2+2\right)y=10\left(e^z+e^{-z}\right)$$

which is a linear equation with constant coefficients.

It's A.E. is

$$D^3-D^2+2=0 \quad \Rightarrow \quad (D+1)\left(D^2-2D+2\right)=0$$

which on solving gives

$$D=-1,\frac{2\pm\sqrt{4-8}}{2}=-1,1\pm i$$

$$\therefore \quad \text{C.F.}=c_1e^{-z}+e^z\left(c_2\cos z+c_3\sin z\right)$$

$$=\frac{c_1}{x}+x\left[c_2\cos(\log x)+c_3\sin(\log x)\right]$$

Now
$$\text{P.I.}=10\frac{1}{D^3-D^2+2}\left(e^z+e^{-z}\right)=10\left(\frac{1}{D^3-D^2+2}e^z+\frac{1}{D^3-D^2+2}e^{-z}\right)$$

$$=10\left(\frac{e^z}{1^3-1^2+2}+z\frac{1}{\frac{d}{dz}\left(D^3-D^2+2\right)}e^{-z}\right) \qquad \text{(Case of failure in 2nd Part)}$$

$$=10\left(e^z\frac{1}{2}+z\frac{1}{3D^2-2D}e^{-z}\right)$$

$$=10\left(e^z\frac{1}{2}+z\frac{1}{3D^2-2D}e^{-z}\right)=10\left(e^z\frac{1}{2}+z.e^{-z}\frac{1}{3(-1)^2-2(-1)}\right)$$

$$=10\left(e^z\frac{1}{2}+\frac{1}{5}z.e^{-z}\right)=5e^z+2ze^{-z}$$

$\therefore$ general solution is

$$y=c_1e^{-z}+(c_2\cos z+c_3\sin z)e^z+5e^z+2ze^{-z}$$

$$\Rightarrow \quad y=c_1e^{-\log x}+\left[c_2\cos(\log x)+c_3\sin(\log x)\right]e^{\log x}+5e^{\log x}+2(\log x)e^{-\log x}$$

$$\Rightarrow \quad y=c_1x^{-1}+\left[c_2\cos(\log x)+c_3\sin(\log x)\right]x+5x+2x^{-1}\log x$$

$$\Rightarrow \quad y=(c_1+2\log x)x^{-1}+x\left[c_2\cos(\log x)+c_3\sin(\log x)+5\right]$$

Example 2.27: Solve the differential equation $x^3\frac{d^3y}{dx^3}+6x^2\frac{d^2y}{dx^2}+4x\frac{dy}{dx}-4y=(\log x)^2$.

Solution: Given differential equation is

$$x^3\frac{d^3y}{dx^3}+6x^2\frac{d^2y}{dx^2}+4x\frac{dy}{dx}-4y=(\log x)^2$$

which is Cauchy's homogeneous linear equation.

Put $\quad x=e^z$ i.e., $z=\log x$

Substituting $x\frac{dy}{dx}=Dy,\ x^2\frac{d^2y}{dx^2}=D(D-1)y,\ x^3\frac{d^3y}{dx^3}=D(D-1)(D-2)y$, we get

$$\left[D(D-1)(D-2)+6D(D-1)+4D-4\right]y=z^2$$

$$\Rightarrow \quad \left(D^3+3D^2-4\right)y=z^2$$

It's A. E. is

$$D^3+3D^2-4=0 \quad \Rightarrow \quad (D-1)(D+2)^2=0$$

which on solving gives $\quad D=1,-2,-2$

$\therefore$ The C.F. is

$$y_c = c_1 e^z + (c_2 + c_3 z)e^{-2z}$$

The P.I. is

$$y_p = \frac{1}{D^3 + 3D^2 - 4} z^2 = -\frac{1}{4} \frac{1}{\left[1 - \frac{D^3 + 3D^2}{4}\right]} z^2$$

$$= -\frac{1}{4}\left(1 - \frac{3D^2 + D^3}{4}\right)^{-1} z^2 = -\frac{1}{4}\left(1 + \frac{3D^2 + D^3}{4} + \ldots\right) z^2$$

$$= -\frac{1}{4}\left(z^2 + \frac{3}{4}(2)\right) = -\frac{1}{4}\left(z^2 + \frac{3}{2}\right)$$

$\therefore$ General solution is

$$y = y_c + y_p = c_1 e^z + (c_2 + c_3 z)e^{-2z} - \frac{1}{4}\left(z^2 + \frac{3}{2}\right)$$

$$\Rightarrow \qquad y = c_1 x + (c_2 + c_3 \log x)x^{-2} - \frac{1}{4}\left((\log x)^2 + \frac{3}{2}\right)$$

Example 2.28: Solve the differential equation $x^2 \frac{d^2 y}{dx^2} + 4x\frac{dy}{dx} + 2y = \sin x$.

Solution: Given differential equation is

$$x^2 \frac{d^2 y}{dx^2} + 4x\frac{dy}{dx} + 2y = \sin x$$

Putting $x = e^z \Rightarrow z = \log x$ and

$$x\frac{dy}{dx} = Dy,\ x^2\frac{d^2 y}{dx^2} = D(D-1)y \qquad \text{where } D = \frac{d}{dz},$$

$$\left[D(D-1) + 4D + 2\right] y = \sin e^z \qquad \Rightarrow \qquad \left(D^2 + 3D + 2\right) y = \sin e^z$$

It's A.E. is

$$D^2 + 3D + 2 = 0 \qquad \Rightarrow \qquad (D+1)(D+2) = 0$$

which on solving gives $D=-1,-2$

$\therefore$ The C.F. is

$$y_c = c_1 e^{-z} + c_2 e^{-2z}$$

The P.I. is

$$y_p = \frac{1}{D^2+3D+2}\sin e^z = \frac{1}{(D+2)(D+1)}\left(\sin e^z\right)$$

Now $\quad \frac{1}{(D+1)}\left(\sin e^z\right) = e^{-z}\int\left(\sin e^z\right)e^z dz \qquad \left[\because \frac{1}{D-a}f(x) = e^{ax}\int f(x)e^{-ax}dx\right]$

$$= e^{-z}\int \sin u\,du \qquad \text{where } u = e^z,\ du = e^z dz$$

$$= e^{-z}(-\cos u) = -e^{-z}\cos e^z$$

$\therefore$
$$y_p = \frac{1}{D+2}\left(-e^z\cos e^z\right) = -\frac{1}{D+2}\left(e^{-z}\cos e^z\right)$$

$$= -e^{-2z}\int\left(e^{-z}\cos e^z\right)e^{2z}dz = -e^{-2z}\int\left(\cos e^z\right)e^z dz,$$

$$= -e^{-2z}\int \cos\mu\, d\mu \qquad \text{where } \mu = e^z,\ d\mu = e^z dz$$

$$= -e^{2z}(\sin\mu) = -e^{-2z}\sin e^z$$

$\therefore$ the general solution is

$$y = y_c + y_p = c_1 e^{-z} + c_2 e^{-2z} - e^{-2z}\sin e^z$$

$$\Rightarrow \qquad y = \frac{c_1}{x} + \frac{c_2}{x^2} - \frac{\sin x}{x^2}$$

EXERCISE 2.4

Solve the following differential equations:

1. $x^2\frac{d^2y}{dx^2} + 9x\frac{dy}{dx} + 25y = 50$
2. $x^2\frac{d^2y}{dx^2} - 3x\frac{dy}{dx} + 4y = 2x^2$

3. $x^3 \dfrac{d^3 y}{dx^3}+3x^2 \dfrac{d^2 y}{dx^2}+x\dfrac{dy}{dx}+8y=65\cos(\log x)$

4. $x^2 \dfrac{d^2 y}{dx^2}+3x\dfrac{dy}{dx}+y=\dfrac{1}{(1-x^2)}$ [PTU May 2009]

5. $x^2 \dfrac{d^2 y}{dx^2}-x\dfrac{dy}{dx}+y=2\log x$

6. $x^2 \dfrac{d^2 y}{dx^2}-3x\dfrac{dy}{dx}+5y=x^2\sin(\log x)$

7. $x^2 \dfrac{d^2 y}{dx^2}-x\dfrac{dy}{dx}+2y=x\log x$ [PTU May 2010]

8. $x^2 \dfrac{d^2 y}{dx^2}-x\dfrac{dy}{dx}-3y=x^2\log x$

9. $x^2 \dfrac{d^2 y}{dx^2}+2x\dfrac{dy}{dx}-12y=x^3\log x$

10. (*a*) $x^2 \dfrac{d^2 y}{dx^2}-2x\dfrac{dy}{dx}-4y=x^2+2\log x$

    (*b*) $(x^2D^2+xD+1)y=\log x\sin(\log x)$

11. Solve $x^2D^2-(2m-1)xD+(m^2+n^2)y=n^2x^m\log x$

12. $x^2 \dfrac{d^2 y}{dx^2}-2x\dfrac{dy}{dx}-4y=x^4$

13. $(x^2D^2-xD+4)y=\cos(\log x)+\sin(\log x)$

14. $x^2 \dfrac{d^2 y}{dx^2}+x\dfrac{dy}{dx}-y=x^2e^x$

15. $x\dfrac{d^2 y}{dx^2}+\dfrac{dy}{dx}=x$

ANSWERS

1. $y=x^{-4}\left[c_1\cos(3\log x)+c_2\sin(3\log x)\right]+2$

2. $y=(c_1+c_2\log x)x^2+(\log x)^2x^2$

3 $y=c_1x^{-2}+x\left[c_2\cos(\sqrt{3}\log x)+c_2\sin(\sqrt{3}\log x)\right]+8\cos c(\log x)-\sin(\log x)$

4. $y=\frac{1}{x}\left[c_1+c_2\log x-\log\left|\frac{1}{x}-1\right|\right]$

5. $y=x(c_1+c_2\log x)+2\log x+4$

6. $y=\left[c_1\cos(\log x)+c_2\sin(\log x)\right]x^2-\frac{x^2}{2}\log x\cos(\log x)$

7. $y=x\left[c_1\cos(\log x)+c_2\sin(\log x)\right]+x\log x$

8. $y=c_1x^3+\frac{c_2}{x}-\frac{1}{9}x^2(3\log x+2)$

9. $y=\frac{c_1}{4}+c_2x^3+\frac{1}{14}(\log x)^2x^2-\frac{1}{49}(\log x)x^3$

10. (*a*) $y=\frac{1}{x}+c_2x^4-\frac{x^2}{6}-\frac{1}{2}\log x+\frac{3}{8}$

 (*b*) $y=c1\cos\left(\log x+c2\sin(\log x)+\frac{1}{4}(\log x.\sin(\log x)-\log x\cos(\log x)\right)$

11. $y=x^m\left(c_1\sin\log x^n+c_2\cos\log x^n+\log x\right)$

12. $y=c_1x^4+\frac{c_2}{x}+\frac{x^4}{5}\log x$

13. $y=x\left(c_1\cos\sqrt{3}\log x\right)+c_2\sin\left(\sqrt{3}\log x\right)+\frac{3}{13}\cos(\log x)+\left(\frac{x}{2}-\frac{2}{13}\right)+\sin(\log x)$

14. $y=c_1x+\frac{c_2}{x}+e^x\left(1-\frac{1}{x}\right)$

15. $y=c_1+c_2\log x+\frac{1}{4}x^2$

2.9 LEGENDRE'S LINEAR EQUATION

[PTU Dec 2005, May 2007]

An equation of the form

$$p_0(ax+b)^n\frac{d^ny}{dx^n}+p_1(ax+b)^{n-1}\frac{d^{n-1}}{dx^{n-1}}+...+p_{n-1}(ax+b)\frac{dy}{dx}+p_ny=f(x) \qquad ...(1)$$

i.e., $$\left[p_0(ax+b)^nD^n+p_1(ax+b)^{n-1}D^{n-1}+...+p_{n-1}(ax+b)D+p_n\right]y=f(x) \qquad ...(2)$$

where $p_0, p_1, p_2, ..., p_n$ are real constants and $f(x)$ is either a constant or a function of x, is called *Legendre's linear equation*. Here Eq. (2) is the symbolic form of the Eq. (1).

Let us reduce this equation to the homogeneous linear equation.

Put $ax+b=e^z \quad \Rightarrow \quad z=\log(ax+b)$

Differentiating w. r. t. x, we get

$$\frac{dz}{dx}=\frac{a}{ax+b}$$

$$\frac{dy}{dx}=\frac{dy}{dz}\cdot\frac{dz}{dx}=\frac{a}{ax+b}\frac{dy}{dz}$$

$$\Rightarrow \quad (ax+b)\frac{dy}{dx}=a\frac{dy}{dz} \qquad \ldots (2)$$

Also, $$\frac{d^2y}{dx^2}=\frac{d}{dx}\left(\frac{dy}{dx}\right)=\frac{d}{dx}\left(\frac{a}{ax+b}\frac{dy}{dz}\right)$$

$$=-\frac{a^2}{(ax+b)^2}\frac{dy}{dz}+\frac{a}{ax+b}\frac{d^2y}{dz^2}\cdot\frac{dz}{dx}=-\frac{a^2}{(ax+b)^2}\frac{dy}{dz}+\frac{a}{ax+b}\frac{d^2y}{dz^2}\cdot\frac{a}{ax+b}$$

$$=\frac{a^2}{(ax+b)^2}\left(\frac{d^2y}{dz^2}-\frac{dy}{dz}\right)$$

$$\Rightarrow \quad (ax+b)^2\frac{d^2y}{dx^2}=a^2\left(\frac{d^2y}{dz^2}-\frac{dy}{dz}\right) \qquad \ldots (3)$$

Similarly, again on differentiating w. r. t. x, we get

$$(ax+b)^3\frac{d^3y}{dx^3}=a^3\left(\frac{d^3y}{dz^3}-3\frac{d^2y}{dz^2}+2\frac{dy}{dz}\right) \qquad \ldots (4)$$

and so on.

Putting $\frac{d}{dz}=D$, in Eq. (2), (3) and (4), we get

$$(ax+b)\frac{dy}{dx}=aDy$$

$$(ax+b)^2\frac{d^2y}{dx^2}=a^2(D^2y-Dy)=a^2\left[D(D-1)\right]y$$

$$(ax+b)^3\frac{d^3y}{dx^3}=a^3(D^3y-3D^2y+2Dy)=a^3(D^3-3D^2+2D)y$$

$$= a^3\left[D(D-1)(D-2)\right]y$$

and so on.

In general,

$$(ax+b)^n\frac{d^ny}{dx^n} = a^n\left[D(D-1)(D-2)...(D-(n-1))\right]y$$

Substituting values of $(ax+b)\frac{dy}{dx}$, $(ax+b)^2\frac{d^2y}{dx^2}$, $(ax+b)^3\frac{d^3y}{dx^3}$,...., $(ax+b)^n\frac{d^ny}{dx^n}$, we get

$$\left[p_o a^n D(D-1)...(D-n+1) + p_1 a^{n-1}D(D-1)...(D-n+2)+...+p_{n-1}aD+p_n\right]y = f(x)$$

which is linear equation with constant coefficients, and can be solved for y in term of z.

Steps to solve Legendre's linear equation with variable coefficients:

1. Write given equation in the symbolic form
$$p_o(ax+b)^n\frac{d^ny}{dx^n}+p_1(ax+b)^{n-1}\frac{d^{n-1}y}{dx^{n-1}}+...+p_{n-1}(ax+b)\frac{dy}{dx}+p_n y = f(x)$$
2. Put $ax+b=e^z$ so that $z=\log(ax+b)$.
3. Replace $(ax+b)\frac{dy}{dx}$ by $aD, (ax+b)^2\frac{d^2y}{dx^2}$ by $a^2D(D-1),...$
$(ax+b)^n\frac{d^ny}{dx^n}$ by $a^nD(D-1)...(D-n+1)$, in step (1) where $D=\frac{d}{dz}$.
4. Then the given equation will be reduced to the form of linear differential equation with constant coefficient.
5. Solve it for y in term z and then put $z=\log(ax+b)$ to get the required solution.

ILLUSTRATIVE EXAMPLES 2.9

Example 2.29: Solve the differential equation $(x+1)^2y''+(x+1)y'=(2x+3)(2x+4)$.

Solution: Given differential equation is

$$(x+1)^2\frac{d^2y}{dx^2}+(x+1)\frac{dy}{dx}=(2x+3)(2x+4)$$

This equation is reducible to homogenous linear form with constant coefficients.

Put $x+1=e^z \Rightarrow z=\log(x+1)$

$\therefore$ Equation reduces to the form

$$[D(D-1)y+Dy]=\left(2(e^z-1)+3\right)\left(2(e^z-1)+4\right)=(2e^z+1)(2e^z+2)$$

$$\Rightarrow \quad D^2y=(4e^{2z}+6e^z+2)$$

It's A.E. is

$$D^2=0 \quad \Rightarrow \quad D=0,0$$

$$\therefore \quad \text{C.F.}=(c_1+c_2z)e^{0z}=c_1+c_2z$$

$$\text{It's P.I.}=\frac{1}{D^2}\left[4e^{2z}+6e^z+2\right]=\frac{1}{D}\int\left[4e^{2z}+6e^z+2\right)dz$$

$$=\frac{1}{D}(2e^{2z}+6e^z+2z)=e^{2z}+6e^z+z^2$$

$\therefore$ the C.S. is given by

$$\Rightarrow \quad y=C.F.+P.I.=c_1+c_2z+e^{2z}+6e^z+z^2$$

But $z=\log(x+1)$

$\therefore$ the C.S. becomes

$$y=c_1+c_2\log(x+1)+(x+1)^2+6(x+1)+\left[\log(x+1)\right]^2$$

Example 2.30: Solve the differential equation $(5+2x)^2\dfrac{d^2y}{dx^2}-6(5+2x)\dfrac{dy}{dx}+8y=0.$

Solution: The given differential equation is

$$(5+2x)^2\frac{d^2y}{dx^2}-6(5+2x)\frac{dy}{dx}+8y=0$$

Putting $5+2x=e^z$

$$(5+2x)\frac{dy}{dx}=2Dy \quad (5+2x)^2\frac{d^2y}{dx^2}=2^2D(D-1)y \quad \text{where } D=\frac{d}{dz}$$

Given equation reduces to the form

$$2^2D(D-1)y-6\times 2Dy+8y=0$$

$$\Rightarrow \quad (4D^2-4D-12D+8)y=0$$

$$\Rightarrow \quad (4D^2-16D+8)y=0 \quad \Rightarrow \quad (D^2-4D+2)y=0$$

It's A.E. is

$$D^2 - 4D + 2 = 0 \quad \Rightarrow \quad D = \frac{4 \pm \sqrt{16-8}}{2} = \left(2 \pm \sqrt{2}\right) \quad \Rightarrow D = 2+\sqrt{2}, 2-\sqrt{2}$$

$$\therefore \quad C.F. = c_1 e^{\left(2+\sqrt{2}\right)z} + c_2 e^{\left(2-\sqrt{2}\right)z}$$

The *C. S.* is given by

$$y = C.F. \qquad [\because P.I. = 0]$$

$$\Rightarrow \qquad y = c_1 e^{\left(2+\sqrt{2}\right)z} + c_2 e^{\left(2-\sqrt{2}\right)z}$$

But $\quad z = \log(5 + 2x)$

$\therefore$ The C.S. becomes

$$y = c_1 (5+2x)^{2+\sqrt{2}} + c_2 (5+2x)^{2-\sqrt{2}}$$

Example 2.31: Solve the differential equation $(2x-1)^3 \dfrac{d^3 y}{dx^3} + (2x-1)\dfrac{dy}{dx} - 2y = x.$

Solution: The given differential equation is

$$(2x-1)^3 \frac{d^3 y}{dx^3} + (2x-1)\frac{dy}{dx} - 2y = x$$

Putting $\;2x - 1 = e^z$, equation reduces to

$$\left[2^3 D(D-1)(D-2) + 2D - 2\right] y = \frac{e^z + 1}{2}$$

$$\Rightarrow \qquad \left[8D(D-1)(D-2) + 2D - 2\right] y = \frac{1}{2} + \frac{1}{2} e^z$$

$$\Rightarrow \qquad \left[4D^3 - 12D^2 + 9D - 1\right] y = \frac{1}{4} + \frac{1}{4} e^z$$

It's A.E. is

$$4D^3 - 12D^2 + 9D - 1 = 0 \qquad \Rightarrow \qquad (D-1)\left(4D^2 - 8D + 1\right) = 0$$

$$\Rightarrow \qquad D = 1, \; \frac{8 \pm \sqrt{64-16}}{8} = 1, \; 1 \pm \frac{\sqrt{3}}{2}$$

$$\therefore \qquad C.F. = c_1 e^z + e^z\left(c_2 e^{\frac{\sqrt{3}}{2}z} + c_3 e^{-\frac{\sqrt{3}}{2}z}\right)$$

Also $P.I. = \dfrac{1}{(D-1)(4D^2-8D+1)}\left(\dfrac{1}{4}+\dfrac{1}{4}e^z\right)$

$$= \frac{1}{(D-1)(4D^2-8D+1)}\left(\frac{1}{4}e^{0z}+\frac{1}{4}e^{1z}\right)$$

$$= \frac{1}{4}\frac{e^{0z}}{(0-1)(0-0+1)} + \frac{1}{4}\frac{1}{(D-1)(4D^2-8D+1)}e^{1z}$$

$$= \frac{-1}{4}+\frac{1}{4}z\frac{1}{\frac{d}{dz}(4D^3-12D^2+9D-1)}e^z \qquad [\because \text{Case failed, II part}]$$

$$= \frac{-1}{4}+\frac{1}{4}z\frac{1}{12D^2-24D+9}e^z$$

$$= \frac{-1}{4}+\frac{1}{4}z\frac{e^z}{12-24+9} = -\frac{1}{4}-\frac{z}{12}e^z$$

$\therefore$ The general solution is

$$y=(2x-1)\left[c_1+c_2(2x-1)^{\frac{\sqrt{3}}{2}}+c_3(2x-1)^{-\frac{\sqrt{3}}{2}}\right]-\frac{1}{4}-\frac{\log(2x-1)}{12}(2x-1)$$

EXERCISE 2.5

1. $(2x-1)^3\dfrac{d^3y}{dx^3}+(2x-1)\dfrac{dy}{dx}-2y=0$

2. $(x+a)^2\dfrac{d^2y}{dx^2}-4(x+a)\dfrac{dy}{dx}+6y=x$

3. $\left[(3x+2)^2D^2+3(3x+2)D-36\right]y=3x^2+4x+1$

4. $(2x+1)^2\dfrac{d^2y}{dx^2}-6(2x+1)\dfrac{dy}{dx}+16y=8(2x+1)^2$

5. $(1+x)^2\dfrac{d^2y}{dx^2}+(1+x)\dfrac{dy}{dx}+y=4\cos\left[\log(1+x)\right]$

6. $(x+1)^2y''+(x+1)y'+y=\sin(2\log(1+x))$ [PTU May 2012]

7. $(1+x^2)y''+(1+x)y'+y=2\sin\log(1+x)$ [PTU Dec 2012]

ANSWERS

1. $y=(2x-1)\left[c_1+c_2(2x-1)^{\frac{\sqrt{3}}{2}}+c_3(2x-1)^{-\frac{\sqrt{3}}{2}}\right]$

2. $y=c_1(x+9)^2+c_2(x+9)^3+\dfrac{1}{2}(x+a)\dfrac{-a}{6}$

3. $y=c_1(3x+2)^2+c_2(3x+2)^3+\dfrac{1}{108}\left[(3x+2)^2\log(3x+2)+1\right]$

4. $y=(2x+1)^2\left[c_1+c_2\log(2x+1)+\{\log(2x+1)\}^2\right]$

5. $y=c_1\cos(\log(1+x))+c_2\sin(\log(1+x))+2\log(1+x)+\sin(\log(1+x))$

6. $y=c_1\cos(\log(1+x))+c_2\sin(\log(1+x))-\dfrac{1}{3}\sin 2(\log(1+x))$

7. $y=c_1\cos(\log(1+x))+c_2\sin(\log(1+x))+\sin(\log(1+x))-\log(1+x)\cos(\log(1+x))$

2.10 SIMULTANEOUS LINEAR EQUATIONS WITH CONSTANT COEFFICIENTS

Now, we discuss differential equations in which one independent variable and two or more than two are dependent variables. Such equations are called simultaneous linear equations. To solve such equations completely, we must have as many simultaneous equations as the number of dependent variables. Here we shall consider simultaneous linear equation with constant coefficients only.

Let x, y be the two dependent variables and t the independent variable. Consider the simultaneous equations

$$f_1(D)x+f_2(D)y=T_1 \qquad \ldots(1)$$

$$\phi_1(D)x+\phi_2(D)y=T_2 \qquad \ldots(2)$$

where $D=\dfrac{d}{dt}$ and T_1,T_2 are functions of t.

To eliminate y, operating on both sides of Eq. (1) by $\phi_2(D)$ and on both sides of Eq. (2) by $f_2(D)$ and subtracting, we get

$$\left[f_1(D)\phi_2(D)-\phi_1(D)f_2(D)\right]x=\phi_2(D)T_1-f_2(D)T_2$$

or $\quad f(D)x=T$

which is a linear equation in x and t and can be solved by the method already discussed. Substituting the value of x in either Eq. (1) or Eq. (2), we get the value of y.

Note: We can also eliminate x to get a linear equation in y and t.

ILLUSTRATIVE EXAMPLES 2.10

Example 2.32: Solve following simultaneous equations

$$\frac{dx}{dt}+4x+3y=t,\ \frac{dy}{dt}+2x+5y=e^t.$$

Solution: Writing D for $\frac{d}{dt}$, the given equations become

$$(D+4)x+3y=t \qquad \ldots(1)$$

$$2x+(D+5)y=e^t \qquad \ldots(2)$$

To eliminate y, operating on both sides of Eq. (1) by $(D+5)$ and of Eq. (2) by 3, and subtracting, we get

$$\left[(D+4)(D+5)-6\right]x=(D+5)t-3e^t$$

$$\Rightarrow \quad \left(D^2+9D+14\right)x=1+5t-3e^t \qquad \left[\because Dt=\frac{dt}{dt}=1\right]$$

It's A.E. is

$$D^2+9D+14=0$$

$$\Rightarrow \quad (D+2)(D+7)=0$$

$$\therefore \quad D=-2,-7$$

$$\text{C.F} = c_1e^{-2t} + c_2e^{-7t}$$

$$\text{P.I.} = \frac{1}{D^2+9D+14}\left(1+5t-3e^t\right) = \frac{1}{D^2+9D+14}\left(e^{0t}+5t-3e^t\right)$$

$$= \frac{e^{0t}}{0^2+9(0)+14} + 5\frac{1}{14\left(1+\frac{9D}{14}+\frac{D^2}{14}\right)}t - 3\frac{e^t}{1^2+9(1)+14}$$

$$= \frac{1}{14} + \frac{5}{14}\left[1+\left(\frac{9D}{14}+\frac{D^2}{14}\right)\right]^{-1} t - \frac{1}{8}e^t = \frac{1}{14} + \frac{5}{14}\left[1-\left(\frac{9D}{14}+\frac{D^2}{14}\right)+\ldots\right]t - \frac{1}{8}e^t$$

$$= \frac{1}{14} + \frac{5}{14}\left(t - \frac{9}{14}\right) - \frac{1}{8}e^t = \frac{1}{14} + \frac{5}{14}t - \frac{45}{196} - \frac{1}{8}e^t = \frac{5}{14}t - \frac{31}{196} - \frac{1}{8}e^t$$

The C.S. = C.F. + P.I.

$$\therefore \quad x = c_1e^{-2t} + c_2e^{-7t} + \frac{5}{14}t - \frac{31}{196} - \frac{1}{8}e^t$$

Now, $\dfrac{dx}{dt} = -2c_1e^{-2t} - 7c_2e^{-7t} + \dfrac{5}{14} - \dfrac{1}{8}e^t$

Substituting the values of x and $\dfrac{dx}{dt}$ in Eq. (1), we get

We have $\quad 3y = t - \dfrac{dx}{dt} - 4x$

$$= t + 2c_1e^{-2t} + 7c_2e^{-7t} - \frac{5}{14} + \frac{1}{8}e^t - 4c_1e^{-2t} - 4c_2e^{-7t} - \frac{10}{7}t + \frac{31}{49} + \frac{1}{2}e^t$$

$$y = \frac{1}{3}\left[-2c_1e^{-2t} + 3c_2e^{-7t} - \frac{3}{7}t + + \frac{27}{98} + \frac{5}{8}e^t\right]$$

Hence $\quad x = c_1e^{-2t} + c_2e^{-7t} + \dfrac{5}{14}t - \dfrac{31}{196} - \dfrac{1}{8}e^t$

$$y = -\frac{2}{3}c_1e^{-2t} + c_2e^{-7t} - \frac{1}{7}t + \frac{9}{98} + \frac{5}{24}e^t$$

Example 2.33: Solve the following simultaneous equations

$$\frac{dx}{dt}+2y=e^t,\ \frac{dy}{dt}-2x=e^{-t}$$

Solution: $\frac{dx}{dt}+2y=e^t$ ... (1)

$$\frac{dy}{dt}-2x=e^{-t} \quad \ldots (2)$$

Differentiating Eq. (1) w. r. t. t, we get

$$\frac{d^2x}{dt^2}+2\frac{dy}{dt}=e^t$$

From Eq. (2), we get

$$\frac{d^2x}{dt^2}+4x+2e^{-t}=e^t$$

$$\Rightarrow \qquad \frac{d^2x}{dt^2}+4x=e^t-2e^{-t}$$

which is linear differential equation with constant coefficients.

It's S.F. is $\left(D^2+4\right)x=e^t-2e^{-t}$ where $D=\frac{d}{dt}$

It's A.E. is $D^2+4=0$

$\Rightarrow \qquad D=\pm 2i$

C.F. $=c_1\cos 2t+c_2\sin 2t$

P.I. $=\frac{1}{D^2+4}\left(e^t-2e^{-t}\right)=\frac{1}{D^2+4}e^t-2\frac{1}{D^2+4}e^{-t}=\frac{1}{5}e^t-\frac{2}{5}e^{-t}$

$\therefore$ C.S. is

$$x=c_1\cos 2t+c_2\sin 2t+\frac{1}{5}e^t-\frac{2}{5}e^{-t}$$

$$\frac{dx}{dt}=-2c_1\sin 2t+2c_2\cos 2t+\frac{1}{5}e^t+\frac{2}{5}e^{-t}$$

From Eq. (1), we have

$$2y = 2c_1 \sin 2t - 2c_2 \cos 2t - \frac{1}{5}e^{t} - \frac{2}{5}e^{-t} + e^{t}$$

$$y = c_1 \sin 2t - 2c_2 \cos 2t + \frac{2}{5}e^{t} - \frac{1}{5}e^{-t}$$

Hence, $x = c_1 \cos 2t + c_2 \sin 2t + \frac{1}{5}e^{t} - \frac{2}{5}e^{-t}$, and

$$y = c_1 \sin 2t - c_2 \cos 2t + \frac{2}{5}e^{t} - \frac{1}{5}e^{-t}$$

Example 2.34: Solve the following simultaneous equations

$$\frac{d^2x}{dt^2} + 4x + 5y = t^2, \quad \frac{d^2y}{dt^2} + 5x + 4y = t + 1.$$

Solution: Writing D for $\frac{d}{dt}$, the given equations becomes

$$\left(D^2 + 4\right)x + 5y = t^2 \quad \text{... (1)}$$

$$5x + \left(D^2 + 4\right)y = t + 1 \quad \text{... (2)}$$

To eliminate y, multiply on both side of Eq. (1) by $\left(D^2 + 4\right)$, and on both side of Eq. (2) by 5 and subtracting, we get

$$\left[\left(D^2 + 4\right)^2 - 25\right]x = \left(D^2 + 4\right)t^2 - 5(t + 1)$$

$$\Rightarrow \quad \left(D^4 + 8D^2 - 9\right)x = 2 + 4t^2 - 5t - 5$$

$$\Rightarrow \quad \left(D^4 + 8D^2 - 9\right)x = 4t^2 - 5t - 3$$

It's A.E. is

$$D^4 + 8D^2 - 9 = 0 \quad \Rightarrow \quad \left(D^2 + 9\right)\left(D^2 - 1\right) = 0$$

which on solving gives

$$D = \pm 1, \pm 3i$$

$$\text{C.F.} = c_1 e^{t} + c_2 e^{-t} + c_3 \cos 3t + c_4 \sin 3t$$

$$\text{P.I.} = \frac{1}{D^4+8D^2-9}\left(4t^2-5t-3\right) = \frac{1}{-9\left(1-\frac{8D^2}{9}-\frac{D^4}{9}\right)}\left(4t^2-5t-3\right)$$

$$= -\frac{1}{9}\left[1-\left(\frac{8D^2}{9}+\frac{D^4}{9}\right)\right]^{-1}\left(4t^2-5t-3\right)$$

$$= -\frac{1}{9}\left[1+\left(\frac{8D^2}{9}+\frac{D^4}{9}\right)+\ldots\right]\left(4t^2-5t-3\right)$$

$$= -\frac{1}{9}\left[4t^2-5t-3+\frac{8}{9}(8)\right] = -\frac{1}{9}\left(4t^2-5t+\frac{37}{9}\right)$$

∴ The C.S. is

$$x = c_1e^t + c_2e^{-t} + c_3\cos 3t + c_4\sin 3t - \frac{4}{9}t^2 + \frac{5}{9}t - \frac{37}{81}$$

Now, $$\frac{dx}{dt} = c_1e^t - c_2e^{-t} - 3c_3\sin 3t + 3c_4\cos 3t - \frac{8}{9}t + \frac{5}{9}$$

$$\frac{d^2x}{dt^2} = c_1e^t + c_2e^{-t} - 9c_3\cos 3t - 9c_4\sin 3t - \frac{8}{9}$$

Substituting the value of x and $\frac{d^2x}{dt^2}$ in Eq. (1), we have

$$5y = t^2 - 4x - \frac{d^2x}{dt^2}$$

$$= t^2 - 4c_1e^t - 4c_2e^{-t} - 4c_3\cos 3t - 4c_4\sin 3t + \frac{16}{9}t^2$$

$$-\frac{20}{9}t + \frac{148}{81} - c_1e^t - c_2e^{-t} + 9c_3\cos 3t + 9c_4\sin 3t + \frac{8}{9}$$

$$\therefore\; y = \frac{1}{5}\left[-c_1e^t - c_2e^{-t} + 5c_3\cos 3t + 5c_4\sin 3t + \frac{25}{9}t^2 - \frac{20}{9}t + \frac{220}{81}\right]$$

Hence $$x = c_1e^t + c_2e^{-t} + c_3\cos 3t + c_4\sin 3t - \frac{1}{9}\left[4t^2 - 5t + \frac{37}{9}\right]$$

$$y = \frac{1}{5}\left[-c_1e^t - c_2e^{-t} + 5c_3\cos 3t + 5c_4\sin 3t + \frac{25}{9}t^2 - \frac{20}{9}t + \frac{220}{81}\right]$$

Example 2.35: Solve the following simultaneous equations:

$$t\frac{dx}{dt}+y=0,\ t\frac{dy}{dt}+x=0,\ \text{given } x(1)=1, y(-1)=0.$$

Solution: The given equations are

$$t\frac{dx}{dt}+y=0 \qquad \ldots (1)$$

$$t\frac{dy}{dt}+x=0 \qquad \ldots (2)$$

Differentiating Eq. (1) w. r. t. t, we have

$$t\frac{d^2x}{dt^2}+\frac{dx}{dt}+\frac{dy}{dt}=0$$

Multiplying throughout by t. we get

$$t^2\frac{d^2x}{dt^2}+t\frac{dx}{dt}+t\frac{dy}{dt}=0 \quad \Rightarrow \quad t^2\frac{d^2x}{dt^2}+t\frac{dx}{dt}-x=0 \qquad \ldots (3)$$

which is Cauchy's homogeneous linear equation.

Putting $t=e^z$ *i.e.*, $z=\log t$, so that

$$t\frac{dx}{dt}=Dx,\ t^2\frac{d^2x}{dt^2}=D(D-1)x \qquad \text{where } D=\frac{d}{dz}$$

Eq. (3) becomes

$$\left[D(D-1)+D-1\right]x=0 \quad \Rightarrow \quad \left(D^2-1\right)x=0$$

It's A.E. is

$$D^2-1=0 \quad \Rightarrow \quad D=\pm 1$$

$$x=c_1e^z+c_2e^{-z}=c_1t+\frac{c_2}{t} \qquad \ldots (4)$$

From Eq. (1)

$$y=-t\frac{dx}{dt}=-t\left(c_1-\frac{c_2}{t^2}\right)=-c_1t+\frac{c_2}{t} \qquad \ldots (5)$$

Since $x(1)=1$, $\therefore$ from Eq. (4), we have $1=c_1+c_2$

Also $y(-1)=0$, $\therefore$ from Eq. (5), we have $0=c_1-c_2$

Solving $c_1=c_2=\dfrac{1}{2}$

Hence $x=\dfrac{1}{2}\left(t+\dfrac{1}{t}\right)$, $y=\dfrac{1}{2}\left(-t+\dfrac{1}{t}\right)$

EXERCISE 2.6

Solve the following simultaneous equations:

1. $\dfrac{dx}{dt}=7x-y,\ \dfrac{dy}{dt}=2x+5y$
2. $\dfrac{dx}{dt}+y=\sin t,\ \dfrac{dy}{dt}+x=\cos t,\ y(0)=0,\ x(0)=2.$ [PTU May 2012]
3. $\dfrac{dx}{dt}-2y+5x=t,\ \dfrac{dy}{dt}+2x+y=0.$ Given that $x(0)=0,\ y(0)=0.$

   [PTU May 2012, Dec 2012]
4. $(D+1)x+(2D+1)y=e^t,\ (D-1)x+(D+1)y=1$
5. $\dfrac{dx}{dt}+2x+3y=0,\ 3x+\dfrac{dy}{dt}+2y=2e^{2t}$
6. $(D-1)x+Dy=2t+1,\ (D+1)x+2Dy=t$
7. $\dfrac{d^2x}{dt^2}-3x-4y=0,\ \dfrac{d^2y}{dt^2}+x+y=0$
8. $\dfrac{d^2x}{dt^2}-\dfrac{dy}{dt}=2x+2t,\ \dfrac{dx}{dt}+\dfrac{4dx}{dt}=3y$
9. $\dfrac{d^2y}{dt^2}+\dfrac{dy}{dt}--2y=\sin t,\ \dfrac{dx}{dt}+x-3y=0$
10. A mechanical system with two degrees of freedom satisfies the equations $2\dfrac{d^2x}{dt^2}+3\dfrac{dy}{dt}=4,\ 2\dfrac{d^2y}{dt^2}=3\dfrac{dx}{dt}$ obtained expressions for x and y is ferns of t, given $x, y, \dfrac{dx}{dt}, \dfrac{dy}{dt}$ all vanish at $t=0$.
11. $\dfrac{d^2x}{dt^2}+y=\sin t,\ \dfrac{d^2y}{dt^2}+x=\cos t$

ANSWERS

1. $x=e^{6t}\left(c_1\cos t+c_2\sin t\right),\ y=e^{6t}\left[\left(c_1-c_2\right)\cos t+\left(c_1+c_2\right)\sin t\right]$

2. $x=e^{t}+e^{-t},\ y=e^{-1}-e^{t}+\sin t$

3. $x=\dfrac{-1}{27}(1+6t)e^{-3t}+\dfrac{1}{27}(1+3t),\ y=\dfrac{-2}{27}(2+3t)e^{-3t}+\dfrac{2}{27}(2-3t)$

4. $x=\dfrac{1}{2}\left[e^{t}-1-ac_1e^{at}+bc_2e^{bt}\right],\ y=c_1e^{bt}+c_2e^{bt}+\dfrac{1}{2}$

   where $a=\dfrac{1}{2}\left(3+\sqrt{17}\right),\ b=\dfrac{1}{2}\left(3-\sqrt{17}\right).$

5. $x=c_1e^{t}+c_2e^{-5t}-\dfrac{6}{7}e^{2t},\ y=c_2e^{-5t}-c_1e^{t}+\dfrac{8}{7}e^{2t}$

6. $x=-t-\dfrac{2}{3},\ y=\dfrac{1}{2}t^2+\dfrac{4}{3}t+c$

7. $x=\left(c_1+c_2t\right)e^{-t}+\left(c_3+c_4t\right)e^{t},\ y=-\dfrac{1}{2}\left[c_1+c_2(1+t)\right]e^{-t}+\dfrac{1}{2}\left[c_4(1-t)-c_3\right]e^{t}$

8. $x=\left(c_1+c_2t\right)e^{t}+c_3e^{-3/2t}$

   $x=y=\left[c_2(3-t)-c_1\right]e^{-t}-\dfrac{1}{6}c_3e^{-32t}$

9. $x=\dfrac{3}{2}c_1e^{t}-3c_2e^{-2t}+c_3e^{-t}+\dfrac{3}{10}e^{t}(\cos t-2\sin t)$

   $y=c_1e^{t}+c_2e^{-2t}-\dfrac{1}{10}(\cos t+3\sin t)$

10. $x=\dfrac{8}{9}\left(1-\cos\dfrac{3t}{2}\right),\ y=\dfrac{4t}{3}-\dfrac{8}{9}\sin\dfrac{3t}{3}$

11. $x=c_1e^{t}+c_2e^{-t}+c_3\cos t+c_4\sin t+\dfrac{t}{4}(\sin t-\cos t)$

    $y=-c_1e^{t}+c_2e^{-t}-c_3\cos t+c_4\sin t+\dfrac{1}{4}(2+t)(\sin t-\cos t).$

❑ ❑ ❑

Chapter 3

Applications of Ordinary Differential Equations

Learning Objectives

After reading this chapter, students will be able to

- understand the importance of differential equations, and how these differential equations can be used to solve various kind of problems
- solve problems related to electrical circuits
- solve problems related to deflection of beams
- solve problems of systems that exhibit simple harmonic motion
- solve problems based on simple pendulum

3.1 INTRODUCTION

The significant applications of differentiation equations include the study of those physically problems, such as electrical and mechanical oscillatory systems, that can be modeled using differential equations. Once the physical problem is modeled using differential equation, we can solve it. Using this solution, we can understand the behaviour of the physical systems.

In this chapter, we will study the applications of differential equations to electric *R-L-C* circuits, deflection of beams, simple harmonic motion and simple pendulum.

3.2 ELECTRIC L-R-C CIRUITS

3.2.1 Basics of Electric Circuits

We will be considering circuits made up of

- three passive elements- *resistance, inductance, capacitance*
- an active-voltage source which may be a *battery* or a *generator.*

Table 3.1: Symbols and units for the elements of electric circuits

SNo.	Element	Symbol	Units
1.	Quantity of electricity	Q	Coulombs
2.	Current	i	Ampere (A)
3.	Resistance	+ − i → − R	Ohm (Ω)
4.	Induction	+ − i → − L	Henry (H)
5.	capacitance	+ − i → − C	Farad (F)
6.	Voltage or electromotive force (e.m.f.)	− i → − + Battery, E constant − i → − ~ + Generator, E variable	Volt (V)
7.	Loop	It is any closed path forced by passing through two or more elements in series.	
8.	Nodes	Nodes are terminals of any of these elements	

Basic Relations between Various Quantities of Electric Circuits

- $i = \dfrac{dq}{dt}$ or $q = \int i\,dt$
- The potential voltage drop across
  - resistance R is iR.
  - inductance L is $L\dfrac{di}{dt}$
  - capacitance C is $\dfrac{q}{C}$.

Basic Electric Circuits

Resistive Circuit

It is a series circuit consisting of an electric source and a resistance R which uses energy (like an electric bulb). The voltage drop across resistor (V_R) is proportional to instantaneous current i, i.e., $V_R \propto i$, i.e., $V_R = iR$ (Ohm's law). Constant of proportionality R is called resistance of the resistor, and is measured in *ohms*.

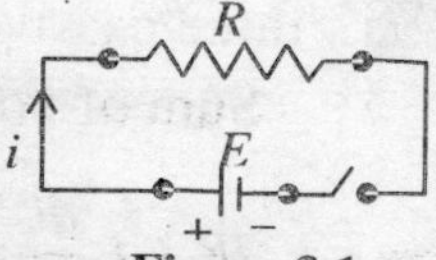

Figure 3.1

Inductive Circuit

It is a series circuit consisting of an electric source and an inductor L (which oppose a change in current). The voltage drop across inductor (V_L) is proportional to rate change of instantaneous current i, i.e, $V_L \propto \dfrac{di}{dt}$, i.e., $V_L = L\dfrac{di}{dt}$. Constant of proportionality L is called inductance of the inductor, and is measured in *henry*.

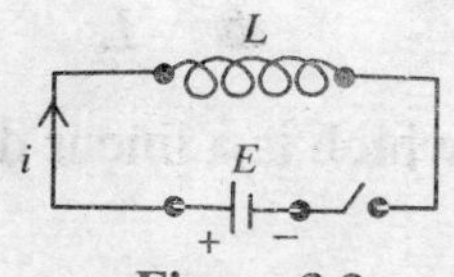

Figure 3.2

Capacitative Circuit

It is a series circuit consisting of an electric source and a capacitor C which is a device that stores energy. The voltage drop across capacitor (V_C) is proportional to instantaneous electric charge Q, i.e, $V_C \propto Q$, i.e., $V_C = \left(\dfrac{1}{C}\right)Q$. Constant of proportionality C is called capacitance of the capacitor, and is measured in *farads*.

Figure 3.3

Kirchoff's Laws

Kirchoff's laws play an important role in the formation of differential equation for an electrical circuit which states as follows:

- The algebraic sum of the voltage drops in any closed circuit is equal to the resultant electromotive force in the circuit. This law is known as *voltage law*.
- The algebraic sum of the currents flowing into any node is zero. This law is known as *circuit law*.

3.2.2 Differential Equation for R-L Series Circuit

Consider a circuit containing resistance R and inductor L in series with a constant voltage source (battery) E. Let i be the current flowing in the circuit at any time t.

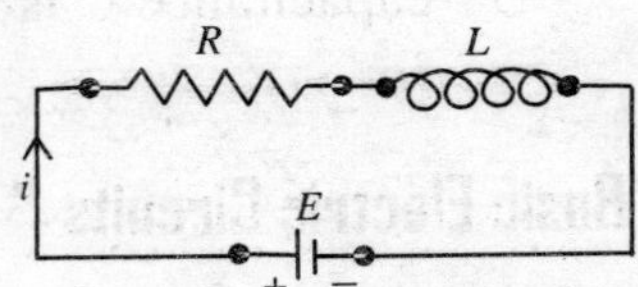

Figure 3.4

By Kirchoff's voltage law, we have

Sum of voltage drops across R and $L = E$

$$\Rightarrow \quad iR + L\frac{di}{dt} = E$$

$$\Rightarrow \quad L\frac{di}{dt} + iR = E$$

$$\Rightarrow \quad \frac{di}{dt} + \frac{R}{L}i = \frac{E}{L} \qquad \ldots(1)$$

which is a linear differential equation in i, *i.e.*, of the type $\frac{dy}{dx} + Py = Q$.

If the constant voltage source (battery) E is replaced with periodic e.m.f. source $E\sin\omega t$, then Eq. (1) takes the form

$$\frac{di}{dt} + \frac{R}{L}i = \frac{E}{L}\sin\omega t \qquad \ldots(2)$$

3.2.3 Differential Equation for L-C Series Circuit

Consider a circuit containing inductor L and resistance capacitor C in series with a constant voltage source (battery) E. Let i be the current flowing in the circuit at any time t.

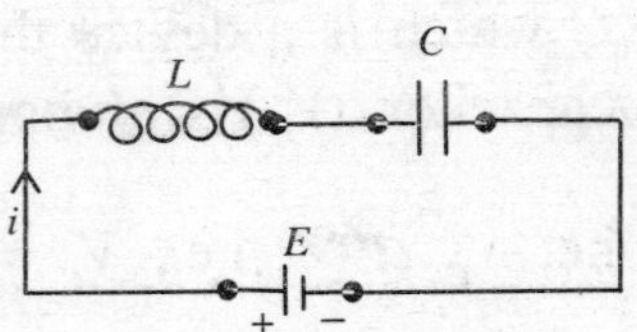

Figure: 3.5

The charge on the capacitor $q = \int i\, dt$, i.e., $i = \frac{dq}{dt}$

By Kirchoff's voltage law, we have

Sum of voltage drops across L and $C = E$

$$\frac{q}{C} + L\frac{di}{dt} = E$$

$$\Rightarrow \quad \frac{q}{C} + L\left[\frac{d}{dt}\left(\frac{dq}{dt}\right)\right] = E$$

$$\Rightarrow \quad \frac{q}{C} + L\frac{d^2q}{d^2t} = E$$

$$\Rightarrow \quad L\frac{d^2q}{d^2t} + \frac{q}{C} = E$$

$$\Rightarrow \quad \frac{d^2q}{d^2t} + \frac{q}{LC} = \frac{E}{L} \qquad \ldots (3)$$

which is a linear differential equation of second order in q.

If the constant voltage source (battery) E is replaced with periodic e.m.f. source $E \sin \omega t$, then Eq. (3) takes the form

$$\frac{d^2q}{dt^2} + \frac{q}{LC} = \frac{E}{L}\sin \omega t \qquad \ldots (4)$$

3.2.4 Differential Equation for R-L-C Series Circuit

Now consider a circuit containing resistance R, induction L and capacitance C all in series with a constant e.m.f. E.

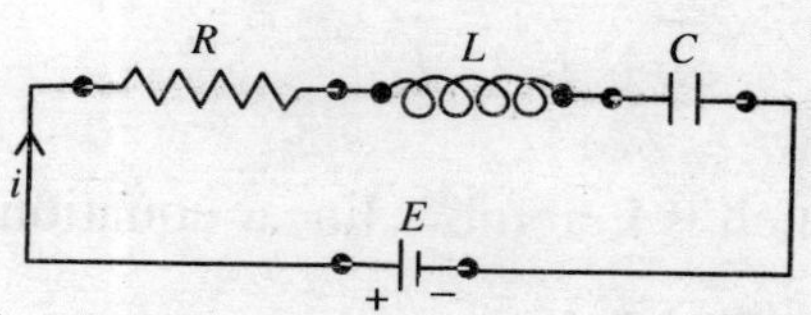

Figure 3.6

If i is the current in the circuit at time t then the charge q on condenser $= \int i dt \quad \Rightarrow \quad i = \frac{dq}{dt}$

Applying Kirchoff's law, we have

Sum of voltage drops across R, L, $C = E$

$$\Rightarrow \quad iR + L\frac{di}{dt} + \frac{q}{C} = E$$

$$\Rightarrow \quad L\frac{d^2q}{dt^2} + R\frac{dq}{dt} + \frac{q}{C} = E$$

$$\Rightarrow \quad \frac{d^2q}{dt^2} + \frac{R}{L}\frac{dq}{dt} + \frac{q}{LC} = \frac{E}{L} \qquad \ldots (5)$$

which is linear differential equation of second order in q.

If the constant voltage source (battery) E is replaced with periodic e.m.f. source $E \sin \omega t$, then Eq. (3) takes the form

$$\frac{d^2q}{dt^2} + \frac{R}{L}\frac{dq}{dt} + \frac{q}{LC} = \frac{E}{L}\sin \omega t \qquad \ldots (6)$$

ILLUSTRATIVE EXAMPLES 3.1

Example 3.1: A constant electromotive force E volts is applied a circuit containing a constant resistance R ohms in series and a constant induction L henries. If the initial current is zero, show that the current builds up to half of its theoretical maximum in $(L \log 2)/R$ seconds.

Solution: Let i be the current in the circuit at any time t.

By Kirchoff's law, we have

$$L\frac{di}{dt} + Ri = E$$

$$\Rightarrow \quad \frac{di}{dt} + \frac{R}{L}i = \frac{E}{L} \qquad \ldots (1)$$

which is Leibnitz's linear equation.

It's I.F. $= e^{\int \frac{R}{L}dt} = e^{\frac{R}{L}t}$

$\therefore$ Solution of Eq. (1) is

$$i(I.F.) = \int \frac{E}{L}(I.F.)\,dt + c$$

$$\Rightarrow \quad ie^{\frac{R}{L}t} = \int \frac{E}{L} e^{\frac{R}{L}t} dt + c = \frac{E}{L} \cdot \frac{L}{R} e^{\frac{R}{L}t} + c$$

$$\Rightarrow \quad i = \frac{E}{R} + ce^{\frac{-Rt}{L}} \quad \ldots (2)$$

Initially, when $t = 0$, $i = 0$, so that $c = \frac{-E}{R}$

Thus, Eq. (2) becomes

$$i = \frac{E}{R}\left(1 - e^{\frac{-Rt}{L}}\right) \quad \ldots (3)$$

This equation gives the current in the circuit at any time t. Clearly, i increases with t and attains the maximum value $\frac{E}{R}$.

Let the current in the circuit be half its theoretical maximum after a time T seconds. Then

$$\frac{1}{2} \cdot \frac{E}{R} = \frac{E}{R}\left(1 - e^{\frac{-RT}{L}}\right)$$

$$\Rightarrow \quad e^{\frac{-RT}{L}} = \frac{1}{2}$$

Taking logarithm on both sides, we get

$$-\frac{RT}{L} = \log \frac{1}{2} = -\log 2$$

$\therefore \quad T = (L \log 2)/R$ Hence the result.

Example 3.2: The initial value problem governing the current i flowing in series R-L circuit when a voltage $v(t) = t$ is applied is given by $iR + L\frac{di}{dt} = t; t \geq 0$; $i(0) = 0$, where R, L are constants. Find the current $i(t)$ at time t. [PTU May 2006]

Solution: Given equation is

$$iR + L\frac{di}{dt} = t \qquad t \geq 0;\ i(0)=0$$

$$\Rightarrow \qquad \frac{di}{dt}+\frac{R}{L}i=\frac{1}{L}t \qquad \ldots (1)$$

which is a linear differential equation in i.

It's $\quad$ $\text{I.F.}=e^{\int \frac{R}{L}dt}=e^{\frac{R}{L}t}$

It's solution is

$$ie^{\frac{R}{L}t}=\int e^{\frac{R}{L}t}\frac{1}{L}t\,dt+c=\frac{1}{L}\int te^{\frac{R}{L}t}\,dt+c \qquad \text{[Integrating by parts]}$$

$$=\frac{1}{L}\left[t\left(e^{\frac{R}{L}t}\left(\frac{L}{R}\right)\right)-(1)\left(\frac{L^2}{R^2}e^{\frac{R}{L}t}\right)\right]+c$$

$$\therefore \qquad ie^{\frac{R}{L}t}=\left(\frac{t}{R}-\frac{L}{R^2}\right)e^{\frac{R}{L}t}+c \qquad \ldots (2)$$

At $t=0,\ i=0$

$\therefore$ $\quad$ Eq. (2) gives $\dfrac{L}{R^2}=c$

Putting the value of c in Eq. (2), it becomes

$$i=\frac{1}{R}t-\frac{L}{R^2}+\frac{L}{R^2}e^{\frac{-R}{L}t}$$

$$\Rightarrow \qquad i=\frac{1}{R}t-\frac{L}{R}\left(1+e^{\frac{-R}{L}t}\right)$$ which is the required answer.

Example 3.3: The equation of *emf* in terms of current i for an electrical circuit having resistance R and a condenser of capacity C, in series, is $E = Ri + \int \frac{i}{C}dt$. Find the current i at any time t, when $E=E_0 \sin \omega t$. [PTU Dec 2006]

Solution: The given equation is

$$E = Ri + \int \frac{i}{C}dt \qquad \ldots (1)$$

which can be written as

$$Ri + \int \frac{i}{C} dt = E_0 \sin \omega t \quad \ldots (2)$$

Differentiating both sides w. r. t. t, we have

$$R\frac{di}{dt} + \frac{i}{C} = \omega E_0 \cos \omega t$$

$$\Rightarrow \quad \frac{di}{dt} + \frac{i}{RC} = \frac{\omega E_0}{R} \cos wt \quad \ldots (3)$$

which is Leibnitz's linear equation.

It's $\quad I.F. = e^{\int \frac{1}{RC} dt} = e^{\frac{t}{RC}}$

$\therefore$ It's solution is

$$ie^{\frac{t}{RC}} = \int \frac{\omega E_0}{R} \cos \omega t.\ e^{\frac{t}{RC}} dt$$

$$= \frac{\omega E_0}{R} \int e^{\frac{t}{RC}} \cos \omega t\, dt$$

$$= \frac{\omega E_0}{R} . \frac{e^{\frac{t}{RC}}}{\sqrt{\left(\frac{1}{RC}\right)^2 + \omega^2}} \cos\left(\omega t - \tan^{-1} \frac{\omega}{\left(\frac{1}{RC}\right)}\right) + c$$

$$\left[\because \int e^{ax} \cos bx\, dx = \frac{e^{ax}}{a^2 + b^2}(a \cos bx + b \sin bx) = \frac{e^{ax}}{\sqrt{a^2 + b^2}} \cos\left(bx - \tan \frac{b}{a}\right)\right]$$

$$= \frac{\omega C E_0}{\sqrt{1 + R^2 C^2 W^2}} e^{\frac{t}{RC}} \cos(\omega t + \phi) + c, \text{ where } \tan \phi = RC\omega$$

$$\Rightarrow \quad i = \frac{\omega C E_0}{\sqrt{1 + R^2 C^2 W^2}} \cos(wt - \phi) + c\, e^{-\frac{t}{RC}}$$

which gives the current at any time t.

Example 3.4: Solve the equation $L\frac{di}{dt} + Ri = E_0 \sin \omega t$ where L, R and E_0 are constants and discuss the case when t increases indefinitely. [PTU May 2007]

Solution: Given equation is

$$L\frac{di}{dt} + Ri = Eo \sin wt \quad \Rightarrow \quad \frac{di}{dt} + \frac{R}{L}i = \frac{E_0}{L}\sin wt$$

which is linear differential equation of first order in t.

It's I.F. $= e^{\int \frac{R}{L}dt} = e^{\frac{R}{L}t}$

It's solution is $ie^{\frac{R}{L}t} = \int e^{\frac{R}{L}t}\frac{E_0}{L}\sin \omega t dt + c = \frac{E_0}{L}\int e^{\frac{R}{L}t} \sin \omega t dt + c$

$$= \frac{E_0}{L} \cdot \frac{e^{\frac{R}{L}t}}{\frac{R^2}{L^2} + \omega^2}\left[\frac{R}{L}\sin \omega t - \omega \cos \omega t\right] + c$$

$$\because \int e^{ax} \sin bx dx = \frac{e^{ax}}{a^2 + b^2}[a \sin bx - b \cos bx]$$

$$= \frac{E_0 L e^{\frac{R}{L}t}}{R^2 + L^2\omega^2}\left[\frac{R}{L}\sin \omega t - \omega \cos \omega t\right] + c$$

$$\therefore \quad i = \frac{E_0 L}{R^2 + L^2\omega^2}\left[\frac{R}{L}\sin \omega t - \omega \cos \omega t\right] + ce^{-\frac{R}{L}t}$$

Let $\frac{R}{L} = a \cos \phi$, $\omega = a \sin \phi$, then $a^2 = \frac{R^2}{L^2} + \omega^2$ and $\tan \phi = \frac{L\omega}{R}$

$$\therefore \quad i = \frac{E_0 L}{R^2 + L^2\omega^2}[a \cos \phi \sin \omega t - a \sin \phi \cos \omega t] + ce^{-\frac{R}{L}t}$$

$$= \frac{E_0 L}{R^2 + L^2\omega^2} a \sin(\omega t - \phi) + ce^{-\frac{R}{L}t}$$

$$= \frac{E_0 L}{R^2 + L^2\omega^2}\sqrt{\frac{R^2 + L^2\omega^2}{L^2}} \sin\left(\omega t - \tan^{-1}\frac{L\omega}{R}\right) + ce^{-\frac{R}{L}t}$$

$$= \frac{E_0}{\sqrt{R^2 + L^2\omega^2}} \sin\left(\omega t - \tan^{-1}\frac{L\omega}{R}\right) + ce^{-\frac{R}{L}t}$$

Now, when t increases indefinitely, *i.e.*, $t \to \infty$, then $e^{-\frac{R}{L}t} \to 0$

$$\therefore \quad i = \frac{E_0}{\sqrt{R^2 + L^2\omega^2}} \sin\left(\omega t - \tan^{-1}\frac{L\omega}{R}\right)$$

EXERCISE 3.1

1. When a resistance R ohms is connected in series with an induction L henries, an emf of E volts, the current i amperes at time t is given by differential equation $L\frac{di}{dt} + Ri = E$. If $E = 10 \sin t$ volts, and $i = 0$, when $t = 0$, find i as a function of t.
2. A voltage Ee^{-at} is applied at $t = 0$ to a circuit containing induction L and resistance R. Show that the current at any time t is $\frac{E}{R - aL}\left(e^{-at} - e^{\frac{-R}{L}t}\right)$.
3. In a circuit containing a battery E, a resistance R, and an induction L, the current i builds up at rate given by differential equation $L\frac{di}{dt} + Ri = E$. Find i as a function of t. How long will it be, before the current has reached one-half of its maximum value if $E = 6$ volts, $R = 100$ ohms and $L = 0.1$ henry?
4. Solve the differential equation $L\frac{di}{dt} + Ri = 200\cos 300t$, when $R = 100$ ohms, $L = 0.05$ henry. Also find i when it is given that $i = 0$ when $t = 0$. What value does i approach after a long time?
5. A resistance R in series with induction L is shunted by an equal resistance R with capacitance C. An alternating e.m.f. $E = \sin pt$ produces currents i_1 and i_2 in the two branches. If initially there is no current, determine i_1 and i_2 from the differential equations $L\frac{di_1}{dt} + Ri_1 = E\sin pt$ and $\frac{i_2}{C} + R\frac{di_2}{dt} = pE\cos pt$. Verify that if $R^2C = L$, then total current $i_1 + i_2$ will be $\frac{E}{R}\sin pt$.
6. Show that the differential equation for the current i in an electrical circuit containing an inductance L and resistance R in series and acted on by an *emf* $E\sin\omega t$ satisfies the differential equation $\frac{di}{dt} + \frac{R}{L}i = \frac{E}{L}\sin\omega t$. Find the value of current at any time t, if initially there is no current in the circuit.

7. A condenser of capacity C is discharging through the inductance L and a resistance R in series and the charge q at any time satisfies the following differential equation

$$\frac{d^2q}{dt^2}+\frac{R}{L}\frac{dq}{dt}+\frac{q}{LC}=0$$

Given that $L = 0.25$ henry, $R = 250$ ohms, $C = 2\times10^{-6}$ farads, $q = 250$ coulombs when $t = 0$, and $\frac{dq}{dt}=0$. Obtain the expression for q in terms of t.

8. For an LCR circuit, the charge q on the condenser is given by

$$\frac{d^2q}{dt^2}+\frac{R}{L}\frac{dq}{dt}+\frac{q}{LC}=E\sin\omega t \quad \text{and the current by } i=\frac{dq}{dt}$$

Given circuit is tuned to resonance so that $\omega^2=\frac{1}{LC}$. If $R^2>\frac{4L}{C}$ and $q=i=0$ at $t=0$, show that

$$q=\frac{E}{R\omega}\left[-\cos\omega t+e^{\frac{R}{2L}t}\left(\cos pt+\frac{R}{2Lp}\sin pt\right)\right], \text{ where } p^2=\frac{1}{LC}-\frac{R^2}{4L^2}.$$

9. An L-C-R circuit with battery e.m.f. '$E\sin pt$' is tuned to resonance so that $p^2=\frac{1}{LC}$. If initially the current i and the charge q be zero, then show that for small value of $\frac{R}{L}$, the current in the circuit at time t is given by $\frac{E}{2L}t\sin pt$. [PTU May 2012]

10. An e.m.f. $E\sin pt$ is applied at $t=0$ to a circuit containing a capacitance C and inductance L. The current i satisfies the equation $L\frac{di}{dt}+\frac{1}{C}\int idt=E\sin pt$. If $p^2=\frac{1}{LC}$ and initially the current and the charge are zero, then find the current at any time t. [PTU Dec 2012]

ANSWERS

1. $i=\frac{10}{L^2+R^2}\left(R\sin t-L\cos t+Le^{\frac{-R}{L}t}\right)$

3. 0.0006931 *secs*

4. $i=\frac{40}{409}[20\cos 300t+3\sin 300t]-\frac{800}{409}e^{-200t};\frac{40}{\sqrt{409}}$

5. $i_1=\frac{E^2}{R^2+L^2p^2}(R\sin pt-pL\cos pt),\ i_2=\frac{pEc}{1+p^2R^2C^2}[\cos pt+RCp\sin pt]$

7. $q = e^{-500t}[0.002\cos 1323t + 0.0008\sin 1323t]$

9. $i = \frac{Et}{2L}\sin pt$

10. $i_1 = \frac{Et}{2p}\sin pt$

3.3 DEFLECTION OF BEAMS

Consider a uniform beam as made up of fibres running lengthwise. When the beam is deflected, the fibers in some region are elongated and those in some region are contracted. In between these two, there is a region of the beam where the fibers are neither compressed nor stretched. This region is called the *neutral surface of the beam* and curve of any particular fibre in this surface is called *elastic curve* or *deflection curve* of the beam. The line in which any plane section of the beam cuts the neutral surface is called the neutral axis of that section.

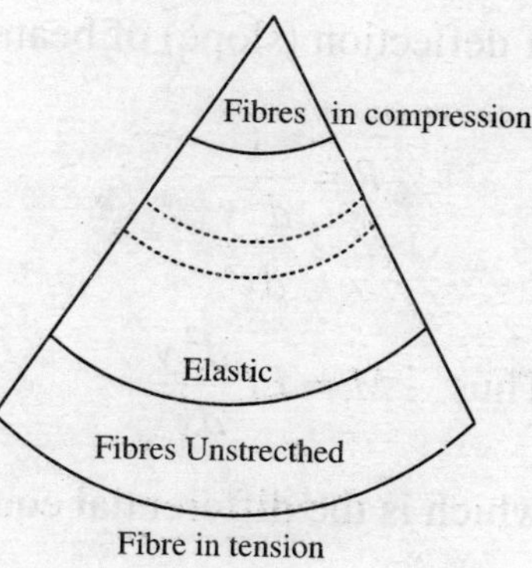

Figure 3.7

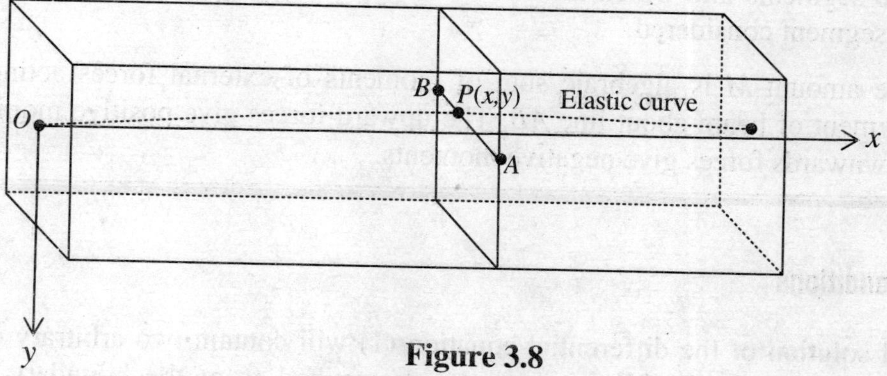

Figure 3.8

Consider a cross-section of bean cutting the elastic curve in P and neutral surface in line AB, the neutral axis of this section.

It is well known from mechanics of structure that the forces above and below the neutral surface acting in opposite directions have a tendency to restore the beam to its original position, creating an interval bending moment M given by $\frac{EI}{R}$,

where E = modulus of elasticity of material of the beam,

I = moment of inertia of cross section of beam about neutral axis AB, and

R = radius of curvature of elastic curve at $P(x,y)$.

Now, $R = \dfrac{\left[1+\left(\dfrac{dy}{dx}\right)^2\right]^{3/2}}{\dfrac{d^2y}{dx^2}}$

If deflection (slope) of beam is small, $\left(\dfrac{dy}{dx}\right)^2$ is very small and can be neglected, so that

$$R = \frac{1}{\dfrac{d^2y}{dx^2}}$$

Thus $\quad M = EI\dfrac{d^2y}{dx^2}$ ... (1)

which is the differential equation of elastic curve.

Note 1: The moment M with respect to AB of all external forces acting on either of the two segments into which the beam is separated by cross-section is independent of segment considered.

Note 2: The amount M is algebraic sum of moments of external forces acting on the segment of beam about line AB. The upward forces give positive moments and downwards forces give negative moments.

Boundary Conditions

The general solution of the differential equation (1) will contain two arbitrary constants which, in any particular problem, are to be determined from the boundary (or end) condition given below:

(*i*) *End freely supported*: At the freely supported end O, there is no deflection of beam, so that $y = 0$. Also there is no bending moment at this end, so that $\dfrac{d^2y}{dx^2} = 0$.

Figure 3.9

(*ii*) *End fixed horizontally*: At the end fixed horizontally, the deflection and slope of beam are zero.

$\therefore \quad y = 0$ and $\dfrac{dy}{dx} = 0$

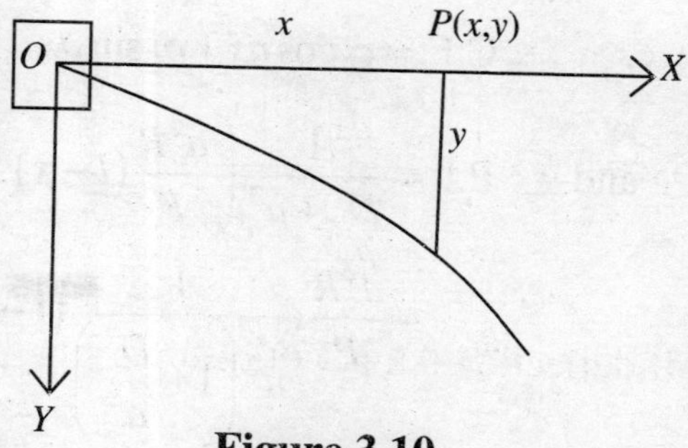

Figure 3.10

(*iii*) *End perfectly free*: At the free end *A* in the above figure, there is no bending moment or shear force, so that $\dfrac{d^2y}{dx^2} = 0$ and $\dfrac{d^3y}{dx^2} = 0$.

ILLUSTRATIVE EXAMPLES 3.2

Example 3.5: The deflection of a strut of length l with one end ($x = 0$) built in and the other supported and subjected to end thrust *P*, satisfies the equation

$$\frac{d^2y}{dx^2} + a^2y = \frac{a^2R}{P}(1-x)$$

Prove that deflection curve is

$$y = \frac{R}{P}\left(\frac{\sin ax}{a} - l\cos ax + l - x\right),$$

where $al = \tan al$.

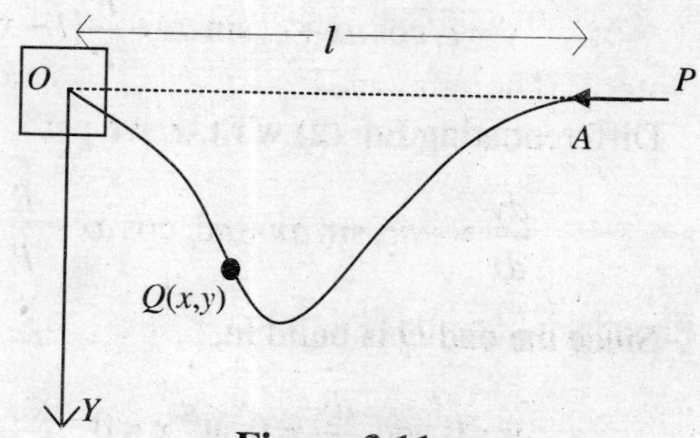

Figure 3.11

Solution: The given differential equation is

$$\frac{d^2y}{dx^2} + a^2y = \frac{a^2R}{P}(l-x)$$

$$\Rightarrow \quad \left(D^2 + a^2\right)y = \frac{a^2R}{P}(l-x) \qquad \ldots (1)$$

It's A.E. is

$$D^2 + a^2 = 0$$

$$\Rightarrow \quad D = \pm ia$$

$\therefore$ $\qquad \text{C.F.} = c_1 \cos ax + c_2 \sin ax$

and $\qquad \text{P.I.} = \dfrac{1}{D^2 + a^2}\left(\dfrac{a^2 R}{P}(l - x)\right) = \dfrac{a^2 R}{P}\dfrac{1}{D^2 + a^2}(l - x)$

$$= \frac{a^2 R}{P} \cdot \frac{1}{a^2\left(1 + \dfrac{D^2}{a^2}\right)}(l - x)$$

$$= \frac{R}{P} \cdot \left(1 + \frac{D^2}{a^2}\right)^{-1}(l - x)$$

$$= \frac{R}{P}\left(1 - \frac{D^2}{a^2} + \ldots\right)(l - x)$$

$$= \frac{R}{P}(l - x)$$

$\therefore$ The C.S. of Eq. (1) is

$$y = c_1 \cos ax + c_2 \sin ax + \frac{R}{P}(l - x) \qquad \ldots (2)$$

Differentiating Eq. (2) w.r.t. x, we get

$$\frac{dy}{dx} = -ac_1 \sin ax + ac_2 \cos ax - \frac{R}{P} \qquad \ldots (3)$$

Since the end O is build in,

$\therefore$ $\qquad y = 0$ and $\dfrac{dy}{dx} = 0$ at $x = 0$

From Eq. (2), we have

$$0 = c_1 + \frac{Rl}{P} \quad \Rightarrow \quad c_1 = -\frac{Rl}{P}$$

From Eq. (3), we have

$$0 = ac_2 - \frac{R}{P} \quad \Rightarrow \quad c_2 = \frac{R}{aP}$$

Substituting the values of c_1 and c_2 in Eq. (2), we have

$$y=\left(-\frac{Rl}{P}\right)\cos ax+\left(\frac{R}{aP}\right)\sin ax+\frac{R}{P}(l-x)$$

$$\Rightarrow \quad y=\frac{R}{P}\left(\frac{\sin ax}{a}-l\cos ax+l-x\right) \qquad \ldots (4)$$

Hence the proof.

Also, at the end A, $y=0$ when $x=l$,

$\therefore$ From Eq. (4), we have

$$0=\frac{R}{P}\left(\frac{\sin al}{a}-l\cos al\right)$$

$$\Rightarrow \quad \frac{\sin al}{a}=l\cos al$$

$$\Rightarrow \quad \frac{\sin al}{\cos al}=al$$

$$\therefore \quad al=\tan al$$

Example 3.6: A light horizontal strut AB of length l is freely pinned at A and B and is under the action of equal and opposite compressive forces P at each of its ends and carries a load W at its centre.

Prove that the deflection at the centre is

$$\frac{W}{2P}\left(\frac{1}{n}\tan\frac{nl}{2}-\frac{l}{2}\right)$$

where $n^2=\frac{P}{EI}$.

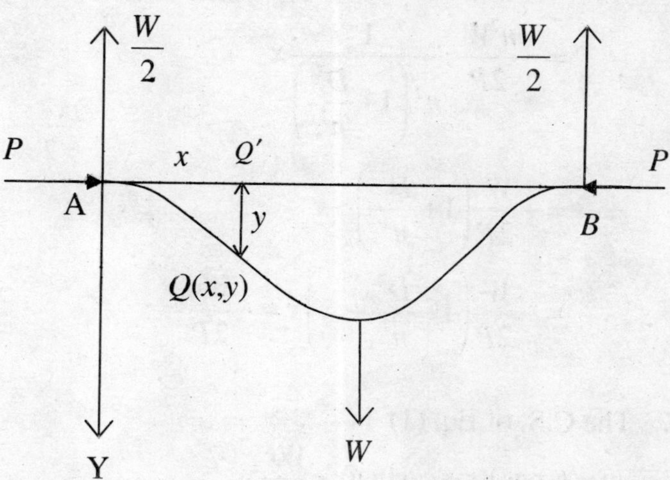

Figure 3.12

Solution: At each end, there is a vertical reaction $\frac{W}{2}$. At any point $Q(x, y)$ of the beam, the internal bending moment is $\left(EI\frac{d^2y}{dx^2}\right)$ and this must be equal to moment of external forces taken to left (or right) of the section QQ'.

$\therefore$ The differential equation for elastic curve of the beam is

$$EI\frac{d^2y}{dx^2} = -\frac{W}{2}x - Py$$

$$\Rightarrow \quad \frac{d^2y}{dx^2} + \frac{P}{EI}y = -\frac{W}{2EI}x$$

$$\Rightarrow \quad \frac{d^2y}{dx^2} + n^2y = -\frac{n^2W}{2P}x \qquad \left[\because \frac{P}{EI} = n^2\right] \qquad \ldots (1)$$

It's A.E. is

$$D^2 + n^2 = 0$$

$$\Rightarrow \quad D = \pm in.$$

$\therefore$ It's C.F. $= c_1 \cos nx + c_2 \sin nx$, and

$$\text{P.I.} = \frac{1}{D^2 + n^2}\left(-\frac{n^2W}{2P}x\right)$$

$$= -\frac{n^2W}{2P} \cdot \frac{1}{n^2\left(1 + \frac{D^2}{n^2}\right)}x$$

$$= -\frac{W}{2P}\left(1 + \frac{D^2}{n^2}\right)^{-1} x$$

$$= -\frac{W}{2P}\left(1 - \frac{D^2}{n^2} + \ldots\right)x = -\frac{Wx}{2P}$$

$\therefore$ The C.S. of Eq. (1) is

$$y = c_1 \cos nx + c_2 \sin nx - \frac{Wx}{2P} \qquad \ldots (2)$$

Differentiating Eq. (2) w.r.t. x, we have

$$\frac{dy}{dx} = -nc_1 \sin nx + nc_2 \cos nx - \frac{W}{2P} \qquad \ldots (3)$$

At the end A, $x = 0$, $y = 0$

$\therefore$ From Eq. (2), we have $c_1 = 0$

$\therefore$ At $x = \frac{l}{2}$, $\frac{dy}{dx} = 0$.

$\therefore$ From Eq. (3), we have

$$0 = nc_2 \cos\frac{nl}{2} - \frac{W}{2P} \qquad [\because c_1 = 0]$$

$$\Rightarrow \quad c_2 = \frac{W}{2Pn\cos\frac{nl}{2}}$$

Substituting the values of c_1 and c_2 in Eq. (2), we have

$$y = \frac{W}{2Pn\cos\frac{nl}{2}} \sin nx - \frac{W}{2P} x$$

$$\Rightarrow \quad y = \frac{W}{2P}\left[\frac{\sin nx}{n\cos\frac{nl}{2}} - x\right]$$

Deflection at the centre

$$y\left(at\ x = \frac{l}{2}\right) = \frac{W}{2P}\left(\frac{\sin\frac{nl}{2}}{n\cos\frac{nl}{2}} - \frac{l}{2}\right)$$

$$= \frac{W}{2P}\left(\frac{1}{n}\tan\frac{nl}{2} - \frac{l}{2}\right)$$

Hence the proof.

EXERCISE 3.2

1. A horizontal tie-rod is freely pinned at each end. It carries a uniform load w kg per cent length and has a horizontal pull P. Find the central deflection and maximum bending moment, taking the origin at one of its ends.
2. Find the equation of elastic curve and its maximum deflection for the horizontal, simply supported, uniform beam of length $2l$ meters, having uniformly distributed load w kg/meter.
3. A horizontal sturt of length l is clamped horizontally at one end and carries a vertical load w at other end. If the horizontal thrust be P, prove that deflection at the free end is $\frac{w}{nP}(\tan nl - nl)$, where $n^2 = \frac{P}{EI}$.
4. A long uniform sturt of length l is clamped at one end, other end being free. If a thrust P be applied at the free end, at a distance a from the neutral axis, prove that deflection at free end is $a(\sec nl - 1)$, where $n^2 = \frac{P}{EI}$.
5. A horizontal tie-rod of length $2l$ with concentrated load w at the centre and ends freely hinged, satisfies the differential equations $EI\frac{d^2y}{dx^2} = Py - \frac{w}{2}x$.

   With conditions $x = 0,\ y = 0$ and $x = l,\ \frac{dy}{dx} = 0$, prove that the deflection s and the bending moment M at the centre $(x=l)$ are given by

   $$s = \frac{w}{2Pn}(nl - \tanh nl) \text{ and } M = \frac{-w}{2n}\tanh nl, \text{ where } n^2EI = P.$$
6. A light horizontal strut of length l is clamped at one end and carries vertical load W at the free end. If the horizontal thrust at the free end is P, show that the strut satisfies the differential equation $EI\frac{d^2y}{dx^2} = (\delta - y)P - W(l - x)$, where y is the displacement of a point at a distance x from the fixed end and δ be the deflection at the free end. Also prove that the deflection at the free end is given by $\frac{W}{nP}(\tan nl - nl)$, where $n^2EI = P$.

ANSWERS

1. $\frac{w}{Pa^2}\left(\sec h\frac{al}{2} - 1\right) + \frac{wl^2}{8p}$; $\quad \frac{w}{a^2}\left(\sec h\frac{al}{2} - 1\right)$
2. $y = \frac{w}{24EI}(4lx^3 - x^4 - 8l^3x)$; $\quad \frac{5wl^4}{24EI}$.

3.4 SIMPLE HARMONIC MOTION

A particle is said to execute *simple harmonic motion* if it moves in a straight line such that its acceleration is always directed towards a fixed point in the line and is proportional to the distance of the particle from the fixed point.

Simple harmonic motion is briefly written as S.H.M.

3.4.1 Expression for Velocity and Position

To find the expression for the velocity and position of the particle at any time, let O be the fixed point in the line $A'OA$ and let P denote the particle position after time t from O moving with a velocity v in the positive direction from O to A.

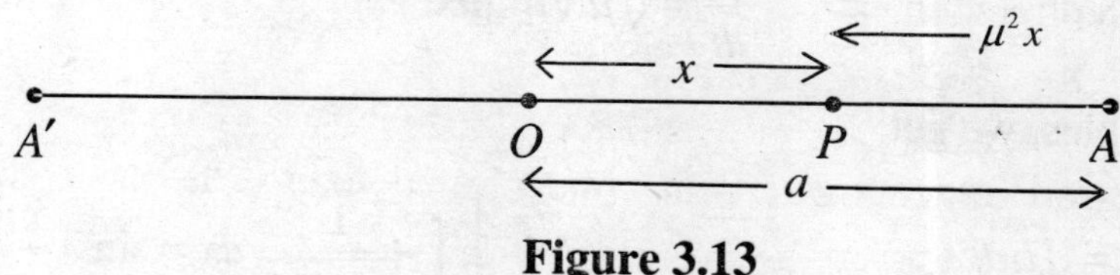

Figure 3.13

Let $OP = x$, then the acceleration is μx, where μ is constant.

Since the acceleration is in the direction opposite to that in which x increase, the equation of motion of the particle is

$$v\frac{dv}{dx} = -\mu x \qquad \ldots (1)$$

Separating the variable, we get

$$vdv = -\mu x dx$$

Integrating, we get

$$\frac{v^2}{2} = -\frac{\mu x^2}{2} + c_1 \qquad \ldots (2)$$

where c_1 is constant of integration

As particle P is supposed to be moving in the direction OA and as the acceleration is given to be taking place in the opposite direction, the particle P must come to rest at some point in OA, say at A such that $OA = a$.

$\therefore \quad v = 0$ when $x = a$

$\therefore$ from Eq. (2), $0 = -\frac{1}{2}\mu a^2 + c_1 \quad \Rightarrow \quad c_1 = \frac{1}{2}\mu a^2$

Putting $c_1 = \frac{1}{2}\mu a^2$ in Eq. (2), we get

$$\frac{v^2}{2} = -\frac{\mu x^2}{2} + \frac{1}{2}\mu a^2 \quad \Rightarrow \quad v^2 = \mu(a^2 - x^2)$$

$$\Rightarrow \quad v = \pm\sqrt{\mu}\sqrt{a^2 - x^2} \qquad \ldots (3)$$

This equation gives the value of the velocity v for any displacement x. Since P is moving in the opposite direction

$$\therefore \quad v = \sqrt{\mu}\sqrt{a^2 - x^2} \quad \Rightarrow \quad \frac{dx}{dt} = \sqrt{\mu}\sqrt{a^2 - x^2}$$

Separating the variables, we get

$$\frac{1}{\sqrt{a^2 - x^2}}dx = \sqrt{\mu}dt \qquad \left[\int \frac{1}{\sqrt{a^2 - x^2}}dx = \sin^{-1}\frac{x}{a}\right]$$

Integrating, we get

$$\sin^{-1}\frac{x}{a} = \sqrt{\mu}t + c_2 \qquad \ldots (4)$$

where c_2 is constant of integration.

(*i*) If the time t is measured from the instant when P is at O, *i.e.*, if $x = 0$ when $t = 0$ then from Eq. (4), we get

$$\sin^{-1}\frac{0}{a} = \sqrt{\mu} \times 0 + c_2$$

$$\Rightarrow \quad \sin^{-1}0 = 0 + c_2 \quad \Rightarrow \quad c_2 = 0 \qquad [\because \ \sin^{-1}0 = 0]$$

Putting $c_2 = 0$ in Eq. (4), we get

$$\sin^{-1}\frac{x}{a} = \sqrt{\mu}t \quad \Rightarrow \quad \frac{x}{a} = \sin\sqrt{\mu}t$$

$$\Rightarrow \quad x = a \ \sin\sqrt{\mu}t \qquad \ldots (5)$$

(ii) If the time t is measured from the instant when the particle is at A, *i.e.*, if $x = a$ when $t = 0$, then from Eq. (4), we get

$$\sin^{-1}\frac{a}{a} = \sqrt{\mu}\times 0 + c_2$$

$$\Rightarrow \quad \sin^{-1}1 = 0 + c_2 \qquad \left[\because \ \sin^{-1}1 = \frac{\pi}{2}\right]$$

$$\Rightarrow \quad c_2 = \frac{\pi}{2}$$

Putting $c_2 = \dfrac{\pi}{2}$ in Eq. (3.4), we get

$$\sin^{-1}\frac{x}{a} = \sqrt{\mu}t + \frac{\pi}{2}$$

$$\Rightarrow \quad \frac{x}{a} = \sin\left(\frac{\pi}{2} + \sqrt{\mu}t\right) \qquad \left[\because \ \sin\left(\frac{\pi}{2} + \sqrt{\mu}t\right) = \cos\sqrt{\mu}t\right]$$

$$\Rightarrow \quad \frac{x}{a} = \cos\sqrt{\mu}t \quad \Rightarrow \quad x = a\cos\sqrt{\mu}t \qquad \ldots (6)$$

Eqs. (5) or (6) give the position of the particle at any time t.

Note 1: The fixed point O is called *the centre of attraction* or *oscillation* or *the mean position* or *the position of equilibrium*. In this position, the acceleration is zero and so the force acting on the particle is zero.

Note 2: We have $f = -\mu x$, so it is clear that f is negative when x is positive, and f is positive when x is negative. Thus, in both the cases, the acceleration and hence the force tries to bring the body back to the equilibrium position. Such a force is called *restoring force*.

Note 3: When the particle is on the left hand side of O, the equation of motion is $v\dfrac{dv}{ax}$ and acceleration in the direction of

$$P_1A = \mu OP_1 = \mu(-x) = -\mu x$$

Hence the same equations that works on the right hand side of O, holds also on the left hand side.

Note 4: The points A and A', where the velocity is zero are called the *extreme positions* or *positions of rest*. For these positions, $x = a, -a$

Note 5: The distance $OA = OA' = a$ is called *the amplitude* of S.H.M.

Note 6: $v = \pm\sqrt{\mu}\sqrt{a^2 - x^2}$ shows that the particle has equal and opposite velocities at a point according as it is moving in direction O to A or in direction at A to O.

3.4.2 Maximum and Minimum Velocities

We have $v^2 = \mu(a^2 - x^2)$

$\therefore$ v is maximum when x is least, *i.e.*, $x = 0$

$\therefore$ maximum velocity $= \sqrt{\mu}\sqrt{a^2 - 0} = \sqrt{\mu}a$, which occurs at the mean position.

Again, v is minimum when x is greatest in magnitude, *i.e.*, $x = a, -a$

$\therefore$ Particle has maximum velocity $\sqrt{u}a$ at O and minimum velocity zero at A or A'.

3.4.3 Maximum and Minimum Accelerations

We have $f = \mu x$

$\therefore$ acceleration (in magnitude) is maximum when $x = a$ or $-a$ and maximum acceleration $= \mu a$ and acceleration is minimum, when x is minimum, *i.e.*, $x = 0$ and minimum acceleration = 0.

$\therefore$ particle has maximum acceleration μa at the extreme position A or A' and has minimum acceleration zero at the mean position O.

3.4.4 Nature of the S.H.M.

Let O be the fixed point in the line $A'OA$ and let P be the position of the particle after t from 0 moving with a velocity v in the position direction from O to A at a distance x.

x

A' O P A

Figure 3.14

Let $OP = x$, $OA = a$.

$\therefore$ $v = \pm\sqrt{\mu}\sqrt{a^2 - x^2}$

gives the velocity of P in terms of its distance from O. Initially when $x = 0$ at the point O, the velocity is maximum and equal to $\sqrt{\mu a}$. As the particle proceeds towards A then acceleration being towards O, the velocity goes on decreasing as x increases. At A where $x = a$, it vanishes and the particle is, for an instant at rest. Then owing to the acceleration towards O, the particle moves in negative direction with a velocity which increases numerically as x decreases and is the greatest at O where it is $\sqrt{\mu a}$. Due to this velocity, the particle proceeds further to the negative side of O, the velocity remaining negative and decreases gradually in magnitude till the particle comes to rest at A', where $x = -a$. The acceleration being towards O, the particle then starts and moves towards O with a positive velocity which increases gradually till it is again maximum at O. The same motion is repeated again and again and thus particle goes on oscillating indefinitely between A and A', the two positions of momentary rest.

3.4.5 Periodic Motion

A particle is said to have a *periodic motion* when it moves in such a manner that after a certain fixed interval of time, it occupies the same position and moves, in the same direction with the same velocity.

The fixed minimal interval of time for such a motion is called the *period of the motion.*

Article: Prove that S.H.M. is periodic with period $\dfrac{2\pi}{\sqrt{\mu}}$.

Proof: Let x be the displacement of the particle and v be its velocity at any time t, measured from the centre of oscillation.

$$x = a \sin \sqrt{\mu} t \qquad \ldots (7)$$

$$v = a\sqrt{\mu} \cos \sqrt{\mu} t \qquad \ldots (8)$$

We know that $\sin\theta$ and $\cos\theta$ are periodic functions of θ, the period in each case being 2π. This is to say that the value of $\sin\theta$ and $\cos\theta$ is repeated when θ is increased by 2π. Therefore, the values of x and v are repeated when $\sqrt{\mu} t$ is increased by 2π or when t is increased by $\dfrac{2\pi}{\sqrt{\mu}}$.

Thus, after every interval of $\frac{2\pi}{\sqrt{\mu}}$ we have the same position and same velocity in the same direction. Therefore, the motion is periodic and the period being $\frac{2\pi}{\sqrt{\mu}}$.

Note 1: Since periodic time $= \frac{2\pi}{\sqrt{\mu}}$,

$\therefore$ S.H.M. is periodic and its period is independent of t.

Note 2: Frequency is the number of complete oscillation in one second so that if n denotes the frequency and T the periodic time,

$$nT = 1 \quad \text{or} \quad n = \frac{1}{T} = \frac{\sqrt{\mu}}{2\pi}$$

It should be noted that frequency is reciprocal of the periodic time.

3.4.6 Summary of Results for S.H.M

1. Acceleration = μx.
2. Acceleration is maximum at extreme position, *i.e.*, when $x = a, -a$ and maximum acceleration = μa.
3. Velocity v is given by $v^2 = \mu(a^2 - x^2)$.
4. $x = a\sin\sqrt{\mu}t$, where t is measured from mean position.
5. $x = a\cos\sqrt{\mu}t$, when t is measured from extreme position.
6. Time period $T = \frac{2\pi}{\sqrt{\mu}}$.
7. Frequency $n = \frac{1}{T} = \frac{\sqrt{\mu}}{2\pi}$.
8. Amplitude = a
9. Time from centre to end point = Time from end point to the centre.

ILLUSTRATIVE EXAMPLES 3.3

Example 3.7: A particle moving with S.H.M. of period 12 seconds, travels 10 cm from the position of rest in 2 seconds. Find the amplitude, the maximum velocity and the velocity at the end of 2 seconds.

Solution: Here $T = 12$ secs $\Rightarrow \dfrac{2\pi}{\sqrt{\mu}} = 12 \Rightarrow \sqrt{\mu} = \dfrac{\pi}{6}$

When $t = 2,\ x = (a - 10)$ cm

$$\therefore \quad a - 10 = a\cos\left(\frac{\pi}{6}\times 2\right)$$

$$\Rightarrow \quad a - 10 = a\cos\frac{\pi}{3} \quad \Rightarrow \quad a - 10 = a\times\frac{1}{2}$$

$$\Rightarrow \quad 2a - 20 = a$$

$$\Rightarrow \quad a = 20$$

Maximum velocity $= \sqrt{\mu}a = \dfrac{\pi}{6}\times 20 = \dfrac{10\pi}{3}$ cm/sec.

Now $v^2 = \mu(a^2 - x^2)$

$$\Rightarrow \quad v^2 = \frac{\pi^2}{36}\times(400 - 100) \qquad [\because x = a - 10 = 20 - 10 = 10cm]$$

$$= \frac{\pi^2}{36}\times 300 = \frac{100\pi^2}{36}\times 3$$

$$\Rightarrow \quad v = \frac{10\pi\sqrt{3}}{6} = \frac{5\sqrt{3}}{3}\pi \text{ cm/ sec}$$

Example 3.8: A point moving in a straight line with S.H.M. has velocities u and v, when its distance from the centre are a and b, respectively. Prove period of motion is $2\pi\sqrt{\dfrac{a^2 - b^2}{V^2 - u^2}}$.

Solution: We know that

$$v^2 = u\left(a^2 - x^2\right) \qquad \ldots (9)$$

where v is the velocity of the particle when its distance from center is x and a is amplitude.

Now when $x = a,\ v = u$

$$\therefore \quad \text{From Eq. (9), } u^2 = u\left(A^2 - a^2\right) \qquad \ldots (10)$$

Again when $x = b$, $v = v$

$\therefore$ From Eq. (9), $v^2 = u(a^2 - b^2)$ ... (11)

Subtracting Eq. (10) from Eq. (11) , we get

$$v^2 - u^2 = u(a^2 - b^2) \quad \Rightarrow \quad u = \frac{v^2 - u^2}{a^2 - b^2} \qquad \text{... (12)}$$

$$\text{Period of motion} = \frac{2\pi}{\sqrt{\mu}} = \frac{2\pi}{\sqrt{\dfrac{v^2 - u^2}{a^2 - b^2}}} = 2\pi\sqrt{\frac{a^2 - b^2}{v^2 - u^2}}$$

Example 3.9: At the end of three successive seconds, the distance of a point moving with S.H.M. from its mean position measured in the same direction are 1,5,5. Show that the period of one complete oscillation is $\dfrac{2\pi}{\theta}$ seconds, we have $\cos\theta = \dfrac{3}{5}$.

Solution: Let 1, 5, 5 be the distances from mean position at times t, t+1, t+2, respectively.

$$\therefore \quad 1 = a\sin\sqrt{\mu}t \qquad \text{... (13)}$$

$$5 = a\sin\sqrt{\mu}(t+1) \qquad \text{... (14)}$$

$$5 = a\sin\sqrt{\mu}(t+2) \qquad \text{... (15)}$$

Adding Eq. (13) and Eq. (15), we get

$$6 = a\left[\sin\sqrt{\mu}(t+2) + \sin\sqrt{\mu}t\right]$$

$$\Rightarrow \quad 6 = a\left[2\sin\sqrt{\mu}(t+1) \times \cos\sqrt{\mu}t\right] \qquad [\because \sin A + \sin B = 2\sin A\cos B]$$

$$\Rightarrow \quad 3 = a\sin\sqrt{\mu}(t+1)\cos\sqrt{\mu}t$$

$$\therefore \quad 3 = a\sin\sqrt{\mu}(t+1) \times \cos\sqrt{\mu}t$$

$$\Rightarrow \quad 3 = 5\cos\sqrt{\mu} \qquad [\because \text{ of Eq. (14)}]$$

$$\Rightarrow \quad \cos\sqrt{\mu} = \frac{3}{5} \quad \Rightarrow \quad \sqrt{\mu} = \cos^{-1}\left(\frac{3}{5}\right)$$

$$\therefore \quad \text{time period } T = \frac{2\pi}{\sqrt{\mu}} = \frac{2\pi}{\theta}$$

where $\theta = \sqrt{\mu} = \cos^{-1}\dfrac{3}{5}$ or $\cos\theta = \dfrac{3}{5}$

Example 3.10: A particle moves with S.H.M. in a straight line. In the first second after starting from rest it travels a distance a and in the next second, it travels a distance b. Prove that the amplitude of the motion is $\dfrac{2a^2}{3a-b}$.

Solution: Let O be the mean position and A' be the extreme position such that $OA' = A$ is the amplitude.

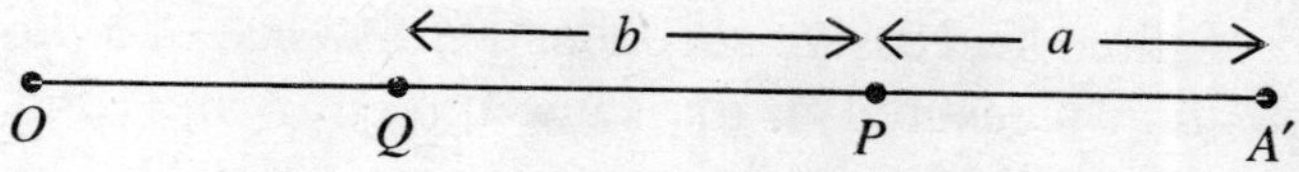

Figure 3.15

Since, time is measured from extreme position

$$\therefore \quad x = A\cos\sqrt{\mu}t \qquad \ldots (16)$$

where x is the distance from mean position.

When $t = 1$, distance $A'P$ from extreme position $= a$

$$\therefore \quad x = A - a \qquad \ldots (17)$$

$\therefore$ from Eqs. (16) and (17), we get

$$A - a = A\cos\sqrt{\mu}t \qquad \ldots (18)$$

When $t = 2$ distance $A'Q$ from extreme position $= a + b$

$$\therefore \quad x = A - a - b$$

$\therefore$ from Eq. (16), we get

$$A - a - b = A\cos^2\sqrt{\mu}t$$

$$\therefore \quad A - a - b = A(2\cos^2\sqrt{ut} - 1) \qquad [\because \cos 2\theta = 2\cos\theta - 1]$$

$$\therefore \quad A - a - b = A\left[2\left(\frac{A-a}{A}\right)^2 - 1\right] \qquad [\because \text{Eq. (18)}]$$

$$\therefore \quad A-a-b=A\left[\frac{2A^2+2a^2-4Aa}{A^2}-1\right]$$

$$\Rightarrow \quad A-a-b=\frac{1}{A}\left[A^2+2a^2-4Aa\right]$$

$$\Rightarrow \quad A^2-Aa-Ab=A^2+2a^2-4Aa$$

$$\Rightarrow \quad 3Aa-Ab=2a^2$$

$$\Rightarrow \quad (3a-b)A=2a^2$$

$$\Rightarrow \quad A=\frac{2a^2}{3a-b}$$

EXERCISE 3.3

1. The amplitude of a particle executing S.H.M. is 2*m*, the time period is 5 secs, calculate:
   (*i*) the maximum acceleration
   (*ii*) maximum velocity
2. A point moving with S.H.M. has a period of oscillation of π seconds and its greatest acceleration is 5*m/sec*^{2}. Find the amplitude and the velocity when the particle is at a distance of 1*m* from the center of oscillation.
3. A particle executing S.H.M. has amplitude *a*. Show that the distance of the point from the center of which the velocity is half of the minimum velocity is $\frac{\sqrt{3}}{2}a$.
4. A particle is executing S.H.M. with amplitude 20*cm* and time 4 seconds. Find the time required by the particle in passing between points which are distance 15*cm* and 5*cm* from the center of force and are on the same side of it.
5. A particle moving in a straight line with S.H.M. has velocities v_1 and v_2 when its distances from the centres are x_1 and x_2, respectively. Show that the period of motion is $2\pi\sqrt{\frac{x_2^2-x_2^2}{v_2^2-v_2^2}}$.
6. At the end of three seconds, the distances of a point moving with S.H.M. from its mean position are x_1, x_2 and x_3, respectively. Show that the time of a complete oscillation is $\left[2\pi / \cos^{-1}\left(\frac{x_1+x_3}{2x_2}\right)\right]$.

7. The period of S.H.M. is 8 *secs.*, and the amplitude is 6*m*. For the motion from one extreme in the path to the centre, what is the average velocity and average acceleration?
8. The acceleration of a particle is defined by the relation $f = -kx$ where x is the displacement. Find the value of k such that $v = 0.15 m/\sec$ when $x = 0$, and $v = 0$ when $x = 0.3m$.
9. A particle moves *a* along a straight line and its velocity at any time t is given by equation $v = \pi \cos\frac{\pi}{2}t$. Show that the motion is S.H.M. and find its time period. Also compute the total distance covered by the particular during the interval from t =0, $t = 3$ *secs.*
10. A point moving in a straight line with S.H.M. has velocity v_1 and v_2 when its distance from the center are x_1 and x_2. Show that the period of motion is $2\pi\sqrt{\frac{x_1^2 - x_2^2}{v_2^2 - v_1^2}}$.
11. A point execute S.H.M. in a straight line such that intro of its position and the velocities are u, v and the corresponding accelerations α, β such that

    (*i*) the distance between two position is $\dfrac{v^2 - u^2}{\alpha + \beta}$

    (*ii*) the time period is $2\pi\sqrt{\dfrac{u^2 - v^2}{\beta^2 - \alpha^2}}$
12. Prove that in S.H.M., where acceleration is n^2 time the displacement, if the displacement be x_0 and initial velocity be v_0, the amplitude will be $\sqrt{x_0^2 + \frac{v_0^2}{x^2}}$.

ANSWERS

1. (*i*) $3.2\pi^2 m/\sec^2$

   (*ii*) $0.8\pi m/\sec$
2. $\left(\frac{5}{4}m, \frac{3}{4}m/\sec.\right)$
3. 0.38 *secs*

7. $3m/\sec, \frac{3\pi}{4} m/\sec^2$

8. $k = \frac{1}{4}$

9. 4 *secs*, 6 *units*

3.5 SIMPLE PENDULUM

If a heavy particle is attached to one end of light inextensible string, the other end of which is fixed, and oscillates under gravity in a vertical plane, the system is called *simple pendulum.*

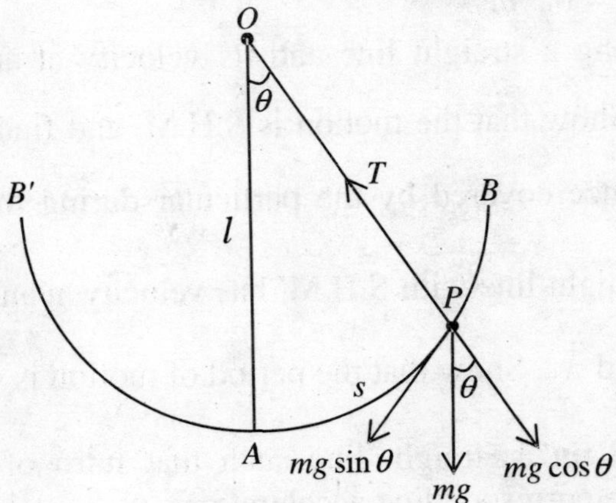

Figure 3.16

Let O be the fixed point, l the length of string and m, the mass of bob. Let P be position of the bob at any time t, let arc $AP = S$ and angle $\angle AOP = \theta$, where OA is vertical line through O.

The forces acting on the bob are

- its weight mg acting vertically downward, and
- the tension T in the string.

The components of weight along and perpendicular to path of motion are $mg\sin\theta$ and $mg\cos\theta$, respectively. The component $mg\cos\theta$ is balanced by tension in the string.

$\therefore$ The equation of motion of bob along tangent is

$$m\frac{d^2 s}{dt^2} = -mg\sin\theta$$

$$\Rightarrow \quad \frac{d^2(l\theta)}{dt^2} = -g\sin\theta \qquad [\because\ s = l\theta\ ]$$

$$\Rightarrow \quad m\frac{d^2\theta}{dt^2} = -\frac{g}{l}\sin\theta = -\frac{g}{l}\left(\theta - \frac{\theta^3}{3!} + \ldots\right) \qquad \left[\because \sin\theta = \theta - \frac{\theta^3}{3!} + \ldots\right]$$

$$= -\frac{g}{l}\theta \qquad \text{to a first approximation}$$

$$\Rightarrow \quad \frac{d^2\theta}{dt^2} + \omega^2\theta = 0 \qquad \text{where } \omega^2 = \frac{g}{l}$$

It's A.E. is

$$D^2 + \omega^2 = 0$$

$$\Rightarrow \quad D = \pm i\omega.$$

$\therefore$ It's solution is

$$\theta = c_1 \cos \omega t + c_2 \sin \omega t$$

$$\Rightarrow \quad \theta = c_1 \cos\sqrt{\frac{g}{l}}t + c_2 \sin\sqrt{\frac{g}{l}}t$$

The motion of bob is simple harmonic and time of an oscillation is $\frac{2\pi}{\omega} = 2\pi\sqrt{\frac{l}{g}}$.

The motion of bob from one extreme position B to other extreme position B' is called *a beat* or *a swing*.

Therefore, the time of one beat = $\frac{1}{2}$(time of one oscillation) = $\pi\sqrt{\frac{l}{g}}$.

A simple pendulum which beats once every second is called a *second's pendulum*. Thus, a second's pendulum beats 86400 times a day, the number of seconds in 24 hours.

Since the time for one beat in a second's pendulum is 1 second, taking g = 981 cm/sec^2, we have

$$1 = \pi\sqrt{\frac{l}{981}} \qquad \Rightarrow \qquad l = \frac{981}{\pi^2} = 99.4 \, cm$$

Hence, it means that the length of second's pendulum is 99.4 *cm*.

Gain or Loss of Beats

Let a simple pendulum of length l makes n beats in a time t so that

$$t = n\pi\sqrt{\frac{l}{g}} \quad \Rightarrow \quad n = \frac{t}{\pi}\sqrt{\frac{g}{l}}$$

Taking logarithm on both sides, we get

$$\log n = \log\left(\frac{t}{\pi}\sqrt{\frac{g}{l}}\right) = \log\left(\frac{t}{\pi}\right) + \log\left(\sqrt{\frac{g}{l}}\right) = \log\left(\frac{t}{\pi}\right) + \frac{1}{2}\log\left(\frac{g}{l}\right)$$

$$\Rightarrow \quad \log n = \log\left(\frac{t}{\pi}\right) + \frac{1}{2}(\log g - \log l)$$

Taking differentials on both sides, we get

$$\frac{dn}{n} = \frac{1}{2}\left(\frac{dg}{g} - \frac{dl}{l}\right) \quad \text{or} \quad dn = \frac{n}{2}\left(\frac{dg}{g} - \frac{dl}{l}\right)$$

which is the change in number of beats.

Now following two cases can arise:

- If g changes while l remains constant, then $dn = \frac{n}{2}.\frac{dg}{g}$
- If l changes while g remains constant, then $dn = -\frac{n}{2}.\frac{dl}{l}$

ILLUSTRATIVE EXAMPLES 3.4

Example 3.11: Find how many seconds a clock would lose per day, if the length of its pendulum increased in the ratio 900:901.

Solution: Let l be the original length and $(l+dl)$ the increased length of the pendulum, then

$$\frac{l}{l+dl} = \frac{900}{901} \quad \Rightarrow \quad \frac{l+dl}{l} = \frac{901}{900}$$

$$\Rightarrow \quad 1 + \frac{dl}{l} = \frac{901}{900} \quad \Rightarrow \quad \frac{dl}{l} = \frac{901}{900} - 1 = \frac{1}{900}$$

Let n be number of beats per day, then $n = 86400$.

If dn is change in number of beats, then

$$dn = -\frac{n}{2}\cdot\frac{dl}{l} = -\frac{86400}{2}\times\frac{1}{900} = -48 \qquad \left[\because \text{since } l \text{ changes, } g \text{ remaining constant, then } dn = -\frac{n}{2}\frac{dl}{l}\right]$$

Since dn is negative, the clock will lose 48 seconds per day.

Example 3.12: A second's pendulum which gains 10 seconds per day at one place loses 10 seconds per day at another. Compare the accelerations to gravity at the two places.

Solution: Let g be the acceleration due to gravity when pendulum beats seconds.

Let $g + g_1$ be acceleration due to gravity at place where it gains 10 seconds per day, then $dn = 10$, $n = 86400$.

Since $dn = \frac{n}{2}\cdot\frac{g_1}{g}$ (here dg has been replaced by g_1)

$$\therefore \quad 10 = \frac{86400}{2}\cdot\frac{g_1}{g}$$

$$\Rightarrow \quad \frac{g_1}{g} = \frac{1}{4320}$$

Adding 1 to both sides, we get

$$\frac{g_1}{g} + 1 = \frac{1}{4320} + 1$$

$$\Rightarrow \quad \frac{g + g_1}{g} = \frac{4321}{4320} \qquad \ldots (1)$$

Let $g + g_2$ be acceleration due to gravity at place where it loses 10 seconds per day, then

$$dn = -10$$

$$\therefore \quad -10 = \frac{86400}{2}\cdot\frac{g_2}{g}$$

$$\Rightarrow \quad \frac{g_2}{g} = -\frac{1}{4320}$$

Adding 1 to both sides, we get

$$\frac{g_2}{g}+1=-\frac{1}{4320}+1$$

$$\Rightarrow \qquad \frac{g+g_2}{g}=\frac{4319}{4320} \qquad \ldots (2)$$

Dividing (1) by (2) we have

$$\frac{g+g_1}{g+g_2}=\frac{4321}{4319}$$

$\therefore$ the accelerations due to gravity at the two places are in ratio 4321:4319.

Example 3.13: A pendulum of length l hangs against a wall inclined at an angle θ to horizontal. Prove that the time of one complete oscillation is $2\pi\sqrt{\frac{l}{g\sin\theta}}$.

Solution: Let the position of bob mass m, at any time t, be P and O be the point of suspension so that $OP = l$ and $\angle AOP = \alpha$, where OA is line of greatest slope through O.

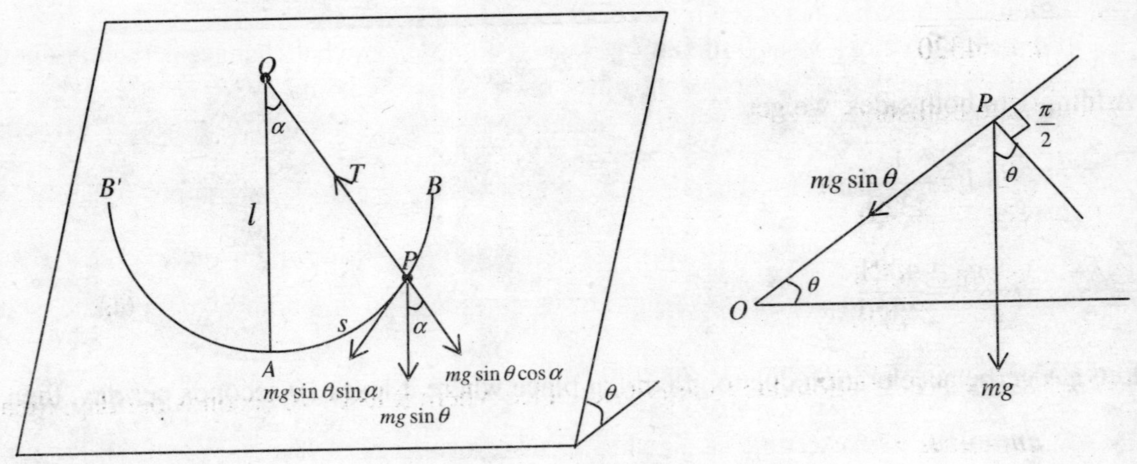

Figure 3.17

The component of weight of bob along plane is $mg\sin\theta$. The equation of motion of bob along tangent at P is

$$m\frac{d^2 s}{dt^2} = -mg \sin\theta \sin\alpha$$

$$\Rightarrow \quad \frac{d^2(l\alpha)}{dt^2} = -g\sin\theta\left(\alpha - \frac{\alpha^3}{3!} + \ldots\right) \qquad \left[\because \sin\alpha = \left(\alpha - \frac{\alpha^3}{3!} + \ldots\right)\right]$$

$$\Rightarrow \quad l\frac{d^2\alpha}{dt^2} = -g\sin\theta.\alpha \quad \text{to a first approximation.}$$

$$\Rightarrow \quad \frac{d^2\alpha}{dt^2} = -\frac{g\sin\theta}{l}\alpha$$

$$\Rightarrow \quad \frac{d^2\alpha}{dt^2} = -\omega^2\alpha \quad \text{where } \omega^2 = \frac{g\sin\theta}{l}$$

$$\Rightarrow \quad \frac{d^2\alpha}{dt^2} + \omega^2\alpha = 0$$

$\therefore$ motion of bob is simple harmonic and time of one oscillation $= \dfrac{2\pi}{\omega} = 2\pi\sqrt{\dfrac{l}{g\sin\theta}}$.

EXERCISE 3.4

1. A clock with a second's pendulum is gaining 2 minutes a day. Prove that length of pendulum must be increased by 0.28cm to make it correctly.
2. If a pendulum clock loses 9 minutes per week, find in *mm* what changes is required in the length of pendulum in order that the clock may keep correct time?
3. A simple pendulum has a period *T*. When the string is lengthened by a small fraction $\dfrac{1}{n}$ of its length, the period becomes T'. Show that $\dfrac{1}{n} = \dfrac{2(T'-T)}{T}$, approximately.
4. If l_1 be length of imperfectly adjusted second's pendulum which gains *n* seconds in one hour and l_2 length of one which loses *n* seconds in one hour at same place, show that true length of second's pendulum is $\dfrac{4l_1l_2}{l_1 + l_2 + 2\sqrt{l_1l_2}}$.
5. A pendulum whose length is *l* makes *m* oscillations in 24 hours. When the length is slightly altered it makes (*m*+*n*) oscillations in 24 hours. Show that diminution in length is $\dfrac{2nl}{m}$ nearly.

6. A clock loses five seconds a day, find the alternation required in length of its pendulum in order that it may keep correct time.
7. A second's pendulum which gains 10 seconds per day at one place loses 10 seconds per day at another. Compare the acceleration due to gravity at the two places.
8. The differential equation of simple pendulum is $\dfrac{d^2x}{dt^2} + \omega_0^2 x = F_0 \sin nt$, where ω_0 and F_0 are constants. If initially $x = 0$, $\dfrac{dx}{dt} = 0$, determine the motion when $\omega_0 = n$.

ANSWERS

1. 1.7 mm
6. $\left(\dfrac{1}{8640} \text{ of its length}\right)$
7. $\dfrac{4321}{4319}$
8. $x = \dfrac{F_0}{2n^2}(\sin nt - nt \cos nt)$

❑ ❑ ❑

Chapter 4

Linear Algebra

Learning Objectives

After reading this chapter, students will be able to

- define a matrix and explain its various types
- explain various operations that can be performed on matrices
- define a determinant and describe its properties
- find the rank of a matrix
- find the inverse of a matrix using Gauss Jordon method
- check the consistency of the system of linear equations
- find the solution of given system of linear equations
- find eigen values and eigen vectors of a given matrix
- check the linear independence and dependence of vectors
- define and prove Cayley-Hamilton theorem
- find the inverse of a matrix using Cayley-Hamilton theorem
- diagonalize a given matrix
- describe the similarity of matrices
- describe the various orthogonal transformations
- describe important theorem of hermitian and unitary matrices

4.1 INTRODUCTION

The various kinds of problems occurring in the field of science, engineering, and applied mathematics can be solved efficiently by expressing them in the form of matrices and determinants. Therefore, the working knowledge of matrices and determinants is very essential.

Elementary aspects of matrices and determinants are having already been covered in XII standard. In this chapter, we will discuss about some advanced aspects of matrices. But before we move on the journey to advanced topics of matrices and determinants, let us look back and review the terminology of the basic aspects of matrices and determinants.

4.2 RECAPITULATION OF FUNDAMENTAL CONCEPTS OF MATRICES

Matrix Defined: A set of $m{\cdot}n$ numbers (real or complex) arranged in a rectangular form having m rows (horizontal lines) and n columns (vertical lines), the numbers being enclosed by bracket [] or (), is called an $m{\times}n$ matrix (read as m by n).

An $m{\times}n$ matrix is written as

$$\begin{pmatrix} a_{11} & a_{12} & a_{13} & \cdots & a_{1n} \\ a_{21} & a_{22} & a_{23} & \cdots & a_{2n} \\ a_{31} & a_{32} & a_{33} & \cdots & a_{3n} \\ \vdots & \vdots & \vdots & \cdots & \vdots \\ a_{m1} & a_{m2} & a_{m3} & \cdots & a_{mn} \end{pmatrix}$$

Elements of the matrix are denoted by a_{ij} where i represent the row and j represent the column.

In terms of its elements, and the matrix is written as $\left[a_{ij}\right]_{(m,n)}$ or $\left[a_{ij}\right]_{m\times n}$.

The suffixes i and j in a_{ij} denotes i^{th} row and j^{th} column respectively to which the element a_{ij} belongs.

Notations: Matrices are denoted by capital letters of alphabet such as A, B, C, … and their respective elements are denoted by corresponding small letters of alphabet, viz. $a_{ij}, b_{ij}, c_{ij}, \ldots$

Types of Matrices

Row matrix - *Row matrix* is a matrix having a single row. This type of matrix is also known as a *row vector.*

Column matrix - *Column matrix* is a matrix having a single column. This type of matrix is also known as a *column vector.*

Square Matrix - A *square matrix* is a matrix with equal number of rows and columns. If square matrix has n rows and n columns, then it is called a square matrix of order n.

The elements whose row and column number is same constitute the principal diagonal.

Diagonal Matrix - *Diagonal matrix* is a square matrix whose all non-diagonal elements are zero.

Diagonal matrix of order n with diagonal elements $a_{11}, a_{22}, a_{33}, ..., a_{nn}$ can also be denoted by

$$A = \text{diag}\left(a_{11}, a_{22}, a_{33}, ..., a_{nn}\right)$$

Identity (Unit) Matrix - *Identity (Unit) matrix* is a square matrix whose elements on the principal diagonal elements are equal to 1 and rest all are zero.

Symmetric Matrix - A square matrix is a *symmetric matrix* if

$a_{ij} = a_{ji}$ for all values of i and j.

Skew-symmetric Matrix - A square matrix is a *skew-symmetric matrix* if

$$a_{ij} = -a_{ji}$$

for all values of i and j, except for $i = j$. All elements on the principal diagonal are zero.

Zero or NULL matrix - If every element of an $m \times n$ matrix is 0, the matrix is called a *zero matrix* or *NULL matrix* of order $m \times n$ and is denoted by $0_{m \times n}$ or by letter O.

Scalar Matrix - A diagonal matrix whose diagonal elements are equal is called a *scalar matrix.*

Triangular Matrix - If every element above or below the leading diagonal of a square matrix is 0, the matrix is called a *triangular matrix.*

If the 0 elements lie below the leading diagonal, the matrix is called *upper triangular matrix*, and if the 0 elements lie above the leading diagonal, the matrix is called *lower triangular matrix.*

Singular and Non-singular Matrices - A Square matrix A is said to be singular if $|A|=0$ and non-singular if $|A| \neq 0$.

Involuntary Matrix - A square matrix A is said to be involuntary if $A^2 = I$.

Orthogonal Matrix - A square matrix is called an orthogonal matrix if $AA' = A'A = I$.

Complex Matrix - A matrix is said to be a complex matrix if at least one element of the matrix is a complex number.

Conjugate of a Matrix - If in a complex matrix A the elements are replaced by their conjugate then matrix so obtained is called conjugate of a matrix and is denoted by $\overline{A}$.

Conjugate Transpose of a Matrix - A matrix A obtained by replacing the elements by their conjugates and then taking transpose of the new matrix is called conjugate transpose of the matrix and is denoted by $\left(\overline{A}\right)' = A^\theta = \left(\overline{A'}\right)$.

Hermitian Matrix - A square matrix A is said to be hermitian if $A^\theta = A$, i.e., $\overline{a}_{ij} = a_{ji}$ $\forall i, j$.

$\therefore$ When $i = j$, then $\overline{a}_{ii} = a_{ii} \Rightarrow a_{ii}$ is purely real.

$\therefore$ In hermitian matrix, all diagonal elements are purely real.

Skew Hermitian Matrix - A square matrix A is said to be skew hermitian if $A^\theta = -A$ i.e., $\overline{a}_{ij} = -a_{ij} \Rightarrow a_{ij}$ is purely imaginary

$\therefore$ In a skew hermitian matrix all diagonal elements are purely imaginary.

Unitary Matrix - A square matrix A is called an unitary matrix if $AA^\theta = A^\theta A = I$.

Equality of Matrices - Two matrices $A = [a_{ij}]$ and $B = [b_{ij}]$ are said to be equal, if

- They have same order.
- They have equal corresponding elements throughout.

The equality of matrices A and B are denoted by writing $A = B$.

Sub-Matrix - A matrix which is obtained from a given matrix by deleting any number of rows or columns is called a sub-matrix.

Echelon Form of a Matrix - A matrix is said to be in echelon form if

- Number of zeros before the first non-zero element in a row is more than the number of zeros before the first non-zero element in the previous row, and
- Rows whose all the elements are zero occur after the non-zero rows.

For example, following matrices

$$A = \begin{pmatrix} 5 & 1 & 4 \\ 0 & 8 & 4 \\ 0 & 0 & 3 \end{pmatrix} \text{ and } B = \begin{pmatrix} -2 & 1 & 0 \\ 0 & 0 & 3 \\ 0 & 0 & 0 \end{pmatrix} \text{ are in echelon form.}$$

Note: If a matrix A is in echelon form, then the rank of a matrix A is equal to the number of non-zero rows in the matrix. Therefore, to find the rank of a given matrix, reduce the matrix to echelon form by applying elementary transformations.

Characteristics of Matrices

1. A matrix is a rectangular array of numbers arranged in rows and columns.
2. The numbers in a matrix are called as entries or elements.
3. Matrices are denoted by capital letters $A, B, C, \ldots$ and the elements by small letters a, b, $c\ldots$
4. The order of a matrix is determined by the number of rows and columns.
5. A matrix can have any number of rows and columns. In other words, it can have any order. The general order is $m\times n$.
6. The first letter of $m\times n$ always denote rows and second denotes columns.
7. a_{ij} symbol is used for elements, where ith refers to row and jth refers to column.
8. a_{ij} or a_{ji} may or may not be equal.
9. $A_{m\times 1}$ matrix is called a column matrix or a column vector, while $A_{1\times n}$ is called a row matrix or a row vector.

Operations on Matrices

Addition of Matrices

Two matrices A and B can be added, if they are comparable, i.e., B has same number of rows and column as A. Their sum, denoted by $A+B$, is defined to be the matrix obtained by adding the corresponding elements of A and B.

Properties of Matrix Addition

Suppose A, B, C are three matrices of same order $m \times n$. Then the matrix addition has the following properties:

- **Commutative –** The addition of matrices is commutative, i.e., $A+B=B+A$.
- **Associative –** The addition of matrices is associative, i.e., $A+(B+C)=(A+B)+C$
- **Distributive law –** Here $m(A+B)=mA+mB$, m being an arbitrary scalar.
- **Existence of additive identity –** Let A be any $m \times n$ matrix, and O the $m \times n$ null matrix, then we have $A + O = O + A = A$, i.e., the null matrix is the identity for the matrix addition
- **Existence of additive inverse –** The matrix $(-A)$ is the additive inverse of A, because $(-A)+A=A+(-A)=O$. Thus, for any matrix A, there exist an unique additive inverse $(-A)$.
- **Cancellation law –** If $A + B = A + C$, then $B = C$. This is said to be left cancellation law. Similarly, if $B + A = C + A$, then $B = C$. This is said to be right cancellation law.

Subtraction of Matrices

A matrix A can be subtracted from matrix B, if they are comparable, *i.e.*, B has the same number of rows and columns as A. Their subtraction, denoted by A – B, is defined to be matrix obtained by subtracting the corresponding element of B from A.

Note that Subtraction of matrices does not follow the commutative law and associative law. It only follows the distributive and cancellation laws.

Multiplication of Matrix by a Scalar – Let $A = [a_{ij}]$ be an $m \times n$ matrix and let k be any real number (called scalar). Then, the product of k and A denoted by kA is defined to be the $m \times n$ matrix whose (i, j)th element is ka_{ij}, i.e., each element of the given matrix is multiplied by the given scalar.

Multiplication of Matrices – The product $A \times B$ or simply as AB, of two matrices $A_{m \times n}$ and $B_{p \times q}$ is defined (feasible) only when the number of columns of matrix A equals number of row of matrix B, i.e., $n = p$.

If product defined, the order of resultant matrix, say C, will be $m \times q$, whose element appearing in ith row and jth column, i.e., c_{ij} is computed as

c_{ij} = sum of the products of the entries in the ith row of matrix A with the corresponding entries in the jth column of matrix B

$$= a_{i0} \times b_{0j} + a_{i1} \times b_{1j} + a_{i2} \times b_{2j} + \ldots + a_{i(n-1)} \times b_{(n-1)j} = \sum_{k=0}^{n-1} a_{ik} b_{kj}$$

Transpose of a Matrix - The matrix obtained by interchanging the rows and columns of a given matrix A is called the transpose of the matrix A and is denoted by A' or A^T

Properties of Transpose:

1. The transpose of the transpose of a matrix is the original matrix. i.e., $(A')' = A$.
2. The transpose of the sum of matrix is the sum of their individual transpose, i.e., $(A+B)' = A' + B'$, A and B being comparable
3. If k is any scalar, then $(kA)' = kA'$.
4. The transpose of the product of matrices is equal to the product of the transposes of individual matrices taken in the reverse order, i.e., $(AB)' = B'A'$.

Trace of a Matrix

Trace of a square matrix A is defined as the sum of the elements in its principal diagonal.

Determinant of a Matrix

With every matrix A, there is an associated quantity called *determinant* of A, and is denoted by *det*(A) or $|A|$.

Minors and Co-Factors of Determinants

The *minor* of any element of a given determinant is a determinant obtained by deleting the row and the column to which the element belong, from the original determinant.

If $|A| = \begin{vmatrix} a_{11} & a_{12} & a_{13} \\ a_{21} & a_{22} & a_{23} \\ a_{31} & a_{32} & a_{33} \end{vmatrix}$ then Minor of $a_{11} = \begin{vmatrix} a_{22} & a_{23} \\ a_{32} & a_{33} \end{vmatrix}$

The *co-factor* of any element of the determinant is equal to the corresponding minor with a proper sign attached to it. The general rule is

$$\text{Co-factor of } a_{ij} = (-1)^{i+j}\left|\Delta_{ij}\right|$$

where Δ_{ij} is the minor of a_{ij} and is denoted by A_{ij}.

$$\text{Co-factor of } a_{11} = A_{11} = (-1)^{1+1}\begin{vmatrix} a_{22} & a_{23} \\ a_{32} & a_{33} \end{vmatrix} = \begin{vmatrix} a_{22} & a_{23} \\ a_{32} & a_{33} \end{vmatrix}$$

Some Important Properties of Determinants

- If the rows are changed to columns and columns into rows, the determinant remains unaltered.
- If any two rows of the determinant are interchanged, the determinant retains its value but the sign changes.
- A determinant vanishes if any two rows or columns are identical.
- If each element of a row is multiplied by the same factor, the whole determinant is multiplied by that factor.
- If an element consists of m terms, the determinant can be expressed as the sum of m determinants.
- If to each element of a row, an equal multiple of the corresponding element of one or more rows is added, the determinant remains unaltered.

Adjoint of a Square Matrix

The adjoint of a square matrix is the transpose of the matrix obtained by replacing each element of A by its cofactors in $|A|$, i.e.,

$$A(Adj\ A) = (Adj\ A)A = |A|\,I_n, \ n \text{ being the order of the square matrix } A.$$

Inverse of a Square Matrix

(*a*) Let A be a square matrix of order n. If there exists another matrix B of the same order such that $AB = BA = I$ then matrix A is said to be invertible and B is called inverse of A. Inverse of A is denoted by A^{-1}. Thus $B = A^{-1}$ and $AA^{-1} = A^{-1}A = I$.

As $A(Adj\ A) = |A|\,I_n$

$$\therefore \quad A^{-1} = \frac{Adj A}{|A|}$$

(*b*) The inverse of a square matrix, if it exists is unique.

(*c*) The necessary and sufficient condition for a square matrix A to possess an inverse is that $|A| \neq 0$, i.e., A is non-singular.

(*d*) If A is invertible, then so is A^{-1} and $(A^{-1})^{-1} = A$.

4.3 ELEMENTARY TRANSFORMATIONS

Following operations on a matrix are called elementary transformation (or E–operation):

- Interchange of two rows or columns:
  - Interchange of *i*th and *j*th rows is denoted by $R_i \leftrightarrow R_j$.
  - Interchange of *i*th and *j*th columns is denoted by $C_i \leftrightarrow C_j$.
- Multiplication of each element of a row or columns by a non zero number k.
  - Multiplication of *i*th row by k is denoted by $R_i \to kR_i$.
  - Multiplication of *j*th column by k is denoted by $C_j \to kC_j$.
- Addition of k times the elements of a row (column) to the corresponding elements of another row (or column), $k \neq 0$.
  - Addition of k times the *j*th row to the *i*th row is denoted by $R_i \to R_i + kR_j$.
  - Addition of k time the *j*th column to the *i*th column is denoted by $C_i \to C_i + kC_j$.

If a matrix B is obtained from a matrix A by one or more E-operations, then B is said to be equivalent to A.

Two equivalent matrices A and B are written as $A \sim B$.

Procedure to Reduce Square Matrix to a Unit Matrix *I* by Elementary Transformations

For convenience, we can consider a matrix A of order 4×4.

1. If in the first column, the principle element is not one but one is present some where else in the first column then first of all make one as principle element.
2. Operate R_1 on R_2, R_3, R_4 to make elements of C_1 all zero except first element.
3. Then operate R_2 on R_3, R_4 to make elements of C_2 all zero except first and second elements. Similarly, operate R_5 on R_4 to make elements of C_3 all zero except first, second, and third.
4. Reduce each diagonal element to element one then reverse process starts.

5. Operate R_4 on R_1, R_2, R_3 to make all elements of C_3 zero except last but one. Similarly operate R_2 on R_1 to make all element of C_2 zero except last but second and the matrix is reduced to unit matrix.

Note: We can apply the above rule to any square matrix of any order.

ILLUSTRATIVE EXAMPLES 4.1

Example 4.1: Reduce the matrix $\begin{pmatrix} 1 & 2 & 3 \\ 2 & 5 & 7 \\ 3 & 1 & 2 \end{pmatrix}$ to upper triangular form:

Solution: (Upper triangular matrix) If in square matrix, all the elements below the principal diagonal are zero, the matrix is called *upper triangular*.

Given matrix $\begin{pmatrix} 1 & 2 & 3 \\ 2 & 5 & 7 \\ 3 & 1 & 2 \end{pmatrix}$ $\quad R_2 \to R_2 - 2R_1, R_3 \to R_3 - 3R_1$

$$\begin{pmatrix} 1 & 2 & 3 \\ 0 & 1 & 1 \\ 0 & -5 & -7 \end{pmatrix} \quad R_3 \to R_3 + 5R_2$$

$$\begin{pmatrix} 1 & 2 & 3 \\ 0 & 1 & 1 \\ 0 & 0 & -2 \end{pmatrix}$$

which is the upper triangular form of the given matrix.

Example 4.2: Transform $\begin{pmatrix} 1 & 3 & 3 \\ 2 & 4 & 10 \\ 3 & 8 & 4 \end{pmatrix}$ into a unit matrix.

Solution: Given matrix $\begin{pmatrix} 1 & 3 & 3 \\ 2 & 4 & 10 \\ 3 & 8 & 4 \end{pmatrix}$ $\quad R_2 \to R_2 - 2R_1, R_3 \to R_3 - 3R_1$

$$\begin{pmatrix} 1 & 3 & 3 \\ 0 & -2 & 4 \\ 0 & -1 & -5 \end{pmatrix} \qquad R_2 \to -\frac{1}{2}R_2$$

$$\begin{pmatrix} 1 & 3 & 3 \\ 0 & 1 & -2 \\ 0 & -1 & -5 \end{pmatrix} \qquad R_3 \to R_3 + R_2$$

$$\begin{pmatrix} 1 & 3 & 3 \\ 0 & 1 & -2 \\ 0 & 0 & -7 \end{pmatrix} \qquad R_1 \to R_1 - 3R_2$$

$$\begin{pmatrix} 1 & 0 & 9 \\ 0 & 1 & -2 \\ 0 & 0 & -7 \end{pmatrix} \qquad R_3 \to -\frac{1}{7}R_3$$

$$\begin{pmatrix} 1 & 0 & 9 \\ 0 & 1 & -2 \\ 0 & 0 & 1 \end{pmatrix} \qquad R_1 \to R_1 - 9R_3, R_2 \to R_2 + 2R_3$$

$$\begin{pmatrix} 1 & 0 & 0 \\ 0 & 1 & 0 \\ 0 & 0 & 1 \end{pmatrix}$$

which is unit matrix.

4.4 RANK OF MATRIX

The number r is called the rank of matrix A if

- every minor of order $(r+1)$ of A is zero.
- there exists at least one minor of order r which is non-zero.

Note:

1. The rank of matrix A is denoted by $\rho(A)$.
2. The rank of a null matrix is zero.
3. If $A = \left(a_{ij}\right)_{m\times n}$, then $\rho(A) \le \min(m,n)$

4. $\rho(I_n) = n$, here I_n is an identity matrix of order n.
5. $\rho(O) = 0$ here O is a null matrix.

Working Rule of Rank of Matrix

Method I:

1. Start with the highest order minor of A. Let their order be r. If any one of them is non-zero then $\rho(A) = r$.
2. If all of them are zero, start with minors of next lower order $(r-1)$ and so on.
3. If you get a non-zero minor. The order of that minor is the rank of A.

Method II:

Reduce the matrix to Echelon/triangular matrix by elementary transformations, then number of non-zero rows of Echelon/triangular matrix is equal to rank of the matrix.

Method III:

Reduce the matrix to the normal form $\begin{pmatrix} I_r & O \\ O & O \end{pmatrix}$ by performing by elementary transformations (row & column both), then r is the rank of the matrix.

Properties of Rank of a matrix:

1. Elementary transformations of a matrix do not change the rank of a matrix.
2. $\rho(A') = \rho(A)$; $\rho(A^\theta) = \rho(A)$
3. $\rho(A) =$ number of non zero rows in upper triangular form of the matrix A.

Normal Form of a Matrix

Any non-zero matrix $A_{m \times n}$ can be reduced to any one of the following forms by performing elementary (row, column or both) transformations:

(i) I_r (ii) $(I_r \;\; 0)$ (iii) $\begin{pmatrix} I_r \\ 0 \end{pmatrix}$ (iv) $\begin{pmatrix} I_r & 0 \\ 0 & 0 \end{pmatrix}$

All above forms are known as *normal forms* of the matrix.

Note:

(*i*) The number r is the rank of the matrix A.

(*ii*) The form $\begin{pmatrix} I_r & 0 \\ 0 & 0 \end{pmatrix}$ is called *first canonical form* of A.

(*iii*) The number r so obtained is the rank of A, because rth order minor $|I_r| = 1 \neq 0$ and each $(r+1)$th order minor = 0.

ILLUSTRATIVE EXAMPLES 4.2

Example 4.3: Find the rank of the following matrices

(*a*) $A = \begin{pmatrix} 1 & 4 & 5 \\ 2 & 6 & 8 \\ 3 & 7 & 22 \end{pmatrix}$ (*b*) $\begin{pmatrix} 1 & 3 & 4 & 5 \\ 1 & 2 & 6 & 7 \\ 1 & 5 & 0 & 10 \end{pmatrix}$ (*c*) diag.$(-1\ \ 0\ \ 1\ \ 0\ \ 0\ \ 4)$.

Solution:

(*a*) Let $A = \begin{pmatrix} 1 & 4 & 5 \\ 2 & 6 & 8 \\ 3 & 7 & 22 \end{pmatrix}$

Since A is of order 3×3

$\therefore \quad \rho(A) \leq 3$

Let us reduce the matrix to triangular form:

$$A \sim \begin{pmatrix} 1 & 4 & 5 \\ 0 & -2 & -2 \\ 0 & -5 & 7 \end{pmatrix} \qquad R_2 \to R_2 - 2R_1, R_3 \to R_3 - 3R_1$$

$$\sim \begin{pmatrix} 1 & 4 & 5 \\ 0 & 1 & 1 \\ 0 & -5 & 7 \end{pmatrix} \qquad R_2 \to -\frac{1}{2}R_2$$

$$\sim \begin{pmatrix} 1 & 4 & 5 \\ 0 & 1 & 1 \\ 0 & 0 & 12 \end{pmatrix} \qquad R_3 \to R_3 + 5R_2$$

Now $|A| = \begin{vmatrix} 1 & 4 & 5 \\ 0 & 1 & 1 \\ 0 & 0 & 12 \end{vmatrix} = 12 \neq 0$ Expanding on first column

$\therefore \quad \rho(A) = 3$

(*b*) Let $A = \begin{pmatrix} 1 & 3 & 4 & 5 \\ 1 & 2 & 6 & 7 \\ 1 & 5 & 0 & 10 \end{pmatrix}$ A is of order 3×4

Since A is of order 3×4

$\therefore \quad \rho(A) \leq 3$

Let us reduce the matrix to triangular form:

$$A \sim \begin{pmatrix} 1 & 3 & 4 & 5 \\ 0 & -1 & 2 & 2 \\ 0 & 2 & -4 & 5 \end{pmatrix} \qquad R_2 \to R_2 - R_1, R_3 \to R_3 - R_1$$

$$\sim \begin{pmatrix} 1 & 3 & 4 & 5 \\ 0 & -1 & 2 & 2 \\ 0 & 0 & 0 & 9 \end{pmatrix} \qquad R_3 \to R_3 + 2R_2$$

Minor of order 3, i.e., $\begin{vmatrix} 3 & 4 & 5 \\ -1 & 2 & 2 \\ 0 & 0 & 9 \end{vmatrix}$ Expanding on the last row

$$= 9(6+4) = 90 \neq 0$$

$\therefore \quad \rho(A) = 3$

(*c*) Let $A = \text{diag.}(-1 \;\; 0 \;\; 1 \;\; 0 \;\; 0 \;\; 4) = \begin{pmatrix} -1 & 0 & 0 & 0 & 0 & 0 \\ 0 & 0 & 0 & 0 & 0 & 0 \\ 0 & 0 & 1 & 0 & 0 & 0 \\ 0 & 0 & 0 & 0 & 0 & 0 \\ 0 & 0 & 0 & 0 & 0 & 0 \\ 0 & 0 & 0 & 0 & 0 & 4 \end{pmatrix}$

Since it is a square matrix of order 6×6

$\therefore \quad \rho(A)\le 6$

Also, it is a diagonal matrix which is a special case of triangular matrix

$\therefore \quad \rho(A) =$ Number of non-zero rows of triangular matrix $=3$

$\therefore \quad \rho(A)=3$.

Example 4.4: Find the rank of the following matrices:

$(a) \begin{pmatrix} 3 & -1 & 1 \\ -15 & 6 & -5 \\ 5 & -2 & 2 \end{pmatrix} \qquad (b) \begin{pmatrix} 2 & 1 & -1 \\ 0 & 3 & -2 \\ 2 & 4 & -3 \end{pmatrix}$

Solution: (*a*) Let $A=\begin{pmatrix} 3 & -1 & 1 \\ -15 & 6 & -5 \\ 5 & -2 & 2 \end{pmatrix}$

$$\Rightarrow \quad |A| = \begin{vmatrix} 3 & -1 & 1 \\ -15 & 6 & -5 \\ 5 & -2 & 2 \end{vmatrix} \qquad \text{Expanding on first row}$$

$$=3(12-10)+1(-30+25)+1(30-30)=6-5+0=1$$

$\therefore \quad |A|\ne 0$, *i.e*, A is non-singular

$\Rightarrow \quad \rho(A)=3$

(*b*) Let $A=\begin{pmatrix} 2 & 1 & -1 \\ 0 & 3 & -2 \\ 2 & 4 & -3 \end{pmatrix}$

$$\Rightarrow |A| = \begin{vmatrix} 2 & 1 & -1 \\ 0 & 3 & -2 \\ 2 & 4 & -3 \end{vmatrix} \qquad \text{Expanding on first row}$$

$$=2(-9+8)-1(0+4)-1(0-6)=-2-4+6=0$$

$\therefore \ |A|=0$

Every minor of order 3 is zero.

Now consider $\begin{vmatrix} 2 & 1 \\ 0 & 3 \end{vmatrix} = 6 \neq 0$

$\Rightarrow \quad \rho(A) = 2$

Example 4.5: Reduce the matrix $A = \begin{pmatrix} 2 & 2 & 2 \\ 1 & 2 & 1 \\ 3 & 4 & 3 \end{pmatrix}$ to normal form and hence find its rank.

Solution: Given $A = \begin{pmatrix} 2 & 2 & 2 \\ 1 & 2 & 1 \\ 3 & 4 & 3 \end{pmatrix} \qquad R_1 \to \frac{1}{2} R_1$

$$\sim \begin{pmatrix} 1 & 1 & 1 \\ 1 & 2 & 1 \\ 3 & 4 & 3 \end{pmatrix} \qquad R_2 \to R_2 - R_1, R_3 \to R_3 - 3R_1$$

$$\sim \begin{pmatrix} 1 & 1 & 1 \\ 0 & 1 & 0 \\ 0 & 1 & 0 \end{pmatrix} \qquad R_3 \to R_3 - R_2$$

$$\sim \begin{pmatrix} 1 & 1 & 1 \\ 0 & 1 & 0 \\ 0 & 0 & 0 \end{pmatrix} \qquad R_1 \to R_1 - R_2$$

$$\sim \begin{pmatrix} 1 & 0 & 1 \\ 0 & 1 & 0 \\ 0 & 0 & 0 \end{pmatrix} \qquad C_3 \to C_3 - C_1$$

$$\sim \left(\begin{array}{cc:c} 1 & 0 & 0 \\ 0 & 1 & 0 \\ \hdashline 0 & 0 & 0 \end{array}\right) = \begin{pmatrix} I_{2\times2} & O_{2\times1} \\ O_{1\times2} & O_{1\times1} \end{pmatrix}$$

which is the required normal form and $\rho(A) = 2$.

Example 4.6: Show that no skew symmetric matrix is of rank 1.

Solution: Let us prove the result 4×4 skew matrix. This method of proof can be used for a skew symmetric matrix of any order

$$\text{Let } A=\begin{pmatrix} 0 & h & g & l \\ -h & 0 & f & m \\ -g & -f & 0 & n \\ -l & -m & -n & 0 \end{pmatrix}$$

If f, g, h, l, m, n are all zero then $\rho(A)=0$. If at least one of these elements say g is non zero, then at least one second order $\begin{vmatrix} 0 & g \\ -g & 0 \end{vmatrix}$ of matrix A is $g^2 \neq 0$

$$\Rightarrow \quad \rho(A) \geq 2$$

Hence $\rho(A) \neq 1$

Example 4.7: Find non-singular matrices P and Q such that PAQ is in the normal form for the matrix $A=\begin{pmatrix} 1 & -1 & -1 \\ 1 & 1 & 1 \\ 3 & 1 & 1 \end{pmatrix}$. Also find A^{-1}, if exists.

Solution: Consider $A = IAI$

$$\begin{pmatrix} 1 & -1 & -1 \\ 1 & 1 & 1 \\ 3 & 1 & 1 \end{pmatrix} = \begin{pmatrix} 1 & 0 & 0 \\ 0 & 1 & 0 \\ 0 & 0 & 1 \end{pmatrix} A \begin{pmatrix} 1 & 0 & 0 \\ 0 & 1 & 0 \\ 0 & 0 & 1 \end{pmatrix} \qquad R_2 \to R_2 - R_1, R_3 \to R_3 - 3R_1$$

Note that all row operations on the LHS are also performed on the pre-factor of RHS and all column operations on the LHS are also performed on the post-factor of RHS.

$$\begin{pmatrix} 1 & -1 & -1 \\ 0 & 2 & 2 \\ 0 & 4 & 4 \end{pmatrix} = \begin{pmatrix} 1 & 0 & 0 \\ -1 & 1 & 0 \\ -3 & 0 & 1 \end{pmatrix} A \begin{pmatrix} 1 & 0 & 0 \\ 0 & 1 & 0 \\ 0 & 0 & 1 \end{pmatrix} \qquad C_2 \to C_2 + C_1, C_2 \to C_3 + C_1$$

$$\begin{pmatrix}1&0&0\\0&2&2\\0&4&4\end{pmatrix}=\begin{pmatrix}1&0&0\\-1&1&0\\-3&0&1\end{pmatrix}A\begin{pmatrix}1&1&1\\0&1&0\\0&0&1\end{pmatrix}\qquad R_3\to R_3-2R_2$$

$$\begin{pmatrix}1&0&0\\0&2&2\\0&0&0\end{pmatrix}=\begin{pmatrix}1&0&0\\-1&1&0\\-1&-2&1\end{pmatrix}A\begin{pmatrix}1&1&1\\0&1&0\\0&0&1\end{pmatrix}\qquad C_3\to C_3-C_2$$

$$\begin{pmatrix}1&0&0\\0&2&0\\0&0&0\end{pmatrix}=\begin{pmatrix}1&0&0\\-1&1&0\\-1&-2&1\end{pmatrix}A\begin{pmatrix}1&1&0\\0&1&-1\\0&0&1\end{pmatrix}\qquad R_2\to\frac{1}{2}R_2$$

$$\left(\begin{array}{cc:c}1&0&0\\0&1&0\\ \hdashline 0&0&0\end{array}\right)=\begin{pmatrix}1&0&0\\-\frac{1}{2}&\frac{1}{2}&0\\-1&-2&1\end{pmatrix}A\begin{pmatrix}1&1&0\\0&1&-1\\0&0&1\end{pmatrix}$$

$$\Rightarrow\quad\begin{pmatrix}I_2&O\\O&O\end{pmatrix}=PAQ,\text{ where }P=\begin{pmatrix}1&0&0\\-\frac{1}{2}&\frac{1}{2}&0\\-1&-2&1\end{pmatrix}\text{ and }Q=\begin{pmatrix}1&1&0\\0&1&-1\\0&0&1\end{pmatrix}$$

As $\rho(A)=2$

$\therefore$ A is a singular matrix, and hence A^{-1} does not exist.

Example 4.8: Find a so that the rank of the matrix $A=\begin{pmatrix}a&0&1\\1&2&a\\1&2&3\end{pmatrix}$ is less than 3. Also find the rank for these values of a.

Solution: Given $A=\begin{pmatrix}a&0&1\\1&2&a\\1&2&3\end{pmatrix}$

$\because\quad \rho(A)<3$

$\therefore$ A is a singular matrix.

$\Rightarrow$ $|A| = 0$

$$\Rightarrow \begin{vmatrix} a & 0 & 1 \\ 1 & 2 & a \\ 1 & 2 & 3 \end{vmatrix} = 0$$

Expanding the determinant on first row, we get

$$a\begin{vmatrix} 2 & a \\ 2 & 3 \end{vmatrix} - 0\begin{vmatrix} 1 & a \\ 2 & 3 \end{vmatrix} + 1\begin{vmatrix} 1 & 2 \\ 1 & 2 \end{vmatrix} = 0$$

$\Rightarrow$ $a(6-2a) + 0(3-2a) + 1(2-2) = 0$

$\Rightarrow$ $a(3-a) = 0$

$\Rightarrow$ $a = 0, 3$

Case I: When $a = 0$, $A = \begin{pmatrix} 0 & 0 & 1 \\ 1 & 2 & 0 \\ 1 & 2 & 3 \end{pmatrix}$

Now $\begin{vmatrix} 2 & 0 \\ 2 & 3 \end{vmatrix} = 6 - 0 = 6 \neq 0$

$\therefore$ $\exists$ a minor of order 2 of A which does not vanish.

$\therefore$ $\rho(A) = 2$

Case II: When $a = 3$, $A = \begin{pmatrix} 3 & 0 & 1 \\ 1 & 2 & 3 \\ 1 & 2 & 3 \end{pmatrix}$

Now $\begin{vmatrix} 3 & 0 \\ 1 & 2 \end{vmatrix} = 6 - 0 = 6 \neq 0$

$\therefore$ $\exists$ a minor of order 2 of A which does not vanish.

$\therefore$ $\rho(A) = 2$

Example 4.9: Under what conditions if any, will the rank of the matrix

$$\begin{pmatrix} 1 & 0 & 0 \\ 0 & h-2 & 2 \\ 0 & k-1 & h+2 \\ 0 & 0 & 3 \end{pmatrix}$$

be less than 3 and what that rank will be?

Solution: Let $A = \begin{pmatrix} 1 & 0 & 0 \\ 0 & h-2 & 2 \\ 0 & k-1 & h+2 \\ 0 & 0 & 3 \end{pmatrix}$

Taking Minor $\begin{vmatrix} 1 & 0 & 0 \\ 0 & h-2 & 2 \\ 0 & 0 & 3 \end{vmatrix} = 0$

$\Rightarrow \quad 3(h-2) = 0$ Expanding on the last column.

$\Rightarrow \quad h = 2$

Now taking minor $\begin{vmatrix} 1 & 0 & 0 \\ 0 & h-2 & 2 \\ 0 & k-1 & h+2 \end{vmatrix} = 0$

$\Rightarrow \quad \begin{vmatrix} 1 & 0 & 0 \\ 0 & 0 & 2 \\ 0 & k-1 & 4 \end{vmatrix} = 0$ Substituting the value of h.

$\Rightarrow \quad 2(k-1) = 0$ Now, expanding on the first column.

$\Rightarrow \quad k = 1$

$\therefore$ the rank will be less than 3 iff $h = 2, k = 1$

Now if $h = 2, k = 1,$ then $A = \begin{pmatrix} 1 & 0 & 0 \\ 0 & 0 & 2 \\ 0 & 0 & 4 \\ 0 & 0 & 3 \end{pmatrix}$

Now $\begin{vmatrix} 1 & 0 \\ 0 & 3 \end{vmatrix} \neq 0 \quad \Rightarrow \quad \rho(A) = 2$

EXERCISE 4.1

1. (*a*) Reduce matrix $A = \begin{pmatrix} 1 & 2 & 3 \\ 2 & 5 & 7 \\ 3 & 1 & 2 \end{pmatrix}$ to upper triangular form.

   (*b*) Reduce matrix $B = \begin{pmatrix} 1 & 3 & 3 \\ 2 & 4 & 10 \\ 3 & 8 & 4 \end{pmatrix}$ to a unit matrix.

2. Find the rank of the following matrices:

   (*a*) $\begin{pmatrix} 1 & 4 & 5 \\ 2 & 6 & 8 \\ 3 & 7 & 22 \end{pmatrix}$ [PTU May 2004] (*b*) $\begin{pmatrix} 1 & 3 & 4 & 5 \\ 1 & 2 & 6 & 7 \\ 1 & 5 & 0 & 10 \end{pmatrix}$

   (*c*) $\begin{pmatrix} 0 & 1 & -3 & -1 \\ 1 & 0 & 1 & 1 \\ 3 & 1 & 0 & 2 \\ 1 & 1 & -2 & 0 \end{pmatrix}$ [PTU Dec 2012] (*d*) $\begin{pmatrix} 1 & 1 & -1 \\ 2 & -3 & 4 \\ 3 & -2 & 3 \end{pmatrix}$ [PTU May 2009]

3. Find the rank of the following matrices:

   (*a*) $\begin{pmatrix} 2 & 3 & -1 & -1 \\ 1 & -1 & -2 & -4 \\ 3 & 1 & 3 & -2 \\ 6 & 3 & 0 & -7 \end{pmatrix}$ [PTU May 2007] (*b*) $\begin{pmatrix} 2 & 2 & 2 \\ 1 & 2 & 1 \\ 3 & 4 & 3 \end{pmatrix}$ [PTU May 2002]

   (*c*) $\begin{pmatrix} 1 & 0 & 2 & 1 \\ 0 & 1 & -2 & 1 \\ 1 & -1 & 4 & 0 \\ -2 & 2 & 8 & 0 \end{pmatrix}$ [PTU May 2004] (*d*) $\begin{pmatrix} 1 & 2 & 1 \\ -1 & 0 & 2 \\ 2 & 1 & 3 \end{pmatrix}$ [PTU May 2003]

4. Find non-singular matrices P and Q such that PAQ is in the normal form for given matrix $A = \begin{pmatrix} 1 & -1 & -1 \\ 1 & 1 & 1 \\ 3 & 1 & 1 \end{pmatrix}$. Also find A^{-1}, if it exists. [PTU Dec 2003]

5. Reduce the following matrices to normal form and hence find their ranks.

(*a*) $\begin{pmatrix} 2 & 2 & 2 \\ 1 & 2 & 1 \\ 3 & 4 & 3 \end{pmatrix}$ (*b*) $\begin{pmatrix} 3 & -1 & 2 \\ -6 & 2 & 4 \\ -3 & 1 & 2 \end{pmatrix}$ (*c*) $\begin{pmatrix} 1 & 2 & 3 & 4 \\ 2 & 1 & 4 & 3 \\ 3 & 0 & 5 & -10 \end{pmatrix}$

6. If $A = \begin{pmatrix} 3 & -3 & 4 \\ 2 & -3 & 4 \\ 0 & -1 & 1 \end{pmatrix}$, find two non-singular matrix P and Q such that $PAQ = 1$. Hence find A^{-1}.

7. Given a matrix $A = \begin{pmatrix} 1 & 1 & 2 \\ 1 & 2 & 3 \\ 0 & -1 & -1 \end{pmatrix}$, ffind two non-singular matrix P and Q such that PAQ is in the normal form. Does A^{-1} exists? If yes, find it.

8. Reduce the matrix $\begin{pmatrix} 0 & 1 & -3 & -1 \\ 1 & 0 & 1 & 1 \\ 3 & 1 & 0 & 2 \\ 1 & 1 & -2 & 0 \end{pmatrix}$ to normal form and hence find its rank.

[PTU May 2012]

ANSWERS

1. $\begin{pmatrix} 1 & 2 & 3 \\ 0 & 1 & 1 \\ 0 & 0 & -2 \end{pmatrix}$ 2. (*a*) 3; (*b*) 3; (*c*) 2; (*d*) 2 3. (*a*) 3; (*b*) 2; (*c*) 3; (*d*) 3

4. $P = \begin{pmatrix} 1 & 0 & 0 \\ \frac{-1}{2} & \frac{1}{2} & 0 \\ -1 & -2 & 1 \end{pmatrix}$, $Q = \begin{pmatrix} 1 & 1 & 0 \\ 0 & 1 & -1 \\ 0 & 0 & 1 \end{pmatrix}$, as $\rho(A) = 2$, therefore, matrix A is singular and hence A^{-1} does not exist.

5. (*a*) 2 (*b*) 2 (*c*) 3

6. $P = \begin{pmatrix} 1 & -1 & 0 \\ 0 & 0 & 1 \\ -2 & 3 & -3 \end{pmatrix}$, $Q = \begin{pmatrix} 1 & 0 & 0 \\ 0 & -1 & 1 \\ 0 & 0 & 1 \end{pmatrix}$, $A^{-1} = \begin{pmatrix} 1 & -1 & 0 \\ -2 & 3 & -4 \\ -2 & 3 & -3 \end{pmatrix}$

7. $P=\begin{pmatrix}1 & 0 & 0\\ -1 & 1 & 0\\ -1 & 1 & 1\end{pmatrix}$, $Q=\begin{pmatrix}1 & -1 & -1\\ 0 & 1 & -1\\ 0 & 0 & 1\end{pmatrix}$, A^{-1} does not exist as $\rho(A)=2$.

8. 2

4.5 INVERSE OF A MATRIX USING GAUSS-JORDAN METHOD

Let A be any square non-singular matrix, then

$$A = IA$$

Apply suitable E-row transformation to A on the left hand side so that A is reduced to I. Simultaneously, apply the same E-row transformations to I on the right hand side. Let I reduce to B, so that

$$I = BA$$

Post-multiplying by A^{-1}, we get

$$IA^{-1} = BAA^{-1}$$

$$\Rightarrow \quad A^{-1} = B\left(AA^{-1}\right) \qquad [\because IA = A]$$

$$\Rightarrow \quad A^{-1} = BI \qquad \left[\because AA^{-1} = I\right]$$

$$\Rightarrow \quad B = A^{-1}$$

Elementary E-row operations which reduce a square matrix A to the unit (identity) matrix, when applied to the unit matrix, gives the inverse matrix A^{-1}.

ILLUSTRATIVE EXAMPLES 4.3

Example 4.10: Using Gauss–Jordon method, find the inverse of matrix $\begin{pmatrix}8 & 4 & 3\\ 2 & 1 & 1\\ 1 & 2 & 1\end{pmatrix}$.

Solution: Let $A=\begin{pmatrix}8 & 4 & 3\\ 2 & 1 & 1\\ 1 & 2 & 1\end{pmatrix}$

Consider $A = IA$

$$\Rightarrow \begin{pmatrix} 8 & 4 & 3 \\ 2 & 1 & 1 \\ 1 & 2 & 1 \end{pmatrix} = \begin{pmatrix} 1 & 0 & 0 \\ 0 & 1 & 0 \\ 0 & 0 & 1 \end{pmatrix} A \qquad R_3 \leftrightarrow R_1$$

$$\Rightarrow \begin{pmatrix} 1 & 2 & 1 \\ 2 & 1 & 1 \\ 8 & 4 & 3 \end{pmatrix} = \begin{pmatrix} 0 & 0 & 1 \\ 0 & 1 & 0 \\ 1 & 0 & 0 \end{pmatrix} A \qquad R_2 \to R_2 - 2R_1, R_3 \to R_3 - 8R_1$$

$$\Rightarrow \begin{pmatrix} 1 & 2 & 1 \\ 0 & -3 & -1 \\ 0 & -12 & -5 \end{pmatrix} = \begin{pmatrix} 0 & 0 & 1 \\ 0 & 1 & -2 \\ 1 & 0 & -8 \end{pmatrix} A \qquad R_3 \to R_3 - 4R_2$$

$$\Rightarrow \begin{pmatrix} 1 & 2 & 1 \\ 0 & -3 & -1 \\ 0 & 0 & -1 \end{pmatrix} = \begin{pmatrix} 0 & 0 & 1 \\ 0 & 1 & -2 \\ 1 & -4 & 0 \end{pmatrix} A \qquad R_1 \to R_1 + R_3, R_2 \to R_2 - R_3$$

$$\Rightarrow \begin{pmatrix} 1 & 2 & 0 \\ 0 & -3 & 0 \\ 0 & 0 & -1 \end{pmatrix} = \begin{pmatrix} 1 & -4 & 1 \\ -1 & 5 & -2 \\ 1 & -4 & 0 \end{pmatrix} A \qquad R_2 \to -\frac{1}{3}R_2, R_3 \to -R_3$$

$$\Rightarrow \begin{pmatrix} 1 & 2 & 0 \\ 0 & 1 & 0 \\ 0 & 0 & 1 \end{pmatrix} = \begin{pmatrix} 1 & -4 & 1 \\ 1/3 & -5/3 & 2/3 \\ -1 & 4 & 0 \end{pmatrix} A \qquad R_1 \to R_1 - 2R_2$$

$$\Rightarrow \begin{pmatrix} 1 & 0 & 0 \\ 0 & 1 & 0 \\ 0 & 0 & 1 \end{pmatrix} = \begin{pmatrix} 1/3 & -2/3 & -1/3 \\ 1/3 & -5/3 & 2/3 \\ -1 & 4 & 0 \end{pmatrix} A$$

$$\therefore \quad I = BA, \text{ where } B = A^{-1} = \begin{pmatrix} 1/3 & -2/3 & -1/3 \\ 1/3 & -5/3 & 2/3 \\ -1 & 4 & 0 \end{pmatrix}$$

Example 4.11: Using elementary row transformations, find the inverse of the matrix

$$A=\begin{pmatrix}1 & 2 & 3\\ -3 & 5 & 0\\ 0 & 1 & 1\end{pmatrix}.$$

Solution: Consider $A = IA$

$$\Rightarrow \begin{pmatrix}1 & 2 & 3\\ -3 & 5 & 0\\ 0 & 1 & 1\end{pmatrix}=\begin{pmatrix}1 & 0 & 0\\ 0 & 1 & 0\\ 0 & 0 & 1\end{pmatrix}A \qquad R_2 \to R_2 + 3R_1$$

$$\Rightarrow \begin{pmatrix}1 & 2 & 3\\ 0 & 11 & 9\\ 0 & 1 & 1\end{pmatrix}=\begin{pmatrix}1 & 0 & 0\\ 3 & 1 & 0\\ 0 & 0 & 1\end{pmatrix}A \qquad R_2 \leftrightarrow R_3$$

$$\Rightarrow \begin{pmatrix}1 & 2 & 3\\ 0 & 1 & 1\\ 0 & 11 & 9\end{pmatrix}=\begin{pmatrix}1 & 0 & 0\\ 0 & 0 & 1\\ 3 & 1 & 0\end{pmatrix}A \qquad R_3 \to R_3 - 11R_2$$

$$\Rightarrow \begin{pmatrix}1 & 2 & 3\\ 0 & 1 & 1\\ 0 & 0 & -2\end{pmatrix}=\begin{pmatrix}1 & 0 & 0\\ 0 & 0 & 1\\ 3 & 1 & -11\end{pmatrix}A \qquad R_3 \to -\frac{1}{2}R_3$$

$$\Rightarrow \begin{pmatrix}1 & 2 & 3\\ 0 & 1 & 1\\ 0 & 0 & 1\end{pmatrix}=\begin{pmatrix}1 & 0 & 0\\ 0 & 0 & 1\\ -3/2 & -1/2 & 11/2\end{pmatrix}A \qquad R_1 \to R_1 - 3R_3, R_2 \to R_2 - R_3$$

$$\Rightarrow \begin{pmatrix}1 & 2 & 0\\ 0 & 1 & 0\\ 0 & 0 & 1\end{pmatrix}=\begin{pmatrix}11/2 & 3/2 & -33/2\\ 3/2 & 1/2 & -9/2\\ -3/2 & -1/2 & 11/2\end{pmatrix}A \qquad R_1 \to R_1 - 2R_2$$

$$\Rightarrow \begin{pmatrix}1 & 0 & 0\\ 0 & 1 & 0\\ 0 & 0 & 1\end{pmatrix}=\begin{pmatrix}5/2 & 1/2 & -15/2\\ 3/2 & 1/2 & -9/2\\ -3/2 & -1/2 & 11/2\end{pmatrix}A$$

$\therefore \quad I = BA$, where $B = A^{-1} = \begin{pmatrix} 5/2 & 1/2 & -15/2 \\ 3/2 & 1/2 & -9/2 \\ -3/2 & -1/2 & 11/2 \end{pmatrix} = \frac{1}{2}\begin{pmatrix} 5 & 1 & -15 \\ 3 & 1 & -9 \\ -3 & -1 & 11 \end{pmatrix}$

Example 4.12: Given matrix $A = \begin{pmatrix} 2 & 0 & 3 \\ -3 & -1 & -3 \\ 4 & 1 & 4 \end{pmatrix}$. Find its inverse using Gauss Jordon method.

Solution: Given $A = \begin{pmatrix} 2 & 0 & 3 \\ -3 & -1 & -3 \\ 4 & 1 & 4 \end{pmatrix}$

Consider $A = IA$

$$\Rightarrow \begin{pmatrix} 2 & 0 & 3 \\ -3 & -1 & -3 \\ 4 & 1 & 4 \end{pmatrix} = \begin{pmatrix} 1 & 0 & 0 \\ 0 & 1 & 0 \\ 0 & 0 & 1 \end{pmatrix} A \qquad R_2 \to 2R_2$$

$$\Rightarrow \begin{pmatrix} 2 & 0 & 3 \\ -6 & -2 & -6 \\ 4 & 1 & 4 \end{pmatrix} = \begin{pmatrix} 1 & 0 & 0 \\ 0 & 2 & 0 \\ 0 & 0 & 1 \end{pmatrix} A \qquad R_2 \to R_2 + 3R_1, R_3 \to R_3 - 2R_1$$

$$\Rightarrow \begin{pmatrix} 2 & 0 & 3 \\ 0 & -2 & 3 \\ 0 & 1 & -2 \end{pmatrix} = \begin{pmatrix} 1 & 0 & 0 \\ 3 & 2 & 0 \\ -2 & 0 & 1 \end{pmatrix} A \qquad R_3 \to 2R_3$$

$$\Rightarrow \begin{pmatrix} 2 & 0 & 3 \\ 0 & -2 & 3 \\ 0 & 2 & -4 \end{pmatrix} = \begin{pmatrix} 1 & 0 & 0 \\ 3 & 2 & 0 \\ -4 & 0 & 2 \end{pmatrix} A \qquad R_3 \to R_3 + R_2$$

$$\Rightarrow \begin{pmatrix} 2 & 0 & 3 \\ 0 & -2 & 3 \\ 0 & 0 & -1 \end{pmatrix} = \begin{pmatrix} 1 & 0 & 0 \\ 3 & 2 & 0 \\ -1 & 2 & 2 \end{pmatrix} A \qquad R_1 \to R_1 + 3R_3, R_2 \to R_2 + 3R_3$$

$$\Rightarrow \begin{pmatrix} 2 & 0 & 0 \\ 0 & -2 & 0 \\ 0 & 0 & -1 \end{pmatrix} = \begin{pmatrix} -2 & 6 & 6 \\ 0 & 8 & 6 \\ -1 & 2 & 2 \end{pmatrix} A \qquad R_1 \to \frac{1}{2} R_1, R_2 \to -\frac{1}{2} R_2, R_3 \to -R_3$$

$$\Rightarrow \begin{pmatrix} 1 & 0 & 0 \\ 0 & 1 & 0 \\ 0 & 0 & 1 \end{pmatrix} = \begin{pmatrix} -1 & 3 & 3 \\ 0 & -4 & -3 \\ 1 & -2 & -2 \end{pmatrix} A$$

$$\therefore \quad I = BA, \text{ where } B = A^{-1} = \begin{pmatrix} -1 & 3 & 3 \\ 0 & -4 & -3 \\ 1 & -2 & -2 \end{pmatrix}$$

EXERCISE 4.2

1. Use elementary transformations to find the inverse of the following matrices:

(*a*) $\begin{pmatrix} 2 & 0 & -1 \\ 5 & 1 & 0 \\ 0 & 1 & 3 \end{pmatrix}$ (*b*) $\begin{pmatrix} 2 & 3 & 1 \\ -3 & 5 & 1 \\ 1 & 7 & 2 \end{pmatrix}$ (*c*) $\begin{pmatrix} 8 & 4 & 3 \\ 2 & 1 & 1 \\ 1 & 2 & 1 \end{pmatrix}$

2. Using Gauss Jordan method, find the inverse of the following matrices:

(*a*) $\begin{pmatrix} 2 & 1 & -1 \\ 0 & 2 & 1 \\ 5 & 2 & -3 \end{pmatrix}$ (*b*) $\begin{pmatrix} 2 & 1 & -1 & 2 \\ 1 & 3 & 2 & -3 \\ -1 & 2 & 1 & -1 \\ 2 & -3 & -1 & 4 \end{pmatrix}$ (*c*) $\begin{pmatrix} 1 & 2 & 2 \\ 2 & 1 & 2 \\ 2 & 2 & 1 \end{pmatrix}$

ANSWERS

1. (*a*) $\begin{pmatrix} 3 & -1 & 1 \\ -15 & 6 & -5 \\ 5 & -2 & 2 \end{pmatrix}$ (*b*) $\begin{pmatrix} 3 & 1 & -2 \\ 7 & 3 & -5 \\ -26 & -11 & 19 \end{pmatrix}$ (*c*) $\begin{pmatrix} 1/3 & -2/3 & -1/3 \\ 1/3 & -5/3 & 2/3 \\ -1 & 4 & 0 \end{pmatrix}$

2. (*a*) $\begin{pmatrix} 8 & -1 & -3 \\ -5 & 1 & 2 \\ 10 & -1 & -4 \end{pmatrix}$ (*b*) $\frac{1}{25}\begin{pmatrix} -23 & 29 & -64 & -18 \\ 10 & -12 & 26 & 7 \\ 1 & -2 & 6 & 2 \\ 2 & -2 & 3 & 1 \end{pmatrix}$ (*c*) $\frac{1}{5}\begin{pmatrix} -3 & 2 & 2 \\ 2 & -3 & 2 \\ 2 & 2 & -3 \end{pmatrix}$

4.6 CONSISTENCY AND SOLUTION OF LINEAR ALGEBRAIC EQUATIONS

Consider the system of linear algebraic equation

$$a_1x+b_1y+c_1z=d_1$$
$$a_2x+b_2y+c_2z=d_2$$
$$a_3x+b_3y+c_3z=d_3$$

In the matrix notation, these equations can be written as

$$AX = B$$

where $A=\begin{pmatrix} a_1 & b_1 & c_1 \\ a_2 & b_2 & c_2 \\ a_3 & b_3 & c_3 \end{pmatrix}$ is called the *coefficient matrix*,

$X=\begin{pmatrix} x \\ y \\ z \end{pmatrix}$ is the column matrix of unknowns, and

$B=\begin{pmatrix} d_1 \\ d_2 \\ d_3 \end{pmatrix}$ is the column matrix of constants.

If $d_1 = d_2 = d_3 = 0$, then $B = O$, and matrix equation $AX = B$ reduce to $AX = O$. Such a system of equation is called a system of *homogenous system of linear equation.*

If at least one of the d_1, d_2, d_3 is the non-zero, then $B \neq 0$ such a system of equation is called a system of *non-homogenous system of linear equation.*

Solving the matrix equation $AX =$ B means finding X, *i.e.*, finding a column matrix of type $\begin{pmatrix} \alpha \\ \beta \\ \gamma \end{pmatrix}$ such that $X=\begin{pmatrix} x \\ y \\ z \end{pmatrix}=\begin{pmatrix} \alpha \\ \beta \\ \gamma \end{pmatrix}$, then $x=\alpha, y=\beta, z=\gamma$.

The matrix equation $AX = B$ need not always have a solution. It may have no solution or a unique solution or an infinite number of solutions.

- **Consistent System of Equation -** A system of linear equation having one or more solution is called a *consistent system of equations*.

For example, following system of linear equations

(*a*) $x+3y=7$, $x+2y=5$ have a unique solution.

(*b*) $x+3y=5$, $2x+6y=10$ have infinite solutions.

For the following system of equations

$$a_1x+b_1y=c_1\,,\ a_2x+b_2y=c_2$$

- If $\frac{a_1}{a_2} \neq \frac{b_1}{b_2} \neq \frac{c_1}{c_2} \Rightarrow$ Unique solution
- If $\frac{a_1}{a_2} = \frac{b_1}{b_2} = \frac{c_1}{c_2} \Rightarrow$ Infinite solutions
- If $\frac{a_1}{a_2} = \frac{b_1}{b_2} \neq \frac{c_1}{c_2} \Rightarrow$ No solution

- **Inconsistent System of Equations -** If a system of equation has no *solution*, it is said to be *inconsistent*.

For example, following system of linear equation

$x+2y=4$, $4x+8y=15$ have no solution .

Criteria for Consistency

In terms of rank of the coefficient matrix and rank of the augmented matrix, the criteria for *a unique solution, no solution, infinite number of solutions* of a system of linear equations is as follows:

- For a system of non-homogeneous of linear equations $AX = B$:
  - If $\rho[A:B] \neq \rho(A)$ the system is inconsistent.
  - If $\rho[A:B] = \rho(A) =$ number of the unknowns, the system has a unique solution.
  - If $\rho[A:B] = \rho(A) <$ number of unknowns, the system has a infinite number of solutions.

The matrix [*A*:*B*] in which the element of *A* and *B* are written side by side is called the *augmented matrix*.

- For a system of homogenous linear equations $AX = O$:
  - $X = O$ is always a solution. This solution in which each unknown has the value zero is called the *null solution* or *trivial solution*. Thus, a homogenous system is always consistent.
  - If $\rho(A)$ = number of unknowns, the system has only the trivial solution.
  - If $\rho(A)$ <number of unknowns, the system has an infinite number of non-trivial solutions.

Art.: If A is a non-singular matrix, then the system of non-homogeneous linear equations $AX = B$ has a unique solution.

Proof: Given system of non-homogeneous linear equations is

$$AX = B \qquad \ldots (1)$$

$\because$ A is the non-singular matrix

$\therefore$ A^{-1} exists

Pre-multiplying both side of (1) by A^{-1}, we get

$$A^{-1}AX = A^{-1}B$$

$$\Rightarrow \quad (A^{-1}A)X = A^{-1}B$$

$$\Rightarrow \quad IX = A^{-1}B$$

$$\Rightarrow \quad X = A^{-1}B$$

which is the required unique solution.

Alternatively:

The augmented matrix is [A:B]. By applying E-row operations on A and B, reduce A to a diagonal matrix as

$$[A:B] = \begin{pmatrix} p_1 & 0 & 0 & : & q_1 \\ 0 & p_2 & 0 & : & q_2 \\ 0 & 0 & p_3 & : & q_3 \end{pmatrix}$$

then $p_1x = q_1, p_2y = q_2, p_3z = q_3$ giving $x = \dfrac{q_1}{p_1}, y = \dfrac{q_2}{p_2}, z = \dfrac{q_3}{p_3}$

ILLUSTRATIVE EXAMPLES 4.4

Example 4.13: Discuss the consistency of the following system of equations. Find the solution if consistent:

(*a*) $x+y+z=4$
$2x+5y-2z=3$

(*b*) $5x+3y+7z=4$
$3x+26y+2z=9$
$7x+2y+10z=5.$

Solution:

(*a*) Given system of equations is

$$x+y+z=4$$
$$2x+5y-2z=3$$

which can be written as

$$AX=B$$

where $A=\begin{pmatrix} 1 & 1 & 1 \\ 2 & 5 & -2 \end{pmatrix}$ and $B=\begin{pmatrix} 4 \\ 3 \end{pmatrix}$ $X=\begin{pmatrix} x \\ y \\ z \end{pmatrix}$

Augmented matrix $(A:B)=\begin{pmatrix} 1 & 1 & 1 & : & 4 \\ 2 & 5 & -2 & : & 3 \end{pmatrix}$ $\quad R_2-2R_1$

$$\sim\begin{pmatrix} 1 & 1 & 1 & : & 4 \\ 0 & 3 & -4 & : & -5 \end{pmatrix} \quad R_2\left(\frac{1}{3}\right)$$

$$\sim\begin{pmatrix} 1 & 1 & 1 & : & 4 \\ 0 & 1 & \frac{-4}{3} & : & \frac{-5}{3} \end{pmatrix} \quad R_1-R_2$$

$$\sim\begin{pmatrix} 1 & 0 & \frac{7}{3} & : & \frac{17}{3} \\ 0 & 1 & \frac{-4}{3} & : & \frac{-5}{3} \end{pmatrix}$$

$\rho(A)=2; \rho(A:B)=2$

$\rho(A)=\rho(A:B) <$ number of unknowns

$\therefore$ Given system of equations are consistent and have infinite number of solutions given by

$$x+\frac{7}{3}z=\frac{17}{3}$$

$$y-\frac{4}{3}z=-\frac{5}{3}$$

Taking $z=k$ we have $x=\frac{17-7k}{3}$, $y=\frac{4k-5}{3}$, $z=k$, where k is any arbitrary constant.

Hence solution is $x=\frac{17-7k}{3}$, $y=\frac{4k-5}{3}$, $z=k$, where k is any arbitrary constant.

(*b*) Given system of equations is

$$5x+3y+7z=4$$
$$3x+26y+2z=9$$
$$7x+2y+10z=5$$

which can be written as

$$AX = B$$

where $A=\begin{pmatrix}5 & 3 & 7\\ 3 & 26 & 2\\ 7 & 2 & 10\end{pmatrix}$, $X=\begin{pmatrix}x\\ y\\ z\end{pmatrix}$, $B=\begin{pmatrix}4\\ 9\\ 5\end{pmatrix}$

Augmented matrix $(A\vdots B)=\begin{pmatrix}5 & 3 & 7 & : & 4\\ 3 & 26 & 2 & : & 9\\ 7 & 2 & 10 & : & 5\end{pmatrix}$ $\quad R_2-\frac{3}{5}R_1,\ R_3-\frac{7}{5}R_1$

$$\sim\begin{pmatrix}5 & 3 & 7 & : & 4\\ 0 & \frac{121}{5} & -\frac{11}{5} & : & \frac{33}{5}\\ 0 & -\frac{11}{5} & \frac{1}{5} & : & -\frac{3}{5}\end{pmatrix} \quad R_3+\frac{1}{11}R_2$$

$$\sim\begin{pmatrix}5 & 3 & 7 & : & 4\\ 0 & \frac{121}{5} & -\frac{11}{5} & : & \frac{33}{5}\\ 0 & 0 & 0 & : & 0\end{pmatrix} \quad R_2\left(\frac{5}{11}\right)$$

$$\sim\begin{pmatrix} 5 & 3 & 7 & \vdots & 4 \\ 0 & 11 & -1 & \vdots & 3 \\ 0 & 0 & 0 & \vdots & 0 \end{pmatrix}$$

$\rho(A)=2,\ \rho(A:B)=2$

$\rho(A)=\rho(A:B)=2<$ number of unknowns

$\therefore$ Given equations are consistent and have infinite number of solutions given by

$$5x+3y+7z=4$$

$$11y-z=3$$

Let $z=k;\ \ y=\dfrac{k+3}{11},\ x=\dfrac{4-\dfrac{3(k+3)}{11}-7k}{5}=\dfrac{7-16k}{11}$

Hence solution is $x=\dfrac{7-16k}{11},\ y=\dfrac{k+3}{11},\ z=k.$

Example 4.14: Show that the following system of linear equation is consistent and hence solve it.

$$x+2y-z=3$$

$$3x-y+2z=1$$

$$2x-2y+3z=2$$

Solution: The given system of equations can be written in the matrix form as

$$\begin{pmatrix} 1 & 2 & -1 \\ 3 & -1 & 2 \\ 2 & -2 & 3 \end{pmatrix}\begin{pmatrix} x \\ y \\ z \end{pmatrix}=\begin{pmatrix} 3 \\ 1 \\ 2 \end{pmatrix}$$

The augmented matrix $[A:B]=\begin{pmatrix} 1 & 2 & -1 & : & 3 \\ 3 & -1 & 2 & : & 1 \\ 2 & -2 & 3 & : & 2 \end{pmatrix}$ $\quad R_2\to R_2-3R_1, R_3\to R_3-2R_1$

$$\begin{pmatrix} 1 & 2 & -1 & : & 3 \\ 0 & -7 & 5 & : & -8 \\ 0 & -6 & 5 & : & -4 \end{pmatrix}\qquad R_2\to-\frac{1}{7}R_2, R_3\to-\frac{1}{6}R_3$$

$$\begin{pmatrix} 1 & 2 & -1 & : & 3 \\ 0 & 1 & -\frac{5}{7} & : & \frac{8}{7} \\ 0 & 1 & -\frac{5}{6} & : & \frac{4}{6} \end{pmatrix} \qquad R_3 \to R_3 - R_2$$

$$\begin{pmatrix} 1 & 2 & -1 & : & 3 \\ 0 & 1 & -\frac{5}{7} & : & \frac{8}{7} \\ 0 & 0 & -\frac{5}{42} & : & -\frac{10}{21} \end{pmatrix} \qquad R_3 \to -\frac{42}{5} R_3$$

$$\begin{pmatrix} 1 & 2 & -1 & : & 3 \\ 0 & 1 & -\frac{5}{7} & : & \frac{8}{7} \\ 0 & 0 & 1 & : & 4 \end{pmatrix}$$

Clearly rank of A = rank of $[A:B] = 3$

Hence the equations are consistent

Now the reduced system

$$\begin{pmatrix} 1 & 2 & -1 \\ 0 & 1 & -\frac{5}{7} \\ 0 & 0 & 1 \end{pmatrix}\begin{pmatrix} x \\ y \\ z \end{pmatrix} = \begin{pmatrix} 3 \\ \frac{8}{7} \\ 4 \end{pmatrix}$$

$\therefore \quad x + 2y - z = 3 \qquad \ldots (1)$

$y - \frac{5}{7}z = \frac{8}{7} \qquad \ldots (2)$

$z = 4 \qquad \ldots (3)$

From Eq. (2) $\quad y = \frac{5}{7}(4)) + \frac{8}{7} = \frac{28}{7} = 4$

From Eq. (1) $\quad x + 8 - 4 = 3$

$\therefore \quad x = -1, y = 4, z = 4$

Example 4.15: Solve the following system of non-homogeneous linear equations

$$5x+3y+7z=4$$
$$3x+26y+2z=9$$
$$7x+2y+11z=5$$

with the help of matrix inversion

Solution: In matrix notation, the given system of equations can be written as

$$AX=B \qquad \text{. . . (1)}$$

where $A=\begin{pmatrix}5&3&7\\3&26&2\\7&2&11\end{pmatrix}$, $X=\begin{pmatrix}x\\y\\z\end{pmatrix}$, $B=\begin{pmatrix}4\\9\\5\end{pmatrix}$

$$|A|=\begin{vmatrix}5&3&7\\3&26&2\\7&2&11\end{vmatrix}=5\begin{vmatrix}26&2\\2&11\end{vmatrix}-3\begin{vmatrix}3&2\\7&11\end{vmatrix}+7\begin{vmatrix}3&26\\7&2\end{vmatrix}$$

Expanding on first column

$$=5(26\times11-2\times2)-3(3\times11-7\times2)+7(3\times2-7\times26)$$

$$=121\neq0$$

$\Rightarrow$ A is non-singular

$\therefore$ A^{-1} exists and the unique solution of (1) is

$$X=A^{-1}B \qquad \text{. . . (2)}$$

To find A^{-1}:

Cofactors of the elements of A are as follows:

$$A_{11}=\begin{vmatrix}26&2\\2&11\end{vmatrix}=282,\ A_{12}=-\begin{vmatrix}3&2\\7&11\end{vmatrix}=-19,\ A_{13}=\begin{vmatrix}3&26\\7&2\end{vmatrix}=-176$$

$$A_{21}=-\begin{vmatrix}3&7\\2&11\end{vmatrix}=-19,\ A_{22}=\begin{vmatrix}5&7\\7&11\end{vmatrix}=6,\ A_{23}=-\begin{vmatrix}5&3\\7&2\end{vmatrix}=11$$

$$A_{31}=\begin{vmatrix}3&7\\26&2\end{vmatrix}=-176,\ A_{32}=-\begin{vmatrix}5&7\\3&2\end{vmatrix}=11,\ A_{33}=\begin{vmatrix}5&3\\3&26\end{vmatrix}=121$$

$$\therefore \quad \text{adj } A = \text{transpose of} \begin{pmatrix} A_{11} & A_{12} & A_{13} \\ A_{21} & A_{22} & A_{23} \\ A_{31} & A_{32} & A_{33} \end{pmatrix} = \text{transpose of} \begin{pmatrix} 282 & -19 & -176 \\ -19 & 6 & 11 \\ -176 & 11 & 121 \end{pmatrix}$$

$$\therefore \quad \text{adj } A = \begin{pmatrix} 282 & -19 & -176 \\ -19 & 6 & 11 \\ -176 & 11 & 121 \end{pmatrix}$$

$$A^{-1} = \frac{1}{|A|} \text{ adj } A = \frac{1}{121} \begin{pmatrix} 282 & -19 & -176 \\ -19 & 6 & 11 \\ -176 & 11 & 121 \end{pmatrix}$$

From Eq. (2), we have

$$X = A^{-1}B = \frac{1}{121} \begin{pmatrix} 282 & -19 & -176 \\ -19 & 6 & 11 \\ -176 & 11 & 121 \end{pmatrix} \begin{pmatrix} 4 \\ 9 \\ 5 \end{pmatrix}$$

$$= \frac{1}{121} \begin{pmatrix} 282\times4-19\times9-176\times5 \\ -19\times4+6\times9+11\times5 \\ -176\times4+11\times9+121\times5 \end{pmatrix} = \frac{1}{121} \begin{pmatrix} 77 \\ 33 \\ 0 \end{pmatrix}$$

$$\Rightarrow \quad \begin{pmatrix} x \\ y \\ z \end{pmatrix} = \frac{1}{121} \begin{pmatrix} 77 \\ 33 \\ 0 \end{pmatrix} = \begin{pmatrix} \frac{77}{121} \\ \frac{33}{121} \\ \frac{0}{121} \end{pmatrix} = \begin{pmatrix} \frac{7}{11} \\ \frac{3}{11} \\ 0 \end{pmatrix}$$

Hence $x = \frac{7}{11}$, $y = \frac{3}{11}$, $z = 0$.

Example 4.16: Solve, with the help of matrices, the simultaneous equation:

$$x+y+z=3$$
$$x+2y+3z=4$$
$$x+4y+9z=6.$$

Solution: Since there is no restriction to find the solution using a particular method, we can use any method.

Augmented matrix $(A:B)=\begin{pmatrix} 1 & 1 & 1 & : & 3 \\ 1 & 2 & 3 & : & 4 \\ 1 & 4 & 9 & : & 6 \end{pmatrix}$ $\quad R_2 \to R_2 - R_1, R_3 \to R_3 - R_1$

$$\sim \begin{pmatrix} 1 & 1 & 1 & : & 3 \\ 0 & 1 & 2 & : & 1 \\ 0 & 3 & 8 & : & 3 \end{pmatrix} \quad R_3 \to R_3 - 3R_2$$

$$\sim \begin{pmatrix} 1 & 1 & 1 & : & 3 \\ 0 & 1 & 2 & : & 1 \\ 0 & 0 & 2 & : & 0 \end{pmatrix} \quad R_3 \to \frac{1}{2}R_3$$

$$\sim \begin{pmatrix} 1 & 1 & 1 & : & 3 \\ 0 & 1 & 2 & : & 1 \\ 0 & 0 & 1 & : & 0 \end{pmatrix} \quad R_1 \to R_1 - R_3, R_2 \to R_2 - 2R_3$$

$$\sim \begin{pmatrix} 1 & 1 & 0 & : & 3 \\ 0 & 1 & 0 & : & 1 \\ 0 & 0 & 1 & : & 0 \end{pmatrix} \quad R_1 \to R_1 - R_2$$

$$\sim \begin{pmatrix} 1 & 0 & 0 & : & 2 \\ 0 & 1 & 0 & : & 1 \\ 0 & 0 & 1 & : & 0 \end{pmatrix}$$

$\therefore \; x=2, \; y=1, \; z=0$

Note: This method is especially useful when the number of unknown is 4 or more, since $|A|$ is of order 4 or more and the minors of its various elements are determinants of order 3 or more, which are difficult to evaluate.

Example 4.17: Solve the equations:

$$x_1 + 3x_2 + 2x_3 = 0$$
$$2x_1 - x_2 + 3x_3 = 0$$
$$3x_1 - 5x_2 + 4x_3 = 0$$
$$x_1 + 17x_2 + 4x_3 = 0$$

Solution: In matrix notation, the given system of equations can be written as

$$AX = O$$

$$\text{where } A = \begin{pmatrix} 1 & 3 & 2 \\ 2 & -1 & 3 \\ 3 & -5 & 4 \\ 1 & 17 & 4 \end{pmatrix}, \ X = \begin{pmatrix} x_1 \\ x_2 \\ x_3 \end{pmatrix}$$

Now let work on matrix A:

$$A = \begin{pmatrix} 1 & 3 & 2 \\ 2 & -1 & 3 \\ 3 & -5 & 4 \\ 1 & 17 & 4 \end{pmatrix} \qquad R_2 \to R_2 - 2R_1,\ R_3 \to R_3 - 3R_1,\ R_4 \to R_4 - R_1$$

$$\sim \begin{pmatrix} 1 & 3 & 2 \\ 0 & -7 & -1 \\ 0 & -14 & -2 \\ 0 & 14 & 2 \end{pmatrix} \qquad R_3 \to R_3 - 2R_2,\ R_4 \to R_4 + 2R_2$$

$$\sim \begin{pmatrix} 1 & 3 & 2 \\ 0 & -7 & -1 \\ 0 & 0 & 0 \\ 0 & 0 & 0 \end{pmatrix} \qquad R_1 \to R_1 + 2R_2$$

$$\sim \begin{pmatrix} 1 & -11 & 0 \\ 0 & -7 & -1 \\ 0 & 0 & 0 \\ 0 & 0 & 0 \end{pmatrix}$$

$\rho(A) = 2 <$ number of unknowns

$\Rightarrow$ The system has an infinite number of non-trivial solutions given by

$$x_1 - 11x_2 = 0$$
$$-7x_2 - x_3 = 0$$

i.e., $x_1 = 11k,\ x_2 = k,\ x_3 = -7k,$ where k is any number.

Different values of k give different solutions.

Example 4.18: For what value of k, the equations

$$x + y + z = 1$$
$$2x + y + 4z = k$$
$$4x + y + 10z = k^2$$

have a solution and solve them completely in each case.

Solution: Given system of equations is

$$x + y + z = 1$$
$$2x + y + 4z = k$$
$$4x + y + 10z = k^2$$

which can be put into matrix from as

$$AX = B$$

where $A = \begin{pmatrix} 1 & 1 & 1 \\ 2 & 1 & 4 \\ 4 & 1 & 10 \end{pmatrix},\ x = \begin{pmatrix} x \\ y \\ z \end{pmatrix},\ B = \begin{pmatrix} 1 \\ k \\ k^2 \end{pmatrix}$

Augmented matrix $(A:B) = \begin{pmatrix} 1 & 1 & 11 & : & 1 \\ 2 & 1 & 4 & : & k \\ 4 & 1 & 10 & : & k^2 \end{pmatrix}$ $\quad R_2 \to R_2 - 2R_1,\ R_3 \to R_3 - 4R_1$

$$\sim\begin{pmatrix} 1 & 1 & 1 & : & 1 \\ 0 & -1 & 2 & : & k-2 \\ 0 & -3 & 6 & : & k^2-4 \end{pmatrix} \qquad R_3-3R_2$$

$$\sim\begin{pmatrix} 1 & 1 & 1 & : & 1 \\ 0 & -1 & 2 & : & k-2 \\ 0 & 0 & 0 & : & k^2-3k+2 \end{pmatrix}$$

$\rho(A)=2<$ number of unknowns

$\therefore$ System of equations cannot have a unique solution.

These will have an infinite number of solution only if

$$\rho(A:B)=\rho(A)=2$$

which is only possible if

$$k^2-3k+2=0 \quad \Rightarrow \quad k=1,\ 2$$

When $k=1$, the augmented matrix $=\begin{pmatrix} 1 & 1 & 1 & : & 1 \\ 0 & -1 & 2 & : & -1 \\ 0 & 0 & 0 & : & 0 \end{pmatrix}$

$$\Rightarrow \quad \begin{aligned} x+y+z&=1 \\ -y+2z&=-1 \end{aligned}$$

Let $z=\lambda$, then $y=1+2\lambda$, $x=-3\lambda$, where λ is an arbitrary constant.

When $k=2$, augmented matrix $=\begin{pmatrix} 1 & 1 & 1 & : & 1 \\ 0 & -1 & 2 & : & 0 \\ 0 & 0 & 0 & : & 0 \end{pmatrix}$

$$\Rightarrow \quad \begin{aligned} x+y+z&=1 \\ -y+2z&=0. \end{aligned}$$

Let $z=\lambda'$, then $y=2\lambda'$, $x=1-3\lambda'$, where λ' is any arbitrary constant.

Example 4.19: Find the values of a and b for which the equations

$$x+ay+z=3$$
$$x+2y+2z=b$$
$$x+5y+3z=9$$

are consistent. When will these equations have a unique solution?

Solution: Given equations are

$$x+ay+z=3$$
$$x+2y+2z=b$$
$$x+5y+3z=9$$

which can be put into matrix from as

$$AX = B$$

where $A=\begin{pmatrix}1 & a & 1\\ 1 & 2 & 2\\ 1 & 5 & 3\end{pmatrix}$, $B=\begin{pmatrix}3\\ b\\ 9\end{pmatrix}$, $X=\begin{pmatrix}x\\ y\\ z\end{pmatrix}$

Augmented matrix is $(A:B)=\begin{pmatrix}1 & a & 1 & : & 3\\ 1 & 2 & 2 & : & b\\ 1 & 5 & 3 & : & 9\end{pmatrix}$ $\quad R_2-R_1, R_3-R_1$

$$\sim\begin{pmatrix}1 & a & 1 & : & 3\\ 0 & 2-a & 1 & : & b-3\\ 0 & 5-a & 2 & : & 6\end{pmatrix} \qquad R_3-\frac{5-a}{2-a}R_2$$

$$\sim\begin{pmatrix}1 & a & 1 & : & 3\\ 0 & 2-a & 1 & : & b-3\\ 0 & 0 & 2-\dfrac{5-a}{2-a} & : & 6-\dfrac{5-a}{2-a}(b-3)\end{pmatrix}$$

$$\Rightarrow \begin{pmatrix} 1 & a & 1 & : & 3 \\ 0 & 2-a & 1 & : & b-3 \\ 0 & 0 & \dfrac{-1-a}{2-a} & : & 6-\dfrac{5-a}{2-a}(b-3) \end{pmatrix}$$

Case I: If $\dfrac{-1-a}{2-a}=0$ and $6-\dfrac{5-a}{2-a}(b-3)=0$, then $a=-1$, $b=6$

$\rho(A)=\rho(A:B)=2<$ number of unknowns

$\therefore$ Given equations have infinite number of solutions given by

$$\begin{pmatrix} 1 & -1 & 1 & : & 3 \\ 0 & 3 & 1 & : & 3 \\ 0 & 0 & 0 & : & 0 \end{pmatrix}$$

$\Rightarrow \quad x-y+z=3$

$3y+z=3$

Let $z=k$, then $y=\dfrac{3-k}{3}$, $\quad x=3+\dfrac{3-k}{3}-k=4-\dfrac{4}{3}k$

$\therefore \quad x=4-\dfrac{4}{3}k$, $y=\dfrac{3-k}{3}$, $z=k$ is the solution.

Case II: If $a=-1$, but $b\neq 6$, then

$\rho(A)=2$, but $\rho(A:B)=3$

$\rho(A)\neq\rho(A:B)$

$\therefore$ Equations are inconsistent, i.e., having no solution.

Case III: If $a\neq -1$, b can have any value then

$\rho(A)=3=\rho(A:B)$

$\therefore$ Given equations have a unique solution which is given by the equations

$$x+ay+z=3$$

$$(2-a)y+z=b-3$$

$$\frac{-(1+a)}{2-a}z=6-\frac{5-a}{2-a}(b-3)$$

$$\therefore \quad z=\frac{2-a}{-(1+a)}\left[6-\frac{(5-a)(b-3)}{2-a}\right]=\frac{6(2-a)-(5-a)(b-3)}{-(1+a)}$$

$$\therefore \quad y=\frac{1}{2-a}\left[(6-3)-\frac{6(2-a)-(5-a)(b-3)}{-(1+a)}\right]$$

$$=\frac{1}{(2-a)(1+a)}\left[(b-3)(1+a)+6(2-a)-(5-a)(b-3)\right]=\frac{2(6-b)}{1+a}$$

$$\therefore \quad x=3-a\left(\frac{2(6-b)}{1+a}\right)-\frac{6(2-a)-(5-a)(b-3)}{-(1+a)}=\frac{(5-3a)(6-b)}{1+a}$$

Hence $x=\dfrac{(5-3a)(6-b)}{1+a}$, $y=\dfrac{2(6-b)}{1+a}$, $=\dfrac{6(2-a)-(5-a)(b-3)}{-(1+a)}$ is the unique solution.

Example 4.20: Investigate for what values of λ and μ, the system of equations

$$x+y+z=6$$
$$x+3y+3z=10$$
$$x+2y+\lambda z=\mu$$

has *(i)* No solution *(ii)* Unique solution *(iii)* Infinite solutions.

Solution: The given system of equations can be written in the matrix form

$$\begin{pmatrix}1 & 1 & 1\\ 1 & 2 & 3\\ 1 & 2 & \lambda\end{pmatrix}\begin{pmatrix}x\\ y\\ z\end{pmatrix}=\begin{pmatrix}6\\ 10\\ \mu\end{pmatrix} \qquad \text{i.e., } AX=B$$

$$\text{Augmented Matrix } [A:B] = \begin{pmatrix} 1 & 1 & 1 & : & 6 \\ 1 & 2 & 3 & : & 10 \\ 1 & 2 & \lambda & : & \mu \end{pmatrix} \qquad R_2 \to R_2 - R_1, R_3 \to R_3 - R_2$$

$$\sim \begin{pmatrix} 1 & 1 & 1 & : & 6 \\ 0 & 1 & 2 & : & 4 \\ 0 & 1 & \lambda-3 & : & \mu-10 \end{pmatrix}$$

Case I: If $\lambda = 3, \mu \neq 10$, then

Rank $A = 2$, Rank$[A:B] = 3$

Rank $A \neq$ Rank$[A:B]$

$\therefore$ the given system is inconsistent i.e. it has no solution.

Case II: If $\lambda \neq 3, \mu$ may have any value

Rank $A = 3$, Rank $[A:B] = 3$

Rank $A =$ Rank$[A:B] = 3$ (number of unknowns)

$\therefore$ the given system is consistent and has a unique solution

Case III: If $\lambda = 3, \mu - 10 = 0$

Rank $A =$ Rank$[A:B] = 2 < 3$ (the number of unknowns)

$\therefore$ the given system is consistent and has a infinite number of solution.

EXERCISE 4.3

1. Examine the consistency of the following system of equation and if found consistent, find the complete solution.

   $x + y + z = 6;\; x + 2y + 3z = 14;\; x + 4y + 7z = 30$

2. Test for consistency the following non-homogeneous system of equations and if found consistent, solve it.

   $5x + 3y + 7z = 4;\; 3x + 26y + 2z = 9;\; 7x + 2y + 10z = 5$

3. Solve the system of equations: $2x_1 + x_2 + 2x_3 + x_4 = 6;\; 6x_1 - 6x_2 + 6x_3 + 12x_4 = 36;$

   $4x_1 + 3x_2 + 3x_3 - 3x_4 = -1;\; 2x_1 + 2x_2 - x_3 + x_4 = 10$

4. Using matrix method, show that the equations

$$3x+3y+2z=1;\ x+2y=4;\ 10y+3z=-2;\ 2x-3y-z=5$$

are consistent and hence obtain the solution for x, y and z.

5. Show that the following system of equations

$$3x+4y+5z=a;\ 4x+5y+6z=b;\ 5x+6y+7z=c$$

have solution only when $a+c=2b$.

6. Find the value of k so that the equation

$$x+y+3z=0;\ 4x+3y+kz=0;\ 2x+y+2z=0$$

have the non trivial solution

7. Show that the following system of equations is not consistent.

$$2x+6y+11=0;\ 6x+20y-6z+3=0;\ 6y-18z+1=0$$

8. Solve the following system of equations

$$2x+y+2z+w=6;\ 6x-6y+6z+12w=36$$
$$4x+3y+3z-3w=-1;\ 2x+2y-z+w=10$$

9. Investigate for what values of λ and μ, the system of equations

$$2x-3y+5z=12;\ 3x+y+\lambda z=\mu;\ x-7y+8z=17$$

has (*i*) No solution (*ii*) Unique solution (*iii*) Infinite solutions.

10. For what value of *k*, the following system of equations fails to have a unique solution

$$3x-y+kz=1;\ 2x+y+z=2;\ x+2y-kz=-1$$

11. Using matrix method, investigate the consistency of following system of equations. If found consistent, find its solution.

$$5x+3y+7z=4;\ 3x+26y+2z=9;\ 7x+2y+11z=5$$

12. Use the rank method to test the consistency of the system of equations

$$4x-y=12;\ -x+5y-2z=0;\ -2y+4z=-8;$$

if consistent then solve it completely. **[PTU Dec 2012]**

13. Test the consistency of the system of equations

$$x+2y-z=3;\ 3x-y+2z=1;\ 2x-2y+3z=2;\ x-y+z=-1,$$

and if consistent, then solve it completely. **[PTU May 2012]**

ANSWERS

1. $x = k - 2,\ y = 8 - 2k,\ z = k$, where k is arbitrary, is the general solution.
2. $x = \dfrac{-16}{11}k + \dfrac{7}{11},\ y = \dfrac{k}{11} + \dfrac{3}{11}, z = k,$ where k is arbitrary
3. $x_1 = 2,\ x_2 = 1,\ x_3 = -1,\ x_4 = 3.$
4. Given system of equation is consistent. Its solution is $x = 2,\ y = 1,\ z = -4.$

6. $k = 8$

8. $x = 2,\ y = 1,\ z = -1,\ w = 3$

9. (*i*) No solution for $\lambda = 2, \mu \neq 7$ (*ii*) Unique solution $\lambda \neq 2$
   (*iii*) Infinite solutions for $\lambda = 2, \mu = 7$

10. $k \neq -\dfrac{7}{2}$

11. Consistent; $x = \dfrac{7}{11}, y = \dfrac{3}{11}, z = 0$

12. Consistent; $x = \dfrac{44}{15}, y = -\dfrac{4}{15}, z = -\dfrac{32}{15}$

13. Consistent; $x = -1, y = 4, z = 4$

4.7 EIGEN VALUES AND EIGEN VECTORS

Characteristic Equation

If A is the square matrix of order n, we can form the matrix $A - \lambda I$, where λ is the scalar and I is the unit matrix of order n. The determinant of this matrix equated to zero, *i.e.*,

$$|A - \lambda I| = \begin{vmatrix} a_{11} - \lambda & a_{12} & \cdots & a_{1n} \\ a_{21} & a_{22} - \lambda & \cdots & a_{2n} \\ \vdots & \vdots & \vdots & \vdots \\ a_{n1} & a_{n2} & \cdots & a_{mn} - \lambda \end{vmatrix} = 0$$

is called the *characteristics equation* of A.

On expanding the determinant the characteristics equation can be written as a polynomial equation of degree n in λ of the form

$$(-1)^n \lambda^n + k_1\lambda^{n-1} + k_2\lambda^{n-1} + ... + k_n = 0$$

The roots of this equation are called the *characteristics roots* or *latent roots* or *eigen values* of A.

Note: The sum of the eigen values of a matrix A is equal to trace of A, where the trace of square matrix is the sum of its diagonal elements.

Eigen Values and Eigen Vectors

Let A is the square matrix of order n. A scalar λ is called an *eigen value* of A iff $\exists$ a non-zero $n\times1$ column matrix

$$X = \begin{pmatrix} x_1 \\ x_2 \\ \vdots \\ x_n \end{pmatrix} \quad \text{such that } AX = \lambda X$$

The non-zero column matrix X is called the *eigen vector* of the matrix A corresponding to the eigen value λ of A.

Note:

1. Eigen values are also known as *proper values, characteristic values, latent roots* or *spectral values*. Similarly, eigen vectors are also called *proper vectors, characteristic vectors, latent vectors* or *spectral vectors*.
2. The set of eigen values of a matrix A is called spectrum of the matrix A.

Properties of Eigen Values and Eigen Vectors

Property 1: If λ is an eigen value of matrix A with corresponding eigen vector X, then $\alpha\lambda$ is an eigen value of αA and the corresponding eigen vector remain the same.

Proof: Given that λ is an eigen value of matrix A with corresponding eigen vector X

$\therefore \quad AX = \lambda X$

$\Rightarrow \quad \alpha(AX) = \alpha(\lambda X)$

$\Rightarrow \quad (\alpha A)X = (\alpha\lambda)X$

$\therefore \quad \alpha\lambda$ is an eigen value of αA and the eigen vector is X.

Property 2: If λ is an eigen value of matrix A with corresponding eigen vector X, then λ^m is an eigen value of A^m and the corresponding eigen vector remain the same.

Proof: Given that λ is an eigen value of matrix A with corresponding eigen vector X

$\therefore \quad AX = \lambda X$

$\Rightarrow \quad A(AX) = A(\lambda X)$

$\Rightarrow \quad (AA)X = \lambda(AX) = \lambda(\lambda X)$

$\Rightarrow \quad A^2X = \lambda^2 X$

$\Rightarrow \quad \lambda^2$ is an eigen value of A^2 and the eigen vector is X.

Pre-multiplying successively m times by A, we get

$A^mX = \lambda^m X$

$\Rightarrow \quad \lambda^m$ is an eigen value of A^m and the eigen vector is X.

Property 3: If λ is an eigen value of matrix A with corresponding eigen vector X, then $\lambda - k$ is an eigen value of $A - kI$ and the corresponding eigen vector remain the same.

Proof: Given that λ is an eigen value of matrix A with corresponding eigen vector X

$\therefore \quad AX = \lambda X$

$\Rightarrow \quad AX - kIX = \lambda X - kIX = \lambda X - kX \qquad [\because IX = X]$

$\Rightarrow \quad (A - kI)X = (\lambda - k)X$

$\Rightarrow \quad \lambda - k$ is an eigen value of $A - kI$ and the eigen vector is X.

Property 4: If λ is an eigen value of matrix A with corresponding eigen vector X, then $\frac{1}{\lambda}$ is an eigen value of A^{-1}, if it exists, and the corresponding eigen vector remain the same.

Proof: Given that λ is an eigen value of matrix A with corresponding eigen vector X

$$\therefore \quad AX = \lambda X$$

Pre-multiplying by A^{-1}, we get

$$A^{-1}(AX) = A^{-1}(\lambda X)$$

$$\Rightarrow \quad (A^{-1}A)X = \lambda(A^{-1}X)$$

$$\Rightarrow \quad IX = \lambda(A^{-1}X) \qquad [\because A^{-1}A = AA^{-1} = I]$$

$$\Rightarrow \quad X = \lambda(A^{-1}X) \qquad [\because IX = X]$$

$$\Rightarrow \quad A^{-1}X = \frac{1}{\lambda}X \qquad [\because IX = X]$$

$\Rightarrow \quad \frac{1}{\lambda}$ is an eigen value of A^{-1} and the eigen vector is X.

Property 5: If λ is an eigen value of matrix A with corresponding eigen vector X, then $\frac{1}{\lambda-k}$ is an eigen value of $(A-kI)^{-1}$ and the corresponding eigen vector remain the same.

Proof: Given that λ is an eigen value of matrix A with corresponding eigen vector X

$$\therefore \quad AX = \lambda X$$

$$\Rightarrow \quad AX - kIX = \lambda X - kIX = \lambda X - kX \qquad [\because IX = X]$$

$$\Rightarrow \quad (A-kI)X = (\lambda - k)X$$

Pre-multiplying by $(A-kI)^{-1}$, we get

$$(A-kI)^{-1}(A-kI)X = (A-kI)^{-1}(\lambda-k)X$$

$$\Rightarrow \quad X=(A-kI)^{-1}(\lambda-k)X \qquad \left[\because (A-kI)^{-1}(A-kI)=I\right]$$

$$\Rightarrow \quad (A-kI)^{-1}X=\frac{1}{\lambda-k}X$$

$\Rightarrow$ $\dfrac{1}{\lambda-k}$ **is an eigen value of $(A-kI)^{-1}$ and the eigen vector is X.**

Property 6: **If λ is an eigen value of matrix A with corresponding eigen vector X, then $\dfrac{|A|}{\lambda}$ is an eigen value of adj A and the corresponding eigen vector remain the same.**

Proof: Given that λ is an eigen value of matrix A with corresponding eigen vector X

$$\therefore \quad AX=\lambda X$$

Pre-multiplying by adj A, we get

$$(\text{adj}A)(AX)=(\text{adj}A)(\lambda X)$$

$$\Rightarrow \quad \left[(\text{adj}A)A\right]X=\lambda((\text{adj}A)X)$$

$$\Rightarrow \quad |A|X=\lambda((\text{adj}A)X) \qquad \left[\because (\text{adj}A)A=A(\text{adj}A)=|A|\right]$$

$$\Rightarrow (\text{adj}A)X=\frac{|A|}{\lambda}X$$

$\Rightarrow$ $\dfrac{|A|}{\lambda}$ **is an eigen value of adj A and the eigen vector is X.**

Property 7: **A square matrix A and its transpose A' have the same eigen values.**

Proof: The eigen values of A are given by the characteristic equation

$$|A-\lambda I|=0$$

$$\Rightarrow \quad \left|(A-\lambda I)'\right|=0 \qquad \left[\because |A|=|A'|\right]$$

$$\Rightarrow \quad \left|A'-(\lambda I)'\right|=0 \qquad \left[\because (A-B)'=A'-B'\right]$$

$\Rightarrow \quad |A' - \lambda I| = 0$

$\Rightarrow$ the characteristic equation of A and A' is same

$\Rightarrow$ eigen value of matrix and its transpose is same

Property 8: Eigen values of a triangular matrix are nothing else but diagonal elements.

Proof: Let $A = \begin{pmatrix} a_{11} & a_{12} & \cdots & a_{1n} \\ 0 & a_{22} & \cdots & a_{2n} \\ \vdots & \vdots & \vdots & \vdots \\ 0 & 0 & \cdots & a_{nn} \end{pmatrix}$ be an upper triangular matrix of order n.

Then $|A - \lambda I| = \begin{vmatrix} a_{11} - \lambda & a_{12} & \cdots & a_{1n} \\ 0 & a_{22} - \lambda & \cdots & a_{2n} \\ \vdots & \vdots & \vdots & \vdots \\ 0 & 0 & \cdots & a_{nn} - \lambda \end{vmatrix} = (a_{11} - \lambda)(a_{22} - \lambda)...(a_{nn} - \lambda)$

$\therefore$ eigen values of A are given by $|A - \lambda I| = 0 \Rightarrow (a_{11} - \lambda)(a_{22} - \lambda)...(a_{nn} - \lambda) = 0$

$\therefore \quad \lambda = a_{11}, a_{22}, ..., a_{nn}$ are the eigen values of A. These eigen values of A are just the diagonal elements of the matrix.

Similarly, we can prove this for the lower triangular matrix.

Property 9: Eigen values of a diagonal matrix are nothing else but diagonal elements.

Proof: Let $A = \begin{pmatrix} a_{11} & 0 & \cdots & 0 \\ 0 & a_{22} & \cdots & 0 \\ \vdots & \vdots & \vdots & \vdots \\ 0 & 0 & \cdots & a_{nn} \end{pmatrix}$ be an upper triangular matrix of order n.

Then $|A - \lambda I| = \begin{pmatrix} a_{11} - \lambda & 0 & \cdots & 0 \\ 0 & a_{22} - \lambda & \cdots & 0 \\ \vdots & \vdots & \vdots & \vdots \\ 0 & 0 & \cdots & a_{nn} - \lambda \end{pmatrix} = (a_{11} - \lambda)(a_{22} - \lambda)...(a_{nn} - \lambda)$

$\therefore$ eigen values of A are given by $|A - \lambda I| = 0 \Rightarrow (a_{11} - \lambda)(a_{22} - \lambda)...(a_{nn} - \lambda) = 0$

$\therefore \quad \lambda = a_{11}, a_{22}, ..., a_{nn}$ are the eigen values of A. These eigen values of A are just the diagonal elements of the matrix.

Property 10: If $a_1, a_2, ..., a_n$ are the eigen values of a square matrix A of order n and k is a scalar, then eigen values of matrix $A - kI$ are $a_1 - k, a_2 - k, ..., a_n - k$.

Proof: Since $a_1, a_2, ..., a_n$ are the eigen values of a square matrix A of order n.

$\therefore$ characteristic polynomial of A is

$$|A - \lambda I| = (a_1 - \lambda)(a_2 - \lambda)...(a_n - \lambda)$$

$\therefore$ characteristic polynomial of $A - kI$ is

$$|(A - kI) - \lambda I| = |A - (k + \lambda)I| = [a_1 - (k + \lambda)][a_2 - (k + \lambda)]...[a_n - (k + \lambda)]$$

$$= [(a_1 - k) - \lambda][(a_2 - k) - \lambda]...[(a_n - k) - \lambda]$$

$\therefore$ eigen values of $A - kI$ are $a_1 - k, a_2 - k, ..., a_n - k$.

Property 11: Eigen values of an idempotent matrix A are either zero or unity.

Proof: Let λ be an eigen values of the given idempotent matrix. Therefore, there exists a non-zero vector X such that

$$AX = \lambda X$$

$$\Rightarrow \quad A(AX) = A(\lambda X) = \lambda(AX) = \lambda(\lambda X)$$

$$\Rightarrow \quad A^2 X = \lambda^2 X$$

$$\Rightarrow \quad AX = \lambda^2 X \qquad [\because A^2 = A \text{ as } A \text{ is idempotent}]$$

$$\Rightarrow \quad \lambda X = \lambda^2 X$$

$$\Rightarrow \quad (\lambda^2 - \lambda) X = O$$

$\Rightarrow \quad$ Either $(\lambda^2 - \lambda) = O$ or $X = O$, but $X \neq O$

$$\therefore \quad \lambda^2 - \lambda = 0$$

$$\Rightarrow \quad \lambda(\lambda - 1) = 0$$

$\Rightarrow \qquad \lambda = 0,1$

$\therefore$ eigen values of an idempotent matrix A are either zero or unity.

Property 12: Sum of eigen values of a square matrix A is the sum of diagonal elements and the product of its eigen values is equal to its determinant.

Proof: Let $A = \begin{pmatrix} a_{11} & a_{12} & \cdots & a_{1n} \\ a_{21} & a_{22} & \cdots & a_{2n} \\ \vdots & \vdots & \vdots & \vdots \\ a_{n1} & a_{22} & \cdots & a_{nn} \end{pmatrix}$ is given square matrix of order n.

Then $|A - \lambda I| = \begin{vmatrix} a_{11} - \lambda & a_{12} & \cdots & a_{1n} \\ a_{21} & a_{22} - \lambda & \cdots & a_{2n} \\ \vdots & \vdots & \vdots & \vdots \\ a_{n1} & a_{n2} & \cdots & a_{nn} - \lambda \end{vmatrix}$

$$= (-1)^n \lambda^n + (-1)^{n-1} \lambda^{n-1} (a_{11} + a_{22} + ... + a_{nn}) \qquad ...(1)$$

Let $\lambda_1, \lambda_2, ..., \lambda_n$ be the eigen values of A.

$$\therefore \quad |A - \lambda I| = (-1)^n (\lambda - \lambda_1)(\lambda - \lambda_2)...(\lambda - \lambda_n) \qquad ...(2)$$

Equating coefficients of λ^{n-1} on the RHS of (1) and (2), we get

$$\lambda_1 + \lambda_2 + ... + \lambda_n = a_{11} + a_{22} + ... + a_{nn}$$

Putting $\lambda = 0$ in (2), we get

$$\lambda_1 . \lambda_2, ..., \lambda_n = |A|$$

ILLUSTRATIVE EXAMPLES 4.5

Example 4.21: Find the eigen values of the matrix $\begin{pmatrix} 2 & 3 & 11 \\ 0 & 3 & 17 \\ 0 & 0 & -2 \end{pmatrix}$.

Solution: Let $A=\begin{pmatrix} 2 & 3 & 11 \\ 0 & 3 & 17 \\ 0 & 0 & -2 \end{pmatrix}$

The characteristic equation of the given matrix is

$$|A-\lambda I|=0$$

$$\Rightarrow \quad \begin{vmatrix} 2-\lambda & 3 & 11 \\ 0 & 3-\lambda & 17 \\ 0 & 0 & -2-\lambda \end{vmatrix}=0$$

Expanding determinant on the first column, we get

$$(2-\lambda)(3-\lambda)(-2-\lambda)=0$$

$$\Rightarrow \quad \lambda=2,3,-2$$

Hence, the eigen values are $2,3,-2$.

Example 4.22: Find the eigen values and corresponding eigen vectors of the matrix

$\begin{pmatrix} -2 & 2 & -3 \\ 2 & 1 & -6 \\ -1 & -2 & 0 \end{pmatrix}$. **[PTU May 2012]**

Solution: Let $\begin{pmatrix} -2 & 2 & -3 \\ 2 & 1 & -6 \\ -1 & -2 & 0 \end{pmatrix}$

The characteristic equation of the given matrix is

$$|A-\lambda I|=0$$

$$\Rightarrow \quad \begin{vmatrix} -2-\lambda & 2 & -3 \\ 2 & 1-\lambda & -6 \\ -1 & -2 & 0-\lambda \end{vmatrix}=0$$

Expanding determinant on the first row, we get

$$(-2-\lambda)\left[-\lambda(1-\lambda)-12\right]-2[-2\lambda-6]-3\left[-4+1(1-\lambda)\right]=0$$

$$\Rightarrow \quad \lambda^3+\lambda^2-21\lambda-45=0 \qquad \ldots(1)$$

By trial we find that $\lambda=-3$ satisfies Eq. (1)

$\therefore$ Extracting factor $(\lambda+3)$ from Eq. (1), we can write

$$(\lambda+3)\left(\lambda^2-2\lambda-15\right)=0$$

$$\Rightarrow \quad (\lambda+3)(\lambda+3)(\lambda-5)=0$$

$$\Rightarrow \quad \lambda=-3,-3,5$$

Thus, the eigen values of A are $-3,-3,5$.

Corresponding to $\lambda=-3$, eigen vectors are given by

$$AX=\lambda X$$

$$\Rightarrow \quad (A-\lambda I)X=O$$

$$\Rightarrow \quad (A+3I)X=O$$

$$\Rightarrow \quad \begin{pmatrix} 1 & 2 & -3 \\ 2 & 4 & -6 \\ -1 & -2 & 3 \end{pmatrix}\begin{pmatrix} x_1 \\ x_2 \\ x_3 \end{pmatrix}=O$$

Simplifying, we get only one independent equation

$$x_1+2x_2-3x_3=0$$

Choosing $x_2=0$, we have

$$x_1-3x_3=0$$

$\therefore \quad \dfrac{x_1}{3}=\dfrac{x_2}{0}=\dfrac{x_3}{1}$ giving the eigen vector $(3,0,1)$

Choosing $x_3=0$, we have

$$x_1+2x_2=0 \quad \Rightarrow \quad x_1=-2x_2$$

$\therefore \quad \dfrac{x_1}{2} = \dfrac{x_2}{-1} = \dfrac{x_3}{0}$ giving the eigen vector $(2, -1, 0)$

Any other eigen vector corresponding to eigen value $\lambda = -3$, will be a linear combination of these two eigen vectors.

Corresponding to $\lambda = 5$, eigen vectors are given by

$$\begin{pmatrix} -7 & 2 & -3 \\ 2 & -4 & -6 \\ -1 & -2 & -5 \end{pmatrix} \begin{pmatrix} x_1 \\ x_2 \\ x_3 \end{pmatrix} = O$$

$$\Rightarrow \quad -7x_1 + 2x_2 - 3x_3 = 0$$

$$2x_1 - 4x_2 - 6x_3 = 0$$

$$-1x_1 - 2x_2 - 5x_3 = 0$$

From first two equations, we have

$$\frac{x_1}{-12-12} = \frac{x_2}{-6-42} = \frac{x_3}{28-4} \Rightarrow \frac{x_1}{-24} = \frac{x_2}{-48} = \frac{x_3}{24} \Rightarrow \frac{x_1}{1} = \frac{x_2}{2} = \frac{x_3}{-1}$$

giving the eigen vector $(1, 2, -1)$

EXERCISE 4.4

1. Prove that eigen values of a skew hermetian matrix are either zero or purely imaginary. [PTU Dec 2012]

2. Find the eigen values of the following matrices:

   (a) $\begin{pmatrix} 1 & 2 & 2 \\ 0 & -4 & 2 \\ 0 & 0 & 7 \end{pmatrix}$ [PTU Dec 2006] (b) $\begin{pmatrix} 3 & 1 & 1 \\ 2 & 4 & 2 \\ 1 & 1 & 3 \end{pmatrix}$

3. If eigen values of matrix $\begin{pmatrix} 8 & -6 & 2 \\ -6 & 7 & -4 \\ 2 & -4 & k \end{pmatrix}$ are 0, 3 and 15. Find the value of k.

4. Find the eigen values and the corresponding eigen vectors of the following matrices:

(a) $\begin{pmatrix} 1 & 1 & 3 \\ 1 & 5 & 1 \\ 3 & 1 & 1 \end{pmatrix}$ [PTU Dec 2012] (b) $\begin{pmatrix} 1 & 0 & -1 \\ 1 & 2 & 1 \\ 2 & 2 & 3 \end{pmatrix}$ [PTU May 2006]

(c) $\begin{pmatrix} 8 & -6 & 2 \\ -6 & 7 & -4 \\ 2 & -4 & 3 \end{pmatrix}$ [PTU Jan 2010] (d) $\begin{pmatrix} -2 & 2 & -3 \\ 2 & 1 & -6 \\ -1 & -2 & 0 \end{pmatrix}$ [PTU Dec 2001]

ANSWERS

2. (a) –4, 1, 7 (b) 2, 2, 6
3. 3
4. (a) 3, 6, -2; $(-1, 0, 1)$, $(-1, 1, -1)$, $(1, 2, 1)$
   (b) 1, 2, 3; $(1, -1, 0)$, $(-2, 1, 2)$, $(1, -1, -2)$
   (c) 0, 3, 15; $(1, 2, 2)$, $(2, 1, -2)$, $(2, -2, 1)$
   (d) –3, -3, 5; $(-2, 1, 0)$, $(3, 0, 1)$, $(1, 2, -1)$

4.8 LINEAR INDEPENDENCE AND DEPENDENCE OF VECTORS

Any ordered n-tuple of number is called an n-vector. For example

$$X = (x_1, x_2, x_3, \ldots, x_n)$$

is a vector of n-dimension and the elements

$$x_1, x_2, x_3, \ldots, x_n$$

are called the first, second, third, . . ., nth component of X.

A vector can be written as a *row vector* or as a *column vector*.

For example, the coordinates of a point in space can be represented by a 3-vector (x, y, z) or $\begin{pmatrix} x \\ y \\ z \end{pmatrix}$.

If A be a matrix of order $m \times n$, then each row of A will be an n-vector and each column of A will be an m-vector.

A vector whose all components are zero is called a zero vector, and is denoted by O. Thus $O = (0, 0, 0, ..., 0)$.

Linearly Dependent Set of Vector

A set $(X_1, X_2, X_3, ..., X_r)$ of r-vectors is said to be linearly dependent set, if $\exists$ r scalars not all zero, such that

$$k_1 X_1 + k_2 X_2 + k_3 X_3 + ... + k_r X_r = O$$

Linearly Independent Set of Vector

A set $(X_1, X_2, X_3, ..., X_r)$ of r-vectors is said to be linearly independent set, if the set is not linearly dependent, i.e., if

$k_1 X_1 + k_2 X_2 + k_3 X_3 + ... + k_r X_r = O$ then all scalars are zero,

$$\Rightarrow \; k_1 = 0,\; k_2 = 0, ..., k_r = 0$$

Linear Combination of Vectors

A vector X is said to be a linear combination of vectors $X_1, X_2, X_3, ..., X_n$ if

$$X = k_1 X_1 + k_2 X_2 + k_3 X_3 + ... + k_n X_n$$

where $k_1, k_2, k_3, ..., k_n$ are scalars.

Art. If a set of vectors is linearly dependent, then show that at least one member of the set is a linear combination of the remaining members.

Proof: Let $X_1, X_2, X_3, ..., X_n$ be any linearly dependent set

$\therefore$ the relation

$$k_1 X_1 + k_2 X_2 + k_3 X_3 + ... + k_n X_n = O$$

Implies that at least one of $k_1, k_2, k_3, ..., k_n$ is non-zero.

Let k_1 is non-zero, then

$$X_1 = \left(-\frac{k_2}{k_1}\right)X_2 + \left(-\frac{k_3}{k_1}\right)X_3 + ... + \left(-\frac{k_n}{k_1}\right)X_n$$

which show that X_1 is a linear combination of $X_2, X_3, ..., X_n$.

Procedure to Test Linear Dependence of Vectors

To test the linear dependence of *r* vectors proceed as follows:

- Write these *r* vectors as row-vectors.
- Add suitable multiple of one of the vector so that the resulting $(r-1)$ vectors have their first component zero.
- Choose any one of these vectors and add its multiples to other vectors so that the remaining $(r-1)$ vectors have their second component zero.
- In this way, continue reducing the successive component to zero.

If the final reduction gives a vector whose all components are zero, then the original vectors are linearly dependent. However, if the final reduction gives a vector whose all components are not zero, then the original vectors are linearly independent.

ILLUSTRATIVE EXAMPLES 4.6

Example 4.23: Show that the vectors $(1,0,2,1)$, $(3,1,2,1)$, $(4,6,2,-4)$ and $(-6,0,-3,-4)$ form a linearly dependent.

Solution: Let X_1, X_2, X_3, X_4 be the four vectors such that $X_1 = (1,0,2,1)$, $X_2 = (3,1,2,1)$, $X_3 = (4,6,2,-4)$ and $X_4 = (-6,0,-3,-4)$.

Consider $k_1X_1 + k_2X_2 + k_3X_3 + k_4X_4 = 0$ ... (1)

$\Rightarrow \quad k_1(1,0,2,1) + k_2(3,1,2,1) + k_3(4,6,2,-4) + k_4(-6,0,-3,-4) = (0,0,0,0)$

$\Rightarrow \quad k_1 + 3k_2 + 4k_3 - 6k_4 = 0$

$k_2 + 6k_3 = 0$

$2k_1 + 2k_2 + 2k_3 - 3k_4 = 0$

$k_1 + k_2 - 4k_3 - 4k_4 = 0$

Rewriting, the above system as

$$k_1 + 3k_2 + 4k_3 - 6k_4 = 0$$

$$0k_1 + k_2 + 6k_3 + 0k_4 = 0$$

$$2k_1 + 2k_2 + 2k_3 - 3k_4 = 0$$

$$k_1 + k_2 - 4k_3 - 4k_4 = 0$$

Rewriting the above system in the matrix form as

$$\begin{pmatrix} 1 & 3 & 4 & -6 \\ 0 & 1 & 6 & 0 \\ 2 & 2 & 2 & -3 \\ 1 & 1 & -4 & 4 \end{pmatrix} \begin{pmatrix} k_1 \\ k_2 \\ k_3 \\ k_4 \end{pmatrix} = \begin{pmatrix} 0 \\ 0 \\ 0 \\ 0 \end{pmatrix} \qquad R_3 \to R_3 - 2R_1,\ R_4 \to R_4 - R_1$$

$$\begin{pmatrix} 1 & 3 & 4 & -6 \\ 0 & 1 & 6 & 0 \\ 0 & -4 & -6 & 9 \\ 0 & -2 & -8 & -2 \end{pmatrix} \begin{pmatrix} k_1 \\ k_2 \\ k_3 \\ k_4 \end{pmatrix} = \begin{pmatrix} 0 \\ 0 \\ 0 \\ 0 \end{pmatrix} \qquad R_3 \to R_3 + 4R_2,\ R_4 \to R_4 + 2R_2$$

$$\begin{pmatrix} 1 & 3 & 4 & -6 \\ 0 & 1 & 6 & 0 \\ 0 & 0 & 18 & 9 \\ 0 & 0 & 0 & 0 \end{pmatrix} \begin{pmatrix} k_1 \\ k_2 \\ k_3 \\ k_4 \end{pmatrix} = \begin{pmatrix} 0 \\ 0 \\ 0 \\ 0 \end{pmatrix} \qquad \text{which is echelon matrix}$$

$$\Rightarrow \quad k_1 + 3k_2 + 4k_3 - 6k_4 = 0$$

$$k_2 + 6k_3 = 0$$

$$18k_3 + 9k_4 = 0$$

Let $k_4 = t$

$$18k_3 + 9t = 0 \quad \Rightarrow \quad k_3 = \frac{-t}{2}$$

$$k_2 - \frac{6t}{2} = 0 \quad \Rightarrow \quad k_2 = 3t$$

$$k_1 + 9t - 2t - 6t = 0 \quad \Rightarrow \quad k_1 = -t$$

Putting in (1), we get

$$-tX_1 + 3tX_2 - \frac{t}{2}X_3 + tX_4 = 0$$

$$\Rightarrow \quad 2X_1 - 6X_2 + X_3 - 2X_4 = 0$$

Since k_1, k_2, k_3 and k_4 are non-zero, therefore, vectors (X_1, X_2, X_3, X_4) form a linearly dependent set.

Example 4.24: Show that the column vectors of the matrix $A = \begin{pmatrix} 1 & 2 & 3 \\ -2 & 1 & 2 \end{pmatrix}$ are linearly dependent.

Solution: Let $X_1 = (1,-2), X_2 = (2,1), X_3 = (3,2)$ are the column vectors.

Consider $B = \begin{pmatrix} 1 & -2 \\ 2 & 1 \\ 3 & 2 \end{pmatrix}$ $\quad R_2 \to R_2 - 2R_1, R_3 \to R_3 - 3R_1$

$\sim \begin{pmatrix} 1 & -2 \\ 0 & 5 \\ 0 & 8 \end{pmatrix}$ $\quad R_3 \to R_3 - \frac{8}{5}R_2$

$\sim \begin{pmatrix} 1 & -2 \\ 0 & 5 \\ 0 & 0 \end{pmatrix}$ which is in Echelon form having 2 non-zero rows

$\therefore$ Vectors X_1, X_2, X_3, i.e., column vectors of given matrix are linearly dependent.

Example 4.25: Show that the vectors $(1,2,1)$, $(2,-2,0)$ form a linearly independent set.

Solution: Consider $k_1(1,2,1) + k_2(2,-2,0) = (0,0,0)$

$$\Rightarrow \quad k_1 + 2k_2 = 0$$

$$2k_1 - 2k_2 = 0$$

$$3k_1 = 0$$

$$\therefore \quad k_1 = k_2 = 0$$

Given vectors form a linearly independent set.

Example 4.26: Are the solutions

$$y_1 = \cos x \quad \text{and} \quad y_2 = \sin x$$

are linearly independent?

Solution: Consider the relation

$$k_1 y_1 + k_2 y_2 = 0$$

$$\Rightarrow \quad k_1 \cos x + k_2 \sin x = 0 \qquad \ldots (1)$$

Putting $x = 0$ in (1), we get

$$k_1 = 0$$

Putting $x = \dfrac{\pi}{2}$ in (1), we get

$$k_2 = 0$$

$\therefore$ y_1 and y_2 are linearly dependent.

Example 4.27: Show that the set consisting only of the zero vectors, O, is linearly dependent.

Solution: Let $X = (0, 0, 0, ..., 0)$ be n-vectors whose all components are zero. Thus, the relation $kX = O = 0$ is true for some non-zero value of number k.

For example

$$2X = O \quad \text{and} \quad 2 \neq 0$$

Hence, the vector X is linearly dependent.

EXERCISE 4.5

1. Show that the vectors $(2,3,-1,-1)$, $(1,-1,-2,-4)$, $(3,1,3,-2)$, and $(6,3,0,-7)$ form a linearly dependent set. Also express one of these as a linear combination of others.
2. Show that the vector { (1,0,0,…,0), (0,1,0,…,0), (0,0,1,…,0),…, (0,0,0,…1) } of n vectors is linearly independent
3. Test for what values of 'k' the set of vectors $\{(k,1,1),\ (0,1,1),(k,0,k)\}$ is linearly independent. **[PTU May 2012]**
4. Is the system of vectors $(2,2,1)$, $(1,3,1)$, $(1,2,2)$ linearly dependent?
5. Examine for linear dependence of the vectors $X_1=(1,2,4)$, $X_2=(2,-1,3)$, $X_3=(0,1,2)$, $X_4=(-3,7,2)$. Also find the relation, if it exists.
6. Prove that the system of vectors $(1,2,3)$, $(1,2,-3)$, $(1,-2,3)$, $(2,-1,5)$ are linearly independent.
7. Test whether the set of vectors $\{(1,1,1),(1,-1,1),(3,-1,3)\}$ are LI or LD by giving suitable reasons. **[PTU Dec 2012]**

ANSWERS

3. 0 4. No 5. $X_1-12X_2+5X_3-5X_4$
7. LD

4.9 LINEAR TRANSFORMATION

If $V(F)$ and $W(F)$ are two vector spaces, then a mapping T from V to W, *i.e.*, $T: V \to W$ is said to be a linear transformation iff

(*i*) $T(v+w) = T(v) + T(w) \quad \forall v, w \in V$

i.e., T take the sum $(v + w)$ of V to the sum of the T-images in W.

(*ii*) $T(av) = aT(v)\ \forall\ v \in V$ and $a \in F$.

i.e., T take the scalar multiplication av of V to the scalar multiplication $aT(v)$ in W.

Matrix Associated with a Linear Transformation

Let R^n and R^m be finite dimensional vector spaces of dimensions n, m, respectively.

Take $T: R^n \to R^m$ be a Linear Transformation (L.T.), and $B=\{e_1,e_2,e_3,\ldots,e_i,\ldots,e_n\}$ be standard basis of R^n where $e_i=(0,0,\ldots,1,0,\ldots,0)$ where 1 occurs at i^{th} place.

Take any vector $v=(x_1,x_2,x_3,...,x_n)\in R^n$

$\therefore$ v can be written as $v=x_1e_1+x_2e_2++x_3e_3+...+x_ne_n$

$$\Rightarrow \quad T(v)=T(x_1e_1+x_2e_2+x_3e_3+...+x_ne_n)$$

$$=x_1T(e_1)+x_2T(e_2)+...+x_nT(e_n)$$

$$=x_1T(e_1)+x_2T(e_2)+...+x_nT(e_n) \qquad [\because \text{T is a L.T.}]$$

$$\Rightarrow \quad T(v)=\sum_{i=1}^{n}x_iT(e_i) \qquad ...(1)$$

Now $T(e_i)\in R^m$ as $T(e_i),T(e_2),...,T(e_n)$ are images of $e_1,e_2,...,e_n$ under linear transformations.

So let $T(e_i)=(\lambda_{1i},\lambda_{2i},...,\lambda_{mi})$, $1\le i\le n$

$\therefore$ Eq. (1) becomes

$$T(v)=\sum_{i=1}^{n}x_i(\lambda_{1i},\lambda_{2i},...,\lambda_{mi})$$

$$=x_1(\lambda_{11},\lambda_{21},...,\lambda_{m1})+x_2(\lambda_{12},\lambda_{22},...,\lambda_{m2})+,...,+x_n(\lambda_{1n},\lambda_{2n},...,\lambda_{mn})$$

$$=(\lambda_{11}x_1+\lambda_{12}x_2+...+\lambda_{1n}x_n),(\lambda_{21}x_1+\lambda_{22}x_2+...+\lambda_{2n}x_n),...,(\lambda_{m1}x_1+\lambda_{m2}x_2+...+\lambda_{mn}x_n)$$

If $T(v)=w=(y_1,y_2...y_m)\in R^m$, then

$$(y_1,y_2,...,y_m)=(\lambda_{11}x_1+\lambda_{12}x_2+...\lambda_{1n}x_n),(\lambda_{21}x_1+\lambda_{22}x_2+...+\lambda_{2n}x_n),...,(\lambda_{m1}x_1+\lambda_{m2}x_2+...\lambda_{mn}x_n)$$
$$=(\lambda_{11}x_1+\lambda_{12}x_2+...\lambda_{1n}x_n,\lambda_{21}x_1+\lambda_{22}x_2+...+\lambda_{2n}x_n,...,\lambda_{m1}x_1+\lambda_{m2}x_2+...\lambda_{mn}x_n)$$

$$\therefore \quad y_1=\lambda_{11}x_1+\lambda_{12}x_2+...+\lambda_{1n}x_n$$
$$y_2=\lambda_{21}x_1+\lambda_{22}x_2+...+\lambda_{2n}x_n$$
$$\vdots$$
$$y_m=\lambda_{m1}x_1+\lambda_{m2}x_2+...+\lambda_{mn}x_n$$

$$\Rightarrow \begin{pmatrix} y_1 \\ y_2 \\ \vdots \\ y_m \end{pmatrix} = \begin{pmatrix} \lambda_{11} & \lambda_{12} & \cdots & \lambda_{1n} \\ \lambda_{21} & \lambda_{22} & \cdots & \lambda_{2n} \\ \vdots & \vdots & \vdots & \vdots \\ \lambda_{m1} & \lambda_{m2} & \cdots & \lambda_{mn} \end{pmatrix} \begin{pmatrix} x_1 \\ x_2 \\ \vdots \\ x_n \end{pmatrix}$$

$$\Rightarrow \quad Y = AX\,(say) \qquad \ldots (2)$$

Thus, if we write $v=(x_1,x_2,...,x_n)$ and $w=(y_1,y_2,...,y_m)$ as column matrix X and Y respectively, then Eq. (2) tells us that image w of vector v can be obtained by pre-multiplying the column matrix X with the matrix A whose columns are the images $T(e_i)$ of vector e_i.

Hence T is completely determined by the matrix $A=\left[\alpha_{ij}\right]_{m\times n}$, $\alpha_{ij}\in R$.

Conversely, for a given matrix $B=\left[\beta_{ij}\right]_{m\times n}$, $\beta_{ij}\in R$.

We define a mapping $T:R^n\to R^m$ by

$$T(v)=w$$

where $v=(x_1,x_2,...x_n)\in R^n$, and $w=(y_1,y_2,y_3...y_m)\in R^m$

Such that

$$y_1=\beta_{11}x_1+\beta_{12}x_2+...+\beta_{1n}x_n$$

$$y_2=\beta_{21}x_1+\beta_{22}x_2+...+\beta_{2n}x_n$$

$$\vdots$$

$$y_m=\beta_{m1}x_1+\beta_{m2}x_2+...+\beta_{mn}x_n$$

$$\Rightarrow \quad Y=BX$$

Clearly the above mapping $T:R^n\to R^m$, defined by matrix B is a L.T.

Hence, there corresponds a $m\times n$ matrix to each L.T. from R^n and for each such matrix, there is a unique L.T. from R^n to R^m.

ILLUSTRATIVE EXAMPLES 4.7

Example 4.28: $T : R^2 \to R^3$ given by $T(x_1, x_2) = (2x_1 - x_2, x_1 - x_2, -2x_1)$ is a L.T. Find the associated matrix.

Solution: Given mapping T is a L.T.

To find matrix take $T(x_1, x_2) = (y_1, y_2, y_3)$ where $(x_1, x_2) \in R^2$ and $(y_1, y_2, y_3) \in R^3$

$$\Rightarrow \quad (y_1, y_2, y_3) = T(x_1, x_2)$$

$$= (2x_1 - x_2, x_1 - x_2, -2x_1)$$

$$\therefore \quad \begin{pmatrix} y_1 = 2x_1 - x_2 \\ y_2 = x_1 - x_2 \\ y_3 = -2x_1 + 0x_2 \end{pmatrix} \Rightarrow \begin{pmatrix} y_1 \\ y_2 \\ y_3 \end{pmatrix} = \begin{pmatrix} 2 & -1 \\ 1 & -1 \\ -2 & 0 \end{pmatrix} \begin{pmatrix} x_1 \\ x_2 \end{pmatrix}$$

So the matrix associated with given L.T. is $= \begin{pmatrix} 2 & -1 \\ 1 & -1 \\ -2 & 0 \end{pmatrix}$.

Example 4.29: If $T : R^4 \to R^3$ is a L.T and $\lambda \in R$, show that the function.

$\lambda T : R^4 \to R^3$ defined by $(\lambda T)(X) = \lambda T(X)$ is a L.T. Also write the associated matrix.

Solution: Let $V_1, V_2 \in R^4$ and $\alpha, \beta \in R$

$$\Rightarrow \quad \alpha v_1 + \beta v_2 \in R^4 \qquad (\because R^4 \text{ is a vector space})$$

Now $(\lambda T)(\alpha v_1 + \beta v_2) = \lambda T(\alpha v_1 + \beta v_2)$

$$= \lambda(\alpha T(v_1) + \beta T(v_2))$$

$$= \lambda(\alpha T(v_1)) + \lambda(\beta T(v_2))$$

$$= \alpha \lambda T(v_1) + \beta \lambda T(v_2)$$

$$= \alpha(\lambda T)(v_1) + \beta(\lambda T)(v_2)$$

$$\Rightarrow \quad \lambda T : R^4 \to R^3 \text{ is a L.T.}$$

To find matrix let $A=\left[\alpha_{ij}\right]_{3\times4}$ be a matrix associated with T,

$$T(X)=\begin{pmatrix}\alpha_{11} & \alpha_{12} & \alpha_{13} & \alpha_{14}\\ \alpha_{21} & \alpha_{22} & \alpha_{23} & \alpha_{24}\\ \alpha_{31} & \alpha_{32} & \alpha_{33} & \alpha_{34}\end{pmatrix}\begin{pmatrix}x_1\\x_2\\x_3\\x_4\end{pmatrix} \quad \forall \ (x_1,x_2,x_3,x_4)\in R^4$$

$$\Rightarrow (\lambda T)(X)=\lambda\begin{pmatrix}\alpha_{11} & \alpha_{12} & \alpha_{13} & \alpha_{14}\\ \alpha_{21} & \alpha_{22} & \alpha_{23} & \alpha_{24}\\ \alpha_{31} & \alpha_{32} & \alpha_{33} & \alpha_{34}\end{pmatrix}\begin{pmatrix}x_1\\x_2\\x_3\\x_4\end{pmatrix}$$

$$=\begin{pmatrix}\lambda\alpha_{11} & \lambda\alpha_{12} & \lambda\alpha_{13} & \lambda\alpha_{14}\\ \lambda\alpha_{21} & \lambda\alpha_{22} & \lambda\alpha_{23} & \lambda\alpha_{24}\\ \lambda\alpha_{31} & \lambda\alpha_{32} & \lambda\alpha_{33} & \lambda\alpha_{34}\end{pmatrix}\begin{pmatrix}x_1\\x_2\\x_3\\x_4\end{pmatrix}$$

$\Rightarrow$ Matrix associated with L.T., λT is

$$\begin{pmatrix}\lambda\alpha_{11} & \lambda\alpha_{12} & \lambda\alpha_{13} & \lambda\alpha_{14}\\ \lambda\alpha_{21} & \lambda\alpha_{22} & \lambda\alpha_{23} & \lambda\alpha_{24}\\ \lambda\alpha_{31} & \lambda\alpha_{32} & \lambda\alpha_{33} & \lambda\alpha_{34}\end{pmatrix}=\lambda A$$

Example 4.30: Consider the L.T., $Y = AX$ in R_3 where $A=\begin{pmatrix}1 & 1 & 1\\ 1 & 2 & 3\\ 2 & 5 & 3\end{pmatrix}$

Find (*a*) The image of $X=(3,5,1)$ (*b*) The vector X whose image is $(0,3,4)$.

Solution:

(*a*) The image of $X=(3,5,1)$ is given by

$$Y = AX \quad \Rightarrow \quad \begin{pmatrix}y_1\\y_2\\y_3\end{pmatrix}=\begin{pmatrix}1 & 1 & 1\\ 1 & 2 & 3\\ 2 & 5 & 3\end{pmatrix}\begin{pmatrix}3\\5\\1\end{pmatrix}=\begin{pmatrix}9\\16\\34\end{pmatrix}$$

$\therefore$ Image of $X=(3,5,1)$ is $(9,16,34)$

(*b*) To find X whose image is $(0,3,4)$ can be obtained on solving

$$\begin{pmatrix} 1 & 1 & 1 \\ 1 & 2 & 3 \\ 2 & 5 & 3 \end{pmatrix}\begin{pmatrix} x_1 \\ x_2 \\ x_3 \end{pmatrix} = \begin{pmatrix} 0 \\ 3 \\ 4 \end{pmatrix}$$

or $\quad AX = y$

$$X = A^{-1}y \qquad \ldots (1)$$

$$\left[\because |A| = -5 \neq 0 \text{ so } A^{-1} \text{ exists}\right]$$

$$\text{Now } A^{-1} = \frac{-1}{5}\begin{pmatrix} -9 & 2 & 9 \\ 3 & 1 & -2 \\ 1 & -3 & 1 \end{pmatrix}$$

Students are advised to check it.

Putting in Eq. (1), we get

$$X = \frac{-1}{5}\begin{pmatrix} -9 & 2 & 1 \\ 3 & 1 & -2 \\ 1 & -3 & 1 \end{pmatrix}\begin{pmatrix} 0 \\ 3 \\ 4 \end{pmatrix}$$

$$= \frac{-1}{5}\begin{pmatrix} 10 \\ -5 \\ -5 \end{pmatrix} = \begin{pmatrix} -2 \\ 1 \\ 1 \end{pmatrix}$$

$\therefore \quad X$ is $(-2,1,1)$ whose image is $(0,3,4)$.

EXERCISE 4.6

1. Show that $T : R^2 \to R^3$ given by

   $T(x, y) = (x + y, x - y, y)$ is a L.T., find the associated matrix.

2. Show that $T : R^2 \to R^3$ given by

   $T(x_1, x_2) = (x_1 - x_2, -x + 2x_2, -2x_1)$

   is a L.T. Also find the associated matrix.

3. Write the L.T. for $T : R^4 \rightarrow R^3$ corresponding to

$$\beta = \begin{pmatrix} 1 & 2 & 3 & 4 \\ 0 & -5 & -6 & -7 \\ -1 & 8 & 9 & -2 \end{pmatrix}$$

4. Using the transformation $Y = AX$, where $A = \begin{pmatrix} 1 & 3 & 2 \\ 2 & 1 & 1 \\ 3 & 2 & 3 \end{pmatrix}$, find

   (*a*) image of $X = (1,1,2)$

   (*b*) vector *X* whose image is $(-2,-5,-5)$

5. Using the transformation $Y = AX$, where $A = \begin{pmatrix} 1 & 1 & 2 \\ 1 & 2 & 5 \\ 1 & 3 & 3 \end{pmatrix}$, find

   (*a*) image of $X = (1,4,0)$

   (*b*) vector whose image is $(1,4,0)$.

ANSWERS

1. $\begin{pmatrix} 1 & 1 \\ 1 & -1 \\ 0 & 1 \end{pmatrix}$ 2. $\begin{pmatrix} 1 & -1 \\ -1 & 2 \\ -2 & 0 \end{pmatrix}$

3. $T(x_1, x_2, x_3, x_4) = (x_1 + 2x_2 + 3x_3 + 4x_4, -5x_2 - 6x_3 - 7x_4, -x_1 + 8x_2 + 9x_3 - 2x_4)$

4. (*a*) $(8,5,11)$

   (*b*) $(-2,-5,-5)$

5. (*a*) $(5,9,13)$

   (*b*) $\left(\frac{-3}{5}, \frac{-6}{5}, \frac{7}{5}\right)$

4.10 CAYLEY-HAMILTON THEOREM

State and prove Cayley-Hamilton Theorem.

[PTU Dec 2002, May 2004, May 2006, May 2007, Jan 2009]

Statement: Every square matrix satisfies its own characteristics equation, *i.e.*, if

$$|A-\lambda I| = (-1)^n \lambda^n + p_1\lambda^{n-1} + p_2\lambda^{n-2} + \ldots + p_{n-2}\lambda^2 + p_{n-1}\lambda + p_n = 0$$

is the characteristics equation of the nth order square matrix A ,then

$$(-1)^n A^n + p_1 A^{n-1} + p_2 A^{n-2} + \ldots + p_{n-2}A^2 + p_{n-1}A + p_n I = 0$$

Proof:

We know that $A(Adj.A) = |A|I$

$$(A-\lambda I)Adj.(A-\lambda I) = |A-\lambda I|I$$

Let $Adj.(A-\lambda I) = B_1\lambda^{n-1} + B_2\lambda^{n-2} + B_3\lambda^{n-3} + \ldots + B_{n-2}\lambda^2 + B_{n-1}\lambda + B_n$

$$(A-\lambda I)Adj.(A-\lambda I) = |A-\lambda I|$$

$$(A-\lambda I)\left(B_1\lambda^{n-1} + B_2\lambda^{n-2} + B_3\lambda^{n-3} + \ldots + B_{n-2}\lambda^2 + B_{n-1}\lambda + B_n\right)$$

$$= (-1)^n \lambda^n + p_1\lambda^{n-1} + p_2\lambda^{n-2} + \ldots + p_{n-2}\lambda^2 + p_{n-1}\lambda + p_n$$

Comparing coefficients of like powers of λ, we get

$$-B_1 = (-1)^n$$

$$AB_1 - B_2 = p_1$$

$$AB_2 - B_3 = p_2$$

$$\vdots$$

$$AB_{n-2} - B_{n-1} = p_{n-2}$$

$$AB_{n-1} - B_n = p_{n-1}$$

$$AB_n = p_n$$

Pre-multiplying above equations by $A^n, A^{n-1}, A^{n-2},...., A^2, A, I$ respectively and adding, we get

$$O = (-1)^n A^n + p_1 A^{n-1} + p_2 A^{n-2} + ... + p_{n-2} A^2 + p_{n-1} A + p_n I$$

$$\Rightarrow \quad (-1)^n A^n + p_1 A^{n-1} + p_2 A^{n-2} + ... + p_{n-2} A^2 + p_{n-1} A + p_n I = O$$

Hence the proof.

Alternatively:

Characteristic equation of square matrix A can be written as

$$|A - \lambda I| = p_0 + p_1 \lambda + p_2 \lambda^2 + ... + p_n \lambda^n$$

We need to prove that A satisfies the following equation

$$p_0 I + p_1 A + p_2 A^2 + ... + p_n A^n = O$$

We know that, for a square matrix P, we have

$$P(Adj.P) = |P| I$$

Let $P = (A - \lambda I)$

$$\therefore \quad (A - \lambda I) Adj.(A - \lambda I) = |A - \lambda I|$$

Let $Adj.(A - \lambda I) = B_0 + B_1 \lambda + B_2 \lambda^2 + ... + B_{n-2} \lambda^{n-2} + B_{n-1} \lambda^{n-1}$

$$(A - \lambda I) Adj.(A - \lambda I) = |A - \lambda I|$$

$$(A - \lambda I)\left(B_0 + B_1 \lambda + B_2 \lambda^2 + ... + B_{n-2} \lambda^{n-2} + B_{n-1} \lambda^{n-1}\right) = p_0 + p_1 \lambda + p_2 \lambda^2 + ... + p_n \lambda^n$$

Comparing coefficients of like powers of λ, we get

$$AB_0 = p_0$$

$$AB_1 - B_0 = p_1$$

$$AB_2 - B_1 = p_2$$

$$\vdots$$

$$AB_{n-1} - B_{n-2} = p_{n-1}$$

$$-B_{n-1} = p_n$$

Pre-multiplying above equations by $I, A, A^2, ..., A^n$ respectively and adding, we get

$$O = p_0 I + p_1 A + p_2 A^2 + ... + p_n A^n$$

$$\Rightarrow \quad p_0 I + p_1 A + p_2 A^2 + ... + p_n A^n = O$$

Hence the proof.

4.10.1 To Find A^{-1} Using Cayley-Hamilton Theorem

We know that

$$p_0 I + p_1 A + p_2 A^2 + ... + p_n A^n = O$$

$$\Rightarrow \quad p_0 I = -\left[p_1 A + p_2 A^2 + ... + p_n A^n\right]$$

Post-multiplying both sides by A^{-1}, we get

$$p_0 I A^{-1} = -\left[p_1 A + p_2 A^2 + ... + p_n A^n\right] A^{-1}$$

$$\Rightarrow \quad p_0 A^{-1} = -\left[p_1 A A^{-1} + p_2 A^2 A^{-1} + ... + p_n A^n A^{-1}\right] \qquad [\because IA = A]$$

$$\Rightarrow \quad p_0 A^{-1} = -\left[p_1 I + p_2 A\left(AA^{-1}\right) + ... + p_n A^{n-1}\left(AA^{-1}\right)\right]$$

$$\Rightarrow \quad p_0 A^{-1} = -\left[p_1 I + p_2 AI + ... + p_n A^{n-1} I\right] \qquad \left[\because AA^{-1} = I\right]$$

$$\Rightarrow \quad p_0 A^{-1} = -\left[p_1 I + p_2 A + ... + p_n A^{n-1}\right]$$

$$\Rightarrow \quad A^{-1} = -\frac{1}{p_0}\left[p_1 I + p_2 A + ... + p_n A^{n-1}\right]$$

ILLUSTRATIVE EXAMPLES 4.8

Example 4.31: Verify Cayley-Hamilton theorem for the matrix $A=\begin{pmatrix}1 & 2 & 1\\ -1 & 0 & 3\\ 2 & -1 & 1\end{pmatrix}$.

Solution: Characteristics equation for A is

$$|A-\lambda I|=\begin{vmatrix}1-\lambda & 2 & 1\\ -1 & -\lambda & 3\\ 2 & -1 & 1-\lambda\end{vmatrix}=0$$

which gives $\lambda^3-2\lambda^2+4\lambda-18=0$

Now we have to show that

$$A^3-2A^2+4A-18I=O$$

$$\text{Again } A^2=\begin{pmatrix}1 & 2 & 1\\ -1 & 0 & 3\\ 2 & -1 & 1\end{pmatrix}\begin{pmatrix}1 & 2 & 1\\ -1 & 0 & 3\\ 2 & -1 & 1\end{pmatrix}=\begin{pmatrix}1 & 1 & 8\\ 5 & -5 & 2\\ 5 & 3 & 0\end{pmatrix}$$

$$A^3=\begin{pmatrix}1 & 1 & 8\\ 5 & -5 & 2\\ 5 & 3 & 0\end{pmatrix}\begin{pmatrix}1 & 2 & 1\\ -1 & 0 & 3\\ 2 & -1 & 1\end{pmatrix}=\begin{pmatrix}16 & -6 & 12\\ 14 & 8 & -8\\ 2 & 10 & 14\end{pmatrix}$$

$$\text{LHS}=A^3-2A^2+4A-18I$$

$$=\begin{pmatrix}16 & -6 & 12\\ 14 & 8 & -8\\ 2 & 10 & 14\end{pmatrix}-2\begin{pmatrix}1 & 1 & 8\\ 5 & -5 & 2\\ 5 & 3 & 0\end{pmatrix}+4\begin{pmatrix}1 & 2 & 1\\ -1 & 0 & 3\\ 2 & -1 & 1\end{pmatrix}-18\begin{pmatrix}1 & 0 & 0\\ 0 & 1 & 0\\ 0 & 0 & 1\end{pmatrix}$$

$$=\begin{pmatrix}0 & 0 & 0\\ 0 & 0 & 0\\ 0 & 0 & 0\end{pmatrix}=O=\text{RHS}$$

Hence, A satisfies its characteristics equation. Thus, Cayley-Hamilton theorem is verified.

Example 4.32: Using Cayley-Hamilton theorem, find A^8 if $A=\begin{pmatrix}1 & 2\\ 2 & -1\end{pmatrix}$.

Solution: Characteristics equation of A is

$$|A-\lambda I|=\begin{vmatrix}1-\lambda & 2\\ 2 & -1-\lambda\end{vmatrix}=0$$

which gives $\lambda^2-5=0$ ... (1)

By Cayley-Hamilton theorem

$$A^2-5I=O \quad \text{... (2)}$$

$$\Rightarrow \quad A^2=5I \quad \text{... (3)}$$

Squaring both sides of Eq. (3), we have

$$\left(A^2\right)^2=(5I)^2$$

$$\Rightarrow \quad A^4=25I \quad \text{... (4)}$$

Squaring both sides of Eq. (4), we have

$$\left(A^4\right)^2=(25I)^2 \quad \Rightarrow \quad A^8=625I=625\begin{pmatrix}1 & 0\\ 0 & 1\end{pmatrix}=\begin{pmatrix}625 & 0\\ 0 & 625\end{pmatrix}$$

Hence $A^8=\begin{pmatrix}625 & 0\\ 0 & 625\end{pmatrix}$

Example 4.33: Find the characteristic equation of the matrix $A=\begin{pmatrix}2 & 1 & 1\\ 0 & 1 & 0\\ 1 & 1 & 2\end{pmatrix}$ and hence, find the matrix represented by $A^8-5A^7+7A^6-3A^5+A^4-5A^3+8A^2-2A+I$.

Solution: The characteristics equation of A is

$$|A-\lambda I|=\begin{vmatrix}2-\lambda & 1 & 1\\ 0 & 1-\lambda & 0\\ 1 & 1 & 2-\lambda\end{vmatrix}=0$$

$$\Rightarrow \quad \lambda^3-5\lambda^2+7\lambda-3=0$$

By Cayley-Hamilton theorem

$$A^3 - 5A^2 + 7A - 3I = O \qquad \ldots (1)$$

Now, $A^8 - 5A^7 + 7A^6 - 3A^5 + A^4 - 5A^3 + 8A^2 - 2A + I$

$$= A^5\left(A^3 - 5A^2 + 7A - 3I\right) + A\left(A^3 - 5A^2 + 7A - 3I\right) + 1\left(A^2 + A + I\right)$$

$$= A^2 + A + \text{I} \qquad \text{[Using (1)]}$$

$$= \begin{pmatrix} 2 & 1 & 1 \\ 0 & 1 & 0 \\ 1 & 1 & 2 \end{pmatrix}\begin{pmatrix} 2 & 1 & 1 \\ 0 & 1 & 0 \\ 1 & 1 & 2 \end{pmatrix} + \begin{pmatrix} 2 & 1 & 1 \\ 0 & 1 & 0 \\ 1 & 1 & 2 \end{pmatrix} + \begin{pmatrix} 1 & 0 & 0 \\ 0 & 1 & 0 \\ 0 & 0 & 1 \end{pmatrix}$$

$$= \begin{pmatrix} 5 & 4 & 4 \\ 0 & 1 & 0 \\ 4 & 4 & 5 \end{pmatrix} + \begin{pmatrix} 2 & 1 & 1 \\ 0 & 1 & 0 \\ 1 & 1 & 2 \end{pmatrix} + \begin{pmatrix} 1 & 0 & 0 \\ 0 & 1 & 0 \\ 0 & 0 & 1 \end{pmatrix}.$$

$$= \begin{pmatrix} 8 & 5 & 5 \\ 0 & 3 & 0 \\ 5 & 5 & 8 \end{pmatrix}.$$

Hence $A^8 - 5A^7 + 7A^6 - 3A^5 + A^4 - 5A^3 + 8A^2 - 2A + I = \begin{pmatrix} 8 & 5 & 5 \\ 0 & 3 & 0 \\ 5 & 5 & 8 \end{pmatrix}.$

Example 4.34: Verify Cayley-Hamilton theorem for the matrix $A = \begin{pmatrix} 2 & -1 & 1 \\ -1 & 2 & -1 \\ 1 & -1 & 2 \end{pmatrix}$.

Hence, compute A^{-1}.

Solution: Characteristic equation of A is

$$|A - \lambda I| = \begin{vmatrix} 2-\lambda & -1 & 1 \\ -1 & 2-\lambda & -1 \\ 1 & -1 & 2-\lambda \end{vmatrix} = 0$$

$$\Rightarrow \lambda^3 - 6\lambda^2 + 9\lambda - 4 = 0 \qquad \ldots (1)$$

To very Cayley-Hamilton theorem, we have to show that

$$A^3 - 6A^2 + 9A - 4I = O \qquad \ldots (2)$$

Now, $A^2 = \begin{pmatrix} 2 & -1 & 1 \\ -1 & 2 & -1 \\ 1 & -1 & 2 \end{pmatrix} \times \begin{pmatrix} 2 & -1 & 1 \\ -1 & 2 & -1 \\ 1 & -1 & 2 \end{pmatrix} = \begin{pmatrix} 6 & -5 & 5 \\ -5 & 6 & -5 \\ 5 & -5 & 6 \end{pmatrix}$

$$A^3 = A^2 \times A = \begin{pmatrix} 6 & -5 & 5 \\ -5 & 6 & -5 \\ 5 & -5 & 6 \end{pmatrix} \begin{pmatrix} 2 & -1 & 1 \\ -1 & 2 & -1 \\ 1 & -1 & 2 \end{pmatrix} = \begin{pmatrix} 22 & -21 & -21 \\ -21 & 22 & -21 \\ 21 & -21 & 22 \end{pmatrix}$$

LHS of Eq. (2) is

$$A^3 - 6A^2 + 9A - 4I$$

$$= \begin{pmatrix} 2 & -21 & -21 \\ -21 & 22 & -21 \\ 21 & -21 & 22 \end{pmatrix} - 6 \begin{pmatrix} 6 & -5 & 5 \\ -5 & 6 & -5 \\ 5 & -5 & 6 \end{pmatrix} + 9 \begin{pmatrix} 2 & -1 & 1 \\ -1 & 2 & -1 \\ 1 & -1 & 2 \end{pmatrix} - 4 \begin{pmatrix} 1 & 0 & 0 \\ 0 & 1 & 0 \\ 0 & 0 & 1 \end{pmatrix}$$

$$= \begin{pmatrix} 0 & 0 & 0 \\ 0 & 0 & 0 \\ 0 & 0 & 0 \end{pmatrix} = 0 \qquad \text{RHS of Eq. (2)}$$

This verifies Cayley-Hamilton theorem.

Now, multiplying both sides of Eq. (2) by A^{-1}, we have

$$A^2 - 6A + 9I - 4A^{-1} = O$$

$$\Rightarrow \quad 4A^{-1} = A^2 - 6A + 9I$$

$$= \begin{pmatrix} 6 & -5 & 5 \\ -5 & 6 & -5 \\ 5 & -5 & 6 \end{pmatrix} - 6 \begin{pmatrix} 2 & -1 & 1 \\ -1 & 2 & -1 \\ 1 & -1 & 2 \end{pmatrix} + 9 \begin{pmatrix} 1 & 0 & 0 \\ 0 & 1 & 0 \\ 0 & 0 & 1 \end{pmatrix} = \begin{pmatrix} 3 & 1 & -1 \\ 1 & 3 & 1 \\ -1 & 1 & 3 \end{pmatrix}$$

$$\therefore \quad A^{-1} = \frac{1}{4} \begin{pmatrix} 3 & 1 & -1 \\ 1 & 3 & 1 \\ -1 & 1 & 3 \end{pmatrix}.$$

Example 4.35: Find the inverse of $A=\begin{pmatrix}2&5&3\\3&1&2\\1&2&1\end{pmatrix}$ using Cayley-Hamilton theorem.

Solution: Characteristic equation of A is

$$|A-\lambda I|=\begin{vmatrix}2-\lambda&5&3\\3&1-\lambda&2\\1&2&1-\lambda\end{vmatrix}=0$$

which gives $\lambda^3-4\lambda^2-17\lambda-4=0$ . . . (1)

By Cayley-Hamilton theorem

$$A^3-4A^2-17A-4I=O \qquad \text{. . . (2)}$$

Now, multiplying both sides of Eq. (2) by A^{-1}, we have

$$A^2-4A-17I-4A^{-1}=O$$

$$4A^{-1}=A^2-4A-17I \qquad \text{. . . (3)}$$

Now, $A^2=\begin{pmatrix}2&5&3\\3&1&2\\1&2&1\end{pmatrix}\times\begin{pmatrix}2&5&3\\3&1&2\\1&2&1\end{pmatrix}=\begin{pmatrix}22&21&19\\11&20&13\\9&9&8\end{pmatrix}$

From Eq. (3), we have

Now, $4A^{-1}=\begin{pmatrix}22&21&19\\11&20&13\\9&9&8\end{pmatrix}-4\begin{pmatrix}2&5&3\\3&1&2\\1&2&1\end{pmatrix}-17\begin{pmatrix}1&0&0\\0&1&0\\0&0&1\end{pmatrix}=\begin{pmatrix}-3&1&7\\-1&-1&5\\5&1&-13\end{pmatrix}$

$$\therefore \qquad A^{-1}=\frac{1}{4}\begin{pmatrix}-3&1&7\\-1&-1&5\\5&1&-13\end{pmatrix}.$$

EXERCISE 4.7

1. Verify Cayley-Hamilton theorem for the matrices:

   (*a*) $\begin{pmatrix} 3 & 2 & 4 \\ 4 & 3 & 2 \\ 2 & 4 & 3 \end{pmatrix}$ [PTU Dec 2006] (*b*) $\begin{pmatrix} 2 & 3 & 1 \\ 3 & 1 & 2 \\ 1 & 2 & 3 \end{pmatrix}$ [PTU May 2006]

2. State Cayley-Hamilton theorem. Use it to find the inverse of the matrix

   $\begin{pmatrix} 4 & 3 & 1 \\ 2 & 1 & -2 \\ 1 & 2 & 1 \end{pmatrix}$ [PTU Dec 2012]

3. Using Cayley-Hamilton theorem, find the inverse of $\begin{pmatrix} 2 & -1 & 1 \\ -1 & 2 & -1 \\ 1 & -1 & 2 \end{pmatrix}$.

   [PTU Dec 2005, Jan 2009]

4. Find the inverse of $\begin{pmatrix} 2 & 5 & 3 \\ 3 & 1 & 2 \\ 1 & 2 & 1 \end{pmatrix}$ using Cayley-Hamilton theorem.

5. Verify Cayley-Hamilton theorem for following matrices, and also find inverse of these matrices:

   (*a*) $\begin{pmatrix} 0 & 0 & 1 \\ 1 & 2 & 0 \\ 2 & -1 & 0 \end{pmatrix}$ (*b*) $\begin{pmatrix} 1 & \sqrt{2} & 0 \\ \sqrt{2} & -1 & 0 \\ 0 & 0 & 1 \end{pmatrix}$ (*c*) $\begin{pmatrix} 5 & -6 & -6 \\ -1 & 4 & 2 \\ 3 & -6 & -4 \end{pmatrix}$

ANSWERS

2. $\frac{1}{11}\begin{pmatrix} 5 & -1 & -7 \\ -4 & 3 & 10 \\ 3 & -5 & -2 \end{pmatrix}$ 3. $\frac{1}{4}\begin{pmatrix} 3 & -1 & 1 \\ 1 & 3 & 1 \\ -1 & 1 & 3 \end{pmatrix}$ 4. $\frac{1}{4}\begin{pmatrix} -3 & 1 & 7 \\ -1 & -1 & 5 \\ 5 & 1 & -13 \end{pmatrix}$

5. ((*a*) $\frac{1}{5}\begin{pmatrix} 0 & 1 & 2 \\ 0 & 2 & -1 \\ 5 & 0 & 0 \end{pmatrix}$ (*b*) $\frac{1}{3}\begin{pmatrix} 1 & \sqrt{2} & 0 \\ \sqrt{2} & -1 & 0 \\ 0 & 0 & 3 \end{pmatrix}$ (*c*) $\frac{1}{2}\begin{pmatrix} -2 & 6 & 6 \\ 1 & -1 & -2 \\ -3 & 6 & 7 \end{pmatrix}$

4.11 REDUCTION OF A MATRIX TO A DIAGONAL FORM

If a square matrix A of order n is said to be diagonalizable if there exist an invertible matrix P of order n such that $P^{-1}AP = D$, where D is a diagonal matrix and the elements of D are eigen values of A.

Art. The necessary and sufficient condition that a square matrix A be diagonalizable is that A has n linearly independent vectors.

Proof: Necessary Condition: Let A be diagonalizable

$\Rightarrow$ A is similar to a diagonal matrix D, say

$\Rightarrow$ $\exists$ an invertible matrix P such that $P^{-1}AP = D$

$\Rightarrow$ $AP = PD$ . . . (1)

If $\lambda_1, \lambda_2, \lambda_3, ..., \lambda_n$ are the elements of the diagonal matrix, then $\lambda_1, \lambda_2, \lambda_3, ..., \lambda_n$ are all eigen values of D.

But similar matrices have the same eigen values, therefore, $\lambda_1, \lambda_2, \lambda_3, ..., \lambda_n$ are the only eigen values of A.

Let $P_1, P_2, P_3, ..., P_n$ are the column vectors of the matrix P where $P_1, P_2, P_3, ..., P_n$ are n linearly independent vectors.

Equating corresponding columns on both sides of Eq. (1), we get

$$AP_i = \lambda_i P_i \quad \forall\, i = 1, 2, 3, ..., n$$

$\therefore$ for each $i = 1, 2, 3, ..., n$; P_i is an eigen vector of A.

$\therefore$ A has n linearly independent vectors.

Sufficient Condition: If $P_1, P_2, P_3, ..., P_n$ are n linearly independent vectors (written as column vectors), then there are n scalars $\lambda_1, \lambda_2, \lambda_3, ..., \lambda_n$ such that

$$AP_i = \lambda_i P_i \quad \forall\, i = 1, 2, 3, ..., n \qquad ...(2)$$

Let P be a square matrix of order n with columns as $P_1, P_2, P_3, ..., P_n$; therefore P is invertible. $[\because\ P_1, P_2, P_3, ..., P_n$ are linearly indepndent$]$

From Eq. (2), we get

$$AP = PD \quad \text{where } D = \text{diag}(\lambda_1, \lambda_2, \lambda_3, ..., \lambda_n)$$

$\therefore \quad P^{-1}AP = D \qquad \Rightarrow \qquad$ A is similar to a diagonal matrix D

$\therefore$ A is diagonalizable.

Note:

1. The matrix P which diagonalizes matrix A is called the *modal matrix* of A, obtained by grouping the eigen values of A into square matrix and the matrix D is called the *spectral matrix* of A.

2. We have $\quad A = PDP^{-1}$

$$A^2 = A.A = \left(PDP^{-1}\right)\left(PDP^{-1}\right) = PD\left(P^{-1}P\right)\left(DP^{-1}\right)$$

$$= PD(I)\left(DP^{-1}\right) = P(DID)\left(P^{-1}\right) = P(D.D)\left(P^{-1}\right)$$

$$= PD^2P^{-1}$$

Repeating this process m times, we get

$$A^m = PD^mP^{-1}$$

$\therefore$ If A is diagonalizable so is A^m.

Steps to Diagonalize Square Matrix

1. Find the eigen values $\lambda_1, \lambda_2, \lambda_3, ..., \lambda_n$ of matrix A.
2. Find the corresponding eigen vectors $X_1, X_2, X_3, ..., X_n$. If the number of eigen vectors $< n$, then A is not diagonalizable.
3. Find $P = (X_1, X_2, X_3, ..., X_n)$ and P^{-1}.
4. $P^{-1}AP = \text{diag}(\lambda_1, \lambda_2, \lambda_3, ..., \lambda_n)$ is the required diagonal matrix.

Note: A is diagonalizable iff A has n linearly independent vectors.

ILLUSTRATIVE EXAMPLES 4.9

Example 4.36: Reduce the matrix $A=\begin{pmatrix}-1 & 2 & -2\\ 1 & 2 & 1\\ -1 & -1 & 0\end{pmatrix}$ to the diagonal form.

Solution: Characteristic equation of A is

$$|A-\lambda I|=\begin{vmatrix}-1-\lambda & 2 & -2\\ 1 & 2-\lambda & 1\\ -1 & -1 & -\lambda\end{vmatrix}=0 \quad\Rightarrow\quad \lambda^3-\lambda^2-5\lambda+5=0$$

Solving it, we get $\lambda=1,\ \pm\sqrt{5}$.

When $\lambda=1$, the corresponding eigen vector is given by

$$\begin{pmatrix}-2 & 2 & -2\\ 1 & 1 & 1\\ -1 & -1 & -1\end{pmatrix}\begin{pmatrix}x\\ y\\ z\end{pmatrix}=0$$

$$\Rightarrow\quad -2x+2y-2z=0 \quad\text{or}\quad -x+y-z=0$$
$$x+y+z=0$$
$$-x-y-z=0$$

Solving first two $\frac{x}{2}=\frac{y}{0}=\frac{z}{-2}$ giving the eigen vector $(1,0,-1)$

When $\lambda=\sqrt{5}$, the corresponding eigen vector is given by

$$\begin{pmatrix}-1-\sqrt{5} & 2 & -2\\ 1 & 2-\sqrt{5} & 1\\ -1 & -1 & -\sqrt{5}\end{pmatrix}\begin{pmatrix}x\\ y\\ z\end{pmatrix}=0$$

$$\Rightarrow\quad \left(-1-\sqrt{5}\right)x+2y-2z=0;\ x+\left(2-\sqrt{5}\right)y+z=0;\ x-y-\sqrt{5}z=0$$

Solving the last two, we get

$$\frac{x}{6-2\sqrt{5}}=\frac{y}{-1+\sqrt{5}}=\frac{z}{1-\sqrt{5}} \quad\text{or}\quad \frac{x}{\left(\sqrt{5}-1\right)^2}=\frac{y}{\sqrt{5}-1}=\frac{z}{1-\sqrt{5}}$$

giving the eigen vector $\left(\sqrt{5}-1,1,-1\right)$

Similarly, the eigen vector corresponding to $\lambda=-\sqrt{5}$ is $\left(\sqrt{5}+1,-1,1\right)$.

$\therefore$ The modal matrix $B=\begin{pmatrix} 1 & \sqrt{5}-1 & \sqrt{5}+1 \\ 0 & 1 & -1 \\ -1 & -1 & 1 \end{pmatrix}$

To Find B^{-1}:

We know that $B^{-1}=\dfrac{Adj.\text{B}}{|\text{B}|}$

$$\text{B}^{-1}=\frac{1}{2\sqrt{5}}\begin{pmatrix} 0 & -2\sqrt{5} & -2\sqrt{5} \\ 1 & +2\sqrt{5} & 1 \\ 1 & 2-\sqrt{5} & 1 \end{pmatrix}$$

Diagonal matrix $D=B^{-1}AB$

$$=\frac{1}{2\sqrt{5}}\begin{pmatrix} 1 & -2\sqrt{5} & -2\sqrt{5} \\ 1 & +2\sqrt{5} & 1 \\ 1 & 2-\sqrt{5} & 1 \end{pmatrix}\begin{pmatrix} -1 & 2 & -2 \\ 1 & 2 & 1 \\ -1 & -1 & 0 \end{pmatrix}\begin{pmatrix} 1 & \sqrt{5}-1 & \sqrt{5}+1 \\ 0 & 1 & -1 \\ -1 & -1 & 1 \end{pmatrix}$$

$$=\begin{pmatrix} 1 & 0 & 0 \\ 0 & \sqrt{5} & 0 \\ 0 & 0 & -\sqrt{5} \end{pmatrix}, \text{ i.e., diag}\left(1,\sqrt{5},-\sqrt{5}\right)$$

is the required diagonal matrix, which is same as $\text{diag}\left(\lambda_1,\lambda_2,\lambda_3\right)$.

Example 4.37: Find a matrix P that transforms the matrix $A=\begin{pmatrix} 1 & 1 & 3 \\ 1 & 5 & 1 \\ 3 & 1 & 1 \end{pmatrix}$ into a diagonal form.

[PTU Dec 2003]

Solution: Given $A=\begin{pmatrix}1 & 1 & 3\\ 1 & 5 & 1\\ 3 & 1 & 1\end{pmatrix}$

Characteristic equation of A is $|A-\lambda I|=\begin{vmatrix}1-\lambda & 1 & 3\\ 1 & 5-\lambda & 1\\ 3 & 1 & 1-\lambda\end{vmatrix}=0$

$$\Rightarrow \quad \lambda^3-7\lambda^2+36=0$$

$$\Rightarrow \quad (\lambda+2)(\lambda^2-9\lambda+18)=0$$

$$\Rightarrow \quad (\lambda+2)(\lambda-3)(\lambda-6)=0$$

$\therefore \quad \lambda=-2,3,6$ are the eigen values.

When $\lambda=-2$, corresponding eigen vector is given by

$$\begin{pmatrix}3 & 1 & 3\\ 1 & 7 & 1\\ 3 & 1 & 3\end{pmatrix}\begin{pmatrix}x\\ y\\ z\end{pmatrix}=0$$

$\Rightarrow \quad 3x+y+3z=0;\ x+7y+z=0;\ 3x+y+3z=0$

Solve first and second equations (third is same as is first equation), we get

$$\frac{x}{-20}=\frac{y}{0}=\frac{z}{20} \quad \Rightarrow \quad \frac{x}{-2}=\frac{y}{0}=\frac{z}{2}=\frac{k}{2}\text{ (say)} \qquad \therefore \quad X_1=k(-1,0,1)$$

When $\lambda=3$, corresponding eigen vector is given by

$$\begin{pmatrix}2 & 1 & 3\\ 1 & 2 & 1\\ 3 & 1 & -2\end{pmatrix}\begin{pmatrix}x\\ y\\ z\end{pmatrix}=0$$

$\Rightarrow \quad 2x+y+3z=0;\ x+2y+z=0;\ 3x+y-2z=0$

Solve first and second equations, we get

$$\frac{x}{-5}=\frac{y}{5}=\frac{z}{-5} \quad \Rightarrow \quad \frac{x}{-1}=\frac{y}{1}=\frac{z}{-1}=k\text{ (say)} \qquad \therefore \quad X_2=k(-1,1,-1)$$

When $\lambda = 6$, corresponding eigen vector is given by

$$\begin{pmatrix} -5 & 1 & 3 \\ 1 & -1 & 1 \\ 3 & 1 & -5 \end{pmatrix}\begin{pmatrix} x \\ y \\ z \end{pmatrix} = 0$$

$$\Rightarrow \qquad -5x + y + 3z = 0\,;\ x - y + z = 0\,;\ 3x + y - 5z = 0$$

Solve first and second equations, we get

$$\frac{x}{4} = \frac{y}{8} = \frac{z}{4} \qquad \Rightarrow \quad \frac{x}{1} = \frac{y}{2} = \frac{z}{1} = k \text{ (say)} \qquad \therefore \ X_2 = k(1, 2, 1)$$

$$\therefore \ \text{Modal matrix } P = \begin{pmatrix} X_1 & X_2 & X_3 \end{pmatrix} = \begin{pmatrix} -1 & -1 & 1 \\ 0 & 1 & 2 \\ 1 & -1 & 1 \end{pmatrix}$$

$$P^{-1} = \frac{Adj.P}{|P|} = -\frac{1}{6}\begin{pmatrix} 3 & 2 & -1 \\ 0 & -2 & -2 \\ -3 & 2 & -1 \end{pmatrix} = -\frac{1}{6}\begin{pmatrix} 3 & 0 & -3 \\ 2 & -2 & 2 \\ -1 & -2 & -1 \end{pmatrix}$$

$\therefore$ Required diagonal form $D = P^{-1}AP$

$$= -\frac{1}{6}\begin{pmatrix} 3 & 0 & -3 \\ 2 & -2 & 2 \\ -1 & -2 & -1 \end{pmatrix}\begin{pmatrix} 1 & 1 & 3 \\ 1 & 5 & 1 \\ 3 & 1 & 1 \end{pmatrix}\begin{pmatrix} -1 & -1 & 1 \\ 0 & 1 & 2 \\ 1 & -1 & 1 \end{pmatrix}$$

$$= -\frac{1}{6}\begin{pmatrix} 12 & 0 & 0 \\ 0 & -18 & 0 \\ 0 & 0 & -36 \end{pmatrix} = \begin{pmatrix} -2 & 0 & 0 \\ 0 & 3 & 0 \\ 0 & 0 & 6 \end{pmatrix}, \text{ i.e., diag}(-2, 3, 6)$$

is the required diagonal matrix, which is same as $\text{diag}(\lambda_1, \lambda_2, \lambda_3)$

Example 4.38: Diagonalize the matrix $A = \begin{pmatrix} -2 & 2 & -3 \\ 2 & 1 & -6 \\ -1 & -2 & 0 \end{pmatrix}$ and obtain its modal matrix.

Solution: Characteristics equation of A is $|A-\lambda I| = \begin{vmatrix} -2-\lambda & 2 & -3 \\ 2 & 1-\lambda & -6 \\ -1 & -2 & 0 \end{vmatrix} = 0$

$\Rightarrow \quad (-2-\lambda)[-\lambda(1-\lambda)-12]-2[2\lambda-6]-3[-4+1-\lambda]=0$

$\Rightarrow \quad -(2+\lambda)(\lambda^2-\lambda-12)+4(\lambda+3)+3(\lambda+3)=0$

$\Rightarrow \quad -(2+\lambda)(\lambda+3)(\lambda-4)+4(\lambda+3)+3(\lambda+3)=0$

$\Rightarrow \quad (\lambda+3)(-\lambda^2+2\lambda+8+7)=0$ or $(\lambda+3)(\lambda^2-2\lambda-15)=0$

$\Rightarrow \quad (\lambda+3)(\lambda+3)(\lambda-5)=0$

$\therefore \quad \lambda=-3,-3,5$ are the eigen values.

When $\lambda=-3$, corresponding eigen vectors are given by

$$\begin{pmatrix} 1 & 2 & -3 \\ 2 & 4 & -6 \\ -1 & -2 & 3 \end{pmatrix}\begin{pmatrix} x \\ y \\ z \end{pmatrix}=0 \qquad R_2 \to R_2-2R_1, R_2 \to R_3+R_1$$

$$\begin{pmatrix} 1 & 2 & -3 \\ 0 & 0 & 0 \\ 0 & 0 & 0 \end{pmatrix}\begin{pmatrix} x \\ y \\ z \end{pmatrix}=0$$

$\Rightarrow \quad x+2y-3z=0$ or $x=-2y+3z$

$y=1.y+0.z$

$z=0.y+1.z$

$$\therefore \quad \begin{pmatrix} x \\ y \\ z \end{pmatrix}=y\begin{pmatrix} -2 \\ 1 \\ 0 \end{pmatrix}+z\begin{pmatrix} 3 \\ 0 \\ 1 \end{pmatrix}$$

$\therefore$ eigen vectors corresponding to $\lambda=-3$ are $(-2,1,0)$ and $(3,0,1)$

When $\lambda = 5$, corresponding eigen vector is given by

$$\begin{pmatrix} -7 & 2 & -3 \\ 2 & -4 & -6 \\ -1 & -2 & -5 \end{pmatrix}\begin{pmatrix} x \\ y \\ z \end{pmatrix} = 0 \qquad R_3 \leftrightarrow R_2$$

$$\begin{pmatrix} -7 & 2 & -3 \\ -1 & -2 & -5 \\ -2 & -4 & -6 \end{pmatrix}\begin{pmatrix} x \\ y \\ z \end{pmatrix} = 0 \qquad R_1 \leftrightarrow R_1$$

$$\begin{pmatrix} -1 & -2 & -5 \\ -7 & 2 & -3 \\ -2 & -4 & -6 \end{pmatrix}\begin{pmatrix} x \\ y \\ z \end{pmatrix} = 0 \qquad R_2 \to R_2 - 7R_1, R_3 \to R_3 + 2R_1$$

$$\begin{pmatrix} -1 & -2 & -5 \\ 0 & 16 & 32 \\ 0 & -8 & -16 \end{pmatrix}\begin{pmatrix} x \\ y \\ z \end{pmatrix} = 0 \qquad R_3 \to R_3 + \frac{1}{2}R_2$$

$$\begin{pmatrix} -1 & -2 & -5 \\ 0 & 16 & 32 \\ 0 & 0 & 0 \end{pmatrix}\begin{pmatrix} x \\ y \\ z \end{pmatrix} = 0$$

$$\Rightarrow \quad -x - 2y - 5z = 0 \quad \text{or} \quad x + 2y + 5z = 0$$
$$16y + 32z = 0 \quad \text{or} \quad y + 2z = 0$$

$$\Rightarrow \quad x = -z\,,\; y = -2z$$

$$\therefore \quad \begin{pmatrix} x \\ y \\ z \end{pmatrix} = \begin{pmatrix} -z \\ -2z \\ z \end{pmatrix} = -z\begin{pmatrix} 1 \\ 2 \\ -1 \end{pmatrix}$$

$\therefore$ Eigen vector corresponding to $\lambda = 5$ is $(1, 2, -1)$

$\therefore$ Modal matrix is $P = \begin{pmatrix} -2 & 3 & 1 \\ 1 & 0 & 2 \\ 0 & 1 & -1 \end{pmatrix}$

$$P^{-1} = \frac{\text{adj}.P}{|P|} = \frac{1}{8}\begin{pmatrix} -2 & 4 & 6 \\ 1 & 2 & 5 \\ 0 & 2 & -3 \end{pmatrix}$$

Now, Diagonal matrix $D = P^{-1}AP$

$$= \frac{1}{8}\begin{pmatrix} -2 & 4 & 6 \\ 1 & 2 & 5 \\ 1 & 2 & -3 \end{pmatrix}\begin{pmatrix} -2 & 2 & 3 \\ 2 & 1 & -6 \\ -1 & -2 & -0 \end{pmatrix}\begin{pmatrix} -2 & 3 & 1 \\ 1 & 0 & 2 \\ 0 & 1 & -1 \end{pmatrix}$$

$$= \frac{1}{8}\begin{pmatrix} -24 & 0 & 0 \\ 0 & -24 & 0 \\ 0 & 0 & 40 \end{pmatrix} = \begin{pmatrix} -3 & 0 & 0 \\ 0 & -3 & 0 \\ 0 & 0 & 5 \end{pmatrix}, \text{ i.e., diag}(-3,-3,5)$$

is the required diagonal matrix, which is same as $\text{diag}(\lambda_1, \lambda_2, \lambda_3)$

EXERCISE 4.8

1. Diagonalize the following, if possible:

   (*a*) $\begin{pmatrix} 2 & 1 \\ 2 & 3 \end{pmatrix}$ (*b*) $\begin{pmatrix} 3 & 0 \\ 0 & 4 \end{pmatrix}$ (*c*) $\begin{pmatrix} 2 & 1 \\ 4 & 2 \end{pmatrix}$

2. Diagonalize, if possible, the matrix $\begin{pmatrix} 6 & 0 & 9 \\ 0 & 7 & 1 \\ 0 & -4 & 3 \end{pmatrix}$.

3. Diagonalize the matrix $A = \begin{pmatrix} 1 & 1 & 3 \\ 1 & 5 & 1 \\ 3 & 1 & 1 \end{pmatrix}$.

4. Prove that the characteristic roots of a diagonal matrix are the diagonal elements of the matrix.

5. Diagonalize the following matrices and obtain the modal matrix in each case.

   (*a*) $\begin{pmatrix} -1 & 1 & 2 \\ 0 & -2 & 1 \\ 0 & 0 & -3 \end{pmatrix}$ (*b*) $\begin{pmatrix} 9 & -1 & -9 \\ 3 & -1 & 3 \\ -7 & 1 & -7 \end{pmatrix}$

6. Show that the following matrices are similar to diagonal matrices. Also find the transforming matrices and the diagonal matrices.

(*a*) $\begin{pmatrix} -9 & 4 & 4 \\ -8 & 3 & 4 \\ -16 & 8 & 7 \end{pmatrix}$ (*b*) $\begin{pmatrix} 6 & -2 & 2 \\ -2 & 3 & -1 \\ 2 & -1 & 3 \end{pmatrix}$

7. Show that the matrix $A = \begin{pmatrix} 3 & 1 & -1 \\ -2 & 1 & 2 \\ 0 & 1 & 2 \end{pmatrix}$ is diagonalizable. Hence, find P such that $P^{-1}AP$ is a diagonal matrix [PTU May 2008]

8. Diagonalize the matrix $A = \begin{pmatrix} -2 & 2 & -3 \\ 2 & 1 & -6 \\ -1 & -2 & 0 \end{pmatrix}$, and obtain its modal matrix. [PTU Dec 2002]

9. Reduce the matrix $\begin{pmatrix} 1 & 0 & -1 \\ 1 & 2 & 1 \\ 2 & 2 & 3 \end{pmatrix}$ to diagonal form. [PTU Dec 2012]

10. Test whether the matrix $\begin{pmatrix} 3 & 1 & -1 \\ -2 & 1 & 2 \\ 0 & 1 & 2 \end{pmatrix}$ is diagonalizable or not? [PTU May 2012]

ANSWERS

1. (*a*) Diagonalizable (*b*) Diagonalizable (*c*) Diagonalizable
2. Not diagonalizable
3. Diagonalizable

5. (*a*) $\begin{pmatrix} 1 & 1 & 1 \\ 0 & -1 & 2 \\ 0 & 0 & -2 \end{pmatrix}$ (*b*) $\begin{pmatrix} 0 & 1 & 4 \\ 1 & 0 & 1 \\ -1 & -1 & -3 \end{pmatrix}$

6. (*a*) $P = \begin{pmatrix} 1 & 1 & 1 \\ 1 & 1 & 1 \\ 1 & -1 & 2 \end{pmatrix}$, $D = \begin{pmatrix} -1 & 0 & 0 \\ 0 & -1 & 0 \\ 0 & 0 & 0 \end{pmatrix}$; (*b*) $P = \begin{pmatrix} -1 & 1 & 2 \\ 0 & 2 & -1 \\ 2 & 0 & 1 \end{pmatrix}$, $D = \begin{pmatrix} 2 & 0 & 0 \\ 0 & 2 & 0 \\ 0 & 0 & 8 \end{pmatrix}$

7. $P=\begin{pmatrix}1&1&0\\-1&0&1\\1&1&1\end{pmatrix}, D=\begin{pmatrix}1&0&0\\0&2&0\\0&0&3\end{pmatrix}$

8. $P=\begin{pmatrix}-2&3&1\\1&0&2\\0&1&-1\end{pmatrix}, D=\begin{pmatrix}-3&0&0\\0&-3&0\\0&0&5\end{pmatrix}$

9. $D=\begin{pmatrix}1&0&0\\0&2&0\\0&0&3\end{pmatrix}$

10. Diagonalizable; $P=\begin{pmatrix}1&1&0\\-1&0&1\\1&1&1\end{pmatrix}, D=\begin{pmatrix}1&0&0\\0&2&0\\0&0&3\end{pmatrix}$

4.12 SIMILAR MATRICES

Let A and B be the square matrices of the same order. The matrix A is said to be similar to B if there exist an invertible matrix P such that

$$A=P^{-1}BP \quad \text{or} \quad PA=BP$$

Consider

$$PA=BP$$

Post multiply both sides by P^{-1}, we have

$$PAP^{-1}=BPP^{-1}=B\left(PP^{-1}\right)=BI=B$$

$$\therefore \quad B=PAP^{-1}$$

Therefore, A is similar to B if and only if B is similar to A. The matrix P is called the *similarity matrix*.

Art.: Similar matrices have the same characteristics equations (and hence the same eigen values). Also if X is any eigen vector of à corresponding to eigen value λ then $\left(P^{-1}.X\right)$ is an eigen vector of B corresponding to the eigen value of λ, where P is similarity matrix.

Proof: $\because$ B is similar to A and P is the similarity matrix

$$\therefore \quad AP=PB \quad \text{or} \quad P^{-1}AP=B$$

Let λ be the eigen value and X be the corresponding eigen vector of A.

$$\therefore \quad AX = \lambda X \qquad \ldots (1)$$

Now $B - \lambda I = P^{-1}AP - \lambda I = P^{-1}AP - P^{-1}(\lambda I)P = P^{-1}(A - \lambda I)P$

$$\therefore \quad |B - \lambda I| = |P^{-1}(A - \lambda I)P|$$

$$= |P^{-1}||A - \lambda I||P|$$

$$= |A - \lambda I||P^{-1}P|$$

$$= |A - \lambda I||I|$$

$$= |A - \lambda I|$$

Therefore, similar matrices have same characteristic polynomial.

Pre-multiplying Eq. (1) both sides by an invertible matrix P^{-1}, we get

$$P^{-1}(AX) = P^{-1}(\lambda X) = \lambda P^{-1}X$$

Let $X = PY$

$$\therefore \quad P^{-1}(APY) = \lambda P^{-1}(PY)$$

$$\Rightarrow \quad (P^{-1}AP)Y = \lambda(P^{-1}P)Y \qquad [\because P^{-1}P = I]$$

$$\Rightarrow \quad BY = \lambda Y \quad \text{where } B = PAP^{-1} \qquad \ldots (2)$$

$\therefore$ B has the same eigen value λ as that of A which shows that eigen values of similar matrices are same.

$\therefore$ similar matrices have the same characteristic equation and hence the same eigen values.

Now, from Eq. (2), Y is an eigen vector of B corresponding to λ, the eigen values of B.

$\therefore$ eigen vector of B is $Y = P^{-1}X$ Hence the result.

Note:

- Converse of the above theorem is not always true, *i.e.*, two matrices which have the same characteristic equation need not always be similar.
- If A is similar to B, B is similar to C, then A is also similar to C.

  Let there be two invertible matrices P and Q, such that

  $$A = P^{-1}BP \text{ and } B = Q^{-1}CQ$$

  $$\text{Thus } A = P^{-1}\left(Q^{-1}CQ\right)P$$

  $$= \left(P^{-1}Q^{-1}\right)C\left(QP\right)$$

  $$= R^{-1}CR \qquad \text{where } R = QP$$

  Hence A is similar to C.

Art.: The necessary and sufficient condition for an n-rowed matrix A to be similar to a diagonal matrix is that the set of eigen vectors of A includes a set of n linearly independent vectors.

Proof:

Necessary Condition: Matrix A is similar to a diagonal matrix D (say)

$\therefore$ $\exists$ a non-singular matrix P such that

$$P^{-1}AP = D = \text{diag}\left(\lambda_1, \lambda_2, \lambda_3, \ldots, \lambda_n\right)$$

Pre-multiplying both sides by P, we get

$$PP^{-1}AP = PD$$

$$\Rightarrow \quad AP = PD \qquad \left[\begin{aligned} \because\ PP^{-1}AP &= \left(PP^{-1}\right)AP \\ &= IAP = \left(IA\right)P = AP \end{aligned}\right]$$

Let $P = [C_1, C_2, C_3, ..., C_n]$

$$\therefore \quad A[C_1, C_2, C_3, ..., C_n] = [C_1, C_2, C_3, ..., C_n][\text{diag}(\lambda_1, \lambda_2, \lambda_3, ..., \lambda_n)]$$

$$\Rightarrow \quad AC_1 = \lambda_1 C_1,\ AC_2 = \lambda_2 C_2,\ AC_3 = \lambda_3 C_3, \ldots,\ AC_n = \lambda_n C_n$$

which shows that $C_1, C_2, C_3, ..., C_n$ are n eigen vectors corresponding to eigen values $\lambda_1, \lambda_2, \lambda_3, ..., \lambda_n$ of *A*. As $C_1, C_2, C_3, ..., C_n$ are columns of a non-singular matrix, therefore, they form a linearly independent set of vectors.

Sufficient Condition: Let $C_1, C_2, C_3, ..., C_n$ be n linearly independent set of eigen vectors and let $\lambda_1, \lambda_2, \lambda_3, ..., \lambda_n$ be the corresponding eigen values.

We have $AC_1 = \lambda_1 C_1,\ AC_2 = \lambda_2 C_2,\ AC_3 = \lambda_3 C_3, \ldots,\ AC_n = \lambda_n C_n$ ... (1)

If we take $P = [C_1, C_2, C_3, ..., C_n]$, then system of Eq. (1) is equivalent to

$$AP = PD \qquad \ldots (2)$$

where $D = \text{diag}(\lambda_1, \lambda_2, \lambda_3, ..., \lambda_n)$

Also matrix P is non-singular as its columns are linearly independent

$\therefore \quad P^{-1}$ exists

Pre-multiplying both sides of (2) by P^{-1}, we get

$$P^{-1}AP = P^{-1}PD$$

$$\Rightarrow \quad P^{-1}AP = D \qquad [\because PP^{-1} = P^{-1}P = I]$$

Hence *A* is similar to *D*.

ILLUSTRATIVE EXAMPLES 4.10

Example 4.39: Examine whether *A* is similar to *B* where

(*a*) $A = \begin{pmatrix} 1 & 0 \\ 0 & 1 \end{pmatrix}$ and $B = \begin{pmatrix} 1 & 1 \\ 0 & 1 \end{pmatrix}$

(*b*) $A = \begin{pmatrix} 5 & 5 \\ -2 & 0 \end{pmatrix}$ and $B = \begin{pmatrix} 1 & 2 \\ -3 & 4 \end{pmatrix}$

Solution: We know that A will be similar to B iff $\exists$ a non singular matrix P such that

$$A = P^{-1}BP \quad \text{or} \quad PA = BP$$

Let $P = \begin{pmatrix} a & b \\ c & d \end{pmatrix}$

Now $PA = BP$

(a) $$\begin{pmatrix} a & b \\ c & d \end{pmatrix}\begin{pmatrix} 1 & 0 \\ 0 & 1 \end{pmatrix} = \begin{pmatrix} 1 & 1 \\ 0 & 1 \end{pmatrix}\begin{pmatrix} a & b \\ c & d \end{pmatrix}$$

$$\begin{pmatrix} a & b \\ c & d \end{pmatrix} = \begin{pmatrix} a+c & b+d \\ c & d \end{pmatrix}$$

$$\therefore \quad a = a + c \quad \Rightarrow \quad c = 0$$

$$b = b + d \quad \Rightarrow \quad d = 0$$

$$\therefore \quad P = \begin{pmatrix} a & b \\ c & d \end{pmatrix} = \begin{pmatrix} a & b \\ 0 & 0 \end{pmatrix} \quad \text{which is singular.}$$

Hence, A is not similar to B.

(b) $$\begin{pmatrix} a & b \\ c & d \end{pmatrix}\begin{pmatrix} 5 & 5 \\ -2 & 0 \end{pmatrix} = \begin{pmatrix} 1 & 2 \\ -3 & 4 \end{pmatrix}\begin{pmatrix} a & b \\ c & d \end{pmatrix}$$

$$\begin{pmatrix} 5a-2b & 5a \\ 5c-2d & 5c \end{pmatrix} = \begin{pmatrix} a+2c & b+2d \\ -3a+4c & -3b+4d \end{pmatrix}$$

$$\therefore \quad 5a - 2b = a + 2c \qquad 5a = b + 2d$$

$$5c - 2d = -3a + 4c \qquad 5c = -3b + 4d$$

These four equations on simplification and re-arrangement of terms becomes

$$2a - b - c + 0d = 0$$

$$3a + 0b + c - 2d = 0$$

$$5a - b + 0c - 2d = 0$$

$$0a + 3b + 5c - 4d = 0$$

which is a set of homogenous equations and can be written as

$$\begin{pmatrix} 2 & -1 & -1 & 0 \\ 3 & 0 & 1 & -2 \\ 5 & -1 & 0 & -2 \\ 0 & 3 & 5 & -4 \end{pmatrix}\begin{pmatrix} a \\ b \\ c \\ d \end{pmatrix} = 0 \qquad C_1 \leftrightarrow C_2$$

$$\begin{pmatrix} -1 & 2 & -1 & 0 \\ 0 & 3 & 1 & -2 \\ -1 & 5 & 0 & -2 \\ 3 & 0 & 5 & -4 \end{pmatrix}\begin{pmatrix} a \\ b \\ c \\ d \end{pmatrix} = 0 \qquad R_3 = R_3 - R_1,\ R_4 = R_4 + 3R_1$$

$$\begin{pmatrix} -1 & 2 & -1 & 0 \\ 0 & 3 & 1 & -2 \\ 0 & 3 & 1 & -2 \\ 0 & 6 & 2 & -4 \end{pmatrix}\begin{pmatrix} a \\ b \\ c \\ d \end{pmatrix} = 0 \qquad R_3 = R_3 - R_2,\ R_4 = R_4 - 2R_2$$

$$\begin{pmatrix} -1 & 2 & -1 & 0 \\ 0 & 3 & 1 & -2 \\ 0 & 0 & 0 & 0 \\ 0 & 0 & 0 & 0 \end{pmatrix}\begin{pmatrix} a \\ b \\ c \\ d \end{pmatrix} = 0$$

$\Rightarrow$ $-a + 2b - c = 0$

$3b + c - 2d = 0$

$\therefore$ if $a = 1$, $b = 1$, then we get

$c = 1$ and $d = 2$

$\therefore$ $P = \begin{pmatrix} 1 & 1 \\ 1 & 2 \end{pmatrix}$ which is non-singular.

Hence, A is similar to B.

Example 4.40: Examine whether the given matrix

$$A=\begin{pmatrix} 2 & 3 & 4 \\ 0 & 2 & -1 \\ 0 & 0 & 1 \end{pmatrix}$$ is similar to a diagonal matrix.

Solution:

Characteristic equation of A is $|A-\lambda I|=0$

$$\Rightarrow \quad \begin{vmatrix} 2-\lambda & 3 & 4 \\ 0 & 2-\lambda & -1 \\ 0 & 0 & 1-\lambda \end{vmatrix}=0$$

Expanding the determinant on first column, we get

$$(2-\lambda)^2(1-\lambda)=0$$

which on solving gives

$$\lambda=1,2,2$$

Eigen vector corresponding to $\lambda=1$ is

$$\begin{pmatrix} 1 & 3 & 4 \\ 0 & 1 & -1 \\ 0 & 0 & 0 \end{pmatrix}\begin{pmatrix} x \\ y \\ z \end{pmatrix}=0$$

$$\Rightarrow \quad x+3y+4z=0$$

$$y-z=0$$

Solving these equations, we get

$$y=z,\ x=-7z$$

$$\therefore \quad X=\begin{bmatrix} x \\ y \\ z \end{bmatrix}=\begin{bmatrix} -7z \\ z \\ z \end{bmatrix}=z\begin{bmatrix} -7 \\ 1 \\ 1 \end{bmatrix}$$

$\therefore$ Single eigen vector corresponding to $\lambda = 1$ is $\begin{bmatrix} -7 \\ 1 \\ 1 \end{bmatrix}$

Eigen vector corresponding to $\lambda = 2$ is

$$\begin{pmatrix} 0 & 3 & 4 \\ 0 & 0 & -1 \\ 0 & 0 & 1 \end{pmatrix}\begin{pmatrix} x \\ y \\ z \end{pmatrix} = 0$$

$$\Rightarrow \quad 3y + 4z = 0$$

$$y = 0$$

$$z = 0$$

$$\therefore \qquad X = \begin{bmatrix} x \\ y \\ z \end{bmatrix} = \begin{bmatrix} x \\ 0 \\ 0 \end{bmatrix} = x\begin{bmatrix} 1 \\ 0 \\ 0 \end{bmatrix}$$

Corresponding to $\lambda = 2$, we get only one vector $\begin{bmatrix} 1 \\ 0 \\ 0 \end{bmatrix}$

As there are only two linearly independent eigen vectors corresponding to three eigen values, therefore, there does not exist any non singular matrix *P*.

Hence, *A* is not similar to diagonal matrix.

Example 4.41: Prove that if *A* is similar to a diagonal matrix, then *A* is similar to A'.

Solution: Let *A* be similar to diagonal matrix *D*, then there exists a non singular matrix *P* such that.

$$P^{-1}AP = D$$

$$\Rightarrow \quad A = PDP^{-1}$$

Now $A' = \left(PDP^{-1}\right)'$

$$= \left(P^{-1}\right)' D' P'$$

$$= \left(P^{-1}\right)' DP' \qquad \left[\begin{array}{l}\because D \text{ is a diagonal matrix,} \\ \therefore D' = D\end{array}\right]$$

$$A' = \left(P^{-1}\right)' DP'$$

$\Rightarrow$ A' is similar to D

$\Rightarrow$ D is similar to A'

Now A is similar to D and D is similar to A'

Hence A is similar to A', i.e., A' is similar to A.

Hence proved.

Example 4.42: Show that rank of every matrix similar to A is the same as that of A.

Solution: Let matrix B be the similar to matrix A .Then there exist a non-singular matrix P such that

$$B = P^{-1}AP$$

Now $\rho(B) = \rho(P^{-1}AP)$

$$= \rho(A)$$

$\because$ We know that rank of a matrix does not change on multiplication by a non singular matrix, hence

$$\rho(B) = \rho(A)$$

4.13 ORTHOGONAL TRANSFORMATIONS

The linear transformation $Y = AX$, where

$$Y = \begin{pmatrix} y_1 \\ y_2 \\ \vdots \\ y_n \end{pmatrix}, A = \begin{pmatrix} a_{11} & a_{12} & \dots & a_{1n} \\ a_{21} & a_{22} & \dots & a_{2n} \\ \vdots & \vdots & \vdots & \vdots \\ a_{n1} & a_{n2} & \dots & a_{nn} \end{pmatrix}, X = \begin{pmatrix} x_1 \\ x_2 \\ \vdots \\ x_n \end{pmatrix}$$

is said to be orthogonal if it transforms $y_1^2 + y_2^2 + \dots + y_n^2$ into $x_1^2 + x_2^2 + \dots + x_n^2$.

Unitary, Hermitian and Orthogonal Matrices

Unitary matrix: A square matrix A of order n is said to be unitary if $A^\theta A = I$, where $A^\theta = (\bar{A})'$.

Hermitian matrix: A square matrix A is said to be hermitian if $A^\theta = A$.

Orthogonal matrix: A square matrix A is said to be orthogonal if $A'A = I$.

Theorem 1: Prove that if A is a unitary matrix of order n, then it is non-singular and $|A| = \pm 1$.

Proof: Since A is a unitary matrix.

$$\therefore \quad A^\theta . A = I$$

Taking determinant on both sides, we get

$$\therefore \quad |A^\theta . A| = |I| \quad \Rightarrow \quad |A^\theta|.|A| = 1 \qquad [\because \ |I| = 1]$$

But $A^\theta = (\bar{A})'$ and $|A^\theta| = |(\bar{A})'| = |\bar{A}|$

$$\therefore \quad |\bar{A}|.|A| = 1 \quad \Rightarrow \quad |A|^2 = 1 \qquad \dots (1)$$

Eq. (1) is only true when $|A| = \pm 1 \quad \Rightarrow \quad |A| \neq 0$.

Thus, A is non-singular and $|A| = \pm 1$.

Theorem 2: Given A and B are two orthogonal matrices of order n, then prove that AB and BA are also orthogonal matrices.

Proof: Since A and B are orthogonal matrices.

$$A'A = I \quad \ldots (1)$$

$$B'B = I \quad \ldots (2)$$

Also, $(AB)'(AB) = B'(A'A)B$

$$= B'(I)B = B'(IB) = B'B \quad [\because IB = B]$$

$$= I$$

Thus, $(AB)'(AB) = I$

$\therefore$ AB is also an orthogonal matrix. Similarly, we can prove for BA.

Theorem 3: A real matrix is unitary if and only if it is orthogonal.

Proof: Suppose that A is a real matrix, then

$$A^{\theta} = (\bar{A})' = A' \quad [\because \text{For real matrix, } \bar{A} = A]$$

Now, A is unitary

$\Rightarrow$ $A^{\theta}A = I$ $\Rightarrow$ $A'A = I$ $[\because A' = A^{\theta}]$.

$\Rightarrow$ A is orthogonal.

Theorem 4: If P is orthogonal, then P' and P^{-1} are also orthogonal.

Proof: Given that P is orthogonal

$\Rightarrow$ $P'P = I$ $\Rightarrow$ $(P'P)' = I'$

$\Rightarrow$ $P'(P')' = I$ $\Rightarrow$ P' is orthogonal.

Again, $P'P = I$ $\Rightarrow$ $(P'P)^{-1} = I^{-1} \Rightarrow P^{-1}(P')^{-1} = I$

$\Rightarrow$ $P^{-1}(P^{-1})' = I$ $\Rightarrow$ P^{-1} is orthogonal.

Theorem 5: (*a*) If P is unitary, then $P', \overline{P}, P^{\theta}$ and P^{-1} are also unitary.

(*b*) If P and Q are unitary, then PQ is also unitary.

Proof:

(*a*) P is unitary $\Rightarrow P^{\theta}P = I \Rightarrow \left(P^{\theta}P\right)' = I' \Rightarrow P'\left(P^{\theta}\right)' = I$

$\Rightarrow P'\left(P'\right)^{\theta} = I$

Hence P' is unitary.

Again, P is unitary, $\Rightarrow P^{\theta}P = I$

$\Rightarrow \overline{P^{\theta}P} = \overline{I} \Rightarrow \overline{P^{\theta}}\,\overline{P} = I$

$\Rightarrow \overline{\left(\left(\overline{P}\right)'\right)}\overline{P} = I$

$\Rightarrow \left(\overline{P}\right)^{\theta}\overline{P} = I$

Hence , $\overline{P}$ is unitary.

P is unitary, $P^{\theta}P = I$

$\Rightarrow \left(P^{\theta}P\right)^{-1} = I^{-1} \qquad \Rightarrow P^{-1}\left(P^{\theta}\right)^{-1} = I$

$\Rightarrow P^{-1}\left(P^{-1}\right)^{\theta} = I$ Hence P^{-1} is unitary.

(*b*) Since P and Q are unitary,

$$P^{\theta}P = I \text{ and } Q^{\theta}Q = I \qquad \ldots (1)$$

Now, $(PQ)^{\theta}(PQ) = \left(Q^{\theta}P^{\theta}\right)(PQ) = Q\left(P^{\theta}P\right)Q$

$= Q^{\theta}IQ$

$= Q^{\theta}Q$

$= I$

Hence, PQ is unitary.

ILLUSTRATIVE EXAMPLES 4.11

Example 4.43: Verify that the matrix

$$A = \begin{pmatrix} \cos\theta & \sin\theta & 0 \\ -\sin\theta & \cos\theta & 0 \\ 0 & 0 & 1 \end{pmatrix} \text{ is orthogonal.}$$

Solution: Given $A = \begin{pmatrix} \cos\theta & \sin\theta & 0 \\ -\sin\theta & \cos\theta & 0 \\ 0 & 0 & 1 \end{pmatrix}$.

$$A' = \begin{pmatrix} \cos\theta & -\sin\theta & 0 \\ \sin\theta & \cos\theta & 0 \\ 0 & 0 & 1 \end{pmatrix}$$

Now, $A'A = \begin{pmatrix} \cos\theta & -\sin\theta & 0 \\ \sin\theta & \cos\theta & 0 \\ 0 & 0 & 1 \end{pmatrix}\begin{pmatrix} \cos\theta & \sin\theta & 0 \\ -\sin\theta & \cos\theta & 0 \\ 0 & 0 & 1 \end{pmatrix}$

$$= \begin{pmatrix} \cos^2\theta + \sin^2\theta & \cos\theta\sin\theta - \sin\theta\cos\theta & 0 \\ \sin\theta\cos\theta - \cos\theta\sin\theta & \sin^2\theta + \cos^2\theta & 0 \\ 0 & 0 & 1 \end{pmatrix}$$

$$= \begin{pmatrix} 1 & 0 & 0 \\ 0 & 1 & 0 \\ 0 & 0 & 1 \end{pmatrix} = I$$

Hence, A is an orthogonal matrix.

Example 4.44: Show that if A is Hermitian and P is unitary, then $P^{-1}AP$ is Hermitian.

Solution: Suppose A is Hermitian

$\Rightarrow \quad A^\theta = A$ . . . (1)

Also P is unitary

$\Rightarrow \quad P^{-1} = P^\theta$ . . . (2)

Now, $\left(P^{-1}AP\right)^{\theta} = \left(P^{\theta}AP\right)^{\theta}$ [by (2)]

$$= P^{\theta}A^{\theta}\left(P^{\theta}\right)^{\theta} = P^{\theta}AP$$ [by (1)]

Thus, $\left(P^{-1}AP\right)^{\theta} = P^{\theta}AP = P^{-1}AP$ [by(2)]

$\therefore$ $P^{-1}AP$ is Hermitian.

EXERCISE 4.9

1. Show that the matrix $\begin{pmatrix} \cos\theta & \sin\theta \\ -\sin\theta & \cos\theta \end{pmatrix}$ is orthogonal.
2. Show that if A is symmetric and P is orthogonal, then $P^{-1}AP$ is symmetric.
3. Show that $A = \begin{pmatrix} 0 & 2m & n \\ l & m & -n \\ l & -m & +n \end{pmatrix}$

   $l = \dfrac{1}{\sqrt{2}},\ m = \dfrac{1}{\sqrt{6}},\ n = \dfrac{1}{\sqrt{3}}$ is orthogonal.
4. If A and B are similar matrices, then prove that $\lambda I - A$ and $\lambda I - B$ are also similar λ being any scalar.
5. Prove that $A = \dfrac{1}{2}\begin{pmatrix} -1 & 1 & 1 & 1 \\ 1 & -1 & 1 & 1 \\ 1 & 1 & -1 & 1 \\ 1 & 1 & 1 & -1 \end{pmatrix}$ is orthogonal.
6. Prove that inverse of a non-singular symmetric matrix is symmetric.
7. If A is invertible, prove that A^{-1} is also invertible and $\left(A^{-1}\right)^{-1} = A$.
8. Let A be a square matrix such that $A^3 = I$. Prove that A is non-singular and find adj. A.
9. Prove that A is non-singular iff adj. A is non-singular.
10. If $F(\alpha) = \begin{pmatrix} \cos\alpha & -\sin\alpha & 0 \\ \sin\alpha & \cos\alpha & 0 \\ 0 & 0 & 1 \end{pmatrix}$, prove that $\left(F(\alpha)\right)^{-1} = F(-\alpha)$.

❑ ❑ ❑

Chapter 5

Infinite Series

Learning Objectives

After reading this chapter, students will be able to

- explain the basic concepts of series and sequences
- describe and prove Cauchy's general principle of convergence of series
- describe various comparison tests for series of positive terms
- describe and use following methods to check the convergence/divergence of series
    - p-Test for series of type $\sum\left(1/n^{p}\right)$
    - Cauchy's integral test
    - D' Alembert's ratio test
    - Raabe's test
    - Logarithmic test
    - Gauss test
    - Cauchy's root test
- explain uniform convergence of series
- describe and use Weierstrass's M-test to check the uniform convergence of series

5.1 RECAPITULATION OF THE CONCEPTS OF SERIES AND SEQUENCE

Before moving on the topic of infinite series as per the requirements of the prescribed syllabus, let us quickly recapitulate (review) the concepts about sets, functions, series and sequence learned in earlier classes.

Set – A set is a collection of well-defined and distinct objects, and is denoted by a capital letter such as *A*, *N*, *R*, etc.

Function – Let *X* and *Y* be two non-empty sets. Then a function *f* from *X* to *Y* is a rule that associates each element $x \in X$ to a unique element $y \in Y, i.e., f(x) = y$, and is denoted as $f : X \rightarrow Y$.

Sequence – A sequence is a function whose domain is the set *N* of all natural numbers where as the range may be any set *S*. In other words, a sequence in a set S is a rule which assign to each natural number a unique element of *S*.

Real Sequence – A real sequence is a function whose domain is the set *N* of all natural numbers and range a subset of the set *R* of real numbers.

Symbolically, $f : N \rightarrow R$ or $(x : N \rightarrow R \text{ or } a : N \rightarrow R)$ is a real sequence.

Note: If $x : N \rightarrow R$ be a sequence, the image of $n \in N$ instead of denoting it by *x*(*n*), we shall generally denote it by x_n. Thus, x_1, x_2, x_3, etc. are the real numbers associated to 1, 2, 3, etc. by this mapping. Also, the sequence $x : N \rightarrow R$ is denoted by $\{x_n\}$ or (x_n).

Here $x_1, x_2, x_3, \ldots$ are called first, second and third terms of the sequence. The *m*th and *n*th terms x_m and x_n for $m \neq n$ are treated as distinct terms. If $x_m = x_n$, i.e., the terms at different positions are treated as distinct terms even if they have the same value.

Range of a Sequence – The set of all distinct terms of a sequence is called its *range*.

Note: In a sequence $\{x_n\}$ since $n \in N$ and *N* is an infinite set, the number of terms of a sequence is always infinite. The range of a sequence may be a finite. The range of a sequence may be a finite set, *e.g.*, if $x_n = (-\text{i})^n$, then $\{x_n\} = \{-1, 1, -1, 1, -1, 1, \ldots\}$.

The range of a sequence $\{x_n\} = \{-1, 1\}$, which is a finite set.

Bounded and Unbounded Sequences

Bounded above sequence – A sequence $\{a_n\}$ is said to be bounded above if $\exists$ a real number K such that $a_n \le K \;\; \forall\, n \in N$.

Bounded below sequence – A sequence $\{a_n\}$ is said to be bounded above if $\exists$ a real number k such that $a_n \ge k \;\; \forall\, n \in N$.

Bounded Sequence – A sequence $\{a_n\}$ is said to be bounded when it is bounded both above and below.

$\Rightarrow$ A sequence $\{a_n\}$ is bounded if $\exists$ two real numbers k and K $(k < K)$ such that $k \le a_n \le K \;\; \forall\, n \in N$.

Choosing $M = \text{Maximum}(|k|, |K|)$, we can also define a sequence $\{a_n\}$ to be bounded if

$$|a_n| \le M \;\; \forall\, n \in N.$$

Example: Consider the sequence $\{a_n\}$ defined by $a_n = \dfrac{1}{n}$.

Here $\{a_n\} = \left\{\dfrac{1}{1}, \dfrac{1}{2}, \dfrac{1}{3}, \ldots\right\}$

$\because \quad 0 < a_n \le 1 \;\; \forall\, n \in N$

$\therefore \quad \{a_n\}$ is bounded.

Unbounded Sequence – If $\exists$ no real numbers M such that $|a_n| \le M \;\; \forall\, n \in N$, then the sequence $\{a_n\}$ is said to be unbounded.

Example: Consider the sequence $\{a_n\}$ defined by $a_n = 2^{n-1}$.

Here $\{a_n\} = \{1, 2, 2^2, 2^3, \ldots\}$

Although $a_n \ge 1, \;\; \forall\, n \in N$, $\exists$ no real number K such that $a_n \le K$.

$\therefore$ the sequence $\{a_n\}$ is bounded below but unbounded above.

Convergent Sequence – A sequence $\{a_n\}$ is said to be convergent if $\underset{n\to\infty}{\text{Lt}}\, a_n$ is finite.

Example: Consider the sequence $\frac{1}{2}, \frac{1}{2^2}, \frac{1}{2^3}, \ldots, \frac{1}{2^n}, \ldots$

Here $a_n = \frac{1}{2^n}$, $\underset{n\to\infty}{\text{Lt}}\, a_n = \underset{n\to\infty}{\text{Lt}}\, \frac{1}{2^n} = 0$, which is finite.

$\Rightarrow$ A sequence $\{a_n\}$ is convergent.

Divergent Sequence – A sequence $\{a_n\}$ is said to be divergent if $\underset{n\to\infty}{\text{Lt}}\, a_n$ is not finite, i.e., if $\underset{n\to\infty}{\text{Lt}}\, a_n = +\infty$ or $-\infty$

Example: (*a*) Consider the sequence $\{n^2\}$

Here $a_n = n^2$, $\underset{n\to\infty}{\text{Lt}}\, a_n = \underset{n\to\infty}{\text{Lt}}\, n^2 = +\infty$, which is infinite.

$\Rightarrow$ A sequence $\{a_n\}$ is divergent.

(*b*) Consider the sequence $\{-2^n\}$

Here $a_n = -2^n$, $\underset{n\to\infty}{\text{Lt}}\, a_n = \underset{n\to\infty}{\text{Lt}}\, (-2^n) = -\infty$, which is infinite.

$\Rightarrow$ A sequence $\{-2^n\}$ is divergent.

Oscillatory Sequence – If a sequence $\{a_n\}$ neither converges to a finite number nor diverges to $+\infty$ or $-\infty$, it is called an *oscillatory sequence.* An oscillatory sequence can oscillate finitely or infinitely.

- A bounded sequence which does not converge is said to be oscillate finitely.

Example: Consider the sequence $\{(-1)^n\}$, which is bounded.

Here $a_n = (-1)^n$

$\underset{n\to\infty}{\text{Lt}}\, a_{2n} = \underset{n\to\infty}{\text{Lt}}\, (-1)^{2n} = 1$

$\underset{n\to\infty}{\text{Lt}}\, a_{2n+1} = \underset{n\to\infty}{\text{Lt}}\, (-1)^{2n+1} = -1$

Thus, $\underset{n\to\infty}{\text{Lt}}\, a_n$ does not exist.

$\Rightarrow$ the sequence $\{(-1)^n\}$ does not converge.

Hence, the sequence $\{(-1)^n\}$ oscillate finitely.

- An unbounded sequence which does not diverge is said to be oscillate infinitely.

Example: Consider the sequence $\{(-1)^n n\}$, which is unbounded.

Here $a_n = (-1)^n n$

$$\underset{n\to\infty}{\text{Lt}}\, a_{2n} = \underset{n\to\infty}{\text{Lt}}\left[(-1)^{2n}.2n\right] = \underset{n\to\infty}{\text{Lt}}\,(2n) = +\infty$$

$$\underset{n\to\infty}{\text{Lt}}\, a_{2n+1} = \underset{n\to\infty}{\text{Lt}}\left[(-1)^{2n+1}.(2n+1)\right] = \underset{n\to\infty}{\text{Lt}}\left[-(2n+1)\right] = -\infty$$

Thus, the sequence $\{(-1)^n n\}$ does not diverge.

Hence, the sequence $\{(-1)^n n\}$ oscillate infinitely.

Note: When we say $\underset{n\to\infty}{\text{Lt}}\, a_n = l$, it means $\underset{n\to\infty}{\text{Lt}}\, a_{2n} = \underset{n\to\infty}{\text{Lt}}\, a_{2n+1} = l$.

Similarly, $\underset{n\to\infty}{\text{Lt}}\, a_n = +\infty$, it means $\underset{n\to\infty}{\text{Lt}}\, a_{2n} = \underset{n\to\infty}{\text{Lt}}\, a_{2n+1} = +\infty$.

Monotonic Sequences

- A sequence $\{a_n\}$ is said to be *monotonically increasing* if $a_{n+1} \geq a_n$, $\forall\, n \in N$, i.e., if $a_1 \leq a_2 \leq a_3 \leq \ldots \leq a_n \leq a_{n+1} \leq \ldots$
- A sequence $\{a_n\}$ is said to be *monotonically decreasing* if $a_{n+1} \leq a_n$, $\forall\, n \in N$, i.e., if $a_1 \geq a_2 \geq a_3 \geq \ldots \geq a_n \geq a_{n+1} \geq \ldots$
- A sequence $\{a_n\}$ is said to be *monotonic* if it is either monotonically increasing or monotonically decreasing.
- A sequence $\{a_n\}$ is said to be *strictly monotonically increasing* if $a_{n+1} > a_n$, $\forall\, n \in N$, i.e., if $a_1 < a_2 < a_3 < \ldots < a_n < a_{n+1} < \ldots$
- A sequence $\{a_n\}$ is said to be *monotonically decreasing* if $a_{n+1} < a_n$, $\forall\, n \in N$, i.e., if $a_1 > a_2 > a_3 > \ldots > a_n > a_{n+1} > \ldots$
- A sequence $\{a_n\}$ is said to be *strictly monotonic* if it is either strictly monotonically increasing or strictly monotonically decreasing.

Limit of a Sequence – A sequence $\{a_n\}$ is said to approach to the limit l (say), when $n \to \infty$, if for each $\varepsilon > 0$, $\exists$ a positive integer m (depending upon ε) such that $|a_n - l| < \varepsilon \quad \forall\, n \geq m.$

In symbolic form, we write $\underset{n\to\infty}{\text{Lt}}\, a_n = l$

Note: $|a_n - l| < \varepsilon \quad \forall\, n \geq m \quad \Rightarrow \quad l - \varepsilon < a_n < l + \varepsilon \quad \forall\, n \geq m$

Theorem: Every Convergent Sequence is Bounded

Proof: Let the sequence $\{a_n\}$ be convergent. Let it tend to the limit l.

Then for given $\varepsilon > 0$, $\exists$ a positive integer m such that

$$|a_n - l| < \varepsilon \quad \forall\, n \geq m$$

$$\Rightarrow \quad l - \varepsilon < a_n < l + \varepsilon \quad \forall\, n \geq m$$

Let k and K be the least and greatest of $a_1, a_2, a_3, \ldots, a_{m-1}, l - \varepsilon, l - \varepsilon$

Then $k \leq a_n \leq K \quad \forall\, n \in N$

$\Rightarrow$ the sequence $\{a_n\}$ is bounded.

Basic Theorems of Convergence of Monotonic Sequences (Without proof)

- The necessary and sufficient condition for the convergence of a monotonic sequence is that it is bounded. A monotonic increasing sequence which is bounded above converges while a monotonic decreasing sequence which is bounded below converges.
- If a monotonic increasing sequence is not bounded above, it diverges to $+\infty$.
- If a monotonic decreasing sequence is not bounded below, it diverges to $-\infty$.
- If $\{a_n\}$ and $\{b_n\}$ are two convergent sequences, then
  - sequence $\{a_n + b_n\}$ is also convergent and converges to $(A+B)$, *i.e.*, if $\underset{n\to\infty}{\text{Lt}}\, a_n = A$ and $\underset{n\to\infty}{\text{Lt}}\, b_n = B$, then $\underset{n\to\infty}{\text{Lt}}\, (a_n + b_n) = (A+B)$.

- sequence $\{a_n.b_n\}$ is also convergent and converges to $(A.B)$, *i.e.*, if $\underset{n\to\infty}{\text{Lt}}\ a_n = A$ and $\underset{n\to\infty}{\text{Lt}}\ b_n = B$, then $\underset{n\to\infty}{\text{Lt}}\ (a_n.b_n) = (A.B)$.
- sequence $\left\{\dfrac{a_n}{b_n}\right\}$ is also convergent and converges to $\left(\dfrac{A}{B}\right)$, *i.e.*, if $\underset{n\to\infty}{\text{Lt}}\ a_n = A$ and $\underset{n\to\infty}{\text{Lt}}\ b_n = B$, then $\underset{n\to\infty}{\text{Lt}}\left(\dfrac{a_n}{b_n}\right) = \left(\dfrac{A}{B}\right)$.

- The sequence $\{|a_n|\}$ converges to zero iff the sequence $\{a_n\}$ converges to zero.
- If a sequence $\{a_n\}$ converges to a and $a_n \geq 0,\ \forall\ n$, then $a \geq 0$.
- If $a_n \to a$, $b_n \to b$ and $a_n \leq b_n,\ \forall\ n$, then $a \leq b$.
- (*Squeeze principle* or *Sandwitch theorem*) If $a_n \to l$, $b_n \to l$ and $a_n \leq c_n \leq b_n,\ \forall\ n$, then $c_n \to l$.

ILLUSTRATIVE EXAMPLES 5.1

Example 5.1: Give an example of a monotonic increasing sequence which is

(*a*) convergent
(*b*) divergent

[PTU Dec 2004]

Solution:

(*a*) Consider the following sequence

$$\frac{1}{2}, \frac{2}{3}, \frac{3}{4}, \ldots, \frac{n}{n+1}, \ldots$$

Since $\dfrac{1}{2} < \dfrac{2}{3} < \dfrac{3}{4}, \ldots$ $\Rightarrow$ given sequence is monotonic increasing.

Now $a_n = \dfrac{n}{n+1}$

$$\therefore \quad \underset{n\to\infty}{\text{Lt}}\ a_n = \underset{n\to\infty}{\text{Lt}}\left(\frac{n}{n+1}\right) = \underset{n\to\infty}{\text{Lt}}\left(\frac{1}{1+\frac{1}{n}}\right) = \left(\frac{1}{1+0}\right) = 1, \text{ which is finite.}$$

$\therefore$ the sequence is convergent.

(*b*) Consider the following sequence

$$1, 2, 3, ..., n, ...$$

Since $1 < 2 < 3 < ...$ $\Rightarrow$ given sequence is monotonic increasing.

Now $a_n = n$

$\therefore$ $\underset{n\to\infty}{\text{Lt}}\, a_n = \underset{n\to\infty}{\text{Lt}}\, n = +\infty$, which is infinite.

$\therefore$ the sequence is divergent.

Example 5.2: Given an example of a monotonic decreasing sequence which is

(*a*) convergent
(*b*) divergent

Solution:

(*a*) Consider the following sequence

$$1, \frac{1}{2}, \frac{1}{3}, ..., \frac{1}{n}, ...$$

Since $1 > \frac{1}{2} > \frac{1}{3}, ...$ $\Rightarrow$ given sequence is monotonic decreasing.

Now $a_n = \frac{1}{n}$

$\therefore$ $\underset{n\to\infty}{\text{Lt}}\, a_n = \underset{n\to\infty}{\text{Lt}} \left(\frac{1}{n}\right) = 0$, which is finite.

$\therefore$ the sequence is convergent.

(*b*) Consider the following sequence

$$-1, -2, -3, ..., -n, ...$$

Since $-1 > -2 > -3 > ...$ $\Rightarrow$ given sequence is monotonic decreasing.

Now $a_n = -n$

$\therefore$ $\underset{n\to\infty}{\text{Lt}}\, a_n = \underset{n\to\infty}{\text{Lt}}\, (-n) = -\infty$, which is infinite.

$\therefore$ the sequence is divergent.

Example 5.3: Discuss the convergence of the following sequences $\{a_n\}$, where

(a) $a_n = \frac{n+1}{n}$ (b) $a_n = \frac{n}{n^2+1}$ (c) $a_n = 1 + \frac{1}{3} + \frac{1}{3^2} + \ldots + \frac{1}{3^n}$

Solution:

(a) Given $a_n = \frac{n+1}{n} \Rightarrow a_{n+1} = \frac{(n+1)+1}{n+1} = \frac{n+2}{n+1}$

$$a_{n+1} - a_n = \frac{n+2}{n+1} - \frac{n+1}{n} = -\frac{1}{n(n+1)} < 0 \;\forall\, n$$

$\Rightarrow$ $a_{n+1} < a_n \;\forall\, n$

$\Rightarrow$ sequence $\{a_n\}$ is a decreasing sequence.

Also $a_n = \frac{n+1}{n} = 1 + \frac{1}{n} > 1 \;\forall\, n$

$\Rightarrow$ sequence $\{a_n\}$ is bounded below by 1.

$\because$ sequence $\{a_n\}$ is decreasing and bounded below, it is also convergent since

$$\underset{n\to\infty}{\text{Lt}}\, a_n = \underset{n\to\infty}{\text{Lt}}\left(\frac{n+1}{n}\right) = \underset{n\to\infty}{\text{Lt}}\left(1 + \frac{1}{n}\right) = 1 + 0 = 1$$

(b) Given $a_n = \frac{n}{n^2+1} \Rightarrow a_{n+1} = \frac{n+1}{(n+1)^2+1} = \frac{n+1}{n^2+2n+2}$

$$a_{n+1} - a_n = \frac{n+1}{n^2+2n+2} - \frac{n}{n^2+1} = \frac{(n+1)(n^2+1) - n(n^2+2n+2)}{(n^2+2n+2)(n^2+1)}$$

$$= \frac{-n^2-n+1}{(n^2+2n+2)(n^2+1)} < 0 \;\forall\, n$$

$\Rightarrow$ $a_{n+1} < a_n \;\forall\, n$

$\Rightarrow$ sequence $\{a_n\}$ is a decreasing sequence.

Also $a_n = \frac{n}{n^2+1} > 0 \;\forall\, n$

$\Rightarrow$ sequence $\{a_n\}$ is bounded below by 0.

$\because$ sequence $\{a_n\}$ is decreasing and bounded below, it is also convergent since

$$\underset{n\to\infty}{\text{Lt}}\, a_n = \underset{n\to\infty}{\text{Lt}}\left(\frac{n}{n^2+1}\right) = \underset{n\to\infty}{\text{Lt}}\left(\frac{\frac{1}{n}}{1+\frac{1}{n^2}}\right) = \left(\frac{0}{1+0}\right) = 0$$

(*c*) Given $\quad a_n = 1+\frac{1}{3}+\frac{1}{3^2}+\ldots+\frac{1}{3^n}$

Note that the given sequence with (n+1) terms is a GP whose first term is 1 and common ratio is $\frac{1}{3}$.

$$a_n = \frac{1\left(1-\frac{1}{3^{n+1}}\right)}{1-\frac{1}{3}} = \frac{\left(1-\frac{1}{3^{n+1}}\right)}{\frac{2}{3}} = \frac{3}{2}\left(1-\frac{1}{3^{n+1}}\right) \qquad \left[\because S_n = \frac{a(1-r^n)}{1-r}\right]$$

$$a_{n+1} = 1+\frac{1}{3}+\frac{1}{3^2}+\ldots+\frac{1}{3^n}+\frac{1}{3^{n+1}}$$

$$a_{n+1}-a_n = \frac{1}{3^{n+1}} > 0 \;\forall\, n$$

$\Rightarrow \quad a_{n+1} > a_n \quad \forall\, n$

$\Rightarrow$ sequence $\{a_n\}$ is an increasing sequence.

Also $a_n = \frac{3}{2}\left(1-\frac{1}{3^{n+1}}\right) < \frac{3}{2} \quad \forall\, n$

$\Rightarrow$ sequence $\{a_n\}$ is bounded above by $\frac{3}{2}$.

$\because$ sequence $\{a_n\}$ is increasing and bounded above, it is also convergent since

$$\underset{n\to\infty}{\text{Lt}}\, a_n = \underset{n\to\infty}{\text{Lt}}\left[\frac{3}{2}\left(1-\frac{1}{3^{n+1}}\right)\right] = \frac{3}{2}(1-0) = \frac{3}{2}$$

5.2 INFINITE SERIES

- If $\{u_n\}$ is a sequence of real numbers, then the expression

$$u_1 + u_2 + u_3 + ... + u_n + ...$$

i.e., the sum of the terms of the sequence, which are infinite in number, is called an *infinite series*. The above series is usually denoted by $\sum_{n=1}^{\infty} u_n$ or simply as Σu_n.

- If all the terms of the series $\Sigma u_n = u_1 + u_2 + u_3 + ... + u_n + ...$ are positive, *i.e.*, $u_n > 0$, $\forall$ *n*, then the series Σu_n is called a *series of positive terms.*

A series in which the terms are positive and negative alternatively, is called an *alternating series*. Thus, the series $\Sigma(-1)^{n-1} u_n = u_1 - u_2 + u_3 - u_4 + ... + (-1)^{n-1} u_n + ...$, where $u_n > 0$, $\forall$ *n*, is an *alternating series.*

- If $\Sigma u_n = u_1 + u_2 + u_3 + ... + u_n + ...$ is an infinite series, where the terms may be positive or negative, then

$$S_n = u_1 + u_2 + u_3 + ... + u_n$$

is called the nth partial sum of Σu_n. Thus, the nth partial sum of an infinite series is the sum of its first *n* terms.

Therefore, $S_1, S_2, S_3, ...$ are the first, second, third, …, partial sums of the series.

Further, since $n \in N$, $\{S_n\}$ is a sequence called the sequence of partial sums of the infinite series Σu_n.

Therefore, to every infinite series Σu_n, there corresponds a sequence $\{S_n\}$ of its partial sums.

Behaviour of an Infinite Series

An infinite series Σu_n converges, diverges or oscillates (finitely or infinitely) according as the sequence $\{S_n\}$ of its partial sums converges, diverges or oscillates (finitely or infinitely).

- Infinite series Σu_n converges (or said to be convergent) if the sequence $\{S_n\}$ of its partial sums converges.

  Thus, Σu_n is convergent if

  $$\underset{n\to\infty}{\text{Lt}}\ S_n = \text{finite}$$

- Infinite series Σu_n diverges (or said to be divergent) if the sequence $\{S_n\}$ of its partial sums diverges.

  Thus, Σu_n is divergent if

  $$\underset{n\to\infty}{\text{Lt}}\ S_n = +\infty \text{ or } -\infty$$

- Infinite series Σu_n oscillates finitely if the sequence $\{S_n\}$ of its partial sums oscillates finitely. Thus, Σu_n oscillates finitely if $\{S_n\}$ is bounded and neither converges nor diverges.

- Infinite series Σu_n oscillates infinitely if the sequence $\{S_n\}$ of its partial sums oscillates infinitely. Thus, Σu_n oscillates infinitely if $\{S_n\}$ is unbounded and neither converges nor diverges.

ILLUSTRATIVE EXAMPLES 5.2

Example 5.4: Discuss the behaviour of the following series

$$\frac{1}{1\cdot 2}+\frac{1}{2\cdot 3}+\frac{1}{3\cdot 4}+\ldots+\frac{1}{n\cdot(n+1)}+\ldots$$

Solution:

Here $u_n = \dfrac{1}{n\cdot(n+1)} = \dfrac{1}{n} - \dfrac{1}{n+1}$

Putting $n = 1, 2, 3, \ldots, n$

$$u_1 = \frac{1}{1} - \frac{1}{2}$$

$$u_2 = \frac{1}{2} - \frac{1}{3}$$

$$u_3 = \frac{1}{3} - \frac{1}{4}$$

$$\vdots$$

$$u_n = \frac{1}{n} - \frac{1}{n+1}$$

On adding, we get

$$S_n = \frac{1}{1} - \frac{1}{n+1} = 1 - \frac{1}{n+1}$$

$\displaystyle\mathop{\text{Lt}}_{n\to\infty} S_n = 1 - 0 = 1$, which is finite.

$\Rightarrow$ $\{S_n\}$ converges to 1.

$\Rightarrow$ Σu_n converges to 1.

Example 5.5: Discuss the behaviour of the following series

$$1^2 + 2^2 + 3^2 + \ldots + n^2 + \ldots$$

OR

Show that the series

$$1^2 + 2^2 + 3^2 + \ldots + n^2 + \ldots$$

diverges to $+\infty$.

Solution: Given series is

$$1^2+2^2+3^2+...+n^2+...$$

Here $u_n = n^2$

And $S_n = 1^2+2^2+3^2+...+n^2$

$$= \frac{n(n+1)(2n+1)}{6}$$ [sum of squares of first n natural numbers]

$$\underset{n\to\infty}{\text{Lt}}\, S_n = \underset{n\to\infty}{\text{Lt}}\, \frac{n(n+1)(2n+1)}{6} = +\infty$$, which is infinite.

$\Rightarrow \{S_n\}$ diverges to $+\infty$.

$\Rightarrow \Sigma u_n$ diverges to $+\infty$.

Nature of Geometric Series

A geometric series in x is of the form

$$1+x+x^2+x^3+...+x^n+...$$

(*a*) It converges if $-1<x<1$, i.e., $|x|<1$

(*b*) It diverges if $x\geq 1$

(*c*) It oscillates finitely if $x=-1$

(*d*) It oscillates infinitely if $x<-1$

Proof:

(*a*) When $|x|<1$

Since $|x|<1$, $x^n \to 0$ as $n\to\infty$

$S_n = 1+x+x^2+x^3+...+x^n$

$$= \frac{1(1-x^n)}{1-x}$$ [Series is GP with $a = 1$ and $r = x$]

$$= \frac{1-x^n}{1-x} = \frac{1}{1-x} - \frac{x^n}{1-x}$$

$$\underset{n\to\infty}{\text{Lt}}\ S_n = \underset{n\to\infty}{\text{Lt}}\left(\frac{1}{1-x} - \frac{x^n}{1-x}\right)$$

$$= \underset{n\to\infty}{\text{Lt}}\left(\frac{1}{1-x}\right) - \underset{n\to\infty}{\text{Lt}}\left(\frac{x^n}{1-x}\right)$$

$$= \frac{1}{1-x} - \left(\frac{0}{1-x}\right)$$

$$= \frac{1}{1-x} \text{ which is finite.}$$

$\Rightarrow$ the sequence $\{S_n\}$ is convergent.

$\Rightarrow$ the given series is convergent.

(*b*) When $x \geq 1$

Sub Case I: When $x = 1$

$$S_n = 1+1+1+1+\ldots \text{ to } n \text{ terms} = n$$

$$\underset{n\to\infty}{\text{Lt}}\ S_n = \underset{n\to\infty}{\text{Lt}}\ n = \infty$$

$\Rightarrow$ the sequence $\{S_n\}$ is divergent and diverges to ∞.

$\Rightarrow$ the given series is divergent and diverges to ∞.

Sub Case II: When $x > 1,\ x^n \to \infty$ as $n \to \infty$,

$$S_n = 1 + x + x^2 + x^3 + \ldots + x^n = \frac{1(x^n-1)}{x-1} = \frac{(x^n-1)}{x-1}$$

$$\underset{n\to\infty}{\text{Lt}}\ S_n = \underset{n\to\infty}{\text{Lt}}\left(\frac{(x^n-1)}{x-1}\right) = \infty$$

$\Rightarrow$ the sequence $\{S_n\}$ is divergent and diverges to ∞.

$\Rightarrow$ the given series is divergent and diverges to ∞.

(*c*) When $x=-1$

$S_n = 1-1+1-1+\ldots$ to n terms

$= 1$ or 0 according as n is odd or even

$\underset{n\to\infty}{\text{Lt}}\ S_n = 1$ or 0

$\Rightarrow$ the sequence $\{S_n\}$ oscillates finitely.

$\Rightarrow$ the given series oscillates finitely.

(*d*) When $x<-1$

$x<-1 \quad \Rightarrow \quad -x>1$

Let $r=-x$, then $r>1$

$\therefore \quad r^n \to \infty$ as $n\to\infty$

$S_n = 1+x+x^2+x^3+\ldots+x^n$

$$= \frac{1(1-x^n)}{1-x} = \frac{1-x^n}{1-x} = \frac{1-(-r)^n}{1-(-r)} \qquad [\because x=-r]$$

$$= \frac{1-r^n}{1+r} \text{ or } \frac{1+r^n}{1+r} \text{ according to } n \text{ is even or odd}$$

$$\underset{n\to\infty}{\text{Lt}}\ S_n = \frac{1-\infty}{1+r} \text{ or } \frac{1+\infty}{1+r} = -\infty \text{ or } +\infty$$

$\Rightarrow$ the sequence $\{S_n\}$ oscillates infinitely.

$\Rightarrow$ the given series oscillates infinitely.

Art. Necessary condition for convergence of a positive term series is that if a positive term series Σu_n is convergent, then $\underset{n\to\infty}{\text{Lt}}\ u_n = 0$.

Proof: Let S_n denote the nth partial sum of the positive term series Σu_n.

Then Σu_n is convergent $\Rightarrow$ $\{S_n\}$ is convergent.

$\Rightarrow \quad \underset{n\to\infty}{\text{Lt}}\ S_n$ is finite and unique $= s$ (say)

$$\Rightarrow \quad \underset{n\to\infty}{\text{Lt}}\ S_n = s$$

Now $S_n - S_{n-1} = u_n$

$$\therefore \quad \underset{n\to\infty}{\text{Lt}}\ u_n = \underset{n\to\infty}{\text{Lt}}\ (S_n - S_{n-1}) = \underset{n\to\infty}{\text{Lt}}\ S_n - \underset{n\to\infty}{\text{Lt}}\ S_{n-1} = s - s = 0$$

Hence Σu_n is convergent $\quad \Rightarrow \quad \underset{n\to\infty}{\text{Lt}}\ u_n = 0$

The converse of the above theorem is not always true, *i.e.*, the *n*th term may tend to zero as $n \to \infty$ even if the series is not convergent.

For example, consider the series

$$1 + \frac{1}{2} + \frac{1}{3} + \ldots + \frac{1}{n} + \ldots$$

It diverges though

$$\underset{n\to\infty}{\text{Lt}}\ u_n = \underset{n\to\infty}{\text{Lt}}\ \frac{1}{n} = 0$$

Hence, $\underset{n\to\infty}{\text{Lt}}\ u_n = 0 \quad \Rightarrow \quad \Sigma u_n$ may or may not be convergent.

Note: $\underset{n\to\infty}{\text{Lt}}\ u_n \neq 0 \quad \Rightarrow \quad \Sigma u_n$ is not convergent.

Art. A positive term series either converges or diverges to $+\infty$.

Proof: Let Σu_n be a positive term series and S_n be its *n*th partial sum, *i.e.*,

$S_n = u_1 + u_2 + u_3 + \ldots + u_n$, then

$$S_{n+1} = u_1 + u_2 + u_3 + \ldots + u_n + u_{n+1}$$

$$= S_n + u_{n+1}$$

$$\Rightarrow \quad S_{n+1} - S_n = u_{n+1} > 0 \ \forall\ n \qquad [\because\ u_n > 0\ \forall\ n]$$

$$\Rightarrow \quad S_{n+1} > S_n \quad \forall\ n$$

$\Rightarrow \quad \{S_n\}$ is a monotonic increasing sequence.

Further two cases arise — the sequence $\{S_n\}$ may be bounded or unbounded above.

Case I: When sequence $\{S_n\}$ is bounded above.

Since $\{S_n\}$ is monotonic increasing and bounded above, it is convergent.

$\Rightarrow$ Σu_n is convergent.

Case II: When sequence $\{S_n\}$ is unbounded above.

Since $\{S_n\}$ is monotonic increasing and unbounded above, it diverges to $+\infty$.

$\Rightarrow$ Σu_n diverges to $+\infty$.

Hence a positive term series either converges or diverges to $+\infty$.

Cor. If $u_n > 0 \ \forall \ n$ and $\underset{n\to\infty}{\text{Lt}} u_n \neq 0$, then the series Σu_n diverges to $+\infty$.

Proof: Given $u_n > 0 \ \forall \ n$

$\Rightarrow$ Σu_n is a series of positive terms.

$\Rightarrow$ Σu_n either converges or diverges to $+\infty$.

Since $\underset{n\to\infty}{\text{Lt}} u_n \neq 0$ (given)

$\therefore$ Σu_n does not converge.

Hence Σu_n diverges to $+\infty$.

Theorem: The necessary and sufficient condition for the convergence of a positive term series Σu_n is that the sequence $\{S_n\}$ of its partial sums is bounded above.

Proof:

Necessary Condition: Suppose the sequence $\{S_n\}$ is bounded above. Since the series Σu_n is of positive terms, the sequence $\{S_n\}$ is monotonically increasing. Since every monotonically increasing sequence which is bounded above converges, therefore, $\{S_n\}$ and hence Σu_n converges.

Sufficient Condition: Suppose Σu_n converges. Then the sequence $\{S_n\}$ of its partial sums also converges. Since every convergent sequence is bounded above, therefore, $\{S_n\}$ is bounded. In particular, $\{S_n\}$ is bounded above.

5.3 CAUCHY'S GENERAL PRINCIPLE OF CONVERGENCE OF SERIES

Statement: The necessary and sufficient condition for the infinite series $\sum_{n=1}^{\infty} u_n$ to converge is that given $\varepsilon > 0$, however small, $\exists$ a positive integer p such that

$$\left|S_{n+p} - S_n\right| < \varepsilon, \ \forall\, n \geq m;\ m \text{ and } p \in N, \text{i.e.,}$$

$$\left|u_{n+1} + u_{n+2} + \ldots + u_{n+p}\right| < \varepsilon \text{ for } n \geq m,\ m \text{ and } p \in N.$$

Proof:

Necessary Condition: Given that $\sum_{n=1}^{\infty} u_n$ is convergent.

$\therefore \quad \underset{n\to\infty}{\text{Lt}}\ S_n = \text{finite}$ where $\{S_n\}$ is the sequence of its partial sums.

Let $\underset{n\to\infty}{\text{Lt}}\ S_n = l$ where l is a finite number.

$\therefore$ given $\varepsilon > 0$, however small, $\exists\ m \in N$ such that

$$\left|S_n - l\right| < \frac{\varepsilon}{2} \quad \forall\, n \geq m \qquad \ldots (1)$$

If $p \in N$ and $n \geq m$ then $n + p \geq m$

$\therefore$ from (1), $\left|S_{n+p} - l\right| < \dfrac{\varepsilon}{2} \quad \forall\, n \geq m$

Now $\left|S_{n+p} - S_n\right| = \left|(S_{n+p} - l) - (S_n - l)\right|$

$$\leq \left|\left(S_{n+p} - l\right)\right| + \left|\left(S_n - l\right)\right| \qquad \left[\because\ |a-b| \leq |a| + |b|\right]$$

$$< \frac{\varepsilon}{2} + \frac{\varepsilon}{2} \quad \text{for } n \geq m,\ p \in N$$

Hence $\left|S_{n+p} - S_n\right| < \varepsilon \quad$ for $n \geq m,\ p \in N$

Sufficient Condition: Given $\left|S_{n+p} - S_n\right| < \varepsilon \quad$ for $n \geq m,\ p \in N$

In particular $\left|S_{m+p} - S_m\right| < \varepsilon \ \forall\, p \in N$

Now S_m, being the sum of first m terms of the sequence $\{S_n\}$ and S_{m+p} differs from S_m by a number $< \varepsilon,\ \ \forall\, p \in N$.

$\therefore\quad S_{m+p}$ cannot be infinite when $p \to \infty$, i.e., $\underset{p\to\infty}{\text{Lt}}\ S_{m+p} \neq \infty$

$\therefore\quad \underset{n\to\infty}{\text{Lt}}\ S_n \neq \infty$ (replace $m + p$ by n)

Also $\underset{n\to\infty}{\text{Lt}}\ S_n$ and $\underset{n\to\infty}{\text{Lt}}\ S_{n+p}$ have the same value S.

Now $\left|S_{n+p} - S_n\right| < \varepsilon \quad \forall\, p \in N$

$\Rightarrow\quad \underset{n\to\infty}{\text{Lt}}\ S_{n+p} = \underset{n\to\infty}{\text{Lt}}\ S_n = l$ (say) $\ \forall\, p \in N$

$\therefore\quad \sum_{n=1}^{\infty} u_n$ is convergent.

Theorem: If m is a given positive integer, then the two series

$u_1 + u_2 + u_3 + \ldots + u_m + u_{m+1} + \ldots + u_n$, and

$u_{m+1} + u_{m+2} + u_{m+3} + \ldots + u_n$

converge or diverge together.

Proof: Let S_n and s_n denote the nth partial sums of the two series, then

$$S_n = u_1 + u_2 + u_3 + \ldots + u_n$$

$$s_n = u_{m+1} + u_{m+2} + u_{m+3} + \ldots + u_n$$

$$= (u_1 + u_2 + u_3 + ... + u_n) - (u_1 + u_2 + u_3 + ... + u_m)$$

$$= S_n - S_m \qquad ...(1)$$

S_m being the sum of a finite number of terms of $\sum u_n$ is a fixed finite quantity.

- If $S_n \to$ a finite limit as $n \to \infty$, then from (1), so does s_n.
- If $S_n \to \infty$ as $n \to \infty$, so does s_n.
- If $S_n \to -\infty$ as $n \to \infty$, so does s_n.
- If S_n does not tend to any limit (finite or infinite), so does s_n.

$\Rightarrow$ The two sequences $\{S_n\}$ and $\{s_n\}$ converge or diverge together.

$\Rightarrow$ The two given series converge or diverge together.

Hence the result.

> **Note:** The above theorem shows that the convergence, divergence or oscillation of a series is not affected by addition or omission of a finite number of its terms.

ILLUSTRATIVE EXAMPLES 5.3

Example 5.6: Using Cauchy's general principle of convergence, prove that the series $\sum_{n=1}^{\infty} \frac{1}{n}$ does not converge.

Solution: Let us suppose that $\sum_{n=1}^{\infty} \frac{1}{n}$ is convergent.

$\therefore$ By Cauchy's general principle of convergence

$$|S_{n+p} - S_n| < \varepsilon \quad \forall\, n \geq m, p \in N$$

Taking $\varepsilon = \frac{1}{2}$, we get

$$|S_{n+p} - S_n| < \frac{1}{2} \quad \forall\, n \geq m, p \in N$$

Putting $n = m$, we get

i.e., $$\left|\frac{1}{m+1}+\frac{1}{m+2}+\frac{1}{m+3}+...+\frac{1}{m+p}\right|<\frac{1}{2} \quad \forall\, p \in N \qquad \left[\because S_n = \frac{1}{n}\right]$$

or $$\frac{1}{m+1}+\frac{1}{m+2}+\frac{1}{m+3}+...+\frac{1}{m+p}<\frac{1}{2} \quad \forall\, p \in N$$

Putting $p = m$, we get

$$\frac{1}{m+1}+\frac{1}{m+2}+\frac{1}{m+3}+...+\frac{1}{2m}<\frac{1}{2} \qquad ...(1)$$

But $$\frac{1}{m+1}+\frac{1}{m+2}+\frac{1}{m+3}+...+\frac{1}{2m}>\frac{1}{m+m}+\frac{1}{m+m}+\frac{1}{m+m}+...+\frac{1}{2m}$$

$$=\frac{m}{2m}=\frac{1}{2}$$

$$\therefore \quad \frac{1}{m+1}+\frac{1}{m+2}+\frac{1}{m+3}+...+\frac{1}{2m}>\frac{1}{2} \qquad ...(2)$$

Eq. (2) contradicts Eq. (1), and hence our supposition is wrong.

$\therefore$ Given series does not converge.

Example 5.7: Using Cauchy's general principle of convergence prove that the series $\sum_{n=1}^{\infty}\frac{1}{n^2}$ is convergent.

Solution: Let $S_n = 1+\frac{1}{2^2}+\frac{1}{3^2}+...+\frac{1}{n^2}$

$$S_{n+p} = 1+\frac{1}{2^2}+\frac{1}{3^2}+...+\frac{1}{n^2}+\frac{1}{(n+1)^2}+...+\frac{1}{(n+p)^2}$$

$$\left|S_{n+p}-S_n\right| = \left|\frac{1}{(n+1)^2}+\frac{1}{(n+2)^2}+\frac{1}{(n+3)^2}+...+\frac{1}{(n+p)^2}\right|$$

$$=\frac{1}{(n+1)^2}+\frac{1}{(n+2)^2}+\frac{1}{(n+3)^2}+...+\frac{1}{(n+p)^2}$$

$$< \frac{1}{n(n+1)} + \frac{1}{(n+1)(n+2)} + \frac{1}{(n+2)(n+3)} + \ldots + \frac{1}{(n+p-1)(n+p)}$$

$$= \left(\frac{1}{n} - \frac{1}{n+1}\right) + \left(\frac{1}{n+1} - \frac{1}{n+2}\right) + \left(\frac{1}{n+2} - \frac{1}{n+3}\right) + \ldots + \left(\frac{1}{n+p-1} - \frac{1}{n+p}\right)$$

$$= \frac{1}{n} - \frac{1}{n+p}$$

$$< \frac{1}{n}$$

$$\therefore \quad \left|S_{n+p} - S_n\right| < \frac{1}{n}$$

Let us choose m such that $m > \frac{1}{\varepsilon}$

$\therefore$ for $n \geq m > \frac{1}{\varepsilon}$, we have $\frac{1}{n} < \varepsilon$

$\therefore$ $\left|S_{n+p} - S_n\right| < \varepsilon$ for $n \geq m,\ p \in N$

$\therefore$ by Cauchy's general principle of convergence, $\sum_{n=1}^{\infty} \frac{1}{n^2}$ is convergent.

Theorem: If Σu_n and Σv_n converge to u and v respectively, then $\Sigma(u_n + v_n)$ converges to $(u+v)$.

Proof: Let $U_n = u_1 + u_2 + u_3 + \ldots + u_n$,

$V_n = v_1 + v_2 + v_3 + \ldots + v_n$, and

$S_n = (u_1 + v_1) + (u_2 + v_2) + (u_3 + v_3) + \ldots + (u_n + v_n)$

Then $S_n = (u_1 + u_2 + u_3 + \ldots + u_n) + (v_1 + v_2 + v_3 + \ldots + v_n) = U_n + V_n$

Since Σu_n converges to u, $\underset{n\to\infty}{\text{Lt}}\ U_n = u$

Σv_n converges to v, $\underset{n\to\infty}{\text{Lt}}\ V_n = v$

$$\underset{n\to\infty}{\text{Lt}}\ S_n = \underset{n\to\infty}{\text{Lt}}\ (U_n + V_n) = \underset{n\to\infty}{\text{Lt}}\ U_n + \underset{n\to\infty}{\text{Lt}}\ V_n = u + v$$

$\therefore$ $\Sigma(u_n + v_n)$ converges to $(u+v)$.

EXERCISE 5.1

1. Show that the series $\sum_{n=1}^{\infty} \frac{1}{n^p}$, $p \le 1$ is divergent.
2. Use Cauchy's Criterion to prove the divergence of the series $\sum \frac{1}{2n-1}$.
3. Apply Cauchy's Criterion to prove that the convergence of the series $\sum \frac{1}{n^2}$.
4. Use Cauchy's Criterion to show that the harmonic series $\sum \frac{1}{n}$ diverges.

5.4 COMPARISON TESTS FOR SERIES OF POSITIVE TERMS

Test 1: Let Σa_n and Σb_n be positive terms series. Let k be a positive constant independent of n and m be a fixed positive integers.

(1) If $a_n \le kb_n \ \forall \ n \ge m$ and Σb_n is convergent then Σa_n is convergent.

(2) If $a_n > kb_n \ \forall \ n \ge m$ and Σb_n is divergent then Σa_n is divergent.

Proof: Let $S_n = a_1 + a_2 + a_3 + ... + a_n$

and $S'_n = b_1 + b_2 + b_3 + ... + b_n$

(1) $\because$ Σb_n converges $\quad \therefore$ $S'_n \to S'$ (finite) *as* $n \to \infty$

$\because$ $b_n > 0 \ \forall n$ $\quad \therefore$ $S'_n < S' \ \forall \ n$ $\quad \left[\because \ S'_n < \sum_{n=1}^{\infty} b_n = S'\right]$

$\because$ $a_n \le kb_n \ \forall \ n \ge m$

$\therefore$ $a_{m+1} \le kb_{m+1}, a_{m+2} \le kb_{m+2}, ..., a_n \le kb_n$

On adding, we get

$$a_{m+1} + a_{m+2} + ... + a_n \le k\,(b_{m+1} + b_{m+2} + ... + b_n)$$

$$\Rightarrow S_n - S_m \le k(S'_n - S'_m)$$

$$\Rightarrow S_n < S_m + k(S'_n - S'_m) \quad \forall \ n \ge m \qquad [\because \ S'_n < S' \ \ \forall n]$$

$\Rightarrow$ $S_n <$ (a finite quantity)

$$\left[\begin{array}{l}\because\ S_m \text{ of } S'_m \text{ are finite as these are the sums of finite} \\ \text{number of terms of } \Sigma a_n \text{ and } \Sigma b_n\text{,respectively.}\end{array}\right]$$

$\Rightarrow$ sequence $\{S_n\}$ is bounded.

Also $\{S_n\}$ is increasing $[\because\ a_n > 0\ \forall n]$

$\therefore$ sequence $\{S_n\}$ converges.

$\Rightarrow$ Σa_n converges.

(2) $\because$ $a_n > kb_n \quad \forall\, n \geq m$

$\therefore$ Proceeding as in (1), we get

$$S_n > S_m + k\left(S'_n - S'_m\right)$$

$\because$ Σb_n diverges

$\therefore$ $S'_n \to +\infty$ $\Rightarrow$ $\{S_n\}$ diverges to ∞

$\Rightarrow$ Σa_n is divergent.

Test 2: Let Σa_n and Σb_n be two positive terms series.

(1) If $\underset{n\to\infty}{\text{Lt}} \dfrac{a_n}{b_n} = l$, l being a finite non-zero constant, the Σa_n and Σb_n both converges or diverges together.

(2) If $\underset{n\to\infty}{\text{Lt}} \dfrac{a_n}{b_n} = 0$ and Σb_n converges, then Σa_n is also converges.

(3) If $\underset{n\to\infty}{\text{Lt}} \dfrac{a_n}{b_n} = \infty$ and Σb_n diverges, then Σa_n also diverges.

(4) If $\underset{n\to\infty}{\text{Lt}} \dfrac{a_n}{b_n} = \infty$ and Σa_n converges, then Σb_n also converges.

Proof: (1) Since $\underset{n\to\infty}{\text{Lt}} \frac{a_n}{b_n} = l$

$\therefore$ given $\varepsilon > 0$, however small, $\exists\, m \in N$ such that

$$\left|\frac{a_n}{b_n} - l\right| < \varepsilon \quad \forall\, n \ge m$$

$$\Rightarrow \quad l - \varepsilon < \frac{a_n}{b_n} < l + \varepsilon \quad \forall\, n \ge m$$

$$\Rightarrow \quad (l-\varepsilon)b_n < a_n < (l+\varepsilon)b_n \qquad [\because\ b_n > 0 \quad \forall n\ ]$$

Since Σa_n and Σb_n are both positive $\forall\, n$

$\therefore \quad l > 0$

Choose ε in such a manner that $(l - \varepsilon) > 0$

Let $l - \varepsilon = k_1$, $l + \varepsilon = k_2$ where k_1 and k_2 are positive constant independent of *n*.

$$\therefore \quad k_1 b_n < a_n < k_2 b_n \qquad \forall\, n \ge m$$

Now following cases arise:

Case 1: If Σb_n is convergent, then Σa_n is also convergent. $[\because\ a_n < k_2 b_n,\ \forall\, n \ge m]$

Case 2: If Σb_n is divergent, then Σa_n is also divergent. $[\because\ a_n > k_1 b_n,\ \forall\, n \ge m]$

Case 3: If Σa_n is convergent, then Σb_n is also convergent. $\left[\because b_n < \frac{1}{k_1} a_n,\ \forall\, n \ge m\right]$

Case 4: If Σa_n is divergent, then Σb_n is also divergent. $\left[\because b_n > \frac{1}{k_2} a_n,\ \forall\, n \ge m\right]$

$\therefore$ Σa_n and Σb_n both convergent or divergent together.

(2) Since $\underset{n\to\infty}{\text{Lt}} \frac{a_n}{b_n} = 0$

$\therefore$ given $\varepsilon > 0$, however small, $\exists m \in N$ such that

$$\frac{a_n}{b_n} < \varepsilon \quad \forall\, n \geq m$$

$\Rightarrow$ $a_n < \varepsilon b_n \quad \forall\, n \geq m$ $\qquad [\because\ b_n > 0,\ \forall\, n]$

$\therefore$ $\sum b_n$ is convergent

$\therefore$ by Comparison Test, $\sum a_n$ converges.

(3) Since $\underset{n\to\infty}{\text{Lt}} \frac{a_n}{b_n} = \infty$

$\therefore$ given $\Delta > 0$, however small, $\exists m \in N$ such that

$$\frac{a_n}{b_n} > \Delta \quad \forall\, n \geq m$$

$\therefore$ $a_n > \Delta b_n \quad \forall\, n \geq m$ $\qquad [\because\ b_n > 0,\ \forall\, n]$

$\because$ $\sum b_n$ diverges.

$\therefore$ by Comparison Test, $\sum a_n$ diverges.

(4) Given $\underset{n\to\infty}{\text{Lt}} \frac{a_n}{b_n} = \infty$ and $\sum a_n$ is convergent.

$\therefore$ given $\Delta > 0$, however large, $\exists m \in N$ such that

$$\frac{a_n}{b_n} > \Delta \quad \forall\, n \geq m$$

$\therefore$ $a_n > \Delta b_n \quad \forall\, n \geq m$ $\qquad [\because\ b_n > 0,\ \forall\, n]$

$\therefore$ $b_n < \frac{1}{\Delta} a_n \quad \forall\, n \geq m$ as Δ is large, therefore, $\frac{1}{\Delta}$ is small

$\because$ $\sum a_n$ is convergent.

$\therefore$ by Comparison Test, $\sum b_n$ is also convergent.

Test 3: Let $a_n > 0$, $b_n > 0$ and $c_n > 0$. Let Σb_n be convergent and Σc_n be divergent.

(1) Σa_n converges if $\overline{\text{lt}}\left(\frac{a_n}{b_n}\right) = \lambda \neq \infty$

(2) Σa_n converges if $\overline{\text{lt}}\left(\frac{a_n}{c_n}\right) = \mu \neq \infty$

Proof : (1) Since $\overline{\text{lt}}\left(\frac{a_n}{b_n}\right) = \lambda$

$\therefore$ by the property of limit superior, for every $\varepsilon > 0$ $\exists m \in N$ such that

$$\Rightarrow \quad \frac{a_n}{b_n} < \lambda + \varepsilon \quad \forall\, n \geq m$$

$$\Rightarrow \quad \frac{a_n}{b_n} < \lambda' \quad \forall\, n \geq m, \quad \text{where } \lambda' = \lambda + \varepsilon$$

$\therefore \quad a_n < \lambda' b_n \quad \forall\, n \geq m$

$\therefore \quad \Sigma b_n$ converges.

$\therefore \quad$ by comparison test.

$\therefore \quad \Sigma a_n$ converges.

(2) Since $\mu \neq 0 \quad \therefore \mu > 0$

Now, $\overline{\text{lt}}\left(\frac{a_n}{b_n}\right) = \mu$

$\therefore$ by property of limit inferior, for every $\varepsilon > 0$, $\exists m \in N$ such that

$$\frac{a_n}{c_n} > \mu - \varepsilon \quad \forall\, n \geq m$$

$$\Rightarrow \quad \frac{a_n}{c_n} > \mu' \quad \forall\, n \geq m, \text{where } \mu' = \mu + \varepsilon$$

Choose ε in such a way that $\mu' = \mu - \varepsilon > 0$

$\therefore \quad a_n > \mu' c_n \quad \forall\, n \geq m$

$\therefore$ $\sum c_n$ diverges.

$\therefore$ by Comparison Test, $\sum a_n$ diverges.

Test 4: Let $\sum a_n$ and $\sum b_n$ be two positive term series.

(1) If $\frac{a_{n+1}}{a_n} < \frac{b_{n+1}}{b_n}$ $\forall\, n \geq m, m \in N$ and $\sum b_n$ is convergent, then $\sum a_n$ is also convergent.

(2) If $\frac{a_{n+1}}{a_n} > \frac{b_{n+1}}{b_n}$ $\forall\, n \geq m$, $m \in n$ and $\sum b_n$ is divergent, then $\sum a_n$ is also divergent.

Proof: Let $S_n = a_1 + a_2 + ... + a_n$ and $S'_n = b_1 + b_2 + ... + b_n$

$$\therefore \quad a_{m+1} + a_{m+2} + ... + a_n = (a_1 + a_2 + ... + a_m + a_{m+1} + ... + a_n) - (a_1 + a_2 + ... + a_m)$$

$$= S_n - S_m$$

Similarly $b_{m+1} + b_{m+2} + ... + b_n = S'_n - S'_m$

(1) $S_n - S_m = a_{m+1} + a_{m+2} + ... + a_n$

$$= a_{m+1}\left(1 + \frac{a_{m+2}}{a_{m+1}} + \frac{a_{m+3}}{a_{m+1}} + ...\right) = a_{m+1}\left(1 + \frac{a_{m+2}}{a_{m+1}} + \frac{a_{m+3}}{a_{m+2}} \times \frac{a_{m+2}}{a_{m+1}} + ...\right)$$

$$\leq a_{m+1}\left(1 + \frac{b_{m+2}}{b_{m+1}} + \frac{b_{m+3}}{b_{m+2}} \times \frac{b_{m+2}}{b_{m+1}} + ...\right) \qquad \left[\because \frac{a_{n+1}}{a_n} < \frac{b_{n+1}}{b_n} \;\; \forall\, n \geq m\right]$$

$$= \frac{a_{m+1}}{b_{m+1}}(b_{m+1} + b_{m+2} + ... + b_n)$$

$$= \frac{a_{m+1}}{b_{m+1}}(S'_n - S'_m)$$

$$\therefore \quad S_n - S_m < \frac{a_{m+1}}{b_{m+1}}(S'_n - S'_m)$$

$$\Rightarrow \quad S_n < S_m + \frac{a_{m+1}}{b_{m+1}}(S'_n - S'_m) \qquad ...(1)$$

Now, Σb_n is convergent. Let Σb_n converges to S'.

$\therefore \quad S'_n < S' \quad \forall\, n$

$\therefore$ from (1), $S_n < S_m + \dfrac{a_{m+1}}{b_{m+1}}(S'_n - S'_m)$

$\Rightarrow \quad S_n <$ (a +ve quantity) $\quad [\because\ S'_n > S'_m]$

$\Rightarrow \quad \{S_n\}$ is bounded above.

Also $\{S_n\}$ is an increasing sequence. $\quad [\because\ a_n > 0\ \forall\, n]$

$\therefore \quad \{S_n\}$ is converges

$\Rightarrow \quad \Sigma a_n$ converges.

(2) Since $\dfrac{a_{n+1}}{a_n} > \dfrac{b_{n+1}}{b_n} \quad \forall\, n \geq m$

$\therefore$ Proceeding as in part (1) we get,

$$S_n > S_m + \frac{a_{m+1}}{b_{m+1}}(S'_n - S'_m) \qquad \ldots (2)$$

Also S_m and S'_m are finite as these are the sums of finite number of terms of Σa_n and Σb_n is convergent, respectively.

$\because \quad \Sigma b_n$ is divergent

$\therefore \quad S'_n \to \infty \text{ as } n \to \infty$

$\therefore$ from (2), $S_n \to \infty \text{ as } n \to \infty$

$\Rightarrow \quad \{S_n\}$ is divergent

$\Rightarrow \quad \Sigma a_n$ divergent.

5.5 THE p-TEST FOR THE SERIES OF TYPE $\Sigma\left(1/n^p\right)$

A series of type

$$\sum\frac{1}{n^p}=\frac{1}{1^p}+\frac{1}{2^p}+\frac{1}{3^p}+\ldots+\frac{1}{n^p}+\ldots,\quad p>0$$

converges if $p>1$ and divergent if $p\le 1$.

Proof: The given series is

$$\sum\frac{1}{n^p}=\frac{1}{1^p}+\frac{1}{2^p}+\frac{1}{3^p}+\ldots+\frac{1}{n^p}+\ldots,\quad p>0$$

Let $\sum a_n=\sum\frac{1}{n^p}$ so that $a_n=\frac{1}{n^p},\ \forall n$

Now, $a_n=\frac{1}{n^p}>0\ \forall\ n$

$\therefore\quad \Sigma a_n$ is a +ve term series.

Hence, convergence or divergence of Σa_n is not affected by grouping the terms in any manner, we like.

Case 1: Let $p>1$

Group the terms in such a way that first group contains first term of the series, second group contains next two terms, third group contains next four terms, so on. In other words $\forall\ r\in N$, the rth group contains 2^{r-1} terms.

$$\therefore\ \sum a_n=\frac{1}{1^p}+\left(\frac{1}{2^p}+\frac{1}{3^p}\right)+\left(\frac{1}{4^p}+\frac{1}{5^p}+\frac{1}{6^p}+\frac{1}{7^p}\right)+\left(\frac{1}{8^p}+\frac{1}{9^p}+\ldots+\frac{1}{15^p}\right)+\ldots$$

$$=\sum b_n$$

where, $b_1=\frac{1}{1^p}=1=a_1$

$$b_2=\frac{1}{2^p}+\frac{1}{3^p}<\frac{1}{2^p}+\frac{1}{2^p}=\frac{2}{2^p}=\frac{1}{2^{p-1}}$$

$$b_3 = \frac{1}{4^p} + \frac{1}{5^p} + \frac{1}{6^p} + \frac{1}{7^p} < \frac{1}{4^p} + \frac{1}{4^p} + \frac{1}{4^p} + \frac{1}{4^p}$$

$$= \frac{4}{4^p} = \frac{1}{4^{p-1}} = \frac{1}{2^{2p-2}} = \frac{1}{(2^{p-1})^2}$$

$$\vdots$$

$$b_n < \frac{1}{(2^{p-1})^{n-1}} \quad \forall\, n \geq 1$$

Now, $\sum \frac{1}{(2^{p-1})^{n-1}}$ is an infinite G.P. with common ratio $\frac{1}{2^{p-1}}$.

$\therefore \quad \sum \frac{1}{\left(2^{p-1}\right)^{n-1}}$ convergent $\quad \left[\because\ 0 < \frac{1}{2^{p-1}} < 1 \text{ for } p > 1\right]$

$\therefore$ By Comparison test, $\sum b_n$ is convergent.

$\Rightarrow \sum a_n$ is convergent

Case 2: Let $p = 1$

$\therefore \quad \sum a_n = \sum \frac{1}{n}$, which is divergent.

Case 3: Let $0 < p < 1$

$\therefore \quad \frac{1}{n^p} > \frac{1}{n} \quad \forall\, n \geq 2$

But $\sum \frac{1}{n}$ is divergent

$\therefore$ By Comparison Test, $\sum \frac{1}{n^p}$ is also divergent.

$\therefore$ we see that $\sum \frac{1}{n^p}$, $p > 0$ converges if $p > 1$ and diverges if $p \leq 1$.

Note 1: The series $\frac{1}{1^p}+\frac{1}{2^p}+\frac{1}{3^p}+\ldots+\frac{1}{n^p}+\ldots$ and geometric series

$1+r+r^2+r^3+\ldots+r^n+\ldots$ with $|r|<1$ are called *auxiliary series*.

Note 2: Take auxiliary series $b_n=\frac{1}{n^p}$

where p = power of n in denominator − power of n in nominator

Note 3: Auxiliary series b_n is convergent if $p > 1$, divergent otherwise.

ILLUSTRATIVE EXAMPLES 5.4

Example 5.8: Discuss the convergence or divergence of the series

$$\frac{1}{1\cdot2\cdot3}+\frac{3}{2\cdot3\cdot4}+\frac{5}{3\cdot4\cdot5}+\ldots$$ [PTU May 2009]

Solution: Let $\Sigma a_n=\frac{1}{1\cdot2\cdot3}+\frac{3}{2\cdot3\cdot4}+\frac{5}{3\cdot4\cdot5}+\ldots$ $\left[\begin{array}{l}\because\ 1,3,5,\ldots \text{ is an A.P. series}\\ \quad\text{with } a=1, d=2,\\ \therefore\ T_n=a+(n-1)d\end{array}\right]$

where $$a_n=\frac{1+(n-1)2}{[(1+(n-1)\cdot1)(2+(n-1)\cdot1)(3+(n-1)\cdot1)]}$$

$$=\frac{2n-1}{n(n+1)(n+2)}$$

Let $b_n=\frac{1}{n^2}$ $[\because p=3-1=2]$

$$\frac{a_n}{b_n}=\frac{2n-1}{n(n+1)(n+2)}\times\frac{n^2}{1}=\frac{n(2n-1)}{(n+1)(n+2)}=\frac{2-\frac{1}{n}}{\left(1+\frac{1}{n}\right)\left(1+\frac{2}{n}\right)}$$

$$\therefore\quad \underset{n\to\infty}{\text{Lt}}\frac{a_n}{b_n}=\underset{n\to\infty}{\text{Lt}}\frac{2-\frac{1}{n}}{\left(1+\frac{1}{n}\right)\left(1+\frac{2}{n}\right)}=\frac{2-0}{(1+0)(1+0)}=2$$

which is non-zero and finite.

$\therefore$ By comparison test, Σa_n and Σb_n converges or diverges together.

But $\Sigma b_n = \Sigma \frac{1}{n^2}$ is convergent. $\left[\because \Sigma \frac{1}{n^p} \text{ converges if } p > 1\right]$

$\therefore$ given series Σa_n is convergent.

Example 5.9: Test the following series for convergence.

$$\sqrt{\frac{1}{2^3}} + \sqrt{\frac{2}{3^3}} + \sqrt{\frac{3}{4^3}} + \ldots$$

Solution: Let $\Sigma a_n = \sqrt{\frac{1}{2^3}} + \sqrt{\frac{2}{3^3}} + \sqrt{\frac{3}{4^3}} + \ldots$

$$\text{where } a_n = \sqrt{\frac{n}{(n+1)^3}} = \sqrt{\frac{n}{n^3\left(1+\frac{1}{n}\right)^3}} = \sqrt{\frac{1}{n^2\left(1+\frac{1}{n}\right)^3}}$$

$$= \frac{1}{n\left(1+\frac{1}{n}\right)^{3/2}} = \frac{1}{n}\left(1+\frac{1}{n}\right)^{-3/2} = \frac{1}{n}\left(1-\frac{3}{2}.\frac{1}{n}+\ldots\right)$$

$$= \frac{1}{n} - \frac{3}{2n^2} + \ldots$$

Take $b_n = \frac{1}{n}$

$$\therefore \quad \frac{a_n}{b_n} = 1 - \frac{3}{2n} + \ldots$$

$\therefore \quad \underset{n\to\infty}{\text{Lt}} \frac{a_n}{b_n} = 1$, which is finite and non-zero.

$\therefore \Sigma a_n$ and Σb_n converges or diverges together.

But $\sum b_n = \sum \frac{1}{n}$ diverges by *p-test*. $\left[\because \quad \sum \frac{1}{n^p} \text{ diverges if } p \leq 1\right]$

$\therefore$ by Comparison Test,

$\sum a_n$ diverges.

Example 5.10: Discuss the convergence or divergence of the series $\sum \frac{1}{n} \tan \frac{1}{n}$.

Solution: Let $\sum a_n = \sum \frac{1}{n} \tan \frac{1}{n}$

where $a_n = \frac{1}{n} \tan \frac{1}{n}$

Take $b_n = \frac{1}{n} \cdot \frac{1}{n} = \frac{1}{n^2}$

$$\therefore \quad \frac{a_n}{b_n} = \left(\frac{1}{n} \tan \frac{1}{n}\right) \times \frac{n^2}{1} = \frac{a_n}{b_n} = n \tan \frac{1}{n}$$

$$= n \left(\frac{1}{n} + \frac{\left(\frac{1}{n}\right)^3}{3} + \frac{\left(\frac{1}{n}\right)^5}{5} + \ldots \right) \qquad \left[\because \quad \tan x = x + \frac{x^3}{3} + \frac{x^5}{5} + \ldots\right]$$

$$= 1 + \frac{1}{3} \cdot \frac{1}{n^2} + \frac{1}{5} \cdot \frac{1}{n^4} + \ldots$$

$\therefore \quad \underset{n \to \infty}{\text{Lt}} \frac{a_n}{b_n} = 1$, which is finite and non-zero.

$\therefore \quad \sum a_n$ and b_n converges or diverges together.

But $\sum b_n = \sum \frac{1}{n^2}$ is converges by *p-test*.

$\therefore \quad$ given series $\sum \frac{1}{n} \tan \frac{1}{n}$ converges.

EXERCISE 5.2

1. Test the series $\sum \sqrt{\dfrac{2^n - 1}{3^n - 1}}$ for the convergence.

2. Discuss the convergence or divergence of the following.

   (*a*) $\sum\left(\sqrt{n^2+1}-n\right)$ (*b*) $\sum\left(\sqrt[3]{n^3+1}-n\right)$ [PTU May 2007, Jan 2010, May 2012]

3. Test the series $\sum\left(n\log\dfrac{2n+1}{2n-1}-1\right)$ for the convergence.

4. Test the series $\sum_{n=1}^{\infty}\dfrac{1}{x^n - x^{-n}}$, $x > 0$ for the convergence.

5. Examine the following series for the convergence:

   (*a*) $\dfrac{1}{1\cdot 3}+\dfrac{3}{5\cdot 7}+\dfrac{5}{9\cdot 11}+\ldots$

   (*b*) $1+\dfrac{2^{\frac{1}{100}}}{2}+\dfrac{3^{\frac{1}{100}}}{3}+\ldots$

   (*c*) $1+\dfrac{2^{\frac{1}{10}}}{2^2}+\dfrac{3^{\frac{1}{10}}}{3^2}+\ldots$

   (*d*) $\dfrac{\sqrt{1}}{5}+\dfrac{\sqrt{2}}{7}+\dfrac{\sqrt{3}}{9}+\dfrac{\sqrt{4}}{11}+\ldots$

   (*e*) $\dfrac{2}{1^p}+\dfrac{3}{2^p}+\dfrac{4}{3^p}+\dfrac{5}{4^p}+\ldots$

   (*f*) $\dfrac{2^p}{1^q}+\dfrac{3^p}{2^q}+\dfrac{4^p}{3^q}+\dfrac{5^p}{4^q}+\ldots$

   (*g*) $\dfrac{2^3}{1^k+3^k}+\dfrac{3^3}{3^k+5^k}+\dfrac{4^3}{5^k+7^k}+\ldots$

   (*h*) $1+\dfrac{1}{2^2}+\dfrac{2^2}{3^3}+\dfrac{3^3}{4^4}+\ldots$

   (*i*) $\dfrac{\sqrt{2}-1}{3^3-1}+\dfrac{\sqrt{3}-1}{4^3-1}+\dfrac{\sqrt{4}-1}{5^3-1}+\ldots$

   (*j*) $\dfrac{1}{\sqrt{1}+\sqrt{2}}+\dfrac{1}{\sqrt{2}+\sqrt{3}}+\dfrac{1}{\sqrt{3}+\sqrt{4}}+\ldots$

   [PTU Dec 2003]

6. Test the following series for convergence or divergence:

   (*a*) $\dfrac{1}{2}+\dfrac{1}{3^2}+\dfrac{1}{2^2}+\dfrac{1}{3^4}+\ldots\infty$

   (*b*) $\dfrac{1}{1\cdot 2}+\dfrac{1}{2\cdot 3}+\dfrac{1}{3\cdot 4}+\ldots\infty$

   (*c*) $\dfrac{1}{1\cdot 4}+\dfrac{1}{2\cdot 5}+\dfrac{1}{3\cdot 6}+\ldots\infty$

   (*d*) $\dfrac{1}{1\cdot 2}+\dfrac{2}{3\cdot 4}+\dfrac{3}{5\cdot 6}+\ldots\infty$

   (*e*) $\dfrac{1}{\sqrt{1\cdot 2}}+\dfrac{1}{\sqrt{2\cdot 3}}+\dfrac{1}{\sqrt{3\cdot 4}}+\ldots\infty$

   (*f*) $\dfrac{1}{1^P}+\dfrac{1}{3^P}+\dfrac{1}{5^P}+\ldots\infty$

   (*g*) $\sum_{n=1}^{\infty}\dfrac{n+1}{n(2n-1)}$

   (*h*) $\sum_{n=1}^{\infty}\dfrac{1}{n^p(n+1)^p}$

   (*i*) $\sum_{n=1}^{\infty}\dfrac{\sqrt{n}}{n^2+1}$ (*j*) $\sum_{n=1}^{\infty}\dfrac{2n^3+5}{4n^5+1}$

   (*k*) $\dfrac{1}{1\cdot 3}+\dfrac{2}{3\cdot 5}+\dfrac{3}{5\cdot 7}+\ldots\infty$

7. Test the series $\sum_{n=1}^{\infty}\frac{n^2+1}{n^3+1}$ for the convergence. [PTU May 2006]

8. Discuss the convergence of the series $\sum\frac{n^2+n-1}{n^4+1}$.

9. Discuss the convergence or divergence of following series:

(a) $\sum\frac{1}{3n-1}$ (b) $\sum\frac{1}{n^2+a^2}$

(c) $\sum\frac{1}{n^p(n+1)^p}$ (d) $\sum\frac{1}{n(n+1)}$

(e) $\sum\frac{k+1}{k(2k-1)}$ (f) $\sum\frac{k+3}{k^2-k+1}$

(g) $\sum\frac{n^5-5n^2+1}{n^7+3n+2}$ (h) $\sum\frac{n(n+1)}{(n+2)^2}$

(i) $\sum\frac{a+bn}{a+bn^2}$ (j) $\sum\frac{2+n}{3+2n^2}$

(k) $\sum\frac{n}{n^2+n+1}$ (l) $\sum\frac{\sqrt{n}}{n^2+1}$

(m) $\sum\frac{n}{\sqrt{2n^3+1}}$ (n) $\sum\sqrt{\frac{1}{n^4+1)}}$

(o) $\sum_{n=2}^{\infty}\frac{1}{\sqrt{n(n^2-1)}}$ (p) $\sum\frac{7(n+2)(n+3)}{n^3\sqrt{n}}$

(q) $\sum\frac{\sqrt{x+n-1}}{\sqrt{x^2+n^2+1}}$ (r) $\sum_{n=2}^{\infty}\frac{n+\sqrt{n}}{n^2-n}$

(s) $\sum\frac{n^p}{(n+1)^q}$

10. Discuss the convergence of the series $\sum_{n=2}^{\infty}\frac{1}{n^2\log n}$

11. Discuss the convergence or divergence of the following series:

(a) $\sum\frac{1}{n^{\left(1+\frac{1}{n}\right)}}$ (b) $\sum\frac{1}{n^{\left(p+\frac{q}{n}\right)}}$

(c) $\sum\frac{(\log n)^{65}}{n^{6/5}}$ (d) $\sum\frac{(\log n)^{72}}{n^{4/3}}$

(*e*) $\sum \frac{1}{n\log n}$ (*f*) $\sum \frac{1}{\sqrt{n}}$

(*g*) $\sum r^{\log n},\ r>0$ (*h*) $\sum \frac{1}{(\log\log n)}\log n$

(*i*) $\sum(\sqrt{n^4+1}-\sqrt{n^4-1})$ (*j*) $\sum \frac{\sqrt{n+1}-\sqrt{n-1}}{n}$

(*k*) $\sum(\sqrt{n^3+1}-\sqrt{n^3})$ (*l*) $\sum \frac{\left(\sqrt{n^2+n+1}-\sqrt{n^2-n+1}\right)}{n}$

(*m*) $\sum\left(\sqrt{n^2+1}-\sqrt{n^2-1}\right)$ [PTU Dec 2012] (*n*) $\sum \frac{\left(\sqrt{n+1}-\sqrt{n}\right)}{n^p}$

12. Discuss the convergence of the following series:

(*a*) $\sum\left(\sqrt[3]{n+1}-\sqrt[3]{n}\right)$ (*b*) $\sum\left(\sqrt[3]{n^2+1}-\sqrt[3]{n^2-1}\right)$

(*c*) $\sum\left(\frac{1}{n}-\log\frac{n+1}{n}\right)$ (*d*) $\sum\left(a^{\frac{1}{n}}-1-\frac{1}{n}\log a\right)$

13. Discuss the convergence or divergence of the following series:

(*a*) $\sum \frac{1}{n}\sin\frac{1}{n}$ (*b*) $\sum \frac{1}{n}\sin^{-1}\frac{1}{n}$

(*c*) $\sum \cos\frac{1}{n}$ (*d*) $\sum \tan\frac{1}{n}$

(*e*) $\sum \frac{1}{\sqrt{n}}\sin\frac{1}{n}$ (*f*) $\sum \sin\frac{1}{n}$

(*g*) $\sum \frac{1}{\sqrt{n}}\tan\frac{1}{n}$ (*h*) $\sum \tan^{-1}\frac{1}{n}$ (*i*) $\sum \cot^{-1} n^2$

14. Show that the series $\sum\left(q^{\frac{1}{n}}-1\right)$ is divergent if $q\neq 1,\ q>0$.

15. Does $\sum_{n=1}^{\infty}\frac{\log_e n}{n^{3/2}}$ Converge?

ANSWERS

1. Converges
2. (*a*) Divergent (*b*) Convergent
3. Converges

5. (*a*) Divergent
   (*b*) Convergent
   (*c*) Convergent
   (*d*) Convergent
   (*e*) Convergent for $p > 2$, divergent for $p \leq 2$
   (*f*) Convergent for $q - p > 1$, diverges for $q - p \leq 1$
   (*g*) Converges for k >4, diverges for $k \leq 4$
   (*h*) Divergent
   (*i*) Convergent
   (*j*) Convergent
6. (*a*) Convergent
   (*b*) Convergent
   (*c*) Convergent
   (*d*) Divergent
   (*e*) Divergent
   (*f*) Converges for $p > 1$, diverges for $p \leq 1$
   (*g*) Divergent
   (*h*) Convergent for $p > \frac{1}{2}$, divergent for $p \leq \frac{1}{2}$
   (*i*) Convergent
   (*j*) Convergent
   (*k*) Divergent
7. Divergent
9. (*a*) Divergent
   (*b*) Convergent
   (*c*) Converges for $p > \frac{1}{2}$ and diverges for $p \leq \frac{1}{2}$
   (*d*) Convergent
   (*e*) Divergent
   (*f*) Divergent
   (*g*) Convergent
   (*h*) Divergent
   (*i*) Divergent
   (*j*) Divergent
   (*k*) Divergent
   (*l*) Convergent
   (*m*) Divergent
   (*n*) Convergent
   (*o*) Convergent
   (*p*) Convergent
   (*q*) Divergent

(*r*) Divergent

(*s*) Convergent for $q > p+1$, Divergent for $q \le p+1$

10. Convergent

11. (*a*) Divergent

(*b*) Converges for $p > 1$, diverges for $p \le 1$

(*c*) Convergent

(*d*) Convergent

(*e*) Divergent

(*f*) Divergent

(*g*) Converges for $r < e^{-1}$, diverges for $r \ge e^{-1}$

(*h*) Convergent (*i*) Convergent

(*j*) Convergent (*k*) Convergent

(*l*) Divergent (*m*) Divergent

(*n*) Converges for $p > \frac{1}{2}$, diverges for $p \le \frac{1}{2}$

12. (*a*) Divergent (*b*) Convergent

(*c*) Convergent (*d*) Convergent

13. (*a*) Convergent (*b*) Convergent

(*c*) Divergent (*d*) Divergent

(*e*) Convergent (*f*) Divergent

(*g*) Convergent (*h*) Divergent (*i*) Convergent

5.6 CAUCHY'S INTEGRAL TEST

Statement: If for $x \ge 1$, $f(x)$ is a non negative monotonic decreasing and integral function of x such that

$$f(n) = u_n \quad \forall\, n \in N$$

then the series $\sum_{n=1}^{\infty} u_n$ and the integral $\int_1^{\infty} f(x)dx$, either both converge or both diverge.

Proof: Let $P > 0$ and let x be any real number $\in [P, P+1]$, i.e., $P \le x \le P+1$

Since $f(x)$ is monotonic decreasing function,

$$\therefore \quad f(P) \ge f(x) \ge f(P+1)$$

$$\Rightarrow \int_P^{P+1} f(P)dx \ge \int_P^{P+1} f(x)dx \ge \int_P^{P+1} f(P+1)dx$$

$$\Rightarrow \int_P^{P+1} u_P dx \ge \int_P^{P+1} f(x)dx \ge \int_P^{P+1} u_{P+1}dx \qquad [\because \quad f(P) = u_P \text{ and } f(P+1) = u_{P+1}]$$

$$\Rightarrow u_P [x]_P^{P+1} \ge \int_P^{P+1} f(x)dx \ge u_{P+1} [x]_P^{P+1}$$

$$\Rightarrow u_P [P+1-P] \ge \int_P^{P+1} f(x)dx \ge u_{P+1} [P+1-P]$$

$$\Rightarrow u_P \ge \int_P^{P+1} f(x)dx \ge u_{P+1} \qquad \ldots (1)$$

Putting $P = 1,\ u_1 \ge \int_1^2 f(x)dx \ge u_2$

Putting $P = 2,\ u_2 \ge \int_2^3 f(x)dx \ge u_3$

$\vdots$

Putting $P = n-1,\ u_{n-1} \ge \int_{n-1}^{n} f(x)dx \ge u_n$

Adding, we get

$$u_1 + u_2 + \ldots + u_{n-1} \ge \int_1^n f(x)dx \ge u_2 + u_3 + \ldots + u_n$$

$$\Rightarrow \qquad S_{n-1} > I_n \ge S_n - u_1, \text{ where } I_n = \int_1^n f(x)dx$$

$$\Rightarrow \qquad S_n > I_n \ge S_n - u_1 \qquad [\because \ S_n > S_{n-1} \quad \forall\, n > 1]$$

$$\Rightarrow \qquad S_n - u_1 \le I_n < S_n \qquad \ldots (2)$$

From last two; $I_n < S_n \Rightarrow I_n$ is bounded above if S_n is bounded above.

From first two; $S_n - u_1 \le I_n \Rightarrow S_n$ is bounded above if I_n is bounded above.

$\therefore$ The sequence $\{S_n\}$ and $\{I_n\}$ be have alike.

Hence the Series $\sum_{n=1}^{\infty} u_n$ and the integral $\int_1^{\infty} f(x)dx$ either both converges or both diverge.

[$\because$ the sequence $\{S_n\}$ and $\{I_n\}$ are both monotonic increasing]

Example 5.11: Discuss the convergence or divergence of the series:

$$\sum_{n=1}^{\infty} \frac{1}{n^P}$$

Solution: Let $f(x) = \frac{1}{x^P}$

$$\therefore \quad f(n) = \frac{1}{n^P} \qquad \text{[given]}$$

Since $P > 0$,

$\therefore \quad f(x) > 0$ for $x \ge 1$ and *f(x)* decreases as *x* increasing.

Thus we can apply Cauchy's Integral Test.

Case I: When $P \ne 1$

$$\text{Here } \int_1^{\infty} f(x)dx = \underset{b \to \infty}{\text{Lt}} \int_1^{b} f(x)dx$$

$$= \underset{b \to \infty}{\text{Lt}} \int_1^{b} \frac{1}{x^P} dx = \underset{b \to \infty}{\text{Lt}} \left[\frac{x^{1-P}}{1-P} \right]_1^b$$

$$= \underset{b \to \infty}{\text{Lt}} \frac{1}{1-P} \left[b^{1-P} - 1 \right]$$

Sub Case I: When $P > 1 \quad \Rightarrow \quad P - 1 > 0$

$$\text{Here } \int_1^{\infty} f(x)dx = \underset{b \to \infty}{\text{Lt}} \frac{1}{P-1} \left[1 - \frac{1}{b^{P-1}} \right] = \frac{1}{P-1}, \text{ which is finite}$$

$$\Rightarrow \quad \int_1^{\infty} f(x)dx \text{ is convergent}$$

$$\Rightarrow \quad \Sigma \frac{1}{n^P} \text{ is convergent}$$

Sub Case II: When $P < 1 \quad \Rightarrow \quad 1 - P > 0$

Here $\int_1^\infty f(x)dx = \underset{b\to\infty}{\text{Lt}} \frac{1}{1-P}\left[b^{1-P} - 1\right] = \infty$

$$\Rightarrow \int_1^\infty f(x)dx \text{ is divergent}$$

$$\Rightarrow \sum \frac{1}{n^P} \text{ is divergent.}$$

Case II: When $P = 1$

Here $\int_1^\infty f(x)dx = \underset{b\to\infty}{\text{Lt}} \int_1^b f(x)dx = \underset{b\to\infty}{\text{Lt}} \int_1^b \frac{1}{x} dx$

$$= \underset{b\to\infty}{\text{Lt}} \left[\log x\right]_1^b = \underset{b\to\infty}{\text{Lt}} \left[\log b - log 1\right] = \infty - 1 = \infty$$

$$\Rightarrow \quad \int_1^\infty f(x)dx \text{ is divergent.}$$

$$\Rightarrow \sum \frac{1}{n^P} \text{ is divergent.}$$

Hence $\sum \frac{1}{n^P}$ converges if $P > 1$ and diverges if $P \le 1$.

Example 5.12: Discuss the convergence of $\sum \frac{1}{n^2+1}$ by applying Cauchy's Integral Test.

Solution: Here $f(x) = \frac{1}{x^2+1}$

$$\therefore \quad f(n) = \frac{1}{n^2+1}$$

f(x) decreases as *x* increases.

Thus we can apply Cauchy's Integral Test.

Here $\int_1^\infty f(x)dx = \underset{b\to\infty}{\text{Lt}} \int_1^b f(x)dx = \underset{b\to\infty}{\text{Lt}} \int_1^b \frac{1}{x^2+1} dx$

$$= \underset{b\to\infty}{\text{Lt}}\left[\tan^{-1}x\right]_1^b = \underset{b\to\infty}{\text{Lt}}\left[\tan^{-1}b - \tan^{-1}1\right]$$

$$= \tan^{-1}(\infty) - \tan^{-1}(1)$$

$$= \frac{\pi}{2} - \frac{\pi}{4} = \frac{\pi}{4}, \text{ which is finite}$$

$$\Rightarrow \quad \int_1^\infty f(x)dx \text{ is convergent.}$$

$$\Rightarrow \quad \sum \frac{1}{n^2+1} \text{ is convergent.}$$

Example 5.13: Show that the infinite series

$$\sum_{n=3}^{\infty} \frac{1}{n\log n(\log\log n)^P} \text{ is convergent if } P > 1 \text{ and divergent if } P \le 1.$$

Solution: Let $f(x) = \dfrac{1}{x\log x(\log\log x)^P}$

$$\therefore \quad f(n) = \frac{1}{n\log n(\log\log n)^P}$$

Since $P > 0$,

$\therefore \quad f(x) > 0$ for $x \ge 3$ and $f(x)$ decreases as x increases.

Thus we can apply Cauchy's Integral Test.

Case I: When $P \ne 1$

$$\text{Here } \int_3^\infty f(x)dx = \underset{b\to\infty}{\text{Lt}} \int_3^b f(x)dx = \underset{b\to\infty}{\text{Lt}} \int_3^b \frac{1}{x\log x(\log\log x)^P}dx$$

$$= \underset{b\to\infty}{\text{Lt}}\left[\frac{(\log\log x)^{1-P}}{1-P}\right]_3^b \qquad \left[\because \int [f(x)]^n f'(x)dx = \frac{[f(x)]^{n+1}}{n+1}\right]$$

$$= \underset{b\to\infty}{\text{Lt}} \frac{1}{1-P}\left[(\log\log b)^{1-P} - (\log\log 3)^{1-P}\right]$$

Sub-Case I: When $P > 1 \Rightarrow P-1 > 0$

Here $$\int_3^b f(x)dx = \underset{b\to\infty}{\text{Lt}} \frac{1}{P-1}\left[\frac{1}{(\log\log 3)^{P-1}} - \frac{1}{(\log\log b)^{P-1}}\right]$$

$$= \frac{1}{P-1}\left[\frac{1}{(\log\log 3)^{P-1}} - 0\right]$$

$$= \frac{1}{P-1}\left(\frac{1}{(\log\log 3)^{P-1}}\right), \text{ which is finite}$$

$\Rightarrow$ $\int_1^\infty f(x)dx$ is convergent.

$\Rightarrow$ $\sum_3^\infty \frac{1}{n\log n(\log\log n)^P}$ is convergent.

Sub-Case II: When $P < 1 \Rightarrow 1-P > 0$

Here $$\int_3^b f(x)dx = \underset{b\to\infty}{\text{Lt}} \frac{1}{1-P}\left[(\log\log b)^{1-P} - (\log\log 3)^{1-P}\right] = \infty$$

$\Rightarrow$ $\int_1^\infty f(x)dx$ is divergent.

$\Rightarrow$ $\sum_3^\infty \frac{1}{n\log n(\log\log n)^P}$ is divergent.

Case II: When $P = 1$

Here $$\int_3^\infty f(x)dx = \underset{b\to\infty}{\text{Lt}} \int_3^b f(x)dx$$

$$= \underset{b\to\infty}{\text{Lt}} \int_3^b \frac{1}{x\log x(\log\log x)}dx$$

$$= \underset{b\to\infty}{\text{Lt}} \int_3^b \frac{\frac{1}{x\log x}}{(\log\log x)}dx$$

$$= \underset{b\to\infty}{\text{Lt}} \left[\log(\log\log x)\right]_3^b \qquad \left[\because \int \frac{f'(x)}{f(x)}dx = \log\left(f(x)\right)\right]$$

$$= \underset{b\to\infty}{\text{Lt}} \left[\log(\log\log b) - \log(\log\log 3)\right]$$

$$= \infty - \log(\log\log 3) = \infty$$

$\Rightarrow \quad \int_1^\infty f(x)dx$ is divergent

$\Rightarrow \quad \sum_{n=3}^{\infty} \frac{1}{n\log n(\log\log n)^P}$ is divergent

Hence $\sum_{3}^{\infty} \frac{1}{n\log n(\log\log n)^P}$ is convergent if $P > 1$ and divergent if $P \leq 1$.

EXERCISE 5.3

Using the integral test, discuss the convergence of the following series:

1. $\sum_{n=1}^{\infty} \frac{n}{(n^2+1)^2}$
2. $\sum \frac{1}{\sqrt{n}}$
3. $\sum \frac{1}{n(n+1)}$
4. $\sum ne^{-n^2}$
5. $\sum \frac{1}{n\sqrt{n^2-1}}$
6. $\sum_{n=1}^{\infty} \frac{3n^4}{n^5+3}$
7. Show that the series $\sum_{n=2}^{\infty} \frac{1}{n^p(\log n)^q}$ series converges iff, either $P > 1$ or $P = 1$ and $q > 1$. Hence apply it to discuss the convergence of series $\sum_{n=2}^{\infty} \frac{n^2+5}{3n^3+1}(\log n)^{-6/5}$.

Hint:

Case I: When $P > 1$, Let $P = 1 + 2h$, where h is small but +ve,

Let $V_n = \dfrac{1}{n^{1+h}}$ and use Comparison Test.

Case II: When $P = 1$, use Cauchy's integral test.

Case III: When $0 < P < 1$, Let $P = 1 - 2h$, where h is small but +ve,

Let $V_n = \dfrac{1}{n^{1-h}}$ and use Comparison Test.

Case IV: When $P < 0$, use n^{th} term test.

8. Discuss the convergence of series: $\sum_{n=2}^{\infty} \dfrac{1}{n(\log n)^P}$, $(P > 0)$. Hence show that $\int_2^{\infty} \dfrac{dx}{x(\log x)^P}$, $(P > 0)$ converges if and only if $P > 1$.

9. Test the convergence of the series $\sum_{n=1}^{\infty} \dfrac{8\tan^{-1} n}{1+n^2}$.

ANSWERS

1. Convergent	2. Divergent	3. Convergent
4. Convergent	5. Convergent	6. Divergent
7. Convergent	9. Convergent	

5.7 D' ALEMBERT'S RATIO TEST

Statement: If Σu_n is a positive term series and $\underset{n\to\infty}{\text{Lt}} \left[\dfrac{u_n}{u_{n+1}}\right] = l$, then

(*i*) Σu_n is convergent if $l > 1$ (*ii*) Σu_n is divergent if $l < 1$

Note: If $l = 1$, the test fails, *i.e.*, no conclusion can be drawn about the convergence or divergence of the series. The series may converge, it may diverge.

Proof: Since $\underset{n\to\infty}{\text{Lt}}\left[\dfrac{u_n}{u_{n+1}}\right]=l$

$\therefore$ given $\varepsilon > 0$, however small, there exist a +ve integer m such that

$$\left|\frac{u_n}{u_{n+1}}-l\right|<\varepsilon \quad \forall\, n\geq m$$

$$l-\varepsilon<\frac{u_n}{u_{n+1}}<l+\varepsilon \qquad \forall\, n\geq m \qquad \ldots (1)$$

Case I: When $l>1$, choose $\varepsilon>0$ such that $l-\varepsilon=r>1$

$\therefore$ for $n\geq m$; $\dfrac{u_n}{u_{n+1}}>l-\varepsilon=r$, *i.e.*, $\dfrac{u_n}{u_{n+1}}>r$ for $n\geq m$

Put $n = m, m + 1, m + 2, \ldots, n\text{-}1$ (*i.e.*, $n-m$ terms)

$$\therefore \quad \frac{u_m}{u_{m+1}}>r$$

$$\frac{u_{m+1}}{u_{m+2}}>r$$

$$\frac{u_{m+2}}{u_{m+3}}>r$$

$$\vdots$$

$$\frac{u_{n-1}}{u_n}>r$$

Multiply these inequalities to get

$$\frac{u_m}{u_n}>r^{n-m}$$

$$\frac{u_n}{u_m}<\frac{1}{r^{n-m}} \quad \Rightarrow \quad u_n<\frac{u_m}{r^{n-m}}=(r^m u_m)\frac{1}{r^n}$$

$$u_n < k.\frac{1}{r^n} \quad \forall\, n \geq m \qquad \text{(where } k = r^m.u_m)$$

Let $v_n = \frac{1}{r^n}$ where $r > 1$

$$\therefore \quad \frac{1}{r} < 1$$

Σv_n is a geometric series with common ratio < 1.

$\therefore$ Σv_n is convergent and by comparison test, $\sum_{n=1}^{\infty} u_n$ is also convergent.

Case II: When $l < 1$, choose $\varepsilon > 0$ such that $l + \varepsilon < R < 1$

$$\therefore \text{ from (1)} \qquad \frac{u_n}{u_{n+1}} < R \quad \forall\, n \geq m$$

Put $n = m, m+1, m+2, \ldots, n-1$, and multiply (as in case I), we get

$$\frac{u_m}{u_n} < R^{n-m}$$

$$\Rightarrow \qquad u_n > \frac{u_m}{R^{n-m}} = \left(R^m u_m\right).\frac{1}{R^n} = k'v_n \qquad \text{where } k' = R^m u_m \text{ and } v_n = \frac{1}{R^n}$$

$\Sigma v_n = \sum \frac{1}{R^n}$ which is a G.P. with common ratio $\frac{1}{R} > 1$

$\therefore$ Σv_n is divergent

$\Rightarrow$ Σu_n is also divergent.

Hence, if Σu_n is a positive term series and $\underset{b \to \infty}{\text{Lt}} \left[\frac{u_n}{u_{n+1}}\right] = l$, then

(*i*) Σu_n is convergent if $l > 1$

(*ii*) Σu_n is divergent if $l < 1$

ILLUSTRATIVE EXAMPLES 5.5

Example 5.14: Discuss the convergence of the following series

$$1+\frac{2^2}{2!}+\frac{3^2}{3!}+\frac{4^2}{4!}+\ldots \text{ to } \infty$$

Solution: Given series can be written as

$$\frac{1^2}{1!}+\frac{2^2}{2!}+\frac{3^2}{3!}+\frac{4^2}{4!}+\ldots \text{ to } \infty$$

Here $u_n = \dfrac{n^2}{n!}$

Using $T_n = a+(n-1)d$, where
$T_n = n^{\text{th}}$ term of 1,2,3,4,…
$= 1+(n-1).1 = n$

$$u_{n+1} = \frac{(n+1)^2}{(n+1)!}$$

$\therefore$ Consider $\displaystyle \operatorname*{Lt}_{n\to\infty} \frac{u_n}{u_{n+1}} = \operatorname*{Lt}_{n\to\infty} \left(\frac{n^2}{n!} \cdot \frac{(n+1)!}{(n+1)^2} \right)$

$$= \operatorname*{Lt}_{n\to\infty} \left(\frac{n^2 (n+1) n!}{n!(n+1)^2} \right)$$

$$\left[\because n! = n(n-1)! \\ = n(n-1)(n-2)! \\ = n(n-1)(n-2)\ldots 1 \right]$$

$$= \operatorname*{Lt}_{n\to\infty} \left(\frac{n^2}{n+1} \right)$$

$$= \operatorname*{Lt}_{n\to\infty} \left(\frac{n}{1+\frac{1}{n}} \right) = \frac{\infty}{1+0} = \infty > 1$$

$\therefore$ By D' Alembert's Ratio Test, $\displaystyle\sum_{n=1}^{\infty} u_n$ is convergent.

Example 5.15: Discuss the convergence of the following series

$$1+\frac{2!}{2^2}+\frac{3!}{3^3}+\frac{4!}{4^4}+\ldots \text{ to } \infty$$

Solution: Given series can be written as

$$\frac{1}{1^1}+\frac{2!}{2^2}+\frac{3!}{3^3}+\frac{4!}{4^4}+\ldots \text{ to } \infty$$

Here $u_n = \dfrac{n!}{n^n}$

Using $T_n = a+(n-1)d$, where
$T_n = n^{\text{th}}$ term of 1,2,3,4,…
$= 1+(n-1).1 = n$

$$u_{n+1} = \frac{(n+1)!}{(n+1)^{n+1}}$$

Consider $\displaystyle \operatorname*{Lt}_{n\to\infty}\frac{u_n}{u_{n+1}} = \operatorname*{Lt}_{n\to\infty}\left(\frac{n!}{n^n}.\frac{(n+1)^{n+1}}{(n+1)!}\right)$

$$= \operatorname*{Lt}_{n\to\infty}\left(\frac{n!}{n^n}.\frac{(n+1)^n(n+1)}{(n+1)n!}\right)$$

$$= \operatorname*{Lt}_{n\to\infty}\left(\frac{(n+1)^n}{n^n}\right) = \operatorname*{Lt}_{n\to\infty}\left(1+\frac{1}{n}\right)^n = e > 1 \qquad [\because e = 2.7188]$$

$\therefore$ By D' Alembert's Ratio test $\displaystyle\sum_{n=1}^{\infty} u_n$ is convergent.

Example 5.16: Discuss the convergence of the following series:

$$\frac{2}{1}+\frac{2\cdot5}{1\cdot5}+\frac{2\cdot5\cdot8}{1\cdot5\cdot9}+\frac{2\cdot5\cdot8\cdot11}{1\cdot5\cdot9\cdot13}+\ldots \text{ to } \infty$$

Solution: Neglecting first term, we have

$$u_n = \frac{2\cdot5\cdot8\cdot11\ldots(3n-1)}{1\cdot5\cdot9\cdot13\ldots(4n-3)}$$

Using $T_n = a+(n-1)d$, where
$T_n = n^{\text{th}}$ term of 2,5,8,…
$= 2+(n-1).3 = 3n-1$

$$u_{n+1} = \frac{2\cdot5\cdot8\cdot11\ldots(3n-1)(3(n+1)-1)}{1\cdot5\cdot9\cdot13\ldots(4n-3)(4(n+1)-3)}$$

$$= \frac{2\cdot5\cdot8\cdot11\ldots(3n-1)(3n+2)}{1\cdot5\cdot9\cdot13\ldots(4n-3)(4n+1)}$$

$$\frac{u_n}{u_{n+1}} = \frac{4n+1}{3n+2} = \frac{4+\frac{1}{n}}{3+\frac{2}{n}}$$

$$\underset{n\to\infty}{\text{Lt}} \frac{u_n}{u_{n+1}} = \underset{n\to\infty}{\text{Lt}} \left(\frac{4+\frac{1}{n}}{3+\frac{2}{n}}\right) = \frac{4+0}{3+0} = \frac{4}{3} > 1$$

$\therefore$ By D' Alembert's Ratio test, the given series $\sum_{n=1}^{\infty} u_n$ is convergent.

EXERCISE 5.4

1. Discuss the convergence of the series $\frac{1}{1+2} + \frac{2}{1+2^2} + \frac{3}{1+2^3} + \ldots$ to ∞.
2. Discuss the convergence of the series $\frac{2!}{3} + \frac{3!}{3^2} + \frac{4!}{3^3} + + \ldots$ to ∞.
3. Discuss the convergence of the series $\left(\frac{1}{3}\right)^2 + \left(\frac{1\cdot 2}{3\cdot 5}\right)^2 + \left(\frac{1\cdot 2\cdot 3}{3\cdot 5\cdot 7}\right)^2 + \ldots$ to ∞.
4. Discuss the convergence of the following series:

(*a*) $\sum_{n=1}^{\infty} \frac{2^{n-1}}{3^n + 1}$ (*b*) $\sum_{n=1}^{\infty} \frac{1}{n!}$

(c) $\sum_{n=1}^{\infty} \frac{n^2(n+1)^2}{n!}$ (*d*) $\sum_{n=1}^{\infty} \frac{x^n}{3^n n^2}$, $x>0$

(*e*) $\sum \frac{n^3+1}{2^n + a}$ (*f*) $\sum \frac{n!2^n}{n^n}$

(*g*) $\sum \frac{2^{n-1}}{3^n + 1}$ (*h*) $\sum_{n=1}^{\infty} \frac{n^2(n+1)^2}{n!}$

(*i*) $\sum \frac{n^2}{3^n}$

5. Discuss the convergence of $\frac{1^2\cdot 2^2}{1!} + \frac{2^2\cdot 3^2}{2!} + \frac{3^2\cdot 4^2}{3!} + \ldots$ to ∞.
6. Discuss the convergence of $\frac{1}{2} + \frac{1}{3} + \frac{1}{5} + \ldots + \frac{1}{2^{n-1}+1} + \ldots$ to ∞.

7. Discuss the convergence or divergence of the following series:

(a) $\sum \frac{\sqrt{n}}{\sqrt{n^2+1}} x^n, x>0$ [PTU Dec 2012] (b) $\sum_{n=1}^{\infty} \frac{x^n}{2n!}$ [PTU May 2012]

(c) $\sum_{n=1}^{\infty} \frac{3^n-2}{3^n+1} x^{n-1},\ x>0$ (d) $\sum_{n=1}^{\infty} \frac{x^n}{n},\ x>0$

(e) $\sum_{n=1}^{\infty} \frac{3^{n-2}}{3^{n+1}} x^{n-1},\ x>0$ (f) $\sum_{n=1}^{\infty} \frac{x^n}{(2n)!},\ x>0$

8. Examine the convergence or divergence of the following series:

$\frac{1}{2\sqrt{1}} + \frac{x^2}{3\sqrt{2}} + \frac{x^4}{4\sqrt{3}} + \frac{x^6}{5\sqrt{4}} + \ldots$ [PTU Dec 2002]

9. Discuss the convergence or divergence of the following series:

(a) $x + \frac{3}{5}x^2 + \frac{8}{10}x^3 + \frac{15}{17}x^4 + \ldots + \frac{n^2-1}{n^2+1}x^n + \ldots$ to ∞

(b) $\frac{x}{2\sqrt{3}} + \frac{x^2}{3\sqrt{4}} + \frac{x^3}{4\sqrt{5}} + \ldots$ to ∞

10. Test the convergence the positive term series:

$1 + \frac{\alpha+1}{\beta+1} + \frac{(\alpha+1)(2\alpha+1)}{(\beta+1)(2\beta+1)} + \frac{(\alpha+1)(2\alpha+1)(3\alpha+1)}{(\beta+1)(2\beta+1)(3\beta+1)} + \ldots$

11. Test the convergence of the series $\sum \frac{x^{n+1}}{(n+1)\sqrt{n}}$.

12. Discuss the convergence of the following series.

(a) $1 + \frac{2^2}{2!} + \frac{3^2}{3!} + \frac{4^2}{4!} + \ldots$ to ∞ (b) $1 + \frac{2!}{2^2} + \frac{3!}{3^3} + \frac{4!}{4^4} + \ldots$ to ∞

(c) $\frac{1}{1+2} + \frac{2}{1+2^2} + \frac{3}{1+2^3} + \ldots$ to ∞ (d) $\frac{2!}{3} + \frac{3!}{3^2} + \frac{4!}{3^4} + \ldots$ to ∞

13. Discuss the convergence of the following series:

(a) $\sum \frac{1}{n!}$ (b) $\sum \frac{x^n}{n},\ x>0$

(c) $\sum \frac{n}{n^2+1} x^n,\ x>0$ (d) $\sum \sqrt{\frac{n+1}{n^3+1}}\, x^n,\ x>0$

14. Discuss the convergence of the following series:

(a) $1 + \frac{x}{2} + \frac{x^2}{5} + \frac{x^3}{10} + \ldots + \frac{x^n}{n^2+1} + \ldots$ to ∞ (b) $\frac{x}{1\cdot 3} + \frac{x^2}{3\cdot 5} + \frac{x^3}{5\cdot 7} + \ldots$ to ∞

(c) $1+\frac{x^2}{2}+\frac{x^4}{4}+\frac{x^6}{6}+\ldots$ to ∞

(d) $1+\frac{x}{2^2}+\frac{x^2}{3^2}+\frac{x^3}{4^2}+\ldots$

(e) $1+3x+5x^2+7x^3+9x^4+\ldots$

(f) $\frac{x}{1\cdot 2}+\frac{x^2}{3\cdot 4}+\frac{x^3}{5\cdot 6}+\frac{x^4}{7\cdot 8}+\ldots$

(g) $\frac{1}{x}+\frac{1}{x+1}+\frac{1}{x+2}+\frac{1}{x+3}+\ldots$

(h) $\frac{x}{x+1}+\frac{x^2}{x+2}+\frac{x^3}{x+3}+\ldots$

(i) $\frac{x}{a+\sqrt{1}}+\frac{x^2}{a+\sqrt{2}}+\frac{x^3}{a+\sqrt{3}}+\ldots$

15. Discuss the convergence of the series $1+\frac{(1!)^2}{2!}x^2+\frac{(2!)^2}{4!}x^4+\frac{(3!)^2}{6!}x^6+\ldots$ to ∞, $x>0$

16. Discuss the convergence of the series

$$1+\frac{2}{5}x+\frac{6}{9}x^2+\frac{14}{17}x^3+\ldots+\frac{2^{n+1}-2}{2^{n+1}+1}x^n+\ldots \text{ to } \infty,\ x>0$$ [PTU Jan 2009]

ANSWERS

1. Convergent
2. Divergent
3. Convergent
4. (a) Convergent
   (b) Convergent
   (c) Convergent
   (d) Convergent if $x \le 3$, and Divergent if $x > 3$
   (f) Convergent
   (g) Convergent
   (h) Convergent
   (i) Convergent
   (j) Convergent
5. Convergent
6. Convergent
7. (a) Converges if $x<1$ and diverges if $x \ge 1$
   (b) Convergent
   (c) Convergent for $x<1$ and divergent if $x \ge 1$
   (d) Convergent for $x<1$ and divergent if $\forall\ x \ge 1$
   (e) Convergent for $x \le 1$ and divergent if $x>1$
   (f) Converges $\forall\ x \ge 1$
8. Converges $x^2 \le 1$ and divergent for $x^2 > 1$
9. (a) Converges for $x<1$ and diverges for $x \ge 1$

(*b*) Converges for $x \le 1$ and diverges for $x > 1$

10. Convergent for $\beta > \alpha > 0$ and divergent if $\alpha \ge \beta > 0$

11. Converges if $x \le 1$ and diverges for $x > 1$

12. (*a*) Convergent (*c*) Convergent (*b*) Convergent (*d*) Divergent

13. (*a*) Convergent (*b*) Convergent for $x < 1$ and divergent for $x \ge 1$

(*c*) Convergent for $x < 1$, divergent for $x \ge 1$

(*d*) Convergent for $x < 1$, divergent for $x \ge 1$

14. (*a*) Converges for $x \le 1$, diverges for $x > 1$

(*b*) Converges for $x \le 1$, diverges for $x > 1$

(*c*) Converges for $x^2 < 1$, diverges for $x^2 \ge 1$

(*d*) Converges for $x \le 1$, diverges for $x > 1$

(*e*) Converges for $x < 1$, diverges for $x \ge 1$

(*f*) Converges for $x \le 1$, diverges for $x > 1$

(*g*) Diverges

(*h*) Converges for $x < 1$, diverges for $x \ge 1$

(*i*) Converges for $x < 1$, diverges for $x \ge 1$

15. Converges for $x^2 < 4$, diverges for $x^2 \ge 4$

16. Converges for $x < 1$, diverges for $x \ge 1$

5.8 RAABE'S TEST

Statement: If Σu_n is a series of positive terms and $\underset{n\to\infty}{\text{Lt}}\left(n\left(\frac{u_n}{u_{n+1}}-1\right)\right) = l$, then the series is convergent if $l > 1$ and divergent if $l < 1$.

Proof: Let us compare the given series Σu_n with an auxiliary series $\sum v_n = \sum \frac{1}{n^p}$ which we know converges if $p > 1$ and diverges if $p \le 1$

Now $\dfrac{v_n}{v_{n+1}} = \dfrac{\dfrac{1}{n^p}}{\dfrac{1}{(n+1)^p}} = \left(\dfrac{n+1}{n}\right)^p = \left(1+\dfrac{1}{n}\right)^p$

$$= 1 + \frac{p}{n} + \frac{p(p-1)}{2!}.\frac{1}{n^2} + \dots$$

Case I: Let $\sum v_n = \sum \dfrac{1}{n^p}$ be convergent, so that $p > 1$.

Then $\sum u_n$ will also converges if $\dfrac{u_n}{u_{n+1}} > \dfrac{v_n}{v_{n+1}}$

$$\Rightarrow \text{if } \frac{u_n}{u_{n+1}} > 1 + \frac{p}{n} + \frac{p(p-1)}{2!}.\frac{1}{n^2} + \dots$$

$$\Rightarrow \text{if } n\left(\frac{u_n}{u_{n+1}} - 1\right) > p + \frac{p(p-1)}{2!}.\frac{1}{n} + \dots$$

$$\Rightarrow \text{if } \underset{n\to\infty}{\text{Lt}}\left[n\left(\frac{U_n}{U_{n+1}} - 1\right)\right] > p$$

$$\Rightarrow \text{if } l > p$$

But p is itself greater than 1, therefore Σu_n is convergent if $l > 1$.

Case II: Let Σv_n be divergent, so that $p \le 1$

Then Σu_n will also diverge if $\dfrac{u_n}{u_{n+1}} < \dfrac{v_n}{v_{n+1}}$

$$\Rightarrow \text{if } \frac{u_n}{u_{n+1}} < 1 + \frac{p}{n} + \frac{p(p-1)}{2!}.\frac{1}{n^2} + \dots$$

$$\Rightarrow \text{if } n\left(\frac{u_n}{u_{n+1}} - 1\right) < p + \frac{p(p-1)}{2!}.\frac{1}{n} + \dots$$

$$\Rightarrow \text{if } \underset{n\to\infty}{\text{Lt}}\left[n\left(\frac{u_n}{u_{n+1}} - 1\right)\right] < p$$

$$\Rightarrow \text{if } l < p$$

But p is itself ≤ 1, thus the given series Σu_n is divergent if $l < 1$.

Note 1: If $\lim_{n\to\infty}\left[n\left(\frac{u_n}{u_{n+1}}-1\right)\right]=1$, then Raabe's Test fails.

Note 2: Raabe's test is used when D'Alembert's Ratio test fails and when in the ratio test, $\frac{u_n}{u_{n+1}}$ does not involve the number *e*. When $\frac{u_n}{u_{n+1}}$ involves *e*, we apply logarithmic test after the ratio test and not the Raabe's test.

5.9 LOGARITHMIC TEST

Statement: A positive term series Σu_n converges or diverges according as

$$\operatorname*{Lt}_{n\to\infty}\left(n\log\frac{u_n}{u_{n+1}}\right)>1 \text{ or } <1$$

Proof: Let us compare the given series Σu_n with an auxiliary series $\sum v_n=\sum\frac{1}{n^p}$ which we know converges if $p > 1$ and diverges if $p \le 1$

$$\text{Now } \frac{v_n}{v_{n+1}}=\frac{(n+1)^p}{n^p}=\left(1+\frac{1}{n}\right)^p$$

Case I: Let Σv_n be convergent, so that $p > 1$.

Then Σu_n will also be convergent if $\frac{u_n}{u_{n+1}}>\frac{v_n}{v_{n+1}}$

$$\Rightarrow \text{ if } \frac{u_n}{u_{n+1}}>\left(1+\frac{1}{n}\right)^p$$

$$\Rightarrow \text{ if } \log\frac{u_n}{u_{n+1}}>\log\left(1+\frac{1}{n}\right)^p=p\log\left(1+\frac{1}{n}\right)$$

$$\Rightarrow \text{ if } \log\frac{u_n}{u_{n+1}}>p\left[\frac{1}{n}-\frac{1}{2n^2}+\frac{1}{3n^3}-\ldots\right] \qquad \left[\because \log(1+x)=x-\frac{x^2}{2}+\frac{x^3}{3}-\ldots\right]$$

$$\Rightarrow \text{ if } n\log\frac{u_n}{u_{n+1}}>p\left[1-\frac{1}{2n}+\frac{1}{3n^2}-\ldots\right]$$

$$\Rightarrow \text{ if } \underset{n\to\infty}{\text{Lt}}\left(n\log\frac{u_n}{u_{n+1}}\right) > p$$

$$\Rightarrow \text{ if } \underset{n\to\infty}{\text{Lt}}\left(n\log\frac{u_n}{u_{n+1}}\right) > 1 \qquad [\because p>1]$$

Case II: Let Σv_n be divergent, so that $p \leq 1$

Then Σu_n also divergent if $\frac{u_n}{u_{n+1}} < \frac{v_n}{v_{n+1}}$

$$\Rightarrow \text{ if } \log\frac{u_n}{u_{n+1}} < \log\frac{v_n}{v_{n+1}}$$

$$\Rightarrow \text{ if } \log\frac{u_n}{u_{n+1}} < \log\left(1+\frac{1}{n}\right)^p = p\log\left(1+\frac{1}{n}\right)$$

$$\Rightarrow \text{ if } \log\frac{u_n}{u_{n+1}} < p\left[\frac{1}{n}-\frac{1}{2n^2}+\frac{1}{3n^3}-\ldots\right]$$

$$\Rightarrow \text{ if } n\log\frac{u_n}{u_{n+1}} < p\left[1-\frac{1}{2n}+\frac{1}{3n^2}-\ldots\right]$$

$$\Rightarrow \text{ If } \lim_{n\to\infty}\left(n\log\frac{u_n}{u_{n+1}}\right) < p$$

$$\Rightarrow \text{ if } \lim_{n\to\infty}\left(n\log\frac{u_n}{u_{n+1}}\right) < 1$$

Thus, the series Σu_n converges or diverges according as

$$\underset{n\to\infty}{\text{Lt}}\left(n\log\frac{u_n}{u_{n+1}}\right) > 1 \quad \text{or} \quad < 1$$

Note 1: The test fails if $\underset{n\to\infty}{\text{Lt}}\left(n\log\frac{u_n}{u_{n+1}}\right) = 1$.

Note 2: The test is applied after the failure of Ratio test and generally when in Ratio test, $\frac{u_n}{u_{n+1}}$ involves 'e'.

5.10 GAUSS TEST

Statement: If $a_n > 0$ and $\frac{a_n}{a_{n+1}} = 1 + \frac{\mu}{n} + O\left(\frac{1}{n^2}\right)$, then $\sum a_n$ converges for $\mu > 1$ and diverges for $\mu \le 1$.

Proof: (*i*) Let $\mu \ne 1$

$$\text{Now } \frac{a_n}{a_{n+1}} = 1 + \frac{\mu}{n} + O\left(\frac{1}{n^2}\right) \Rightarrow \frac{a_n}{a_{n+1}} - 1 = \frac{\mu}{n} + O\left(\frac{1}{n^2}\right)$$

$$\Rightarrow n\left(\frac{a_n}{a_{n+1}} - 1\right) = \mu + O\left(\frac{1}{n}\right) \Rightarrow \underset{n\to\infty}{\text{Lt}}\left[n\left(\frac{a_n}{a_{n+1}} - 1\right)\right] = \mu$$

$\therefore$ By Raabe's Test, $\sum a_n$ converges for $\mu > 1$ and diverges for $\mu < 1$.

(*i*) Let $\mu = 1$

$$\therefore n\left(\frac{a_n}{a_{n+1}} - 1\right) = 1 + O\left(\frac{1}{n}\right) \Rightarrow n\left(\frac{a_n}{a_{n+1}} - 1\right) - 1 = O\left(\frac{1}{n}\right)$$

$$\Rightarrow \left[n\left(\frac{a_n}{a_{n+1}} - 1\right) - 1\right]\log n = O\left(\frac{\log n}{n}\right)$$

$$\Rightarrow \underset{n\to\infty}{\text{Lt}}\left[\left\{n\left(\frac{a_n}{a_{n+1}} - 1\right) - 1\right\}\log n\right] = 0 < 1 \qquad \left[\because \underset{n\to\infty}{\text{Lt}}\left(\frac{\log n}{n}\right) = 0\right]$$

$\therefore$ By D' Morgan and Bertrand's Test, $\sum a_n$ converges for $\mu > 1$ and diverges for $\mu \le 1$.

D' Morgan and Bertrand's Test

Statement: If $\sum a_n$ is a positive terms series and $\underset{n\to\infty}{\text{Lt}}\left[\left\{n\left(\frac{a_n}{a_{n+1}} - 1\right) - 1\right\}\log n\right] = l$, then $\sum a_n$ converges if $l > 1$ and diverges if $l < 1$.

Note: The test never fails as we know that the series diverges for $\mu = 1$. Moreover, the test is applied after the failure of Ratio test and when it is possible to expand $\frac{u_n}{u_{n+1}}$ in powers of $\frac{1}{n}$ by Binomial Theorem or by any other method.

[Binomial theorem: $(1+x)^n = 1 + nx + \frac{n(n-1)}{2!}x^2 + \ldots\infty$ where $|x| < 1$]

ILLUSTRATIVE EXAMPLES 5.6

Example 5.17: Discuss the convergence of the series

$$\frac{1}{2} + \frac{1\cdot 3}{2\cdot 4} + \frac{1\cdot 3\cdot 5}{2\cdot 4\cdot 6} + \ldots$$

Solution: Here $u_n = \frac{1\cdot 3\cdot 5\ldots(2n-1)}{2\cdot 4\cdot 6\ldots .2n}$

$$u_{n+1} = \frac{1\cdot 3\cdot 5\ldots(2n-1)(2n+1)}{2\cdot 4\cdot 6\ldots(2n)(2n+2)}$$

$$\therefore \quad \frac{u_n}{u_{n+1}} = \frac{1\cdot 3\cdot 5\ldots 2n-1)}{2\cdot 4\cdot 6\ldots(2n)} \times \frac{2\cdot 4\cdot 6\ldots 2n(2n+2)}{1\cdot 3\cdot 5\ldots(2n-1)(2n+1)}$$

$$= \frac{2n+2}{2n+1} = \frac{1+\frac{1}{n}}{1+\frac{1}{2n}} \to 1 \text{ as } n \to \infty$$

$\therefore$ D' Alembert's Ratio test fails.

$$\text{Now,} \quad n\left(\frac{u_n}{u_{n+1}} - 1\right) = n\left[\frac{2n+2}{2n+1} - 1\right] = \frac{n}{2n+1} = \frac{1}{2+\frac{1}{n}}$$

$$\therefore \quad \underset{n\to\infty}{\text{Lt}}\, n\left(\frac{u_n}{u_{n+1}} - 1\right) = \frac{1}{2} < 1$$

$\therefore$ By Raabe's test, Σu_n diverges.

Example 5.18: Discuss the convergence of the series:

$$\sum \frac{(\lfloor n)^2}{\lfloor 2n} x^n,\ x>0$$

Solution: Here $u_n = \dfrac{(\lfloor n)^2}{\lfloor 2n} x^n$

$$u_{n+1} = \frac{(\lfloor n+1)^2}{\lfloor 2n+2} x^{n+1}$$

$$\therefore \quad \frac{u_n}{u_{n+1}} = \frac{(\lfloor n)^2}{\lfloor 2n} x^n \times \frac{\lfloor 2n+2}{(\lfloor n+1)^2 x^{n+1}}$$

$$= \frac{(\lfloor n)^2}{\lfloor 2n} x^n \times \frac{(2n+2)(2n+1)\lfloor 2n}{((n+1)\lfloor n)^2 x^{n+1}}$$

$$= \frac{(2n+2)(2n+1)}{(n+1)^2}\frac{1}{x} = \frac{2(n+1)(2n+1)}{(n+1)^2}\frac{1}{x}$$

$$= \frac{2(2n+1)}{n+1}.\frac{1}{x} = \frac{2\left(2+\frac{1}{n}\right)}{1+\frac{1}{n}}.\frac{1}{x}$$

$$\therefore \quad \underset{n\to\infty}{\text{Lt}} \frac{u_n}{u_{n+1}} = \frac{2(2+0)}{1+0}.\frac{1}{x} = \frac{4}{x}$$

$\therefore$ By Ratio test, $\sum u_n$ converges for $\dfrac{4}{x} > 1$, i.e., $x < 4$ and diverges for $\dfrac{4}{x} < 1$, i.e., $x > 4$.

When $x = 4$, Ratio test fails.

For $x = 4$ $\qquad \dfrac{u_n}{u_{n+1}} = \dfrac{2(2n+1)}{n+1}.\dfrac{1}{4} = \dfrac{2n+1}{2(n+1)}$

$$\therefore \quad \frac{u_n}{u_{n+1}} - 1 = \frac{2n+1}{2(n+1)} - 1 = \frac{2n+1-2n-2}{2(n+1)} = -\frac{1}{2(n+1)}$$

$$\therefore \quad n\left(\frac{u_n}{u_{n+1}}-1\right)=-\frac{n}{2(n+1)}=-\frac{1}{2\left(1+\frac{1}{n}\right)}$$

$$\therefore \quad \underset{n\to\infty}{\text{Lt}}\left[n\left(\frac{u_n}{u_{n+1}}-1\right)\right]=-\frac{1}{2(1+0)}=-\frac{1}{2}<1$$

$\therefore$ By Raabe's Test, $\sum a_n$ diverges.

Hence, given series $\sum a_n$ converges for $x < 4$ and diverges for $x \geq 4$.

Example 5.19: Show that the series

$$1+\frac{\alpha+1}{\beta+1}+\frac{(\alpha+1)(2\alpha+1)}{(\beta+1)(2\beta+1)}+\ldots+\frac{(\alpha+1)(2\alpha+1)\ldots(n\alpha+1)}{(\beta+1)(2\beta+1)\ldots(n\beta+1)}+\ldots$$

Converges if $\beta > \alpha$ and diverges if $\beta \leq \alpha$.

Solution: Here $u_n=\dfrac{(\alpha+1)(2\alpha+1)\ldots(n\alpha+1)}{(\beta+1)(2\beta+1)\ldots(n\beta+1)}$

$$u_{n+1}=\frac{(\alpha+1)(2\alpha+1)\ldots(n\alpha+1)((n+1)\alpha+1)}{(\beta+1)(2\beta+1)\ldots(n\beta+1)((n+1)\beta+1)}$$

$$\therefore \quad \frac{u_n}{u_{n+1}}=\frac{((n+1)\alpha+1)}{((n+1)\beta+1)}=\frac{n\alpha+\alpha+1}{n\beta+\beta+1}=\frac{\alpha+\frac{\alpha+1}{n}}{\beta+\frac{\beta+1}{n}}$$

$$\therefore \quad \underset{n\to\infty}{\text{Lt}}\frac{u_n}{u_{n+1}}=\frac{\alpha+0}{\beta+0}=\frac{\alpha}{\beta}$$

$\therefore$ by Ratio Test, $\sum a_n$ converges for $\dfrac{\alpha}{\beta}>1$, i.e., $\alpha>\beta$ and diverges for $\dfrac{\alpha}{\beta}<1$, i.e.,

$\alpha<\beta$.

When $\dfrac{\alpha}{\beta}=1$, i.e., $\alpha=\beta$, Ratio Test fails.

$$\frac{u_n}{u_{n+1}}=\frac{n\alpha+\alpha+1}{n\beta+\beta+1}=1$$

$$\therefore \quad \frac{u_n}{u_{n+1}} - 1 = 1 - 1 = 0$$

$$\therefore \quad n\left(\frac{u_n}{u_{n+1}} - 1\right) = 0$$

$$\therefore \quad \underset{n\to\infty}{\text{Lt}}\left[n\left(\frac{u_n}{u_{n+1}} - 1\right)\right] = 0 < 1$$

$\therefore$ by Raabe's Test, $\sum a_n$ diverges.

$\therefore$ given series converges if $\alpha > \beta$ and diverges if $\alpha \le \beta$.

Example 5.20: Discuss the convergence of the series

$$\frac{1^2}{4^2} + \frac{1^2 \cdot 5^2}{4^2 \cdot 8^2} + \frac{1^2 \cdot 5^2 \cdot 9^2}{4^2 \cdot 8^2 \cdot 12^2} + \ldots \text{ to } \infty$$

Solution: Here $\sum u_n = \frac{1^2}{4^2} + \frac{1^2 \cdot 5^2}{4^2 \cdot 8^2} + \frac{1^2 \cdot 5^2 \cdot 9^2}{4^2 \cdot 8^2 \cdot 12^2} + \ldots$ to ∞

$$\therefore \quad u_n = \frac{1^2 \cdot 5^2 \cdot 9^2 \ldots (4n-3)^2}{4^2 \cdot 8^2 \cdot 12^2 \ldots (4n)^2}, \qquad u_{n+1} = \frac{1^2 \cdot 5^2 \cdot 9^2 \ldots (4n-3)^2 (4n+1)^2}{4^2 \cdot 8^2 \cdot 12^2 \ldots (4n)^2 (4n+4)^2}$$

$$\therefore \quad \frac{u_n}{u_{n+1}} = \frac{(4n+4)^2}{(4n+1)^2} = \frac{4^2 (n+1)^2}{4^2\left(n + \frac{1}{4}\right)^2} = \frac{\left(1 + \frac{1}{n}\right)^2}{\left(1 + \frac{1}{4n}\right)^2} = \left(1 + \frac{1}{n}\right)^2 \left(1 + \frac{1}{4n}\right)^{-2}$$

$$= \left[1 + \frac{2}{n} + O\left(\frac{1}{n^2}\right)\right]\left[1 - \frac{2}{4n} + O\left(\frac{1}{n^2}\right)\right]$$

$$= 1 + \frac{2}{n} - \frac{1}{2n} + O\left(\frac{1}{n^2}\right) = 1 + \frac{3}{2n} + O\left(\frac{1}{n^2}\right)$$

$$= 1 + \frac{\mu}{n} + O\left(\frac{1}{n^2}\right) \qquad \text{where } \mu = \frac{3}{2} > 1$$

$\therefore$ By Gauss Test, given Series $\sum u_n$ converges.

Example 5.21: Test the Series

$$1+\frac{\alpha\beta}{1\cdot\gamma}x+\frac{\alpha(\alpha+1)\beta(\beta+1)}{1\cdot2\cdot\gamma(\gamma+1)}x^2+\frac{\alpha(\alpha+1)(\alpha+2)\beta(\beta+1)(\beta+2)}{1\cdot2\cdot3\cdot\gamma(\gamma+1)(\gamma+2)}x^3+\ldots\infty,\ x>0$$

for convergence.

Solution: The given series is

$$1+\frac{\alpha\beta}{1\cdot\gamma}x+\frac{\alpha(\alpha+1)\beta(\beta+1)}{1\cdot2\cdot\gamma(\gamma+1)}x^2+\frac{\alpha(\alpha+1)(\alpha+2)\beta(\beta+1)(\beta+2)}{1\cdot2\cdot3\cdot\gamma(\gamma+1)(\gamma+2)}x^3+\ldots\infty$$

$$\therefore\quad u_n=\frac{\alpha(\alpha+1)(\alpha+2)\ldots(\alpha+n-1).\beta(\beta+1)(\beta+2)\ldots(\beta+n-1)}{1\cdot2\cdot3\ldots n.\gamma(\gamma+1)(\gamma+2)\ldots(\gamma+n-1)}x^n$$

(neglecting first term)

$$u_{n+1}=\frac{\alpha(\alpha+1)(\alpha+2)\ldots(\alpha+n-1)(\alpha+n).\beta(\beta+1)(\beta+2)\ldots(\beta+n-1)(\beta+n)}{1\cdot2\cdot3\ldots n\cdot(n+1)\cdot\gamma(\gamma+1)(\gamma+2)\ldots(\gamma+n-1)(\gamma+n)}x^{n+1}$$

$$\therefore\quad \frac{u_n}{u_{n+1}}=\frac{(n+1)(\gamma+n)}{(\alpha+n)(\beta+n)}\frac{1}{x}=\frac{\left(1+\frac{1}{n}\right)\left(1+\frac{\gamma}{n}\right)}{\left(1+\frac{\alpha}{n}\right)\left(1+\frac{\beta}{n}\right)}\frac{1}{x}$$

$$\therefore\quad \underset{x\to\infty}{\text{Lt}}\frac{u_n}{u_{n+1}}=\frac{(1+0)(1+0)}{(1+0)(1+0)}\frac{1}{x}=\frac{1}{x}$$

$\therefore$ by Ratio Test, Σa_n convergence for $\frac{1}{x}>1$, i.e., $x<1$ and diverges for $\frac{1}{x}<1$,

i.e., $x>1$

For $x=1$, Ratio Test fails

For $x=1$

$$\frac{u_n}{u_{n+1}}=\frac{(n+1)(\gamma+n)}{(n+\alpha)(n+\beta)}=\frac{\left(1+\frac{1}{n}\right)\left(1+\frac{\gamma}{n}\right)}{\left(1+\frac{\alpha}{n}\right)\left(1+\frac{\beta}{n}\right)}$$

$$=\left(1+\frac{1}{n}\right)\left(1+\frac{\gamma}{n}\right)\left(1+\frac{\alpha}{n}\right)^{-1}\left(1+\frac{\beta}{n}\right)^{-1}$$

$$=\left(1+\frac{1}{n}\right)\left(1+\frac{\gamma}{n}\right)\left\{1-\frac{\alpha}{n}+O\left(\frac{1}{n^2}\right)\right\}\left\{1-\frac{\beta}{n}+O\left(\frac{1}{n^2}\right)\right\}$$

$$=1+\frac{1+\gamma-\alpha-\beta}{n}+O\left(\frac{1}{n^2}\right)$$

$$=1+\frac{\mu}{n}+O\left(\frac{1}{n^2}\right) \qquad \text{where } \mu=1+\gamma-\alpha-\beta$$

$\therefore$ by Gauss Test, $\sum u_n$ converges for $1+\gamma-\alpha-\beta>1$, i.e., $\gamma>\alpha+\beta$ and diverges for $1+\gamma-\alpha-\beta<1$, i.e., $\gamma<\alpha+\beta$.

Example 5.22: Discuss the convergence of series.

$$\frac{a+x}{\lfloor 1}+\frac{(a+2x)^2}{\lfloor 2}+\frac{(a+3x)^3}{\lfloor 3}+\ldots$$

Solution: Here $u_n=\dfrac{(a+nx)^n}{\lfloor n}$, $u_{n+1}=\dfrac{(a+(n+1)x)^{n+1}}{\lfloor (n+1)}$

$$\frac{u_n}{u_{n+1}}=\frac{(a+nx)^n}{\lfloor n}\cdot\frac{\lfloor (n+1)}{(a+(n+1)x)^{n+1}}$$

$$=\frac{n^n x^n\left(1+\dfrac{a}{nx}\right)^n}{(n+1)^{n+1}x^{n+1}\left(1+\dfrac{a}{(n+1)x}\right)^{n+1}}\cdot\frac{(n+1)\lfloor n}{\lfloor n}$$

$$=\frac{1}{x}\frac{n^n}{(n+1)^n(n+1)}\cdot\frac{\left(1+\dfrac{a}{nx}\right)^n}{\left(1+\dfrac{a}{(n+1)x}\right)^{n+1}}(n+1)$$

$$=\frac{1}{x}\frac{n^n}{n^n\left(1+\dfrac{1}{n}\right)^n}\cdot\frac{\left(1+\dfrac{a}{nx}\right)^n}{\left(1+\dfrac{a}{(n+1)x}\right)^{n+1}}=\frac{1}{x}\frac{1}{\left(1+\dfrac{1}{n}\right)^n}\cdot\frac{\left(1+\dfrac{a}{nx}\right)^n}{\left(1+\dfrac{a}{(n+1)x}\right)^{n+1}}$$

$$\underset{x\to\infty}{\text{Lt}} \frac{u_n}{u_{n+1}} = \underset{x\to\infty}{\text{Lt}} \left(\frac{1}{x} \cdot \frac{1}{\left(1+\frac{1}{n}\right)^n} \cdot \frac{\left[\left(1+\frac{a}{nx}\right)^{\frac{nx}{a}}\right]^{\frac{a}{x}}}{\left[\left(1+\frac{a}{(n+1)x}\right)^{\frac{(n+1)x}{a}}\right]^{\frac{a}{x}}} \right)$$

$$= \frac{1}{xe} \cdot \frac{e^{\frac{a}{x}}}{e^{\frac{a}{x}}} = \frac{1}{xe}$$

By D' Alembert's Ratio Test, the given series converges if $\frac{1}{xe} > 1$. diverges if $\frac{1}{xe} < 1$, and fails at $\frac{1}{xe} = 1$.

Therefore, by D' Alembert's Ratio Test, the given series converges if $x < \frac{1}{e}$ and diverges if $x > \frac{1}{e}$.

At $x = \frac{1}{e}$, we apply Logarithmic Test because expression of $\frac{u_n}{u_{n+1}}$ involves e.

$$\frac{u_n}{u_{n+1}} = \frac{e\left(1+\frac{ae}{n}\right)^n}{\left(1+\frac{1}{n}\right)^n \left(1+\frac{ae}{(n+1)}\right)^{n+1}}$$

$$= \left[\log e + n\log\left(1+\frac{ae}{n}\right)\right] - \left[n\log\left(1+\frac{1}{n}\right) + (n+1)\log\left(1+\frac{ae}{(n+1)}\right)\right]$$

$$= \left[\log e - n\log\left(1+\frac{1}{n}\right)\right] + \left[n\log\left(1+\frac{ae}{n}\right) - (n+1)\log\left(1+\frac{ae}{(n+1)}\right)\right]$$

$$=1-n\left(\frac{1}{n}-\frac{1}{2n^2}+\frac{1}{3n^3}-...\right)+n\left(\frac{ae}{n}-\frac{a^2e^2}{2n^2}+\frac{a^3e^3}{3n^3}-...\right)$$

$$-(n+1)\left(\frac{ae}{n+1}-\frac{a^2e^2}{2(n+1)^2}+\frac{a^3e^3}{3(n+1)^3}-...\right)$$

$$=\left(\frac{1}{2n}-\frac{1}{3n^2}+...\right)+\left(-\frac{a^2e^2}{2n}+\frac{a^3e^3}{3n^2}+...\right)+\left(\frac{a^2e^2}{2(n+1)}+\frac{a^3e^3}{3(n+1)^2}+...\right)$$

$$\underset{n\to\infty}{\text{Lt}}\left[n\log\left(\frac{u_n}{u_{n+1}}\right)\right]=\frac{1}{2}-\frac{a^2e^2}{2}+\frac{a^2e^2}{2}=\frac{1}{2}<1$$

$\Rightarrow$ Σu_n diverges by Logarithmic Test at $x=\frac{1}{e}$

Hence, Σu_n converges for $x<\frac{1}{e}$ and diverges for $x\geq\frac{1}{e}$.

EXERCISE 5.5

1. Test the convergence or divergence of the following series:

(a) $\frac{1}{2}+\frac{1\cdot3}{2\cdot4}+\frac{1\cdot3\cdot5}{2\cdot4\cdot6}+...$

(b) $1+\frac{2}{1}.\frac{1}{2}+\frac{2\cdot4}{1\cdot3}.\frac{1}{2}+\frac{2\cdot4\cdot6}{1\cdot3\cdot5}.\frac{1}{4}+...$

(c) $\frac{1}{1^2}+\frac{1+2}{1^2+2^2}+\frac{1+2+3}{1^2+2^2+2^2}+...$

2. Test the convergence or divergence of the following series:

(a) $1+\frac{1}{2}.\frac{x^2}{4}+\frac{1\cdot3}{2\cdot4}.\frac{x^4}{8}+\frac{1\cdot3\cdot5}{2\cdot4\cdot6}.\frac{x^6}{12}+...$

(b) $1+\frac{3}{7}x+\frac{3\cdot6}{7\cdot10}x^2+\frac{3\cdot6\cdot9}{7\cdot10\cdot13}x^3+...$ to ∞

3. Test the convergence or divergence of the series

$$1+\frac{1+\alpha}{1+\beta}+\frac{(1+\alpha)(2+\alpha)}{(1+\beta)(2+\beta)}+...$$

4. Examine the convergence or divergence of the following series:

(*a*) $1+\frac{2^2}{3^2}+\frac{2^2\cdot 4^2}{3^2\cdot 5^2}+\frac{2^2\cdot 4^2\cdot 6^2}{3^2\cdot 5^2\cdot 7^2}+\ldots$ [PTU May 2004]

(*b*) $\frac{1^2}{2^2}+\frac{1^2}{2^2}\cdot\frac{2^2}{4^2}+\frac{1^2}{2^2}\cdot\frac{3^2}{4^2}\cdot\frac{5^2}{6^2}+\ldots$

(*c*) $\frac{1^2}{4^2}+\frac{1^2}{4^2}\cdot\frac{5^2}{8^2}+\frac{1^2}{4^2}\cdot\frac{5^2}{8^2}\cdot\frac{9^2}{12^2}+\ldots$ to ∞

5. Examine the convergences or divergence of the following series:

(*a*) $\frac{1^2}{2^2}+\frac{1^2\cdot 3^2}{2^2\cdot 4^2}\cdot x+\frac{1^2\cdot 3^2\cdot 5^2}{2^2\cdot 4^2\cdot 6^2}\cdot x^2+\ldots,\ x>0$

(*b*) $1+\frac{(\lfloor 1)^2}{\lfloor 2}x+\frac{(\lfloor 2)^2}{\lfloor 4}x^2+\frac{(\lfloor 3)^2}{\lfloor 6}x^3+\ldots,\ x>0$

(*c*) $\frac{x}{1}+\frac{1}{2}\cdot\frac{x^3}{3}+\frac{1\cdot 3}{2\cdot 4}\cdot\frac{x^5}{5}+\frac{1\cdot 3\cdot 5}{2\cdot 4\cdot 6}\cdot\frac{x^7}{7}+\ldots$

(*d*) $1+\frac{3}{7}x+\frac{3\cdot 6}{7\cdot 10}x^2+\frac{3\cdot 6\cdot 9}{7\cdot 10\cdot 13}x^3+\ldots,\ x>0$

(*e*) $1+\frac{2x}{\lfloor 2}+\frac{3^2x^2}{\lfloor 3}+\frac{4^3x^3}{2}+\ldots,\ x>0$

(*f*) $\frac{x}{1\cdot 2}+\frac{x^2}{2\cdot 3}+\frac{x^3}{3\cdot 4}+\frac{x^4}{4\cdot 5}+\ldots\ \infty$

(*g*) $\frac{1}{12}+\frac{3}{7}x+\frac{3\cdot 6}{7\cdot 10}x^2+\frac{3\cdot 6\cdot 9}{7\cdot 10\cdot 13}x^3+\ldots$ to ∞

6. Discuss the convergence of the following series:

(*a*) $x+\frac{2^2x^2}{2!}+\frac{3^3x^3}{3!}+\frac{4^4x^4}{4!}+\ldots\ \infty$ [PTU May 2007, May 2008]

(*b*) $1+\frac{2x}{2!}+\frac{3^2x^2}{3!}+\frac{4^3x^3}{4!}+\frac{5^4x^4}{5!}+\ldots\ \infty$ [PTU May 2005, May 2006]

7. Discuss the convergence or divergence of the following series for $x > 0$:

(*a*) $\Sigma\frac{\lfloor 2n}{(\lfloor n)^2}x^n$

(*b*) $\Sigma\frac{\lfloor n}{4\cdot 7\cdot 10\cdot(3n+1)x^n}$

(*c*) $\Sigma\frac{1\cdot 2\cdot 3\ldots n}{3\cdot 5\cdot 7\ldots(2n+1)}x^n$

(*d*) $\Sigma\frac{1\cdot 3\cdot 5\ldots(2n-1)}{2\cdot 4\cdot 6\ldots(2n)}\cdot\frac{x^n}{2n+2}$

(*e*) $\Sigma\frac{(n!)^2}{(2n)!}x^{2n}$

(*f*) $\Sigma\frac{4\cdot 7\cdot 10\ldots(3n+1)}{1\cdot 2\cdot 3\ldots n}x^n$

(*g*) $\Sigma\frac{2^{3n}}{3^{2n}}$

8. Discuss the convergence of the following series:

(a) $\sum \frac{\alpha\beta}{1\cdot\gamma}x+\frac{\alpha(\alpha+1)\beta(\beta+1)}{1\cdot2\cdot\gamma(\gamma+1)}x^2+\frac{\alpha(\alpha+1)(\alpha+2)\beta(\beta+1)(\beta+2)}{1\cdot2\cdot3\cdot\gamma(\gamma+1)(\gamma+2)}x^3+\ldots$ [PTU Dec 2003]

(b) $\sum \frac{n!}{x(x+1)(x+2)\ldots(x+n-1)}$

(c) $1+\frac{\alpha}{1\cdot\beta}x+\frac{\alpha(\alpha+1)^2}{1\cdot2\cdot\beta(\beta+1)}x^2+\frac{\alpha(\alpha+1)^2(\alpha+2)^2}{1\cdot2\cdot3\cdot\beta(\beta+1)(\beta+2)}x^3+\ldots,\ x>0$

Converges when $x<1$ and diverges for $x>1$ and when $x = 1$, the series converges if $\beta>2$ and diverges if $\beta\leq 2\alpha$.

9. Discuss the convergence or divergence of the following series:

(a) $1+\frac{a}{b}+\frac{a(a+1)}{b(b+1)}+\frac{a(a+1)(a+2)}{b(b+1)(b+2)}+\ldots$

(b) $1+\frac{1}{2}\cdot\frac{a}{b}+\frac{1\cdot3}{2\cdot4}\cdot\frac{a(a+1)}{b(b+1)}+\frac{1\cdot3\cdot5}{2\cdot4\cdot6}\cdot\frac{a(a+1)(a+2)}{b(b+1)(b+2)}+\ldots$

(c) $1+a+\frac{a(a+1)}{1\cdot2}+\frac{a(a+1)(a+2)}{1\cdot2\cdot3}+\ldots,\ a>0$

(d) $1\cdot\frac{a}{2}+2\cdot\frac{a(a+1)}{b(b+1)}+3\cdot\frac{a(a+1)(a+2)}{b(b+1)(b+2)}+\ldots,\ a>0$

10. Discuss the convergence or divergence of the following series:

(a) $\sum\left[\frac{1\cdot3\cdot5\ldots(2n-1)}{2\cdot4\cdot6\ldots2n}\right]^P x^n$ (b) $\sum\frac{4\cdot7\cdot10\ldots(3n+1)}{1\cdot2\cdot3\ldots n}x^n$ [PTU Dec 2006]

(c) $\sum\left[\frac{1\cdot3\cdot5\ldots(2n-1)}{2\cdot4\cdot6\ldots(2n)}\right]^P,\ P\in R$

ANSWERS

1. (a) Divergent (b) Divergent (c) Divergent
2. (a) Convergent for $x^2\leq1$ and divergent for $x^2>1$

   (b) Convergent for $x<1$ and divergent for $x>1$
3. Convergent for $\beta-\alpha>1$ and divergent for $\beta-\alpha\leq1$
4. (a) Divergent (b) Divergent (c) Convergent
5. (a) Convergent for $x<1$ and divergent for $x\geq1$

   (b) Converges for $x<4$ and diverges for $x\geq4$

(*c*) Converges for $x^2 \leq 1$ and diverges for $x^2 > 1$

(*d*) Converges for $x^2 \leq 1$ and diverges for $x > 1$

(*e*) Converges for $x \leq \frac{1}{e}$ and diverges for $x > \frac{1}{e}$

(*f*) Convergent for $x \leq 1$ and divergent for $x > 1$

(*g*) Convergent for $x \leq 1$ and divergent for $x > 1$

6. (*a*) Convergent for $x < \frac{1}{e}$ and divergent for x $x \geq \frac{1}{e}$

(*b*) Convergent for $x \leq \frac{1}{e}$ and divergent for x $x > \frac{1}{e}$

7. (*a*) Converges for $x \leq \frac{1}{4}$ and diverges for $x \geq \frac{1}{4}$

(*b*) Converges for $x < 3$ and diverges for $x \geq 3$

(*c*) Converges $x < 2$ and diverges $x \geq 2$

(*d*) Converges $x \leq 1$ and diverges $x > 1$

(*e*) Diverges (*f*) Diverges (*g*) Convergent

8. (*a*) Convergence if $\gamma > \alpha + \beta$ and divergence if $\gamma \leq \alpha + \beta$

(*b*) Converges if $x > 2$ and diverges if $x < 2$

9. (*a*) Converges for $b - a > 1$ and diverges for $b - a \leq 1$

(*b*) Converges for $b - a > \frac{1}{2}$ and diverges for $b - a \leq \frac{1}{2}$

(*c*) Diverges

(*d*) Converges for $b - a > 2$ and diverges for $b - a \leq 2$

10. (*a*) Convergent for $x \leq 1$, $p > 2$ and divergent if $x > 1$, $p \leq 2$

(*b*) Convergent for $x < \frac{1}{3}$ and divergent if $x \geq \frac{1}{3}$

(*c*) Convergent for $p > 2$ and divergent for $p \leq 2$

5.11 CAUCHY'S ROOT TEST

Statement: If $\sum u_n$ is a real series of positive terms and $\underset{n\to\infty}{\text{Lt}} (u_n)^{1/n} = r$, then the series $\sum u_n$ is convergent if $r < 1$ and divergent if $r > 1$.

Proof:

(1) When $r < 1$

Let us select $\varepsilon > 0$, such that $r < r + \varepsilon < 1$

Since $\underset{n\to\infty}{\text{Lt}} (u_n)^{1/n} = r$ [Given]

$\therefore$ corresponding to $\varepsilon > 0$, $\exists m \in n$ such that

$$r - \varepsilon < (u_n)^{1/n} < r + \varepsilon \qquad \forall n \geq m$$

Let $r + \varepsilon = K$

$\therefore 0 < K < 1$

Then, $(u_n)^{1/n} < K \qquad \forall n \geq m$

$\Rightarrow \quad u_n < K^n \qquad \forall n \geq m$

But $\sum K^n$, being G.P. with common Ratio $K < 1$, is Convergent.

$\therefore$ The given series is also convergent. [By Direct comparison Test]

(2) When $r > 1$

Let us select $\varepsilon > 0$ such that $r > r - \varepsilon > 1$

Since $\underset{n\to\infty}{\text{Lt}} (u_n)^{1/n} = r$

Corresponding to $\varepsilon > 0$ $\exists m \in N$ such that

$$r - \varepsilon < (u_n)^{1/n} < r + \varepsilon \qquad \forall n \geq m$$

Let $r - \varepsilon = K$

$\Rightarrow \quad K > 1$

Then $(u_n)^{1/n} > K \quad \forall n \geq m$

$\Rightarrow \quad u_n > K^n \quad \forall n \geq m$

But $\sum K^n$, being G.P. with common ratio $K > 1$, is divergent.

Hence, the given series is also divergent. [By Direct comparison Test]

Remarks: Cauchy's root test is more general than D' Alembert's Ratio Test because *D'* Alembert's ratio test may fail, when Cauchy's root test is applicable.

This is because if $\underset{n\to\infty}{\text{Lt}} \dfrac{u_n}{u_{n+1}}$ exists, then $\underset{n\to\infty}{\text{Lt}} (u_n)^{1/n}$ must exist, while if $\underset{n\to\infty}{\text{Lt}} (u_n)^{1/n}$ exists,

then $\underset{n\to\infty}{\text{Lt}} \dfrac{u_n}{u_{n+1}}$ may not exist.

Note 1: Cauchy's Root Test fails when $r = 1$.

Note 2: This test is applicable when power of every factor of u_n an integral multiple of *n*.

ILLUSTRATIVE EXAMPLES 5.7

Example 5.23: Test the convergence of the following series.

(*a*) $\displaystyle\sum_{n=1}^{\infty}\left(\frac{n}{n+1}\right)^{n^2}$ or $\displaystyle\sum_{n=1}^{\infty}\left(1+\frac{1}{n}\right)^{-n^2}$ [PTU Dec 2012] (*b*) $\displaystyle\sum_{n=2}^{\infty}\frac{1}{(\log n)^n}$ [PTU Dec 2002]

Solution:

(*a*) Given series is $\sum u_n = \sum\left(\dfrac{n}{n+1}\right)^{n^2}$

$$\text{Here } u_n = \left(\frac{n}{n+1}\right)^{n^2} \quad \Rightarrow \quad (u_n)^{1/n} = \left[\left(\frac{n}{n+1}\right)^{n^2}\right]^{1/n}$$

$$= \left(\frac{n}{n+1}\right)^{n} = \left(\frac{n+1}{n}\right)^{-n} = \left[\left(1+\frac{1}{n}\right)^{n}\right]^{-1}$$

$$\underset{n\to\infty}{\text{Lt}}\,(u_n)^{1/n} = \underset{n\to\infty}{\text{Lt}}\left[\left(1+\frac{1}{n}\right)^n\right]^{-1} = e^{-1} = \frac{1}{e} < 1 \qquad [\because e = 2.7]$$

$\therefore$ By Cauchy's Root Test, the given series $\sum u_n$ is Convergent.

(*b*) Given series $\sum u_n = \sum_{n=2}^{\infty} \frac{1}{(\log n)^n}$

Here $u_n = \frac{1}{(\log n)^n}$

$$(u_n)^{1/n} = \left[\frac{1}{(\log n)^n}\right]^{1/n} = \frac{1}{\log n}$$

$$\underset{n\to\infty}{\text{Lt}}\,(u_n)^{1/n} = \underset{n\to\infty}{\text{Lt}}\left(\frac{1}{\log n}\right) = \frac{1}{\log\infty} = \frac{1}{\infty} = 0 < 1$$

$\therefore$ By Cauchy's Root Test, the given series

$\sum u_n = \sum_{n=2}^{\infty} \frac{1}{(\log n)^n}$ is Convergent.

Example 5.24: Solve the convergence or divergence of the series:

$$\left(\frac{2^2}{1^2}-\frac{2}{1}\right)^{-1} + \left(\frac{3^3}{2^3}-\frac{3}{2}\right)^{-2} + \left(\frac{4^4}{3^4}-\frac{4}{3}\right)^{-3} + \ldots$$

Solution: Here $u_n = \left[\frac{(n+1)^{n+1}}{n^{n+1}} - \frac{n+1}{n}\right]^{-n} = \left[\left(\frac{n+1}{n}\right)^{n+1} - \frac{n+1}{n}\right]^{-n}$

$$(u_n)^{1/n} = \left[\left[\left(\frac{n+1}{n}\right)^{n+1} - \frac{n+1}{n}\right]^{-n}\right]^{1/n} = \left[\left(\frac{n+1}{n}\right)^{n+1} - \frac{n+1}{n}\right]^{-1}$$

$$= \left[\left(1+\frac{1}{n}\right)^{n+1} - \left(1+\frac{1}{n}\right)\right]^{-1} = \left[\left(1+\frac{1}{n}\right)^{n}.\left(1+\frac{1}{n}\right) - \left(1+\frac{1}{n}\right)\right]^{-1}$$

$$\underset{n\to\infty}{\text{Lt}}\,(u_n)^{1/n} = \underset{n\to\infty}{\text{Lt}}\left[\left(1+\frac{1}{n}\right)^n . \left(1+\frac{1}{n}\right) - \left(1+\frac{1}{n}\right)\right]^{-1}$$

$$= [e(1+0)-(1+0)]^{-1}$$

$$= (e-1)^{-1} = \frac{1}{e-1} < 1 \qquad [\because\ 2 < e < 3]$$

$\therefore$ by Cauchy's Root test, the series Σu_n is convergent.

EXERCISE 5.6

1. Discuss the convergence or divergence of the following series:

(a) $\sum_{n=2}^{\infty} (\log n)^{-n}$ (b) $\sum \left(\frac{n+1}{3n}\right)^n$

(c) $\sum \left(\frac{n}{n+1}\right)^{n^2}$ (d) $\sum \left(1+\frac{1}{n}\right)^n \left(\frac{1}{2}\right)^n$

(e) $\sum \frac{(n-\log n)^n}{2^n n^n}$ (f) $\sum \left(\frac{n-\log n}{2n}\right)^{2n}$

(g) $\sum \left[\frac{n+1}{n}\left[\left(\frac{n+1}{n}\right)^n - 1\right]\right]^{-n}$ (h) $\sum \left(1+\frac{1}{\sqrt{n}}\right)^{-n^{3/2}}$ [PTU May 2008]

(i) $\sum \left(1-\frac{1}{n}\right)^{n^2}$ (j) $\sum_{n=2}^{\infty} \frac{n^{\log n}}{(\log n)^n}$

2. Test the convergence of the following series:

(a) $\sum \frac{2^{3n}}{3^{2n}}$ (b) $\sum e^{2n}\left(\frac{n}{n+1}\right)^{n^2}$

(c) $\sum \frac{e^n}{n^n}$ (d) $\sum_{n=1}^{\infty} \frac{1}{1+\frac{1}{n}}$

(e) $\sum_{n=1}^{\infty} \left(n^{\frac{1}{n}} - 1\right)^n$ (f) $\sum_{n=1}^{\infty} \left(1+\frac{1}{n}\right)^{-n^2}$

(g) $\sum_{n=1}^{\infty} \left(\frac{n}{n+1}\right)^{n^2}$

3. Test the convergence of the following series:

(a) $\sum \frac{((n+1)x)^n}{n^{n+1}}$

(b) $\frac{2}{3}x+\left(\frac{3}{4}\right)^2 x^2+\left(\frac{4}{5}\right)^3 x^3+\ldots$

4. Discuss the convergence of the following series:

(a) $\sum \frac{n^{10}}{(\log e^3)^n}$

(b) $\sum \left(\frac{n+1}{n}\right)^{3n} \cdot \frac{1}{3n}$

(c) $\sum 5^{-n}-(-1)^n$

(d) $\sum \frac{(n-\log e^n)^n}{2^n n^n}$

(e) $\sum \left(1+\frac{1}{\sqrt{n}}\right)^{-n^{3/2}}$

(f) $\sum \left(\frac{nx}{n+1}\right)^n$

(g) $\sum_{n=1}^{\infty} \frac{n^n}{2^{(n^2)}}$

(h) $\sum_{n=1}^{\infty} \frac{n}{(\log e^n)^n}$

(i) $\sum_{n=1}^{\infty} \frac{(\log e^n)^n}{n^n}$

5. Discuss the convergence or divergence of the following series:

(a) $\sum_{n=2}^{\infty} \frac{1}{[\log(\log n)]^n}$

(b) $\sum q^{n^2} r^n, \quad q, r > 0$

(c) $\sum n^n x^n, \quad x > 0$

(d) $\sum n^K x^n, \quad x > 0$

(e) $\sum \frac{x^n}{n^n}, \quad x > 0$

(f) $\sum \frac{x^n}{3^n n^2}$

(g) $\sum e^{\sqrt{n}} r^n, \quad r > 0$

(h) $\sum e^{-\sqrt{n}} r^n, \quad r > 0$

(i) $\sum \frac{(1+nx)^n}{n^n}$

ANSWERS

1. (a) Convergent (b) Convergent (c) Convergent (d) Convergent
(e) Convergent (f) Convergent (g) Convergent (h) Convergent
(i) Convergent (j) Convergent

2. (*a*) Convergent (*b*) Convergent (*c*) Convergent (*d*) Convergent

   (*e*) Convergent (*f*) Convergent (*g*) Convergent

3. (*a*) Convergent if $x < 1$ and divergent if $x \geq 1$

   (*b*) Convergent if $x < 1$ and divergent if $x \geq 1$

4. (*a*) Convergent (*b*) Convergent (*c*) Convergent (*d*) Convergent

   (*e*) Convergent

   (*f*) Convergent if $x < 1$, divergent if $x \geq 1$

   (*g*) Convergent (*h*) Convergent (*i*) Convergent

5. (*a*) Convergent

   (*b*) Convergent for $0 < r < 1$ and divergent for $r \geq 1$

   (*c*) Divergent

   (*d*) Converges for $k < -1$ and diverges for $k \geq -1$

   (*e*) Convergent

   (*f*) Converges for $0 \leq x \leq 3$ and diverges for $x > 3$

   (*g*) Converges for $0 \leq r \leq 1$ and diverges for $r \geq 1$

   (*h*) Converges for $0 < r \leq 1$ and diverges for $r > 1$

   (*i*) Converges for $x < 1$ and diverges for $x \geq 1$

5.12 ALTERNATING SERIES

A series whose terms are alternatively positive and negative is called an *alternating series*.

For example

$$1 - \frac{1}{2} + \frac{1}{3} - \frac{1}{4} + \ldots \quad \text{or} \quad \sum_{n=1}^{\infty} (-1)^{n-1} \frac{1}{n}$$

is an alternating series.

5.13 LEIBNITZ'S (OR ALTERNATING SERIES) TEST

Statement: If $\{a_n\}$ is monotone decreasing sequence of positive terms, i.e., $a_{n+1} > a_n$, and converges to zero, i.e., $\underset{n\to\infty}{\text{Lt}}\ a_n = 0$, then $\sum_{n=1}^{\infty}(-1)^{n-1}\ a_n$ is convergent.

Proof:

Here $\{a_n\}$ is monotone decreasing sequence of positive terms and converges to zero.

$\therefore$ We have

(a) $a_n > 0 \quad \forall n,$

(b) $a_n \geq a_{n+1} \quad \forall n,$ and

(c) $a_n \to 0 \quad$ as $n \to \infty$

Let $\{S_n\}$ be the sequence of partial sums of $\sum(-1)^{n-1}a_n$

$\therefore \quad S_{2n} = (a_1 - a_2) + (a_3 - a_4) + ... + (a_{2n-1} - a_{2n})$

$\Rightarrow \quad S_{2n} \geq 0 \ \forall n$ ... (1)

[$\because$ content of each bracket ≥ 0]

Again $S_{2n} = a_1 - [(a_2 - a_3) + (a_4 - a_5) + ... + (a_{2n-2} - a_{2n-1}) + a_{2n}]$

$= a_1 - $[some positive quantity]

$< a_1$

$\therefore \quad S_{2n} < a_1 \quad \forall n$ ... (2)

From (1) and (2), we get

$0 \leq S_{2n} < a_1 \quad \forall n$

$\Rightarrow$ $\{S_{2n}\}$ is bounded sequence.

Also $\quad S_{2n+2} - S_{2n} = a_{2n+1} - a_{2n+2} \geq 0$

$\therefore \quad S_{2n+2} > S_{2n} \quad \forall n$

$\Rightarrow \{S_{2n}\}$ is an increasing sequence.

Also $\{S_{2n}\}$ is bounded above

$\therefore \quad \{S_{2n}\}$ is convergent.

Let $S_{2n} \to s$ as $\quad n \to \infty$

Now $\quad S_{2n+1} = S_{2n} + a_{2n+1}$

$\Rightarrow \quad \text{Lt}\, S_{2n+1} = \text{Lt}\, S_{2n} + \text{Lt}\, a_{2n+1}$

$\Rightarrow \quad \text{Lt}\, S_{2n+1} = s + 0$ $\qquad\qquad$ [$\because a_n \to 0$ as $n \to \infty$]

$\Rightarrow \quad \text{Lt}\, S_{2n+1} = s$

Now $S_{2n} \to s$ as $n \to \infty$ and also $S_{2n+1} \to s$ as $n \to \infty$

$\therefore \quad \underset{n\to\infty}{\text{Lt}}\, S_n = s$

$\therefore$ given series $\sum_{n=1}^{\infty} (-1)^{n-1}\, a_n$ converges to s.

Cor. 1: We have $0 \leq S_{2n} < a_1$

$\Rightarrow \quad 0 \leq \underset{n\to\infty}{\text{Lt}}\, S_{2n} < a_1$ $\qquad\qquad$ [$\because$ of squeeze Principle]

$\Rightarrow \quad 0 \leq s < a_1$

$\therefore$ sum of convergent series $\sum_{n=1}^{\infty} (-1)^{n-1}\, a_n$ lies between 0 and a_1 (both inclusive)

Cor. 2: If $a_n \to a \neq 0 \;\; \forall n$ then $\sum (-1)^n\, a_n$ oscillates finitely.

Proof: We have

$$S_{2n+1} = S_{2n} + a_{2n+1}$$

$$\Rightarrow \underset{n\to\infty}{\text{Lt}} S_{2n+1} = \underset{n\to\infty}{\text{Lt}} S_{2n} + \underset{n\to\infty}{\text{Lt}} a_{2n+1}$$

$$\Rightarrow \underset{n\to\infty}{\text{Lt}} S_{2n+1} = s + a \neq s$$

$$\Rightarrow \underset{n\to\infty}{\text{Lt}} S_{2n+1} \neq \underset{n\to\infty}{\text{Lt}} S_{2n}$$

$\therefore$ $\{S_n\}$ has two sub-sequential limits s and $s + a$, which are finite.

Hence $\{S_n\}$ oscillates finitely.

$\therefore$ $\sum(-1)^r\ a_n$ oscillates finitely between s and $s+a$ which differ by a.

ILUSTRATIVE EXAMPLES 5.6

Example 5.25: Examine the convergence of the following series:

(*a*) $1-\frac{1}{\sqrt{2}}+\frac{1}{\sqrt{3}}-\frac{1}{\sqrt{4}}+...\infty$ [PTU May 2004]

(*b*) $1-\frac{1}{2}+\frac{1}{3}-\frac{1}{4}+...\infty$

Solution:

(*a*) It is an alternating series

$$u_n = \frac{1}{\sqrt{n}}, \qquad u_{n+1} = \frac{1}{\sqrt{n+1}}$$

$$\frac{1}{\sqrt{n}} > \frac{1}{\sqrt{n+1}} \qquad \forall\ n$$

$$\therefore\ u_n > u_{n+1} \qquad \forall\ n$$

$\therefore$ First condition of Leibnitz's test is true.

Now, $\underset{n\to\infty}{\text{Lt}}\ u_n = \underset{n\to\infty}{\text{Lt}} \frac{1}{\sqrt{n}} = 0$

Second condition of Leibnitz's test is also true.

Both the conditions of Leibnitz's test are satisfied.

$\therefore$ The given series is convergent.

(*b*) It is an alternating series

$$u_n = \frac{1}{n}, \qquad u_{n+1} = \frac{1}{n+1}$$

$$\frac{1}{n} > \frac{1}{n+1} \quad \forall\ n$$

$$\therefore \quad u_n > u_{n+1} \quad \forall\ n$$

$\therefore$ First condition of Leibnitz's test is true.

Now $\quad \underset{n\to\infty}{\text{Lt}}\ u_n = \underset{n\to\infty}{\text{Lt}}\ \frac{1}{n} = 0$

Second condition of Leibnitz's test is also satisfied.

Both the conditions of Leibnitz's test are satisfied.

$\therefore$ The given series is convergent.

Example 5.26: Test the convergence of the series $\sum_{n=1}^{\infty} \frac{(-1)^{n-1} n}{2n-1}$.

Solution: The given series is $\sum_{n=1}^{\infty} \frac{(-1)^{n-1} n}{2n-1} = \frac{1}{1} - \frac{2}{3} + \frac{3}{5} - \frac{4}{7} + \ldots$

It is an alternating series.

(*i*) $u_n = \frac{n}{2n-1}, \quad u_{n+1} = \frac{n+1}{2n+1}$

$$u_n - u_{n+1} = \frac{1}{4n^2-1} > 0 \quad \forall\ n$$

$$\Rightarrow \quad u_n > u_{n+1} \quad \forall\ n$$

(*ii*) $\underset{n\to\infty}{\text{Lt}}\ u_n = \underset{n\to\infty}{\text{Lt}}\ \frac{n}{2n-1} = \underset{n\to\infty}{\text{Lt}}\ \frac{1}{2-\frac{1}{n}} = \frac{1}{2} \neq 0$

Here the second condition of Leibnitz's Test is not satisfied.

Hence the given series is not convergent.

Example 5.27: Examine the convergence of the series

$$2-\frac{3}{2}+\frac{4}{3}-\frac{5}{4}+\ldots$$

Solution: It is an alternating series.

$$u_n=\frac{n+1}{n}, \qquad u_{n+1}=\frac{n+2}{n+1}$$

$$u_n-u_{n+1}=\frac{n+1}{n}-\frac{n+2}{n+1}=\frac{(n+1)^2-n(n+2)}{n(n+1)}=\frac{1}{n(n+1)}>0$$

$$\Rightarrow \quad u_n>u_{n+1} \quad \forall\ n$$

Now $\qquad \underset{n\to\infty}{\text{Lt}}\ u_n=\underset{n\to\infty}{\text{Lt}}\ \frac{n+1}{n}=\underset{n\to\infty}{\text{Lt}}\left(1+\frac{1}{n}\right)=1\neq 0$

Here the second condition of Leibnitz's test is not satisfied.

$\therefore$ The given series is not convergent.

EXERCISE 5.7

1. Show that the following series are convergent:

(*a*) $1-\frac{1}{3}+\frac{1}{5}-\frac{1}{7}+\ldots$ or $\sum_{n=1}^{\infty}\frac{(-1)^n}{2n-1}$ [PTU Dec 2012]

(*b*) $1-\frac{1}{\sqrt{2}}+\frac{1}{\sqrt{3}}-\frac{1}{\sqrt{4}}+\ldots$

(*c*) $\frac{1}{\log 2}-\frac{1}{\log 3}+\frac{1}{\log 4}-\frac{1}{\log 5}+\ldots$

(*d*) $1-\frac{1}{2\sqrt{2}}+\frac{1}{3\sqrt{3}}-\frac{1}{4\sqrt{4}}+\ldots$

(*e*) $1-\frac{1}{2^2}+\frac{1}{3^2}-\frac{1}{4^2}+\ldots$

(*f*) $1-\frac{1}{2^p}+\frac{1}{3^p}-\frac{1}{4^p}+\ldots$

(*g*) $\frac{1}{2^3}-\frac{1+2}{3^3}+\frac{1+2+3}{4^3}-\frac{1+2+3+4}{5^3}+\ldots$

(*h*) $\frac{1}{xy}-\frac{1}{(x+1)(y+1)}+\frac{1}{(x+2)(y+2)}+\ldots,\ x>0,\ y>0$

(*i*) $1-\frac{1}{2}+\frac{1}{3}-\frac{1}{4}+\ldots$

(*j*) $\frac{1}{2^3}-\frac{1}{3^3}(1+2)+\frac{1}{4^3}(1+2+3)-\frac{1}{5^3}(1+2+3+4)\ldots$

2. Examine the convergence of the series $1+\frac{1}{2^2}-\frac{1}{3^2}-\frac{1}{4^2}+\frac{1}{5^2}+\frac{1}{6^2}-\frac{1}{7^2}-\frac{1}{8^2}+\ldots\infty$.

3. The series $\frac{1}{3}-\frac{1}{2}+\frac{1}{9}-\frac{1}{4}+\frac{1}{27}-\frac{1}{8}+\ldots\infty$ does not meet one of the conditions of Leibnitz's test which one? Find the sum of this series. [PTU Dec 2004]

4. Show that the following series are convergent

(*a*) $\sum\frac{(-1)^{n-1}}{\lfloor n}$ (*b*) $\sum\frac{(-1)^{n-1}}{\log n}$

(*c*) $\sum\frac{(-1)^{n-1}}{n+a},\quad a\geq 0$ (*d*) $\sum\frac{(-1)^{n-1}}{\sqrt{n}+\sqrt{a}},\quad a\geq 0$

(*e*) $\sum(-1)^{n-1}\frac{\log n}{n^2}$ (*f*) $(-1)^{n-1}\sin\frac{k}{n}$ where k is constant

(*g*) $\sum(-1)^{n-1}\frac{m}{5^n}$ (*h*) $\sum_{n=1}^{\infty}\frac{(-1)^{n-1}}{3n-2}$

5. Examine the convergence of the series $\sum_{n=1}^{\infty}\frac{n(-1)^{n-1}}{5^n}$.

6. Examine the convergence of the series.

$$\frac{1}{\sqrt{2}+1}-\frac{1}{\sqrt{3}+1}+\frac{1}{\sqrt{4}+1}-\frac{1}{\sqrt{5}+1}+\ldots\infty$$

7. Examine the convergence of the series $\sum_{n=1}^{\infty}(-1)^{n-1}\frac{n}{n^2+1}$.

8. Examine the convergence of the series $\sum_{n=1}^{\infty}\frac{\cos n\pi}{n^2+1}$. $\left[\text{Hint: } \cos n\pi=(-1)^n\right]$

[PTU May 2012]

9. Discuss the convergence of the following series

(*a*) $\sum(-1)^n\left(n^{\frac{1}{n}}-1\right)$ (*b*) $\sum(-1)^n\frac{1}{\sqrt{n}}$

(c) $\sum \frac{(-1)^n}{n}$ (d) $\sum_{n=1}^{\infty} \frac{(-1)^n}{\lfloor n}$

ANSWERS

9. (a) Convergent (b) Convergent (c) Convergent (d) Convergent

5.14 POWER SERIES

Statement: A *power series* is a series of the form

$$\sum_{n=0}^{\infty} a_n x^n = a_0 + a_1 x + a_2 x^2 + \ldots + a_n x^n + \ldots$$

where a's are constants and x is a variable. We can also denote power series by $\sum a_n x^n$. This is called *power series* about $x = 0$.

A power series about $x = a$ is a series of the form

$$\sum_{n=0}^{\infty} a_n (x-a)^n = a_0 + a_1(x-a) + a_2(x-a)^2 + \ldots + a_n(x-a)^n + \ldots$$

in which centre a and coefficients $a_0, a_1, a_2, \ldots, a_n$ are constants.

For Example:

(a) $\frac{1}{1+x} = 1 - x + x^2 - x^3 + \ldots$ (b) $e^x = 1 + x + \frac{x^2}{2!} + \frac{x^3}{3!} + \ldots$

(c) $\log(1+x) = x - \frac{x^2}{2} + \frac{x^3}{3} \ldots$ (d) $(1+x)^n = 1 + nx + \frac{n(n-1)}{2!} x^2 + \ldots$

5.15 INTERVAL OF CONVERGENCE

If the power Series is $\sum a_n x^n$ then take

$\therefore$ Series becomes $\sum u_n$ where $u_n = a_n x^n$

As in power series, a's can be +ve as well as –ve.

$\therefore$ for convergence of $\sum u_n$, we test the convergence of $\sum |u_n|$

$\because$ every absolutely convergent series is a convergent series.

$$\therefore \quad \left|\frac{u_n}{u_{n+1}}\right| = \left|\frac{a_n x^n}{a_{n+1} x^{n+1}}\right| = \left|\frac{a_n}{a_{n+1}} \cdot \frac{1}{x}\right| = \left|\frac{a_n}{a_{n+1}}\right| \cdot \frac{1}{|x|}$$

Let $\underset{n\to\infty}{\text{Lt}} \left|\frac{a_n}{a_{n+1}}\right| = l \quad \therefore \quad \underset{n\to\infty}{\text{Lt}} \left|\frac{a_n}{a_{n+1}}\right| = \frac{l}{|x|}$

$\therefore \quad \sum |u_n|$ converges if $\frac{l}{|x|} > 1$ [By Ratio Test]

$\therefore \quad \sum u_n$ converges if $|x| < l$, i.e., for $-l < x < l$

$\therefore$ The power series converges in the interval $(-l, l)$ and diverges outside this interval.

Interval $(-l, l)$ is called the *interval of convergence* of the power series.

Art. To Find Radius of Convergence

Step 1: Use the Ratio Test (or n^{th} Root Test) to the interval where the series converges absolutely. Apply Ratio Test on the series

$$a_0 + a_1 x + a_2 x^2 + \ldots + a_n x^n + \ldots$$

$$\underset{n\to\infty}{\text{Lt}} \frac{|u_n|}{|u_{n+1}|} = \underset{n\to\infty}{\text{Lt}} \frac{|a_n x^n|}{|a_{n+1} x^{n+1}|} = \underset{n\to\infty}{\text{Lt}} \left|\frac{a_n}{a_{n+1}}\right| \frac{1}{|x|}$$

By Ratio Test, the given series converges absolutely or diverges accordingly as

$$\underset{n\to\infty}{\text{Lt}} \frac{|u_n|}{|u_{n+1}|} > 1 \quad \text{or} \quad \underset{n\to\infty}{\text{Lt}} \frac{|u_n|}{|u_{n+1}|} < 1$$

$$\Rightarrow \quad \underset{n\to\infty}{\text{Lt}} \left|\frac{a_n}{a_{n+1}}\right| \frac{1}{|x|} > 1 \quad \text{or} \quad \underset{n\to\infty}{\text{Lt}} \left|\frac{a_n}{a_{n+1}}\right| \frac{1}{|x|} < 1$$

$$\Rightarrow \quad \text{if } |x| < R \quad \text{or} \quad |x| > R$$

where $R = \underset{n\to\infty}{\text{Lt}} \left|\frac{a_n}{a_{n+1}}\right|$ is *radius of convergence.*

Step 2: If the interval of convergence is finite, test for convergence or divergence at each end-point. Use *Comparison Test, Integral Test* or *Alternating Series Test.*

Note 1: To find radius of convergence R (in step 1), we can also apply Root Test on $\sum|u_n| = \sum|a_n x^n|$.

Note 2: Similar steps are followed to find radius of convergence R for series $\sum a_n (x-a)^n$.

Remark 1: A power series can be differentiated term by term at each interior point of its interval of convergence.

Remark 2: A power series can be integrated term by term at each interior point of its interval of convergence.

ILLUSTRATIVE EXAMPLES 5.8

Example 5.28: Prove that the series $\frac{\sin x}{1^3} - \frac{\sin 2x}{2^3} + \frac{\sin 3x}{3^3} \ldots$ converges absolutely.

Solution: The given series is $\sum_{n=1}^{\infty} u_n = \sum_{n=1}^{\infty} \frac{(-1)^{n-1} \sin nx}{n^3}$

Since $|u_n| = \frac{|\sin nx|}{n^3} \leq \frac{1}{n^3} \quad \forall n$

and $\Sigma \frac{1}{n^3}$ converges. [by *p-test*]

$\therefore$ By comparison test, the series $\Sigma|u_n|$ converges

$\Rightarrow$ The given series converges absolutely.

Example 5.29: Prove that the series

$$x-\frac{x^2}{2}+\frac{x^3}{3}-\frac{x^4}{4}+\ldots\infty$$

is convergent for $-1<x\leq 1$. Also write the interval of convergence.

Solution: The given series is

$$\Sigma u_n=\Sigma(-1)^{n-1}\frac{x^n}{n}$$

$$|u_n|=\left|\frac{x^n}{n}\right|=\frac{|x^n|}{n},\quad |u_{n+1}|=\frac{|x^{n+1}|}{n+1}=\frac{|x|^{n+1}}{n+1}$$

$$\frac{|u_n|}{|u_{n+1}|}=\frac{|x^n|}{n}\cdot\frac{n+1}{|x|^{n+1}}=\frac{n+1}{n}\cdot\frac{1}{|x|}$$

$$\underset{n\to\infty}{\text{Lt}}\frac{|u_n|}{|u_{n+1}|}=\underset{n\to\infty}{\text{Lt}}\left(\frac{n+1}{n}\cdot\frac{1}{|x|}\right)=\underset{n\to\infty}{\text{Lt}}\left[\left(1+\frac{1}{n}\right)\cdot\frac{1}{|x|}\right]$$

$$=(1+0)\cdot\frac{1}{|x|}=\frac{1}{|x|}$$

∴ By Ratio test $\Sigma|u_n|$ converges when

$$\frac{1}{|x|}>1,\text{ i.e., }|x|<1,\text{ i.e., }-1<x<1$$

and diverges when $\frac{1}{|x|}<1$, i.e., $|x|>1$, i.e., for $x>1$ or $x<-1$

Ratio test fails when $|x|=1$, i.e., when $x=\pm 1$

∴ When x = 1 the series become

$$1-\frac{1}{2}+\frac{1}{3}-\frac{1}{4}+\ldots\infty$$

which is an alternating series and is convergent.

When $x=-1$, the series becomes

$$-1-\frac{1}{2}-\frac{1}{3}-\frac{1}{4}-\ldots\infty=-\left(1+\frac{1}{2}+\frac{1}{3}+\frac{1}{4}+\ldots\infty\right)=-\Sigma\frac{1}{n}$$

which is of the type $\Sigma\frac{1}{n^p}$ where $p=1$

$\therefore$ by *p-test*; it is divergent.

$\therefore$ given series converges for $-1<x\leq 1$ and diverges for $x>1$ or $x\leq -1$

$\therefore$ The interval of convergence $(-1,1]$.

Example 5.30: Discus the convergence of the series $\sum_{n=1}^{\infty}(-1)^n\frac{\tan^{-1}n}{1+n^2}$. [PTU May 2003]

Solution: Here $u_n=\frac{(-1)^n\tan^{-1}n}{1+n^2}$

$$|u_n|=\left|\frac{(-1)^n\tan^{-1}n}{1+n^2}\right|=\frac{\tan^{-1}n}{1+n^2}=f(x)\quad\text{(say)}$$

$\therefore$ By Cauchy's integral test $\sum_{n=1}^{\infty}|u_n|$ and $\int_1^{\infty}f(x)dx$ converge or diverge together

$$\int_1^{\infty}f(x)dx=\int_1^{\infty}\frac{\tan^{-1}x}{1+x^2}dx=\left[\frac{(\tan^{-1}x)^2}{2}\right]_1^{\infty}$$

$$=\frac{1}{2}\left[(\tan^{-1}\infty)^2-(\tan^{-1}1)^2\right]=\frac{1}{2}\left[\left(\frac{\pi}{2}\right)^2-\left(\frac{\pi}{4}\right)^2\right]$$

$$=\frac{3}{8}\pi^2\qquad\text{which is finite}$$

$\therefore$ $\int_1^{\infty}f(x)dx$ is convergent. $\therefore$ $\sum_{n=1}^{\infty}|u_n|$ is also convergent

Hence $\sum_{n=1}^{\infty}(-1)^n\frac{\tan^{-1}n}{1+n^2}$ is convergent series.

Example 5.31: Discuss the radius of convergence and interval of convergence of series

$$\sum_{n=1}^{\infty} \underline{|2n-1}\, x^n$$

Solution: The given power series is $\sum_{n=1}^{\infty} \underline{|2n-1}\, x^n$

Here $a_n = \underline{|2n-1}$

$$a_{n+1} = \underline{|2(n+1)-1} = \underline{|2n+1}$$

$$\therefore \operatorname*{Lt}_{n\to\infty} \left|\frac{a_{n+1}}{a_n}\right| = \operatorname*{Lt}_{n\to\infty} \left|\frac{\underline{|2n+1}}{\underline{|2n-1}}\right| = \operatorname*{Lt}_{n\to\infty} \frac{(2n+1)2n\underline{|2n-1}}{\underline{|2n-1}} = \operatorname*{Lt}_{n\to\infty} [2n(2n+1)] = \infty$$

$$\text{Using } \frac{1}{R} = \operatorname*{Lt}_{n\to\infty} \left|\frac{a_{n+1}}{a_n}\right| = \frac{1}{R} = \infty$$

$\Rightarrow$ Radius of Convergence = $R = 0$

$\therefore$ The given series converges only at $x = 0$

$\Rightarrow$ There is no interval of convergence.

EXERCISE 5.8

1. For what values of x the following series converges?

   (*a*) $x - \frac{x^2}{\sqrt{2}} + \frac{x^3}{\sqrt{3}} - \frac{x^4}{\sqrt{4}} + \ldots$ (*b*) $1 + x + \frac{x^2}{2!} + \frac{x^3}{3!} + \ldots$

   (*c*) $x - \frac{x^3}{3} + \frac{x^5}{5} - \frac{x^7}{7} + \ldots$ (*d*) $x - \frac{x^2}{2^2} + \frac{x^3}{3^2} - \frac{x^4}{x^2} + \ldots$

2. For what value of x does the series $\sum_{n=0}^{\infty} (-1)^n (4x+1)^n$ converges absolutely?

3. For what value of x the power series

   $$1 - \frac{1}{2}(x-2) + \frac{1}{4}(x-2)^2 - \frac{1}{8}(x-2)^3 + \ldots + \left(-\frac{1}{2}\right)^n (x-2)^n + \ldots\infty$$ [PTU May 2002]

   converges? What is its Sum?

4. Prove that the series $\sum_{n=0}^{\infty} \frac{(-1)^n \cos^2 nx}{\sqrt[n]{n}}$ convergence absolutely.

5. Test the absolute convergence of the series $\sum_{n=2}^{\infty} \frac{(-1)^n}{n(\log n)^2}$.

6. Find the radius of convergence of the following power series.

(a) $\sum_{n=0}^{\infty} \frac{(2n)!}{(n!)^2} x^n$ (b) $\sum \frac{n!}{n^n} x^n$

(c) $\sum \frac{x^n}{n^2}$ (d) $\sum_{n=1}^{\infty} \lfloor n \, x^n$

(e) $\sum (3+4i) x^n$ (f) $\sum_{n=1}^{\infty} \frac{x^n}{n}$

ANSWERS

1. (a) $-1 < x \le 1$ (b) all x (c) $-1 < x < 1$ (d) $-1 < x \le 1$

2. $-\frac{1}{2} < x < 0$ 3. $0 < x < 4; \frac{2}{x}$

5. Absolutely convergent

6. (a) $R = \frac{1}{4}$ (b) $R = e$ (c) $R = 1$ (d) $R = 0$ (e) $R = 1$ (f) $R = 1$

5.16 UNIFORM CONVERGENCE

Definition: A sequence $\{(u_n(x))\}$, $n \in N$, $\forall n \in I$ is said to converge uniformly to $u(x)$ in I if for every $\varepsilon > 0$, however small $\exists$ a positive integer m (independent of $x \; \forall \; x \in I$) such that $|u_n(x) - u(x)| < \varepsilon \quad \forall n \ge m$ and $x \in I$.

Remark: For different values of $x \in I$, we have different values of m, say $m_1, m_2, m_3, \ldots$

Let $m = \max(m_1, m_2, m_3, \ldots)$ Such m is independent of x.

5.17 WEIERSTRASS'S M-TEST

Statement: A series $\sum_{n=1}^{\infty} u_n(x)$ of functions converges uniformly and absolutely in I if $\exists$ a convergent series $\sum_{n=1}^{\infty} M_n$ of positive constants such that

$$|u_n(x)| \le M_n \qquad \forall \; n \in N \text{ and } \forall \; x \in I$$

Proof: Since $|u_n(x)| \le M_n$ and $\sum_{n=1}^{\infty} M_n$ converges.

$\therefore \quad \Sigma|u_n(x)|$ also converges $\quad \Rightarrow \Sigma u_n(x)$ converges absolutely

Since $\sum_{n=1}^{\infty} M_n$ Converges

$\therefore$ given $\varepsilon > 0$, however small $\exists$ a + *ve* integer m such that

$$|S - S_n| < \varepsilon \quad \forall n \ge m \qquad \ldots (1)$$

where $S = M_1 + M_2 + \ldots + M_n + M_{n+1} + \ldots$

$$S_n = M_1 + M_2 + \ldots + M_n$$

$$S - S_n = M_{n+1} + M_{n+2} + \ldots$$

Hence (1) implies

$$|M_{n+1} + M_{n+2}, \ldots| < \varepsilon \quad \forall n \ge m$$

Also $|u_n(x)| \le M_n$

$$|u_{n+1}(x) + u_{n+2}(x) + \ldots| \le |u_{n+1}(x)| + |u_{n+2}(x)| + \ldots \le M_{n+1} + M_{n+2} + \ldots \quad < \varepsilon \quad \forall n \ge m$$

$$\therefore \quad |S(x) - S_n(x)| < \varepsilon \quad \forall n \ge m$$

Hence Proved.

ILLUSTRATIVE EXAMPLES 5.9

Example 5.32: Show that the series $\sum_{n=1}^{\infty} \frac{\sin(x^2 + nx)}{n(n+2)}$, for all real values of *x*, converges uniformly. [PTU May 2009]

Solution: Here $u_n(x) = \frac{\sin(x^2 + nx)}{n(n+2)}$

$$\therefore \quad |u_n(x)| = \left|\frac{\sin(x^2 + nx)}{n(n+2)}\right| = \frac{|\sin(x^2 + nx)|}{n(n+2)} \le \frac{1}{n(n+2)} < \frac{1}{n^2} (= M_n) \qquad \forall \; x \in R$$

$$[\because \; \sin\theta \le 1]$$

Since $\sum_{n=1}^{\infty} M_n = \sum_{n=1}^{\infty} \frac{1}{n^2}$ is convergent by *p-test*, therefore, by M-test, the given series is uniformly convergent for all real values of x.

Example 5.33: Prove that the series

$$\cos x + \frac{\cos 2x}{2^2} + \frac{\cos 3x}{3^2} + \ldots$$

converges uniformly on R.

Solution: Here $u_n(x) = \frac{\cos nx}{n^2}$

$$\therefore \quad |u_n(x)| = \left|\frac{\cos nx}{n^2}\right| = \frac{|\cos nx|}{n^2} \leq \frac{1}{n^2} (= M_n) \quad \forall \ x \in R \qquad [\because \cos nx \leq 1]$$

Since $\sum_{n=1}^{\infty} M_n = \sum_{n=1}^{\infty} \frac{1}{n^2}$ is convergent by *p-test*, therefore, by *M*-test, the given series is uniformly convergent for all values of x.

Example 5.34: Show that the series

$$\sum_{n=1}^{\infty} \frac{\cos nx}{n^p} \quad \text{for all real } x \text{ and } p > 1 \qquad \text{[PTU Dec 2005]}$$

converges uniformly.

Solution: Here $u_n(x) = \frac{\cos nx}{n^p} \quad \therefore \quad |u_n(x)| = \left|\frac{\cos nx}{n^p}\right| = \frac{|\cos nx|}{n^p} \leq \frac{1}{n^p} (= M_n) \quad \forall \ x \in R$

Since $\sum_{n=1}^{\infty} M_n = \sum_{n=1}^{\infty} \frac{1}{n^p}$ is convergent for $p > 1$ by *p-test*, therefore, by *M*-test, the given series is uniformly convergent for all values of x and for $p > 1$.

Example 5.35: Show that the series

$$\sum_{n=1}^{\infty} \frac{1}{n^p + n^q x^2}, \text{ for all real values of } x \text{ and } p > 1;$$

converges uniformly.

Solution: Here $u_n(x)=\dfrac{1}{n^p+n^q x^2}$

Since $x^2 \ge 0$ for all real x, $\quad \therefore \; n^q x^2 \ge 0$

$$\Rightarrow \quad n^p + n^q x^2 \ge n^p \quad \Rightarrow \quad \frac{1}{n^p+n^q x^2} \le \frac{1}{n^p}$$

$$\therefore \; |u_n(x)| = \frac{1}{n^p+n^q x^2} \le \frac{1}{n^p} \; (= M_n) \qquad \forall \; x \in R$$

Since $\sum\limits_{n=1}^{\infty} M_n = \sum\limits_{n=1}^{\infty} \dfrac{1}{n^p}$ is convergent for $p > 1$ by *p-test*, therefore, by *M*-test, the given series is uniformly convergent for all values of x and $p > 1$.

EXERCISE 5.9

Test the uniform convergence the series.

1. (*a*) $\dfrac{\sin x}{1^2}+\dfrac{\sin 2x}{2^2}+\dfrac{\sin 3x}{3^2}+\dfrac{\sin 4x}{4^2}+\ldots$ [PTU May 2003]

   (*b*) $\sin x-\dfrac{\sin 2x}{2\sqrt{2}}+\dfrac{\sin 3x}{3\sqrt{3}}-\dfrac{\sin 4x}{4\sqrt{4}}+\ldots$

2. (*a*) $\sum\limits_{n=1}^{\infty} \dfrac{\cos(x^2+n^2x)}{m(n^2+2)}$ (*b*) $\sum\limits_{n=1}^{\infty} \dfrac{1}{n^4+n^3x^2}$

4. Show that if $0 < r < 1$, then each of the following series is uniformly convergent on R;

   (*a*) $\sum\limits_{n=1}^{\infty} r^n \cos nx$ (*b*) $\sum\limits_{n=1}^{\infty} r^n \sin nx$ (*c*) $\sum\limits_{n=1}^{\infty} r^n \cos n^2 x$ (*d*) $\sum\limits_{n=1}^{\infty} r^n \sin a^n x$

5. Prove that $\sum \dfrac{x}{n(1+nx^2)}$ is uniformly convergent for all real values of x.

6. Discuss the uniform convergence of $\sum \dfrac{x}{(n+x^2)^2}$.

ANSWERS

1. (*a*) Uniformly convergent for all real x. (*b*) Uniformly convergent for all real x.
2. (*a*) Uniformly convergent for all real x. (*b*) Uniformly convergent for all real x.
6. Uniformly convergent.

TESTING CRITERIA FOR CONVERGENCE OF AN INFINITE SERIES

Types of infinite series:

- Positive term series
- Alternating series
- Geometric series
- Power series

1. For a positive term series, first find u_n, and if possible evaluate $\underset{n\to\infty}{\text{Lt}}\, u_n$. If $\underset{n\to\infty}{\text{Lt}}\, u_n \neq 0$, the series is divergent. However, if $\underset{n\to\infty}{\text{Lt}}\, u_n = 0$, then compare $\sum u_n$ with $\sum v_n$, where v_n is always of type $\frac{1}{n^p}$, and then

   - Apply Comparison test, if it fails, then
   - Apply Ratio test, if it fails, then
   - Apply Raabe's test, if it fails, then
   - Apply Logarithmic test, if it fails, then
   - Apply Gauss test

   Special Cases:

   (*a*) If in Ratio test, $\frac{u_n}{u_{n+1}}$ in involves *e*, we directly apply Logarithmic test.

   (*b*) If in Ratio test, it is possible to expand $\frac{u_n}{u_{n+1}}$ in powers of $\frac{1}{n}$, then directly apply Gauss Test.

2. For alternating series, apply Leibnitz's rule.
3. Geometric series $1+x+x^2+...$
   - converges if $-1<x<1$, i.e., $|x|<1$
   - diverges if $x\geq 1$
   - oscillates finitely if $x=-1$
   - oscillates infinitely if $x<-1$
4. For power series find
   - interval of convergence & radius of convergence by Ratio test
   - uniform convergence by Weierstrass's M-Test

Chapter 6

Complex Numbers

Learning Objectives

After reading this chapter, students will be able to

- explain the basic concepts of complex numbers
- state and prove De-Moivre's theorem
- demonstrate applications of De-Moivre's theorem
- describe the following functions of a complex variable
  - exponential function
  - logarithmic function
  - circular functions
  - hyperbolic functions
  - inverse trigonometric functions
  - inverse hyperbolic functions
- describe the summation of trigonometric functions –C+iS method
- solve various probems related to complex numbers

6.1 RECAPITULATION OF BASIC CONCEPTS OF COMPLEX NUMBERS

A number of the form $x + iy$, where x and y are real numbers is called a *complex number*. The set of complex numbers is denoted by C.

If $z = x + iy$ is a complex number, then

x is called the real part of z and we write $\text{Re}(z) = x$
y is called the imaginary part of z and we write $\text{Im}(z) = y$

If $x = 0$ and $y \neq 0$, then $z = 0 + iy = iy$ is called a purely imaginary number.

If $y = 0$, then $z = x + i0 = x$ is called a purely real number.

If $x = 0$ and $y = 0$, then $z = 0 + i0 = 0$ is called the zero complex number.

We observe that the system of complex number includes the system of real number, *i.e.*, $R \subseteq C$. Every real number can be represented as a complex number. For example, -2 is a real number can be represented as $-2+0i$.

Also we observe that a complex number $(x + iy)$ can be uniquely determined by an ordered pair (x, y) of real numbers in which the first member in the real part and the second member is the imaginary part of the complex number.

6.1.1 Conjugate Complex Number

If two complex numbers differ only in the sign of imaginary part, they are called *conjugate complex number*. Thus $(x + iy)$ and $(x - iy)$ are conjugate complex numbers.

The conjugate of a complex number is obtained by simply changing the sign of the imaginary part. Clearly the conjugate of a real number is the real number itself because there is no imaginary part in it.

The conjugate of a complex number z is denoted by $\bar{z}$.

6.1.2 Geometrical Representation of Complex Numbers

Let $z = x + iy$ be a complex number. It can be uniquely determined by the ordered pair (x, y) of real numbers. Since there exists one-one correspondence between the ordered pairs of real numbers and the points on a plane, every complex number $z = x + iy$ can be represented by a unique point $P(x,y)$ with regard to two rectangular axes x-axis and y-axis.

The plane representing complex numbers as ordered pairs of real number is called the *Complex plane* or *Argand plane* or *Gaussian plane*. Complex numbers of the form $x + i.0$ (*i.e.*, real numbers) are represented by points $(x,0)$ on the y-axis which is, therefore, called the *axis of imaginaries*.

The length OP is called the *modulus* of the complex number $z = x + iy$ and is denoted by $|Z|$.

Clearly $|Z| = OP = \sqrt{x^2 + y^2} = \sqrt{[\text{Re}(z)^2 + [\text{Im}(z)]^2}$

Now $Z = x + iy$

$\therefore$ if $P(x, y)$ represents z, then $\bar{z}$ is represented by $P'(x, -y)$, which is the image of P in x-axis.

Since $OP' = OP$, $|\bar{z}| = |z|$.

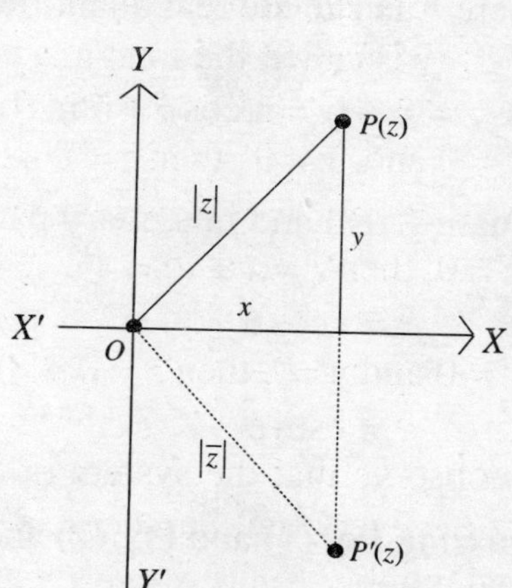

Figure 6.1: Geometrical representation of complex number

6.1.3 Properties of Complex Number

(*i*) The sum, difference, product and quotient of two complex number is also a complex number.

(*ii*) If a complex number is equal to zero, then its real and imaginary parts are separately equal to zero.

Thus $x + iy = 0 \Rightarrow x = 0$ and $y = 0$

(*iii*) If two complex number are equal, then their real and imaginary parts are separately equal.

Thus $x + iy = a + ib \Rightarrow x = a$ and $y = b$

(*iv*) If two complex number are equal, then their conjugates are also equal.

Thus $a + ib = c + id \Rightarrow a - ib = c - id$

In general, $f(x + iy) = g(x + iy) \Rightarrow f(x - iy) = g(x - iy)$ (on changing i to $-i$)

6.1.4 Standard Form of a Complex Number/Polar Representation of a Complex Number

Every non-zero complex number $z = x + iy$ can be put in the form

$$z = r(\cos\theta + i\sin\theta)$$

where r and θ are real quantities.(polar co-ordinates)

Let $z = x + iy = r(\cos\theta + i\sin\theta)$

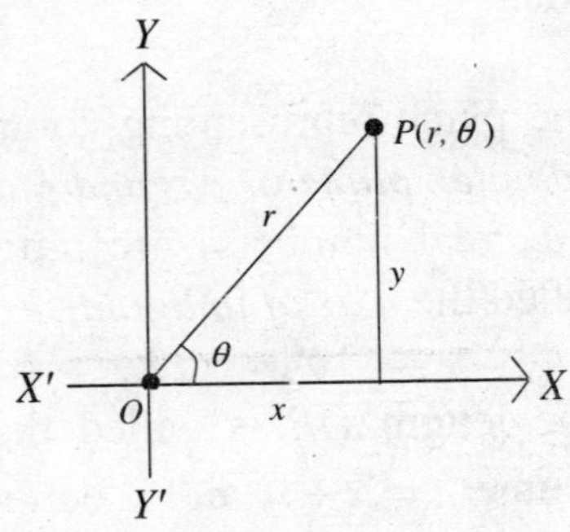

Figure 6.2: Polar coordinates of complex number

Equating real and imaginary parts, we get

$$x = r\cos\theta \qquad \ldots (1)$$

$$y = r\sin\theta \qquad \ldots (2)$$

Squaring Eq. (1) and Eq. (2) and adding, we get

$$x^2 + y^2 = r^2 \text{ or } r = \sqrt{x^2 + y^2} \qquad \text{(taking +ve with the square root)}$$

Substituting this value of r in Eq. (1) and Eq. (2), we get

$$\left.\begin{aligned} \cos\theta &= \frac{x}{r} = \frac{x}{\sqrt{x^2 + y^2}} \\ \sin\theta &= \frac{y}{r} = \frac{y}{\sqrt{x^2 + y^2}} \end{aligned}\right\} \qquad \ldots (3)$$

Since $z = x + iy$ is a non-zero complex number, at least one of x and y is not equal to zero.

Whatever be the value of x and y, it is always possible to find a value of θ satisfying both the equation given by Eq. (3), where $-\pi < \theta \le \pi$.

Hence every non-zero complex number $z = x + iy$ can be put in the form

$$r(\cos\theta + i\sin\theta) = r \text{ cis } \theta \qquad [\cos\theta + i\sin\theta \text{ can be written as } \text{cis}\,\theta \text{ for brevity}]$$

$$= re^{i\theta}$$

Note 1: If $z = x + iy = r(\cos\theta + i\sin\theta)$, then

(*i*) The positive real number $r = \sqrt{x^2 + y^2}$ is called the modulus or norms or an absolute value of the complex number z and is written as mod Z or $|Z|$.

Thus $z = x + iy \Rightarrow |z| = \sqrt{x^2 + y^2}$

If $z = r(\cos\theta + i\sin\theta) \Rightarrow |z| = r$

(*ii*) The number θ is called the amplitude or argument of z and is written as amplitude z or argument z.

The unique value of θ such that $-\pi < \theta \le \pi$, which satisfies the equations given by Eq. (3), is called the *principal value* of the amplitude.

Dividing the Eq. (3*a*) by Eq. (3*b*), we get $\tan\theta = \dfrac{y}{x}$

$$\therefore \text{ amplitude } z = \theta = \tan^{-1}\frac{y}{x} = \tan^{-1}\frac{\text{Im}(z)}{\text{Re}(z)}$$

Note 2: While reducing any complex number to the standard form $r((\cos\theta + \sin\theta)$ always take the principal value of the amplitude.

Note 3: $1 = \cos 0$, $-1 = \text{cis } \pi$, $i = \text{cis } \dfrac{\pi}{2}$, $-i = \text{cis}\left(-\dfrac{\pi}{2}\right)$.

6.1.5 Geometrical Representation of Addition of two Complex Numbers

In the complex piane, let A and B represents the two complex numbers representing complex numbers $z_1 = x_1 + iy_1$ and $z_2 = x_2 + iy_2$. Then

$$z_1 + z_2 = (x_1 + x_2) + i(y_1 + y_2)$$

$$A \leftrightarrow (x_1, y_1),\ B \leftrightarrow (x_2, y_2), \text{ and}$$

$$C \leftrightarrow (x_1 + x_2, y_1 + y_2)$$

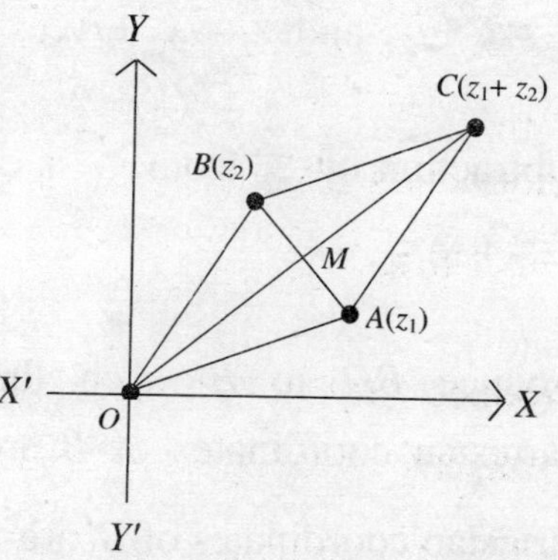

Figure 6.3: Addition of two complex numbers

Complete the parallelogram *OACB*. Let its diagonal intersects at *M*.

Since M is the mid point of AB

$$\therefore \quad M \leftrightarrow \left(\frac{x_1+x_2}{2}, \frac{y_1+y_2}{2}\right) \qquad \text{(mid point Formula)}$$

M is the mid point of OC

$$\therefore \quad C \leftrightarrow (x_1+x_2, y_1+y_2)$$

$\Rightarrow$ C represents the complex number

$$(x_1+x_2)+i(y_1+y_2)=(x_1+iy_1)+(x_2+iy_2)=z_1+z_2$$

Thus, if two points A and B represents the complex numbers z_1 and z_2 in a complex plane, then their sum (z_1+z_2) is represented by the point C of the diagonal OC of parallelogram $OACB$.

Cor. : $\quad |z_1|=OA, \quad |z_2|=OB, \quad |z_1+z_2|=OC$

Since $OC \leq OA + AC = OA + OB$

$$\therefore \quad |z_1+z_2| \leq |z_1|+|z_2|$$

6.1.6 Geometrical Representation of Difference of two Complex Numbers

In the complex plane, let A and B represents the two complex numbers representing complex numbers $z_1 = x_1 + iy_1$ and $z_2 = x_2 + iy_2$.

Subtraction of z_2 from z_1 is equivalent to addition of $(-z_2)$ to z_1.

Produce BO to B' such that OB' = OB. Since, Cartesian coordinates of B are (x_2, y_2), therefore the cartesian coordinates of B' are $(-x_2, -y_2)$,

$\because$ B' is in 3rd quadrant.

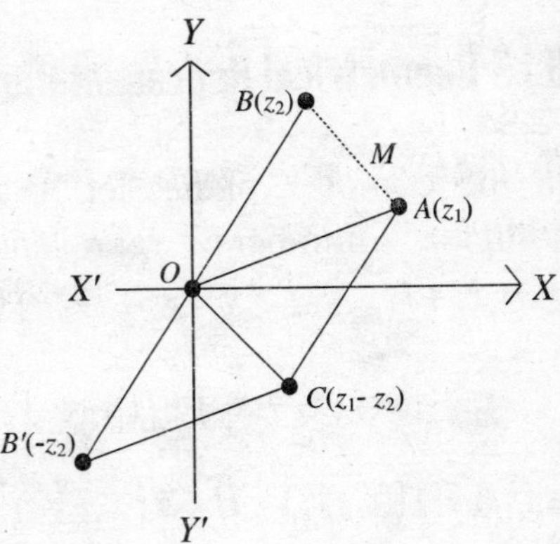

Figure 6.4: Subtraction of two complex numbers

$\therefore$ B' represents the complex number

$$-x_2 - iy_2 = -(x_2 + iy_2) = -z_2$$

Complete the parallelogram $OAB'C$. Then, C represents the sum of the complex numbers z_1 and $(-z_2)$, *i.e.*, C represents the complex number

$$\therefore \quad z_1 + (-z_2) = z_1 - z_2$$

Cor. : $|z_1| = OA, \quad |z_2| = OB, \quad |z_1 - z_2| = OC = BA$

Since $OB + BA \geq OA$

$$\therefore \quad |z_2| + |z_1 - z_2| \geq |z_1| \quad \Rightarrow \quad |z_1 - z_2| \geq |z_1| - |z_2|$$

(sum of two sides of a triangle is greater than the third side)

Cor. : Geometrical meaning of $|z_1 - z_2|$

If two complex numbers z_1 and z_2 be represented by the end points A and B on the complex plane, then

$$|z_1 - z_2| = \left|\overrightarrow{OA} - \overrightarrow{OB}\right| = \left|\overrightarrow{BA}\right| = BA = AB$$

Thus, $A(z_1)$ and $B(z_2)$ $\Rightarrow$ $|z_1 - z_2| = BA = AB$

6.1.7 Geometrical Representation of Product of two Complex Numbers

If z_1 and z_2 are two complex numbers, prove geometrically that

$$|z_1 z_2| = |z_1|.|z_2| \text{ and}$$
$$\text{amp.}(|z_1 z_2|) = \text{amp.}(|z_1|) + \text{amp.}(|z_2|)$$

Proof: Let A and B represents the two complex numbers

$$z_1 = r_1(\cos\theta_1 + i\sin\theta_1) \text{ and}$$
$$z_2 = r_2(\cos\theta_2 + i\sin\theta_2)$$

in the complex plane.

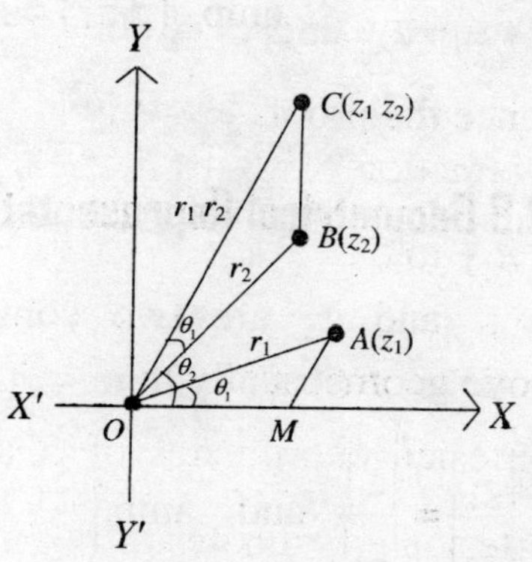

Figure 6.5: Product of two complex numbers

Then $OA = r_1$, $OB = r_2$, $\angle XOA = \theta_1$, $\angle XOB = \theta_2$.

Take $OM = 1$ unit of length on the real axis. Construct $\Delta\, OBC$ similar to $\Delta\, OMA$ on the side of OB away from A such that

$$\angle BOC = \angle MOA = \theta_1 \quad \text{and} \quad \angle OBC = \angle OMA$$

As Δs OMA and OBC are similar (by construction), therefore

$$\frac{OC}{OA} = \frac{OB}{OM} = \frac{OB}{1} = OB \qquad [\because\ OM = 1]$$

$$\therefore\quad OC = OA \times OB = r_1 r_2 \text{ and}$$

$$\angle MOC = \angle MOB + \angle BOC = \angle MOB + \angle MOA = \theta_2 + \theta_1 = \theta_1 + \theta_2$$

$\therefore$ Polar coordinates of C are $(r_1 r_2, \theta_1 + \theta_2)$

$\therefore$ Point C represents the complex number

$$r_1 r_2 (\cos(\theta_1 + \theta_2) + i \sin(\theta_1 + \theta_2)) = r_1 r_2 (\text{cis}(\theta_1 + \theta_2)) = r_1 r_2 (\text{cis}\theta_1 \text{cis}\theta_2)$$

$$[\because\ \text{cis}(\alpha + \beta) = \text{cis}\alpha\, \text{cis}\beta]$$

$$r_1 r_2 (\cos(\theta_1 + \theta_2) + i \sin(\theta_1 + \theta_2)) = r_1 (\cos\theta_1 + i \sin\theta_1) \times r_2 (\cos\theta_2 + i \sin\theta_2) = z_1 z_2$$

$$|z_1 z_2| = OC = r_1 r_2 = OA \times OB = |z_1||z_2|$$

$$\text{amp. } (z_1 z_2) = \angle XOC = \theta_1 + \theta_2 = \text{amp. } (z_1) + \text{amp. } (z_2)$$

Hence the proof.

6.1.8 Geometrical Representation of Quotient of two Complex Numbers

If z_1 and z_2 are two complex numbers, prove geometrically that

$$\left|\frac{z_1}{z_2}\right| = \frac{|z_1|}{|z_2|} \quad \text{and} \quad \text{amp.}\left(\frac{z_1}{z_2}\right) = \text{amp.}(|z_1|) - \text{amp.}(|z_2|)$$

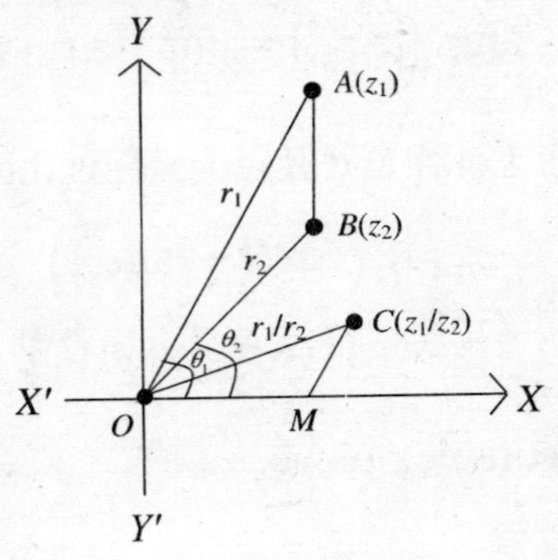

Figure 6.6: Quotient of two complex numbers

Proof: Let A and B represents the two complex numbers

$$z_1 = r_1(\cos\theta_1 + i\sin\theta_1) \text{ and}$$

$$z_2 = r_2(\cos\theta_2 + i\sin\theta_2)$$

in the complex plane.

Then $OA = r_1$, $OB = r_2$, $\angle XOA = \theta_1$, $\angle XOB = \theta_2$.

Take $OM = 1$ unit of length on the real axis.

Construct $\Delta\, OMC$ on OM similar to $\Delta\, OBA$ such that

$$\angle MOC = \angle BOA = \angle XOA = \angle XOB = \theta_1 - \theta_2 \quad \text{and} \quad \angle OMP = \angle OBA.$$

As Δs OMC and OBA are similar (by construction), therefore

$$\frac{OC}{OM} = \frac{OA}{OB}$$

$$= \frac{OC}{1} = \frac{OA}{OB} = \frac{r_1}{r_2} \qquad [\because \; OM = 1]$$

$$\therefore \quad OC = \frac{r_1}{r_2}$$

$\therefore$ Polar coordinates of C are $\left(\frac{r_1}{r_2}, \theta_1 - \theta_2\right)$

$\therefore$ Point C represents the complex number

$$\frac{r_1}{r_2}\left(\cos(\theta_1 - \theta_2) + i\sin(\theta_1 - \theta_2)\right) = \frac{r_1(\cos\theta_1 + i\sin\theta_1)}{r_2(\cos\theta_2 + i\sin\theta_2)}$$

$$= \frac{z_1}{z_2} \qquad \left[\because \quad \text{cis}(\alpha - \beta) = \frac{\cos\alpha}{\cos\beta}\right]$$

$$\left|\frac{z_1}{z_2}\right| = OC = \frac{r_1}{r_2} = \frac{OA}{OB} = \frac{|z_1|}{|z_2|}$$

$$\text{amp.}\left(\frac{z_1}{z_2}\right) = \angle XOC = \theta_1 - \theta_2 = \text{amp.}\,(z_1) - \text{amp.}\,(z_2)$$ Hence the proof.

ILLUSTRATIVE EXAMPLES 6.1

Example 6.1: Express the following in the form $r(\cos\theta + i\sin\theta)$

(*i*) $1-i$ (*ii*) $\dfrac{1+i}{1-i}$

Solution: (*i*) $1-i = r(\cos\theta + i\sin\theta)$ ... (1)

Equating real and imaginary parts:

$$1 = r\cos\theta \text{ and } -1 = r\sin\theta$$

Squaring and adding, we get

$$r^2 = 2 \Rightarrow r = \sqrt{2}$$

Also $\cos\theta = \dfrac{1}{\sqrt{2}}$ and $\sin\theta = -\dfrac{1}{\sqrt{2}}$ $\left[\therefore\ \theta \text{ is in } 4^{th} \text{quadrant}\right]$

Both these equations are satisfied when $\theta = -\dfrac{\pi}{4}$

Putting the value of r and θ in (1), we get

$$1-i = \sqrt{2}\left[\cos\left(-\frac{\pi}{4}\right) + i\sin\left(-\frac{\pi}{4}\right)\right]$$

$$= \sqrt{2}\left[\cos\frac{\pi}{4} - i\sin\frac{\pi}{4}\right] \qquad \left[\begin{array}{l}\because \cos(-\theta) = \cos\theta \\ \sin(-\theta) = -\sin\theta\end{array}\right]$$

(*ii*) $$\frac{1+i}{1-i} \times \frac{1+i}{1+i} = \frac{(1+i)^2}{(1+i)(1-i)} = \frac{(1+i^2+2i)}{(1^2-i^2)} = \frac{(1+(-1)+2i)}{(1-(-1))} = \frac{(1-1+2i)}{(1+1))} = \frac{(2i)}{2}$$

$$= i = \cos\frac{\pi}{2} + i\sin\frac{\pi}{2}$$

Example 6.2: Find the modulus and amplitude of the following

(*i*) $\tan\alpha - i$ (*ii*) $1-\cos\alpha + i\sin\alpha$

Solution: (*i*) $z=\tan\alpha-i=\dfrac{\sin\alpha}{\cos\alpha}-i=\dfrac{1}{\cos\alpha}(\sin\alpha-i\cos\alpha)$

$$z=\sec\alpha\left(\cos\left(\frac{\pi}{2}-\alpha\right)-i\sin\left(\frac{\pi}{2}-\alpha\right)\right)$$

$$=\sec\alpha\left(\cos\left(\alpha-\frac{\pi}{2}\right)+i\sin\left(\alpha-\frac{\pi}{2}\right)\right)$$

On comparing with $z=r(\cos\theta+i\sin\theta)$

$$|z|=\sec\alpha,\quad \text{amp. } z=\alpha-\frac{\pi}{2}$$

(*ii*) $z=1-\cos\alpha+i\sin\alpha$

$$=2\sin^2\frac{\alpha}{2}+2i\sin\frac{\alpha}{2}\cos\frac{\alpha}{2}\quad\left[\because\ 1-\cos\alpha=2\sin^2\frac{\alpha}{2},\ \sin\alpha=2\sin\frac{\alpha}{2}\cos\frac{\alpha}{2}\right]$$

$$=2\sin\frac{\alpha}{2}\left[\sin\frac{\alpha}{2}+i\cos\frac{\alpha}{2}\right]$$

$$=2\sin\frac{\alpha}{2}\left[\cos\left(\frac{\pi}{2}-\frac{\alpha}{2}\right)+i\sin\left(\frac{\pi}{2}-\frac{\alpha}{2}\right)\right]\quad\left[\begin{aligned}\because\ \cos\left(\frac{\pi}{2}-\theta\right)&=\sin\theta\\ \sin\left(\frac{\pi}{2}-\theta\right)&=\cos\theta\end{aligned}\right]$$

On comparing with $z=r(\cos\theta+i\sin\theta)$, we get

$$\Rightarrow\ |z|=2\sin\frac{\alpha}{2},\quad \text{amp. } z=\frac{\pi}{2}-\frac{\alpha}{2}=\frac{\pi-\alpha}{2}$$

Example 6.3: If $a+ib=\dfrac{(x+i)^2}{2x-i}$, prove that $a^2+b^2=\dfrac{(x+1)^2}{4x^2+1}$.

Solution: Given that $a+ib=\dfrac{(x+i)^2}{2x-i}$ . . . (1)

Changing i to $-i$, we get $a-ib=\dfrac{(x-i)^2}{2x+i}$ . . . (2)

Multiplying Eq. (1) by Eq. (2), we get

$$(a+ib)(a-ib)=\frac{(x+i)^2}{2x-i}\cdot\frac{(x-i)^2}{2x+i}$$

$$\Rightarrow\ \left(a^2-(ib)^2\right)=\frac{(x+i)^2(x-i)^2}{(2x-i)(2x+i)}$$

$$\Rightarrow\ a^2-(i^2b^2)=\frac{\left[(x+i)(x-i)\right]^2}{(2x)^2-i^2} \qquad \Rightarrow\ a^2-((-1)b^2)=\frac{\left[(x^2-i^2)\right]^2}{4x^2-(-1)}$$

$$\Rightarrow\ a^2+b^2=\frac{\left[x^2-i^2\right]^2}{4x^2+1}=\frac{\left[x^2-(-1)\right]^2}{4x^2+1}=\frac{\left[x^2+1\right]^2}{4x^2+1}$$ Hence the proof.

Example 6.4: Find square root of $\dfrac{86-27i}{2+3i}$.

Solution: $$\frac{86-27i}{2+3i}=\frac{86-27i}{2+3i}\times\frac{2-3i}{2-3i}=\frac{172-54i-258i+81i^2}{2^2-(3i)^2}$$

$$=\frac{172-54i-258i+81\times(-1)}{4-3^2i^2}=\frac{172-54i-258i-81}{4-9\times(-1)}$$

$$=\frac{91-312i}{4+9}=\frac{91-312i}{13}=7-24i$$

Let $\sqrt{7-24i}=a+ib$

$$\Rightarrow\ 7-24i=(a+ib)^2=a^2+(ib)^2+2abi \qquad \text{[by squaring]}$$

$$=\left(a^2+i^2b^2\right)+2abi$$

$$=\left(a^2-b^2\right)+2abi$$

Comparing real and imaginary parts

$$\left(a^2-b^2\right)=7,\ \ 2ab=-24$$

Now $\left(a^2+b^2\right)^2=\left(a^2-b^2\right)^2+4a^2b^2=49+576=625$

$\Rightarrow \quad a^2+b^2=25$

Also $a^2-b^2=7$

Adding, we get

$$2a^2=32 \Rightarrow a^2=16 \Rightarrow a=\pm 4$$

Subtracting, we get

$$2b^2=18 \Rightarrow b^2=9 \Rightarrow b=\pm 3$$

Either $a=4, b=-3$ or $a=-4, b=3$ $\qquad \because 2ab=-24$

$$\therefore \sqrt{7-24i}=4-3i \quad \text{or} \quad -4+3i$$

6.1.9 Some Useful Results

1. If z is a complex numbers, then

(*i*) $|z|^2=|\overline{z}|^2=z.\overline{z}$ $\qquad$ (*ii*) amp. $(z)+$ amp. $(\overline{z})=0$

Proof: (*i*) Let $z=x+iy$, then $\overline{z}=x-iy$

$$|z|^2=\left(\sqrt{x^2+y^2}\right)^2=x^2+y^2$$

$$|\overline{z}|^2=\left(\sqrt{x^2+(-y)^2}\right)^2=\left(\sqrt{x^2+y^2}\right)^2=x^2+y^2$$

$$z.\overline{z}=(x+iy).(x-iy)=x^2-(iy)^2=x^2-i^2y^2=x^2+y^2 \qquad \left[\because i^2=-1\right]$$

$$\therefore |z|^2=|\overline{z}|^2=z.\overline{z}$$

(*ii*) Let $z=r(\cos\theta+i\sin\theta)$, then

$$\overline{z}=r(\cos\theta-i\sin\theta)=r\left[\cos(-\theta)+i\sin(-\theta)\right]$$

$$\text{amp. } (z) = \theta \quad \text{and} \quad \text{amp. } (\bar{z}) = -\theta$$

$$\therefore \quad \text{amp. } (z) + \text{amp. } (\bar{z}) = \theta + (-\theta) = 0$$

2. If z_1 and z_2 are two complex numbers, then

(*i*) $|z_1 z_2| = |z_1||z_2|$ (*ii*) $\left|\dfrac{z_1}{z_2}\right| = \dfrac{|z_1|}{|z_2|}$

Proof: (*i*) $|z_1 z_2|^2 = (z_1 z_2)\left(\overline{z_1 z_2}\right) = (z_1 z_2)\left(\bar{z}_1 \bar{z}_2\right) = \left(z_1 \bar{z}_1\right)\left(z_2 \bar{z}_2\right)$

$$= |z_1|^2 |z_2|^2 \quad \Rightarrow \quad |z_1 z_2| = |z_1||z_2|$$

(*ii*) $\left|\dfrac{z_1}{z_2}\right|^2 = \left(\dfrac{z_1}{z_2}\right)\left(\overline{\dfrac{z_1}{z_2}}\right) = \dfrac{z_1 \cdot \bar{z}_1}{z_2 \cdot \bar{z}_2} = \dfrac{|z_1|^2}{|z_2|^2}$

$$\Rightarrow \quad \left|\frac{z_1}{z_2}\right| = \frac{|z_1|}{|z_2|}$$

6.1.10 Effect of Rotation on Complex Number

Let us study the effect of rotation, in the anti-clockwise direction, through an angle α on the complex numbers.

Let $z = r(\cos\theta + i\sin\theta)$ . . . (1)

and is represented by point P.

Rotate OP through an angle α in the anti-clockwise direction so that it occupy the position OQ, then

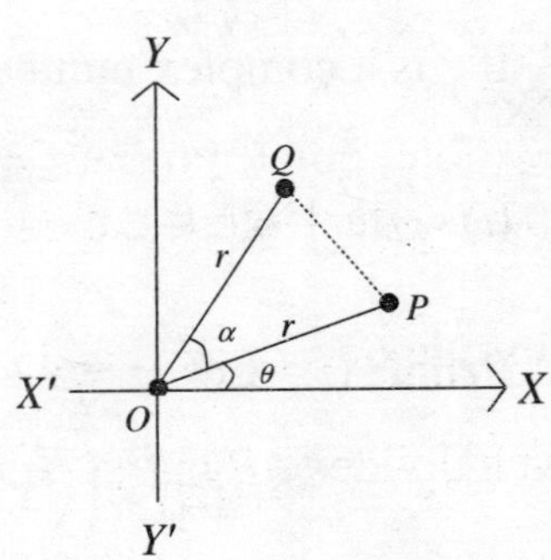

Figure 6.7: Effect of rotation on Complex number

$$OQ = OP = r$$

$$\angle XOQ = \angle XOP + \angle POQ = \theta + \alpha$$

$\therefore$ Point Q represents the complex number

$$r[\cos(\theta+\alpha) + i\sin(\theta+\alpha)] = r \text{ cis}(\theta+\alpha) = r \text{ cis}\theta \text{ cis}\alpha$$

$$= r(\cos\theta + i\sin\theta)(\cos\alpha + i\sin\alpha)$$

$$= z(\cos\alpha + i\sin\alpha) \qquad \text{[Using Eq. (1)]}$$

Hence, the effect of rotation of OP representing the complex number z through an angle α in the anti-clockwise direction can be obtained by multiplying z with a factor $(\cos\alpha + i\sin\alpha)$.

ILLUSTRATIVE EXAMPLES 6.2

Example 6.5: Find the locus of $P(z)$ if z_1 and z_2 are two given complex numbers such that $|z-z_1|=|z-z_2|$.

Solution: Let A and B represent the given complex numbers z_1 and z_2, and P represents the complex number z, as shown in Figure 6.7.

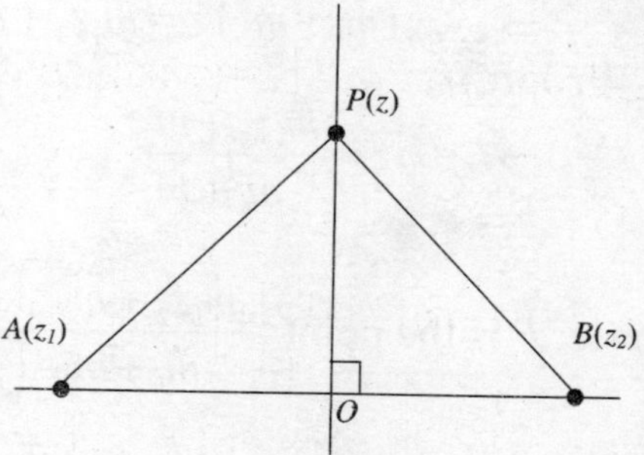

Figure 6.8: Locus of a point $P(z)$

Now $|z-z_1|=|z-z_2|$ (Given)

$$\therefore \quad \left|\overrightarrow{OP}-\overrightarrow{OA}\right| = \left|\overrightarrow{OP}-\overrightarrow{OB}\right|$$

$$\therefore \quad \left|\overrightarrow{OP}+\overrightarrow{AO}\right| = \left|\overrightarrow{OP}-\overrightarrow{BO}\right|$$

$$\Rightarrow \quad \left|\overrightarrow{AP}\right| = \left|\overrightarrow{BP}\right| \quad \Rightarrow \quad AP = BP \qquad \text{[By triangle law]}$$

Therefore, point P moves such that its distance from two fixed points A and B always remains equal. Hence, the locus of P is the right bisector of AB.

Example 6.6: If $A(z_1)$ and $B(z_2)$ be two points in the Argand plane, prove that the point which divides AB in the ratio $m_1{:}m_2$ is $\left(\dfrac{m_1z_2+m_2z_1}{m_1+m_2}\right)$, $(m_1>0,\ m_2>0)$.

Solution: Given that points A and B represent the given complex numbers z_1 and z_2. Let point P divide the join of A and B in the ratio $m_1{:}m_2$. Then, we have

$$\frac{AP}{BP}=\frac{m_1}{m_2}$$

$$\Rightarrow \quad m_2.AP = m_1.BP$$

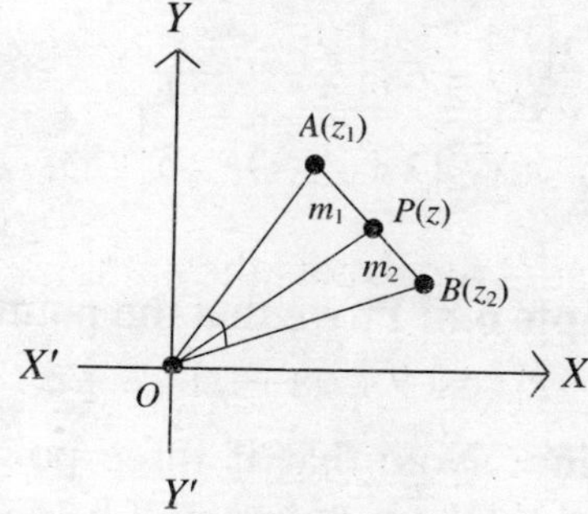

Figure 6.9: Point P dividing the line AB in ratio $m_1{:}m_2$

In vector notation, we can write

$$m_2.\overrightarrow{AP} = m_1.\overrightarrow{BP}$$

$$\Rightarrow \quad m_2\left(\overrightarrow{OP} - \overrightarrow{OA}\right) = m_1\left(\overrightarrow{OB} - \overrightarrow{OP}\right)$$

$$\Rightarrow \quad m_2\left(z - z_1\right) = m_1\left(z_2 - z\right)$$

$$\Rightarrow \quad \left(m_1 + m_2\right)z = m_1 z_2 + m_2 z_1$$

$$\Rightarrow \quad z = \frac{m_1 z_2 + m_2 z_1}{m_1 + m_2}$$

Hence, P is the point $\left(\dfrac{m_1 z_2 + m_2 z_1}{m_1 + m_2}\right)$

Note that mid point of $A(z_1)$ and $B(z_2)$ is represented by $\left(\dfrac{z_1 + z_2}{2}\right)$.

Example 6.7: Show that the centroid of the triangle formed by $A(z_1)$, $B(z_2)$ and $C(z_3)$ is $\frac{1}{3}\left(z_1 + z_2 + z_3\right)$.

Solution: Let D be the mid point of BC. Then D is the point $\left(\dfrac{z_2 + z_3}{2}\right)$. If G is the centroid of ΔABC, then G divides AD in the ratio 2:1.

$\therefore$ G represents the point

$$\frac{2\left(\dfrac{z_2 + z_3}{2}\right) + 1.z_1}{2+1} = \frac{1}{3}\left(z_1 + z_2 + z_3\right)$$

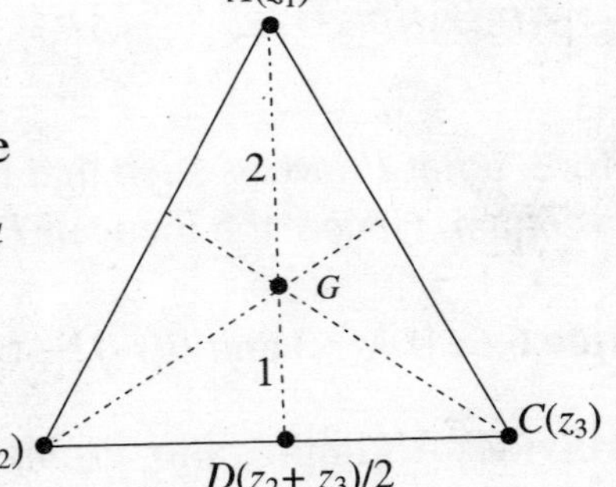

Figure 6.10: Centroid G of a triangle

Example 6.8: Prove that the points $x + iy$ and $\dfrac{1}{-x+iy}$ lie on a straight line through origin.

Solution: Note that if three points lie on a straight line, then the area of the triangle formed by these points will be zero. Therefore, we need to prove that area of the triangle formed by these points is zero.

Let $z_1 = x+iy$

$$z_2 = \frac{1}{-x+iy} = \frac{1}{-x+iy} \times \frac{-x-iy}{-x-iy} = \frac{-x-iy}{(-x+iy)(-x-iy)} = \frac{-x-iy}{(-x)^2-(iy)^2}$$

$$= \frac{-x-iy}{x^2-i^2y^2} = \frac{-x-iy}{x^2+y^2} = \frac{-x}{x^2+y^2} + \frac{-iy}{x^2+y^2}$$

Let O be the origin. Let z_1 and z_2 be represented by A and B. Then, the coordinates of A and B are (x, y) and $\left(\frac{-x}{x^2+y^2}, \frac{-y}{x^2+y^2}\right)$, respectively.

$$\text{Area of } \Delta OAB = \frac{1}{2}\begin{vmatrix} 0 & 0 & 1 \\ x & y & 1 \\ \frac{-x}{x^2+y^2} & \frac{-y}{x^2+y^2} & 1 \end{vmatrix} = \frac{1}{2}\left(-\frac{xy}{x^2+y^2} + \frac{xy}{x^2+y^2}\right) = 0$$

Thus, given points lie on a straight line passing through the origin.

Example 6.9: If amp. $\left(\frac{z_2-z_3}{z_1-z_3}\right)$ = amp. $\left(\frac{z_2-z_4}{z_1-z_4}\right)$, show that z_1, z_2, z_3, z_4 are con-cyclic.

Solution: Let A, B, C, D represent z_1, z_2, z_3, z_4, respectively.

$$\because \text{ amp. }\left(\frac{z_2-z_3}{z_1-z_3}\right) = \text{amp. }\left(\frac{z_2-z_4}{z_1-z_4}\right)$$

$$\therefore \text{ amp. }\left(\frac{\overrightarrow{OB}-\overrightarrow{OC}}{\overrightarrow{OA}-\overrightarrow{OC}}\right) = \text{amp. }\left(\frac{\overrightarrow{OB}-\overrightarrow{OD}}{\overrightarrow{OA}-\overrightarrow{OD}}\right)$$

$$\Rightarrow \text{ amp. }\left(\frac{\overrightarrow{CB}}{\overrightarrow{CA}}\right) = \text{amp. }\left(\frac{\overrightarrow{DB}}{\overrightarrow{DA}}\right)$$

$$\Rightarrow \angle ACB = \angle ADB$$

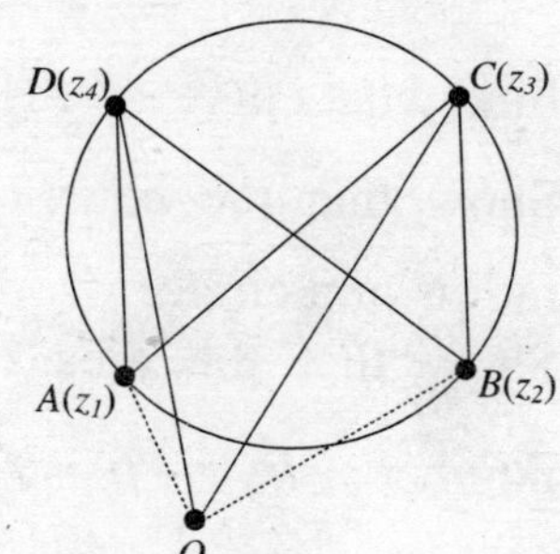

Figure 6.11

Since AB subtends equal angles at two points, C and D, on the same side of it.

$\therefore$ points $A, B, C, D,$ and hence z_1, z_2, z_3, z_4 are con-cyclic.

Example 6.10: If $A(a)$ and $B(b)$ are two fixed points in the Argand plane, find the locus of variable point $P(z)$ if amp. $\left(\frac{z-a}{z-b}\right) = k$ (constant).

Solution: Given amp. $\left(\frac{z-a}{z-b}\right) = k$

$$\Rightarrow \text{amp.}\left(\frac{\overrightarrow{OP}-\overrightarrow{OA}}{\overrightarrow{OP}-\overrightarrow{OB}}\right) = k$$

$$\Rightarrow \text{amp.}\left(\frac{\overrightarrow{AP}}{\overrightarrow{BP}}\right) = k$$

$$\Rightarrow \angle BPA = k$$

A(a) B(b)

Figure 6.12

EXERCISE 6.1

1. If $(x+iy)^{\frac{1}{3}} = a+ib$, show that $4\left(a^2-b^2\right) = \frac{x}{a}+\frac{y}{b}$.
2. Find the real values of x and y for which the following equation is satisfied $(x+iy)(2-3i) = 4+i$
3. Find the value of x, y if $\frac{x-1}{3+i}+\frac{y-1}{3-i} = i$.
4. Find what curve $z\bar{z}+(1+i)z+(1-i)\bar{z} = 0$ represents?
5. Show that the equation $|z-z_1|^2+|z-z_2|^2 = k$ represent a circle. Find centre and radius of a circle.
6. Show that the area of a triangle on the Argand diagram formed by complex number z, iz, $z+iz$ is $\frac{1}{2}|z|^2$.
7. If $z = x+iy$ and z moves on the circle $|z| = 1$, find the locus of $\frac{2}{z}$.
8. If $|z_1+z_2| = |z_1-z_2|$, prove that the difference of the amplitudes of z_1 and z_2 is $\frac{\pi}{2}$.
9. If z_1, z_2, z_3 are the vertices of an isosceles triangle right angled at the vertex z_2, prove

$$z_1^2+2z_2^2+z_3^2 = 2z_2(z_1+z_3)$$

10. What domain of the z-plane is represented by

$$|z+2|+|z-2|<4$$

ANSWERS

2. $x=\dfrac{5}{13}, y=\dfrac{14}{13}$

3. $x=-4, y=6$

4. Centre (-1,1) and radius = $\sqrt{2}$

7. Locus is a circle with centre (0,0) and radius 2

10. Region of complex plane below the real axis $y=0$

6.2 DE MOIVRE'S THEOREM

Statement: (1) If n is any integer, positive or negative, then

$$(\cos\theta+i\sin\theta)^n=\cos n\theta+i\sin n\theta$$

(2) If n is a fraction, positive or negative, then one of the values of

$$(\cos\theta+i\sin\theta)^n \text{ is } \cos n\theta+i\sin n\theta$$

Proof:

Case 1: When n is a positive integer.

We shall prove the theorem by induction method.

When n = 1, the theorem becomes

$$(\cos\theta+i\sin\theta)^1=\cos 1\theta+i\sin 1\theta$$

$$\Rightarrow \cos\theta+i\sin\theta=\cos\theta+i\sin\theta \quad \text{which is true,}$$

Let us suppose that the theorem is true for $n=m$.

$$\Rightarrow (\cos\theta+i\sin\theta)^m=(\cos m\theta+i\sin m\theta) \qquad \ldots(1)$$

Now put $n=m+1$

$$(\cos\theta + i\sin\theta)^{m+1} = (\cos\theta + i\sin\theta)^m (\cos\theta + i\sin\theta)$$

$$= (\cos m\theta + i\sin m\theta)(\cos\theta + i\sin\theta)$$

$$= (\cos m\theta\cos\theta - \sin m\theta\sin\theta) + i(\sin m\theta\cos\theta + \cos m\theta\sin\theta)$$

$$= \cos(m\theta + \theta) + i\sin(m\theta + \theta)$$

$$= \cos(m+1)\theta + i\sin(m+1)\theta$$

$\Rightarrow$ Thus, theorem is true for $n = m + 1$.

Hence, by the principle of mathematical induction, the theorem is true for all positive values of integer n.

Case 2: When n is a negative integer.

Let $n = -m$, where m is a positive integer.

$$\therefore \quad (\cos\theta + i\sin\theta)^n = (\cos\theta + i\sin\theta)^{-m}$$

$$\Rightarrow \quad \frac{1}{(\cos\theta + i\sin\theta)^m} = \frac{1}{\cos m\theta + i\sin m\theta} \qquad \text{[by case 1]}$$

$$= \frac{1}{\cos m\theta + i\sin m\theta} \times \frac{\cos m\theta - i\sin m\theta}{\cos m\theta - i\sin m\theta}$$

$$= \frac{\cos m\theta - i\sin m\theta}{(\cos m\theta)^2 - (i\sin m\theta)^2} = \frac{\cos m\theta - i\sin m\theta}{\cos^2 m\theta - i^2\sin^2 m\theta}$$

$$= \frac{\cos m\theta - i\sin m\theta}{\cos^2 m\theta + \sin^2 m\theta} \qquad \left[\because i^2 = -1\right]$$

$$= \cos m\theta - i\sin m\theta \qquad \left[\because \cos^2 m\theta + \sin^2 m\theta = 1\right]$$

$$= \cos(-m\theta) + i\sin(-m\theta) \qquad \left[\begin{array}{l}\because \cos(-m\theta) = \cos m\theta, \text{ and} \\ \sin(-m\theta) = -\sin m\theta\end{array}\right]$$

$$= \cos(-m)\theta + i\sin(-m)\theta = \cos n\theta + i\sin n\theta$$

This proves that the theorem is true for all negative values of integer n.

Case 3: When n is a fraction (positive or negative).

Let $n=\dfrac{p}{q}$, where q is a positive integer and p is any integer (positive or negative).

$$\therefore \quad \left(\cos\frac{\theta}{q}+i\sin\frac{\theta}{q}\right)^q=\left(\cos q\frac{\theta}{q}+i\sin q\frac{\theta}{q}\right)=\cos\theta+i\sin\theta \qquad \text{[by case 1]}$$

Taking q^{th} root of both sides, we get

$$\cos\frac{\theta}{q}+i\sin\frac{\theta}{q}=(\cos\theta+i\sin\theta)^{\frac{1}{q}}$$

Raising both sides to the pth power, we get

$$\left(\cos\frac{\theta}{q}+i\sin\frac{\theta}{q}\right)^p=(\cos\theta+i\sin\theta)^{\frac{p}{q}}$$

$$\Rightarrow \quad \cos p\frac{\theta}{q}+i\sin p\frac{\theta}{q}=(\cos\theta+i\sin\theta)^{\frac{p}{q}}$$

$$\Rightarrow \quad (\cos\theta+i\sin\theta)^{\frac{p}{q}}=\cos\left(\frac{p}{q}\theta\right)+i\sin\left(\frac{p}{q}\theta\right)$$

This proves that the theorem is true for all $n=\dfrac{p}{q}$, $q>0$.

Cor 1: $(\cos\theta+i\sin\theta)^{-n}=\cos(-n\theta)+i\sin(-n\theta)=\cos(n\theta)-i\sin(n\theta)$

Cor 2: $(\cos\theta-i\sin\theta)^{n}=\left[\cos(-\theta)+i\sin(-\theta)\right]^n=\cos(-n\theta)+i\sin(-n\theta)=\cos(n\theta)-i\sin(n\theta)$

Cor 3: $(\cos\theta-i\sin\theta)^{-n}=\left[\cos(-\theta)+i\sin(-\theta)\right]^{-n}=\cos(-(-n)\theta)+i\sin(-(-n)\theta)=\cos(n\theta)+i\sin(n\theta)$

Cor 4: $\dfrac{1}{\cos\theta+i\sin\theta}=(\cos\theta+i\sin\theta)^{-1}=\cos(-1\times\theta)+i\sin(-1\times\theta)=\cos(-\theta)+i\sin(-\theta)=\cos(\theta)-i\sin(\theta)$

Note: $(\text{cis}\,\theta_1)(\text{cis}\,\theta_2)(\text{cis}\,\theta_3)...(\text{cis}\,\theta_n)=\text{cis}\,(\theta_1+\theta_2+...+\theta_n)$ and $\dfrac{\text{cis}\,\theta_1}{\text{cis}\,\theta_2}=\text{cis}\,(\theta_1-\theta_2)$

ILLUSTRATIVE EXAMPLES 6.3

Example 6.11: Simplify $\dfrac{(\cos 3\theta + i\sin 3\theta)^4}{(\cos 4\theta - i\sin 4\theta)^3}$

Solution: $\dfrac{(\cos 3\theta + i\sin 3\theta)^4}{(\cos 4\theta - i\sin 4\theta)^3} = \dfrac{\cos 12\theta + i\sin 12\theta}{(\cos(-4\theta) + i\sin(-4\theta))^3} = \dfrac{\cos 12\theta + i\sin 12\theta}{\cos(-12\theta) + i\sin(-12\theta)} =$

$$= \frac{\text{cis}\,12\theta}{\text{cis}(-12\theta)} = \text{cis}(12\theta - (-12\theta)) = \text{cis}(12\theta + 12\theta)$$

$$= \text{cis}\,24\theta = \cos 24\theta + i\sin 24\theta$$

Example 6.12: Simplify $\dfrac{(\cos 2\theta + i\sin 2\theta)^4 (\cos\theta - i\sin\theta)^5}{(\cos 6\theta + i\sin 6\theta)^3 (\cos 4\theta - i\sin 4\theta)^7}$

Solution: $\dfrac{(\cos 2\theta + i\sin 2\theta)^4 (\cos\theta - i\sin\theta)^5}{(\cos 6\theta + i\sin 6\theta)^3 (\cos 4\theta - i\sin 4\theta)^7} = \dfrac{\text{cis}(8\theta)(\cos(-5\theta) + i\sin(-5\theta))}{\text{cis}(18\theta)(\cos(-28\theta) + i\sin(-28\theta))}$

$$= \frac{\text{cis}(8\theta)(\text{cis}(-5\theta))}{\text{cis}(18\theta)(\text{cis}(-28\theta))} = \frac{\text{cis}(8\theta - 5\theta)}{\text{cis}(18\theta - 28\theta)} = \frac{\text{cis}(3\theta)}{\text{cis}(-10\theta)}$$

$$= \text{cis}(3\theta - (-10\theta)) = \text{cis}(13\theta) \qquad \left[\because \frac{\text{cis}\,\theta_1}{\text{cis}\,\theta_2} = \text{cis}(\theta_1 - \theta_2)\right]$$

$$= \cos 13\theta + i\sin 13\theta$$

Example 6.13: If $x + \dfrac{1}{x} = 2\cos\theta$, prove that

$$x^n + \frac{1}{x^n} = 2\cos n\theta \quad \text{and} \quad x^n - \frac{1}{x^n} = \pm 2\sin n\theta$$

Solution: $x + \dfrac{1}{x} = 2\cos\theta \quad \Rightarrow \quad x^2 - 2x\cos\theta + 1 = 0$

$$\Rightarrow \quad x = \frac{-(-2\cos\theta) \pm \sqrt{4\cos^2 - 4}}{2} \qquad \left[\begin{array}{l}\because x^2 - 2x\cos\theta + 1 = 0 \text{ is a quadratic} \\ \text{equation in } x \text{ with } a = 1, b = -2\cos\theta, c = 1\end{array}\right]$$

$$= \frac{2\cos\theta \pm \sqrt{4(\cos^2 - 1)}}{2} \qquad \left[\begin{array}{l}\because \cos^2\theta + \sin^2\theta = 1 \\ \Rightarrow \cos^2\theta - 1 = -\sin^2\theta\end{array}\right]$$

$$=\frac{2\cos\theta \pm \sqrt{4\left(-\sin^2\theta\right)}}{2}=\frac{2\cos\theta \pm 2\sqrt{-1\times\sin^2\theta}}{2}$$

$$=\frac{2\cos\theta \pm 2i\sin\theta}{2}=\cos\theta \pm i\sin\theta$$

$$x=\cos\theta+i\sin\theta \qquad \text{[taking positive sign only]}$$

Now $x^n=(\cos\theta+i\sin\theta)^n=\cos n\theta+i\sin n\theta$

$$\frac{1}{x^n}=x^{-n}=(\cos\theta+i\sin\theta)^{-n}=\cos n\theta-i\sin n\theta$$

$$\therefore \quad x^n+\frac{1}{x^n}=(\cos n\theta+i\sin n\theta)+(\cos n\theta-i\sin n\theta)=2\cos n\theta$$

$$\therefore \quad x^n-\frac{1}{x^n}=(\cos n\theta+i\sin n\theta)-(\cos n\theta-i\sin n\theta)=2i\sin n\theta$$

Similarly, the result holds for negative sign also.

Example 6.14: If α, β are the roots of $x^2-2x+2=0$, then prove that

$$\frac{(t+\alpha)^n-(t+\beta)^n}{\alpha-\beta}=\frac{\sin n\phi}{\sin^n\phi}, \text{ where } t+1=\cot\phi.$$

Solution: $x^2-2x+2=0$

It is a quadratic equation with $a = 1$, $b = -2$, $c = 2$.

$$\therefore \quad x=\frac{-(-2)\pm\sqrt{(-2)^2-4\times1\times2}}{2}=\frac{2\pm\sqrt{4-8}}{2}=\frac{2\pm\sqrt{-4}}{2}=\frac{2\pm2i}{2}=1\pm i$$

$$\therefore \quad \alpha=1+i, \quad \beta=1-i$$

$$\text{Now } \frac{(t+\alpha)^n-(t+\beta)^n}{\alpha-\beta}=\frac{[t+(1+i)]^n-[t+(1-i)]^n}{(1+i)-(1-i)}=\frac{[t+1+i]^n-[t+1-i]^n}{2i}$$

$$=\frac{[\cot\phi+i]^n-[\cot\phi-i]^n}{2i} \qquad [\text{Given } t+1=\cot\phi]$$

$$=\frac{\left[\dfrac{\cos\phi}{\sin\phi}+i\right]^n-\left[\dfrac{\cos\phi}{\sin\phi}-i\right]^n}{2i}=\frac{\left[\dfrac{\cos\phi+i\sin\phi}{\sin\phi}\right]^n-\left[\dfrac{\cos\phi-i\sin\phi}{\sin\phi}\right]^n}{2i}$$

$$=\frac{\dfrac{(\cos\phi+i\sin\phi)^n}{\sin^n\phi}-\dfrac{(\cos\phi-i\sin\phi)^n}{\sin^n\phi}}{2i}=\frac{(\cos\phi+i\sin\phi)^n-(\cos\phi-i\sin\phi)^n}{2i\sin^n\phi}$$

$$=\frac{\cos n\phi+i\sin n\phi-(\cos n\phi-i\sin n\phi)}{2i\sin^n\phi}=\frac{\cos n\phi+i\sin n\phi-\cos n\phi+i\sin n\phi}{2i\sin^n\phi}$$

$$=\frac{2i\sin n\phi}{2i\sin^n\phi}=\frac{\sin n\phi}{\sin^n\phi}$$ Hence proved.

Example 6.15: Is $(\sin\theta+i\cos\theta)^n=\sin n\theta+i\cos n\theta$?

Solution: $(\sin\theta+i\cos\theta)^n\neq\sin n\theta+i\cos n\theta$

$$\because\quad(\sin\theta+i\cos\theta)^n=\left[\cos\left(\frac{\pi}{2}-\theta\right)+i\sin\left(\frac{\pi}{2}-\theta\right)\right]^n$$

$$=\cos n\left(\frac{\pi}{2}-\theta\right)+i\sin n\left(\frac{\pi}{2}-\theta\right)\qquad\text{[De Moivre's theorem]}$$

$$=\cos\left(\frac{n\pi}{2}-n\theta\right)+i\sin\left(\frac{n\pi}{2}-n\theta\right)\neq\sin n\theta+i\sin n\theta$$

Example 6.16: If $x_r=\cos\dfrac{\pi}{3^r}+i\sin\dfrac{\pi}{3^r}$, show that $\cos\left[\dfrac{\pi}{2}\left(1-\dfrac{1}{3^3}\right)\right]+i\sin\left[\dfrac{\pi}{2}\left(1-\dfrac{1}{3^3}\right)\right]$.

Also show that $x_1.x_2.x_3 \ldots x_n=i$.

Solution: Given $x_r=\cos\dfrac{\pi}{3^r}+i\sin\dfrac{\pi}{3^r}=\text{cis}\dfrac{\pi}{3^r}$

Put $r = 1, 2, 3, \ldots, n$

$$\therefore\quad x_1.x_2.x_3\ldots x_n=\text{cis}\frac{\pi}{3}\times\text{cis}\frac{\pi}{3^2}\times\text{cis}\frac{\pi}{3^3}\times\ldots\times\text{cis}\frac{\pi}{3^n}$$

$$= \text{cis}\left[\frac{\pi}{3}+\frac{\pi}{3^2}+\frac{\pi}{3^3}+...+\frac{\pi}{3^n}\right] \quad \left[\because \begin{array}{l}(\text{cis}\,\theta_1)(\text{cis}\,\theta_2)(\text{cis}\,\theta_3)...(\text{cis}\,\theta_n)\\ =\text{cis}(\theta_1+\theta_2+...+\theta_n)\end{array}\right]$$

$$= \text{cis}\frac{\pi}{3}\left[1+\frac{1}{3}+\frac{1}{3^2}+...+\frac{1}{3^{n-1}}\right]$$

Here $1+\frac{1}{3}+\frac{1}{3^2}+...+\frac{1}{3^{n-1}}$ is a *G.P.*

with first term = 1, and common ratio = $\frac{1}{3}$

$$\therefore \quad x_1.x_2.x_3...x_n = \text{cis}\frac{\pi}{3}\left[\frac{1.\left(1-\frac{1}{3^n}\right)}{1-\frac{1}{3}}\right] \quad \left[\because \begin{array}{l}\text{for a } G.P. \text{ of } n \text{ terms with intital term}\\ a \text{ and common ratio } r,\ s_n = \dfrac{a(1-r^n)}{1-r}\end{array}\right]$$

$$= \text{cis}\frac{\pi}{3}\left[\frac{\left(1-\frac{1}{3^n}\right)}{\frac{3-1}{3}}\right] = \text{cis}\frac{\pi}{3}\left[\frac{\left(1-\frac{1}{3^n}\right)}{\frac{2}{3}}\right] = \text{cis}\frac{\pi}{2}\left(1-\frac{1}{3^n}\right)$$

$$x_1.x_2.x_3...x_n = \cos\left[\frac{\pi}{2}\left(1-\frac{1}{3^n}\right)\right]+i\sin\left[\frac{\pi}{2}\left(1-\frac{1}{3^n}\right)\right],$$

which proves the first part.

When $n \to \infty,\quad \frac{1}{3^n} \to 0,$ *i.e.*,

$$x_1.x_2.x_3...\infty = \underset{n\to 0}{\text{Lt}}\ \text{cis}\frac{\pi}{2}\left(1-\frac{1}{3^n}\right) = \text{cis}\frac{\pi}{2} = \cos\frac{\pi}{2}+i\sin\frac{\pi}{2} = 0+i\times 1 = i$$

This proves the first part.

Example 6.17: If $x^2-2x\cos\theta+1=0$, show that $x^{2n}-2x^n\cos n\theta+1=0$.

Solution: Given $x^2-2x\cos\theta+1=0$

$\Rightarrow \quad x=\cos\theta \pm i\sin\theta$ [Already proved in Ex. (6.13)]

Let $x = \cos\theta + i\sin\theta$

Now $x^{2n} - 2x^n \cos n\theta + 1 = (\cos\theta + i\sin\theta)^{2n} - 2(\cos\theta + i\sin\theta)^n \cos n\theta + 1$

$$= \cos 2n\theta + i\sin 2n\theta - 2\cos n\theta(\cos n\theta + i\sin n\theta) + 1$$

$$= \cos 2n\theta + i\sin 2n\theta - 2\cos^2 n\theta - 2i\cos n\theta \sin n\theta + 1$$

$$= \left[\cos 2n\theta - 2\cos^2 n\theta + 1\right] + i\left[\sin 2n\theta - 2\cos n\theta \sin n\theta\right]$$

$$= \left[(1 + \cos 2n\theta) - 2\cos^2 n\theta\right] + i\left[\sin 2n\theta - \sin 2n\theta\right]$$

$$= \left[\cos^2 n\theta - 2\cos^2 n\theta\right] + i \times 0 \qquad \left[\begin{array}{l}\because\ 2\sin n\theta \cos n\theta = \cos 2n\theta, \\ 1 + \cos 2n\theta = 2\cos^2 n\theta\end{array}\right]$$

$= 0 + 0 = 0 = R.H.S.$ Hence proved.

Example 6.18: If $x = \cos\theta + i\sin\theta$, and $y = \cos\phi + i\sin\phi$, then show that

$$\frac{x-y}{x+y} = i\tan\frac{\theta-\phi}{2}.$$ [PTU Dec 2012]

Solution:

$$\frac{x-y}{x+y} = \frac{(\cos\theta + i\sin\theta) - (\cos\phi + i\sin\phi)}{(\cos\theta + i\sin\theta) + (\cos\phi + i\sin\phi)} = \frac{\cos\theta + i\sin\theta - \cos\phi - i\sin\phi}{\cos\theta + i\sin\theta + \cos\phi + i\sin\phi}$$

$$= \frac{(\cos\theta - \cos\phi) + i(\sin\theta - \sin\phi)}{(\cos\theta + \cos\phi) + i(\sin\theta + \sin\phi)}$$

$$= \frac{\left(-2\sin\frac{\theta+\phi}{2}\sin\frac{\theta-\phi}{2}\right) + i\left(2\cos\frac{\theta+\phi}{2}\sin\frac{\theta-\phi}{2}\right)}{\left(2\cos\frac{\theta+\phi}{2}\cos\frac{\theta-\phi}{2}\right) + i\left(2\sin\frac{\theta+\phi}{2}\cos\frac{\theta-\phi}{2}\right)}$$

$$= \frac{2\sin\frac{\theta-\phi}{2}\left(-\sin\frac{\theta+\phi}{2} + i\cos\frac{\theta+\phi}{2}\right)}{2\cos\frac{\theta-\phi}{2}\left(\cos\frac{\theta+\phi}{2} + i\sin\frac{\theta+\phi}{2}\right)} = \frac{\sin\frac{\theta-\phi}{2}\left(i^2\sin\frac{\theta-\phi}{2} + i\cos\frac{\theta+\phi}{2}\right)}{\cos\frac{\theta-\phi}{2}\left(\cos\frac{\theta-\phi}{2} + i\sin\frac{\theta+\phi}{2}\right)}$$

$$\left[\because\ i^2 = -1\right]$$

$$= \frac{i\sin\frac{\theta-\phi}{2}\left(\cos\frac{\theta+\phi}{2}+i\sin\frac{\theta+\phi}{2}\right)}{\cos\frac{\theta-\phi}{2}\left(\cos\frac{\theta+\phi}{2}+i\sin\frac{\theta+\phi}{2}\right)} = i\tan\frac{\theta-\phi}{2}$$

EXERCISE 6.2

1. Prove that $(1+\cos\theta+i\sin\theta)^n+(1+\cos\theta-i\sin\theta)^n = 2^{n+1}\cos^n\left(\frac{\theta}{2}\right)\cos\left(\frac{n\theta}{2}\right)$.

2. Prove that $\left(\sqrt{3}+i\right)^n+\left(\sqrt{3}-i\right)^n = 2^{n+1}\cos\left(\frac{n\pi}{6}\right)$.

3. Prove that $(1+i)^n+(1-i)^n = 2^{\frac{n}{2}+1}\cos\left(\frac{n\pi}{4}\right)$.

4. Show that $\frac{(\cos 7\theta - i\sin 7\theta)^2(\cos 3\theta+i\sin 3\theta)^4}{(\cos 5\theta+i\sin 5\theta)^3(\cos\theta - i\sin\theta)^{17}} = 1$.

5. Show that $(\sin\theta - i\cos\theta)^5 = \sin 5\theta - i\cos 5\theta$.

6. If $2\cos\theta = x+\frac{1}{x}$ and $2\cos\phi = y+\frac{1}{y}$, prove that $2\cos(\theta+\phi) = xy+\frac{1}{xy}$.

7. If $a = \text{cis}\,\alpha$, $a = \text{cis}\,\beta$, $c = \text{cis}\,\gamma$, show that $\frac{ab}{c}+\frac{c}{ab} = 2\cos(\alpha+\beta+\gamma)$.

8. If $x+\frac{1}{x} = 2\cos\theta$ and $y+\frac{1}{y} = 2\cos\phi$, show that

   (*a*) $x^m y^n + \frac{1}{x^m y^n} = 2\cos(m\theta+n\phi)$ (*b*) $\frac{x^m}{y^n}+\frac{y^n}{x^m} = 2\cos(m\theta - n\phi)$

9. If $\sin\phi = i\tan\theta$, show that $\text{cis}\,\theta = \pm\tan\left(45^\circ+\frac{\phi}{2}\right)$.

10. If $x_r = \cos\frac{\pi}{2^r}+i\sin\frac{\pi}{2^r}$. Prove that $x_1.x_2.x_3\ldots\infty = -1$.

11. If $\sin\alpha+\sin\beta+\sin\gamma = 0 = \cos\alpha+\cos\beta+\cos\gamma$, prove that

    (*a*) $\cos^2\alpha+\cos^2\beta+\cos^2\gamma = \frac{3}{2}$

    (*b*) $\cos 3\alpha+\cos 3\beta+\cos 3\gamma = 3\cos(\alpha+\beta+\gamma)$

6.3 APPLICATIONS OF DE-MOIVRE'S THEOREM

De-Moivre's theorem can be used to find all *n*-roots of a complex number.

Since $\cos\theta = \cos(2k\pi+\theta)$ and $\sin\theta = \sin(2k\pi+\theta)$, where *k* is an integer.

From De-Moivre's theorem

$$(\cos\theta + i\sin\theta)^{1/n} = [\cos(2k\pi+\theta) + i\sin(2k\pi+\theta)]^{1/n}$$

$$\Rightarrow \quad \boxed{(\cos\theta + i\sin\theta)^{1/n} = \cos\left(\frac{2k\pi+\theta}{n}\right) + i\sin\left(\frac{2k\pi+\theta}{n}\right)}$$

By putting $k = 0, 1, 2, 3, \ldots, n-1,$ we get *n*-roots of the complex number.

Some Standard Results

$$1 = \cos 0 + i\sin 0 \qquad -1 = \cos\pi + i\sin\pi$$

$$i = \cos\frac{\pi}{2} + i\sin\frac{\pi}{2} \qquad -i = \cos\frac{\pi}{2} - i\sin\frac{\pi}{2}$$

$$1+i = \sqrt{2}\left(\cos\frac{\pi}{4} + i\sin\frac{\pi}{4}\right) \qquad -i = \cos\frac{\pi}{2} - i\sin\frac{\pi}{2}$$

Example 6.19: Prove that the *n* *n*th roots of a unity are in geometric progression.

Solution: Consider the equation $z^n = 1$

$$z^n = \cos 0 + i\sin 0 = \cos 2k\pi + i\sin 2k\pi$$

$$\Rightarrow \quad z = (\cos 2k\pi + i\sin 2k\pi)^{1/n} = \cos\frac{2k\pi}{n} + i\sin\frac{2k\pi}{n}, \text{ where } k = 0, 1, 2, \ldots, n-1$$

By putting $k = 0, 1, 2, 3, \ldots, n-1,$ we get *n*-roots of the complex number.

For $k = 0, \quad z_0 = 1$

For $k = 1, \quad z_1 = \cos\frac{2\pi}{n} + i\sin\frac{2\pi}{n} = \omega$ (say)

For $k=2$, $z_2=\cos\frac{4\pi}{n}+i\sin\frac{4\pi}{n}=\left(\cos\frac{2\pi}{n}+i\sin\frac{2\pi}{n}\right)^2=\omega^2$ (say)

For $k=3$, $z_3=\cos\frac{6\pi}{n}+i\sin\frac{6\pi}{n}=\left(\cos\frac{2\pi}{n}+i\sin\frac{2\pi}{n}\right)^3=\omega^3$ (say)

$\vdots$

For $k=n-1$, $z_{n-1}=\cos\frac{2(n-1)\pi}{n}+i\sin\frac{2(n-1)\pi}{n}=\left(\cos\frac{2\pi}{n}+i\sin\frac{2\pi}{n}\right)^{n-1}=\omega^{n-1}$ (say)

Hence, the roots are $1,\omega,\omega^2,\omega^3,\ldots,\omega^{n-1}$ which are in *G.P.*

$\therefore$ The values are $\pm\frac{1}{\sqrt{2}}\pm i\frac{1}{\sqrt{2}}$

Example 6.20: Solve $z^6-i=0$.

Solution: Consider the equation $z^n=1$

$$z^6=i=\cos\frac{\pi}{2}+i\sin\frac{\pi}{2}=\cos\left(2k\pi+\frac{\pi}{2}\right)+i\sin\left(2k\pi+\frac{\pi}{2}\right)$$

$$\Rightarrow \quad z=\left[\cos\left(2k\pi+\frac{\pi}{2}\right)+i\sin\left(2k\pi+\frac{\pi}{2}\right)\right]^{1/6}$$

$$\Rightarrow \quad z=\left[\cos(4k+1)\frac{\pi}{2}+i\sin(4k+1)\frac{\pi}{2}\right]^{1/6}$$

$$\Rightarrow \quad z=\cos(4k+1)\frac{\pi}{12}+i\sin(4k+1)\frac{\pi}{12}=\text{cis}(4k+1)\frac{\pi}{12}$$

By putting $k=0,1,2,3,4,5$ we get the roots as

$$\text{cis}\frac{\pi}{12},\ \text{cis}\frac{5\pi}{12},\ \text{cis}\frac{9\pi}{12},\ \text{cis}\frac{13\pi}{12},\ \text{cis}\frac{17\pi}{12},\ \text{cis}\frac{21\pi}{12}$$

Example 6.21: Solve the equation $z^4-z^3+z^2-z+1=0$ using De-Moivre's theorem.

Solution: Multiplying the equation by $z+1$, we get

$$(z+1)\left(z^4-z^3+z^2-z+1\right)=0$$

$$\Rightarrow \quad z^5-z^4+z^3-z^2+z+z^4-z^3+z^2-z+1=0$$

$$\Rightarrow \quad z^5+1=0$$

$$\Rightarrow \quad z^5=-1 \qquad [\because \cos\pi+i\sin\pi=-1]$$

$$=\cos\pi+i\sin\pi$$

$$=\cos(2k+1)\pi+i\sin(2k+1)\pi$$

$$\Rightarrow \quad z=\left[\cos(2k+1)\pi+i\sin(2k+1)\pi\right]^{1/5}$$

$$\Rightarrow \quad z=\cos(2k+1)\frac{\pi}{5}+i\sin(2k+1)\frac{\pi}{5}=\text{cis}(2k+1)\frac{\pi}{5}$$

By putting $k=0,1,2,3,4$ we get the roots as

$$\text{cis}\frac{\pi}{5},\ \text{cis}\frac{3\pi}{5},\ \text{cis}\frac{5\pi}{5},\ \text{cis}\frac{7\pi}{5},\ \text{cis}\frac{9\pi}{5}$$

Example 6.22: Find all the values of $\left(\frac{1}{2}+\frac{\sqrt{3}}{2}i\right)^{3/4}$. [PTU Dec 2012]

Solution: $\frac{1}{2}+i\frac{\sqrt{3}}{2}=r(\cos\theta+i\sin\theta)=r\cos\theta+ir\sin\theta$

Equating real and imaginary parts, we get

$$r\cos\theta=\frac{1}{2},\quad r\sin\theta=\frac{\sqrt{3}}{2}$$

Squaring and adding, we get

$$r^2(\cos^2\theta+\sin^2\theta)=\frac{1}{4}+\frac{3}{4}=1 \qquad \Rightarrow \qquad r=1$$

Dividing imaginary part by real part, we get

$$\frac{r\sin\theta}{r\cos\theta}=\frac{\sqrt{3}/2}{1/2}=\sqrt{3} \qquad \Rightarrow \qquad \tan\theta=\sqrt{3} \qquad \Rightarrow \qquad \theta=\frac{\pi}{3}$$

$$\therefore \quad \frac{1}{2}+i\frac{\sqrt{3}}{2}=\cos\frac{\pi}{3}+i\sin\frac{\pi}{3}$$

$$\Rightarrow \quad \left(\frac{1}{2}+i\frac{\sqrt{3}}{2}\right)^{3/4}=\left(\cos\frac{\pi}{3}+i\sin\frac{\pi}{3}\right)^{3/4}=\left[\cos\left(2k\pi+\frac{\pi}{3}\right)+i\sin\left(2k\pi+\frac{\pi}{3}\right)\right]^{3/4}$$

$$k=0,1,2,3$$

Apply De-Moivre's theorem, we get

$$=\left[\cos\frac{3}{4}\left(2k\pi+\frac{\pi}{3}\right)+i\sin\frac{3}{4}\left(2k\pi+\frac{\pi}{3}\right)\right] \quad k=0,1,2,3$$

$$=\left[\cos\left(\frac{3}{2}k\pi+\frac{\pi}{4}\right)+i\sin\left(\frac{3}{2}k\pi+\frac{\pi}{4}\right)\right] \quad k=0,1,2,3$$

The values are given below:

For $k=0$, value is $\cos\frac{\pi}{4}+i\sin\frac{\pi}{4}=\frac{1}{\sqrt{2}}+i\frac{1}{\sqrt{2}}$

For $k=1$, value is $\cos\frac{7\pi}{4}+i\sin\frac{7\pi}{4}=\cos\left(2\pi-\frac{\pi}{4}\right)+i\sin\left(2\pi-\frac{\pi}{4}\right)$

$$=\cos\frac{\pi}{4}-i\sin\frac{\pi}{4}=\frac{1}{\sqrt{2}}-i\frac{1}{\sqrt{2}}$$

For $k=2$, value is $\cos\frac{13\pi}{4}+i\sin\frac{13\pi}{4}=\cos\left(3\pi+\frac{\pi}{4}\right)+i\sin\left(3\pi+\frac{\pi}{4}\right)$

$$=-\cos\frac{\pi}{4}-i\sin\frac{\pi}{4}=-\frac{1}{\sqrt{2}}-i\frac{1}{\sqrt{2}}$$

For $k=3$, value is $\cos\frac{19\pi}{4}+i\sin\frac{19\pi}{4}=\cos\left(5\pi-\frac{\pi}{4}\right)+i\sin\left(5\pi-\frac{\pi}{4}\right)$

$$=-\cos\frac{\pi}{4}+i\sin\frac{\pi}{4}=-\frac{1}{\sqrt{2}}+i\frac{1}{\sqrt{2}}$$

$\therefore$ roots are $\pm\frac{1}{\sqrt{2}}\pm i\frac{1}{\sqrt{2}}$

Example 6.23: Use De Moiver's theorem to solve the equation $(z-1)^5+z^5=0$.

Solution: The given equation is $(z-1)^5+z^5=0$

$$\Rightarrow (z-1)^5=-z^5 \quad \Rightarrow \quad \frac{(z-1)^5}{z^5}=-1=\cos\pi+i\sin\pi$$

$$\Rightarrow \left(\frac{z-1}{z}\right)^5=\cos\pi+i\sin\pi \quad \Rightarrow \quad \frac{z-1}{z}=(\cos\pi+i\sin\pi)^{\frac{1}{5}}$$

Adding $2k\pi$ in the angles of both cosine and sin e, we get

$$1-\frac{1}{z}=\left[\cos(2k\pi+\pi)+i\sin(2k\pi+\pi)\right]^{\frac{1}{5}}, \qquad k=0,1,2,3,4$$

Apply De Moiver's theorem, we get

$$1-\frac{1}{z}=\cos\frac{(2k+1)}{5}\pi+i\sin\frac{(2k+1)}{5}\pi, \qquad k=0,1,2,3,4.$$

Putting $\left(\frac{2k+1}{5}\right)\pi=\theta$, we get

$$1-\frac{1}{z}=\cos\theta+i\sin\theta \quad \Rightarrow \quad 1-\cos\theta+i\sin\theta=\frac{1}{z}$$

$$\Rightarrow z=\frac{1}{1-\cos\theta-i\sin\theta}=\frac{1}{2\sin^2\frac{\theta}{2}-i2\sin\frac{\theta}{2}\cos\frac{\theta}{2}}$$

$$\Rightarrow z=\frac{1}{2\sin\frac{\theta}{2}\left(\sin\frac{\theta}{2}-i\cos\frac{\theta}{2}\right)}=\frac{i}{2\sin\frac{\theta}{2}\left(\cos\frac{\theta}{2}+i\sin\frac{\theta}{2}\right)}$$

$$\Rightarrow z=\frac{i}{2\sin\frac{\theta}{2}}\left(\cos\frac{\theta}{2}+i\sin\frac{\theta}{2}\right)^{-1}=\frac{i}{2\sin\frac{\theta}{2}}\left(\cos\frac{\theta}{2}-i\sin\frac{\theta}{2}\right)^{-1}$$

$$\Rightarrow z=\frac{1}{2}\left(i\cot\frac{\theta}{2}+1\right)=\frac{1}{2}\left[i\cot(2k+1)\frac{\pi}{10}+1\right], k=0,1,2,3,4$$

EXERCISE 6.X

1. Find all the roots of the following:

   (a) $(-1)^{1/5}$ (b) $(-i)^{1/3}$

   (c) $(1+i)^{2/3}$ (d) $(-8i)^{1/3}$ [PTU May 2012]

2. Solve the following equations:

   (a) $z^{12}-1=0$ (b) $z^7+z^4+z^3+1=0$

   (c) $z^7-z^4+z^3-1=0$ (d) $z^7-z^4-z^3-1=0$

3. Find all the values of $(1-i\sqrt{3})^{1/4}$

4. Use De-Moivre's theorem to find the roots of $z^5+1=0$.

5. Use De Moivre's theorem to find all the roots of the equation $z^4-(1-z)^4=0$.

6. Solve the equation $z^4+z^3+z^2+z+1=0$ using De Moivre's theorem.
   [**Hint:** Multiply by $(z-1)$]

7. Prove that $\sqrt[n]{a+ib}+\sqrt[n]{a-ib}$ has *n* real values.

8. Prove that $(a+ib)^{m/n}+(a-ib)^{m/n}=2\left(\sqrt{a^2+b^2}\right)^{m/n}\cos\left(\frac{m\theta}{n}\right)$, where $\theta=\tan^{-1}\left(\frac{b}{a}\right)$.

ANSWERS

1. (a) $-1,\ \text{cis}\frac{\pi}{5},\ \text{cis}\frac{3\pi}{5}$ (b) $i,\ \pm\frac{\sqrt{3}}{2}-\frac{i}{2}$

   (c) $2,\ 2\text{cis}\frac{2\pi}{5},\ 2\text{cis}\frac{4\pi}{5}$ (d) $z=2\text{cis}(4k-1)\frac{\pi}{12},\ k=0,1,2$

2. (a) $\pm1,\ \pm i,\ \pm\text{cis}\frac{\pi}{6},\ \pm\text{cis}\frac{\pi}{3}$ (b) $\pm\left(\frac{1}{\sqrt{2}}\pm i\frac{1}{\sqrt{2}}\right),\ \frac{1}{\sqrt{2}}\pm i\frac{\sqrt{3}}{2},\ -1$

   (c) $1,\ -\frac{1}{2}\left(1\pm i\sqrt{3}\right),\ \pm\left(\frac{1}{\sqrt{2}}\pm i\frac{1}{\sqrt{2}}\right)$ (d) $\pm1,\ \pm i,\ -\frac{1}{2}\left(1\pm i\sqrt{3}\right)$

3. $z=2^{1/4}\text{cis}(6k-1)\frac{\pi}{12},\ k=0,1,2,3$

4. $z=\text{cis}\frac{2k+1}{5}\pi$, where $k=0,1,2,3,4$

5. $z=\frac{1}{2}\left[1-i\tan\left(\frac{k\pi}{4}\right)\right],\ k=0,1,2,3$

6. $\cos\frac{2\pi}{5}\pm i\sin\frac{2\pi}{5},\ \cos\frac{4\pi}{5}\pm i\sin\frac{4\pi}{5}$

6.4 EXPONENTIAL FUNCTION OF A COMPLEX VARIABLE

We know that when x is real

$$e^x \text{ or } \exp(x) = 1 + x + \frac{x^2}{2!} + \frac{x^3}{3!} +, \ldots, + \frac{x^n}{n!} + \ldots\infty$$

Similarly, we can define the exponential function of the complex variable $z = x + iy$ as

$$e^z \text{ or } \exp(z) = 1 + z + \frac{z^2}{2!} + \frac{z^3}{3!} +, \ldots, + \frac{z^n}{n!} + \ldots\infty \qquad \ldots (1)$$

Properties:

I. Exponential form of $z(= x + iy) = re^{i\theta}$ $\qquad \left[e^{i\theta} = \cos\theta + i\sin\theta\right]$

Putting $x = 0$ in Eq. (1), we get

$$e^{iy} = 1 + iy + \frac{(iy)^2}{2!} + \frac{(iy)^3}{3!} + \frac{(iy)^4}{4!} + \ldots\infty$$

$$= \left(1 - \frac{y^2}{2!} + \frac{y^4}{4!} \mp \ldots\infty\right) + i\left(y - \frac{y^3}{3!} + \frac{y^5}{5!} \mp \ldots\infty\right) = \cos y + i\sin y$$

Thus $e^z = e^x . e^{iy} = e^x(\cos y + i\sin y)$

Also $x + iy = r(\cos\theta + i\sin\theta) = re^{i\theta}$

II. e^z is a periodic function having period $2\pi i$.

Since $e^z = e^{x+iy} = e^x e^{iy} = e^x(\cos y + i\sin y)$

$$= e^x(\cos(2n\pi + y) + i\sin(2n\pi + y))$$

$$= e^x \text{cis}(2n\pi + y) = e^x e^{i(y+2n\pi)} = e^{x+iy+2n\pi i}$$

$$= e^{z+2n\pi i}$$

III. e^z does not vanish for any value of z, *i.e.*, its value is never zero for any value of z.

Since $e^z = e^{x+iy} = re^{i\theta} \quad \Rightarrow \quad e^x e^{iy} = re^{i\theta}$

$\therefore\ r = e^x > 0,\ y = \theta,\ \left|e^{iy}\right| = 1$

Thus $\left|e^z\right| = \left|e^x\right|\left|e^{iy}\right| = e^x \neq 0$

IV. $e^{\bar{z}} = \overline{e^z}$

Since $e^{\bar{z}} = e^{x-iy} = e^x e^{-iy} = e^x(\cos y - i\sin y) = e^x\overline{\cos y - i\sin y}$

$= \overline{e^z} = \text{R.H.S}$

6.5 LOGARITHMIC FUNCTION OF A COMPLEX VARIABLE

If $z = x + iy$ and $w = u + iv$ be so rotated that $e^w = z$, then w is said to be a logarithm of z to the base e and is written as

$$w = \log_e z \qquad \ldots (1)$$

Also $e^{w+2n\pi i} = e^w . e^{2n\pi i} = z \quad \Rightarrow \quad z = e^{w+2n\pi i}$

$$\therefore \quad \log_e z = w + 2n\pi i \qquad \ldots (2)$$

Which means that logarithm of a complex number has an infinite number of values and is, therefore, a multi-valued function.

The general value of the logarithm of z is written as Log z (beginning with capital L) so as to distinguish it from its principal value which is written as log z. This principal value is obtained by taking $n = 0$ in Log z.

Thus, from Eqs. (1) and (2), we have

$$\boxed{\text{Log}(x+iy) = 2n\pi i + \log(x+iy)}$$

Some Observations

1. If $y = 0$, then $\text{Log}(x) = 2n\pi i + \log x$. This shows that the logarithm of a real quantity is also multi-valued. Its principal value is real while all other values are imaginary.
2. We know that the logarithm of a negative quantity has no real value. But we can now evaluate this, as shown below:

$$\log_e(-z) = \log_e(z(-1)) = \log_e z + \log_e(-1)$$

$$= \log_e z + i\pi \qquad \left[\because\ -1 = \cos\pi + i\sin\pi = e^{i\pi}, \text{ and } \log_e e^{i\pi} = i\pi\right]$$

$$= 0.6931 + i(3.1416)$$

3. Real and imaginary parts of $\log(x+iy)$.

$$\left[\text{Put } x = r\cos,\ y = r\sin\theta, \text{ so that } r = \sqrt{x^2+y^2} \text{ and } \theta = \tan^{-1}\left(\frac{y}{x}\right)\right]$$

$$\text{Log}(x+iy) = 2n\pi i + \log(x+iy)$$

$$= 2n\pi i + \log\left[r(\cos\theta + i\sin\theta)\right] = 2n\pi i + \log\left(re^{i\theta}\right)$$

$$= 2n\pi i + \log r + \log e^{i\theta} = 2n\pi i + \log r + i\theta$$

$$= \log\sqrt{x^2+y^2} + i\left[2n\pi + \tan^{-1}\left(\frac{y}{x}\right)\right]$$

Therefore, $\text{Re}\left[\text{Log}(x+iy)\right] = \log\sqrt{x^2+y^2}$, and $\text{Im}\left[\text{Log}(x+iy)\right] = 2n\pi + \tan^{-1}\left(\frac{y}{x}\right)$

4. Real and imaginary parts of $(\alpha+i\beta)^{x+iy}$.

$$\left[\text{Put } \alpha = r\cos,\ \beta = r\sin\theta, \text{ so that } r = \sqrt{\alpha^2+\beta^2} \text{ and } \theta = \tan^{-1}\left(\frac{\beta}{\alpha}\right)\right]$$

$$(\alpha+i\beta)^{x+iy} = e^{x+iy}\text{Log}(\alpha+i\beta)$$

$$= e^{x+iy}\left[2n\pi i + \log(\alpha+i\beta)\right]$$

$$= e^{x+iy}\left[2n\pi i + \log\left(re^{i\theta}\right)\right] = e^{x+iy}\left[2n\pi i + \log r + \log e^{i\theta}\right]$$

$$= e^x e^{iy}\left[2n\pi i + \log r + i\theta\right] \qquad \left[\because\ e^{iy} = \cos y + i\sin y;\ \log e^x = x\right]$$

$$= e^x(\cos y + i\sin y)[\log r + i(2n\pi + \theta)]$$

$$= e^x\left[\cos y\log r + i(2n\pi+\theta)\cos y + i\sin y\log r + i^2(2n\pi+\theta)\sin y\right]$$

$$= e^x\left[\cos y\log r + i(2n\pi+\theta)\cos y + i\sin y\log r - (2n\pi+\theta)\sin y\right]$$

$$= e^x\left[\cos y\log\sqrt{\alpha^2+\beta^2} - (2n\pi+\theta)\sin y\right]$$

$$+ ie^x\left[(2n\pi+\theta)\cos y + \sin y\log\log\sqrt{\alpha^2+\beta^2}\right]$$

$$\text{Real Part} = e^x\left[\cos y \log\sqrt{\alpha^2+\beta^2} - (2n\pi+\theta)\sin y\right]$$

$$\text{Imaginary Part} = e^x\left[(2n\pi+\theta)\cos y + \sin y \log\log\sqrt{\alpha^2+\beta^2}\right]$$

ILLUSTRATIVE EXAMPLES 6.4

Example 6.24: Show that (*i*) $\exp(2\pm 3\pi i) = -e^2$ (*ii*) $\exp\left(\frac{2+\pi i}{4}\right) = \sqrt{\frac{e}{2}}(1+i)$

Solution: (*i*) $\text{LHS} = \exp(2\pm 3\pi i) = e^{2\pm 3\pi i} = e^2.e^{\pm 3\pi i} = e^2\left[\cos(\pm 3\pi) + i\sin(\pm 3\pi)\right]$

$$= e^2\left[\cos(3\pi) + i\times 0\right] \qquad \left[\because \cos(\pm\theta) = \cos\theta,\ \sin n\theta = 0\right]$$

$$= e^2[-1+0] \qquad \left[\because \cos(3\pi) = -1\right]$$

$$= -e^2 = \text{RHS}$$

(*ii*) $\text{LHS} = \exp\left(\frac{2+\pi i}{4}\right) = e^{\frac{2}{4}+\frac{\pi i}{4}} = e^{\frac{1}{2}}.e^{\frac{\pi i}{4}}$

$$= e^{\frac{1}{2}}\left(\cos\frac{\pi}{4} + i\cos\frac{\pi}{4}\right)$$

$$= e^{\frac{1}{2}}\left[\frac{1}{\sqrt{2}} + i\times\frac{1}{\sqrt{2}}\right] \qquad \left[\because \cos\left(\frac{\pi}{4}\right) = \sin\left(\frac{\pi}{4}\right) = \frac{1}{\sqrt{2}}\right]$$

$$= \frac{1}{\sqrt{2}}e^{\frac{1}{2}}(1+i) = \sqrt{\frac{e}{2}}(1+i) = \text{RHS}$$

Example 6.25: For any integer r, prove that $\sum_{k=1}^{n}\left[\exp\left(\frac{2\pi ik}{n}\right)\right]^r$ is n or zero according as $\frac{r}{\pi}$ is or is not an integer.

Solution: Let $S = \sum_{k=1}^{n}\left[\exp\left(\frac{2\pi ik}{n}\right)\right]^r = \sum_{k=1}^{n}\exp\left(\frac{2\pi ikr}{n}\right)$ $\left[\begin{array}{l}\because (\exp z)^r = (e^z)^r \\ \quad = e^{rz} = (\exp z)^r\end{array}\right]$

$$=\sum_{k=1}^{n}\left[\exp\left(\frac{2\pi ir}{n}\right)\right]^{k} \qquad \left[\because \exp(i\theta)=e^{i\theta} = \cos\theta + i\sin\theta = \text{cis}\,\theta\right]$$

$$=\sum_{k=1}^{n}\left[\text{cis}\left(\frac{2\pi r}{n}\right)\right]^{k}$$

$$=\sum_{k=1}^{n}\alpha^{k} \quad \text{where } \alpha = \text{cis}\left(\frac{2\pi r}{n}\right)$$

$$S=\alpha+\alpha^{2}+\alpha^{3}+\ldots+\alpha^{n} \qquad \ldots (1)$$

Case I: When $\frac{r}{n}$ is an integer.

Let $\frac{r}{n}=m$, where $m \in Z$

$\therefore \quad \alpha = cis(2m\pi) = \cos(2m\pi) + i\sin(2m\pi) = 1 + i\times 0 = 1$

From Eq. (1), we have

$$S = 1+(1)^{2}+(1)^{2}+\ldots+(1)^{n} = n \qquad \ldots (2)$$

Case II: When $\frac{r}{n}$ is not an integer, then $\alpha = \text{cis}\left(\frac{2\pi r}{n}\right)$.

$\therefore \quad 1-\alpha \neq 0 \quad \text{and} \quad \alpha^{\text{n}} = \text{cis}(2\pi r) = 1$

From Eq. (2), we have

$$S = \frac{\alpha(1-\alpha^{n})}{1-\alpha} = \frac{\alpha(1-1)}{1-\alpha} = 0$$

$$\text{Hence } S = \sum_{k=1}^{n}\left[\exp\left(\frac{2\pi ik}{n}\right)\right]^{r} = \begin{cases} n & \text{if } \frac{r}{n} \in Z \\ 0 & \text{if } \frac{r}{n} \notin Z \end{cases}$$

Example 6.26: Find the general value of $\log(-i)$.

Solution: The general value of

$$\text{Log}(-i) = 2n\pi i + \log(-i) \qquad \left[\because -i = -\cos\frac{\pi}{2} - i\sin\frac{\pi}{2} = e^{-\frac{\pi i}{2}}\right]$$

$$= 2n\pi i + \log(-i) = 2n\pi i + \log\left(e^{-\frac{\pi}{2}i}\right)$$

$$= 2n\pi i - \frac{\pi}{2} i \log e \;= 2n\pi i - \frac{\pi}{2} i \qquad [\because \log e = 1]$$

$$= (4n -- 1)\frac{\pi}{2} i$$

Example 6.27: Prove that i^i is wholly real and find its principal value. Also show that value of i^i form a G.P.

or

Prove that $\log i^i = -\left(2n + \frac{1}{2}\right)\pi$.

Solution: Here $i^i = e^{\text{Log}\, i^i} = e^{i\text{Log}\, i} = e^{i(2n\pi i + \log i)}$

$$= e^{i\left(2n\pi i + \log e^{\frac{\pi}{2}i}\right)} = e^{i\left(2n\pi i + \frac{\pi}{2}i\right)} \qquad \left[\because e^{\frac{\pi}{2}i} = \cos\left(\frac{\pi}{2}\right) + i\sin\left(\frac{\pi}{2}\right) = i\right]$$

$$= e^{\left(2n\pi i^2 + \frac{\pi}{2}i^2\right)} = e^{\left(-2n\pi - \frac{\pi}{2}\right)} = e^{-\left(2n\pi + \frac{\pi}{2}\right)} = e^{-4(n+1)\frac{\pi}{2}} \qquad [\because i^2 = -1]$$

Taking logarithm on both sides, we get

$$\log\left(i^i\right) = \log\left(e^{-4(n+1)\frac{\pi}{2}}\right) = -4(n+1)\frac{\pi}{2} = -\left(2n + \frac{1}{2}\right)\pi$$

The principal value is obtained by replacing $n = 0$, i.e.,

$$i^i = e^{-\frac{\pi}{2}}$$

Now putting $n = 0, 1, 2, \ldots$

The values of i^i are $e^{-\frac{\pi}{2}}, e^{-\frac{5\pi}{2}}, e^{-\frac{9\pi}{2}}, \ldots$ which is a G.P. series with common ratio $r = e^{-2\pi}$.

Example 6.28: If $i^{i^{i..\infty}} = A + iB$, prove that $A^2 + B^2 = e^{-\pi B}$ and $\tan\left(\frac{\pi}{2}A\right) = \frac{B}{A}$.

Solution: Here $i^{i^{i..\infty}} = A + iB$

$$\Rightarrow \quad i^{A+iB} = A + iB$$

$$\Rightarrow \quad \left(\cos\frac{\pi}{2} + i\sin\frac{\pi}{2}\right)^{A+iB} = A + iB \qquad \left[\because \cos\frac{\pi}{2} + i\sin\frac{\pi}{2} = i\right]$$

$$\Rightarrow \quad \left(e^{i\frac{\pi}{2}}\right)^{A+iB} = A + iB$$

$$\Rightarrow \quad e^{i\frac{\pi}{2}A - \frac{\pi}{2}B} = A + iB \qquad \Rightarrow \qquad e^{i\frac{\pi}{2}A} \cdot e^{-\frac{\pi}{2}B} = A + iB$$

$$\Rightarrow \quad e^{-\frac{\pi}{2}B}\left[\cos\left(\frac{\pi}{2}A\right) + i\sin\left(\frac{\pi}{2}A\right)\right] = A + iB$$

Comparing real and imaginary parts

$$e^{-\frac{\pi}{2}B}\cos\left(\frac{\pi}{2}A\right) = A \qquad \ldots(1)$$

$$e^{-\frac{\pi}{2}B}\sin\left(\frac{\pi}{2}A\right) = B \qquad \ldots(2)$$

Squaring (1) and (2), and adding, we get

$$A^2 + B^2 = \left[e^{-\frac{\pi}{2}B}\cos\left(\frac{\pi}{2}A\right)\right]^2 + \left[e^{-\frac{\pi}{2}B}\sin\left(\frac{\pi}{2}A\right)\right]^2$$

$$= e^{-\pi B}\cos^2\left(\frac{\pi}{2}A\right) + e^{-\pi B}\sin^2\left(\frac{\pi}{2}B\right)$$

$$= e^{-\pi B}\left[\cos^2\left(\frac{\pi}{2}A\right) + \sin^2\left(\frac{\pi}{2}B\right)\right] = e^{-\pi B}$$

$$\Rightarrow \quad A^2 + B^2 = e^{-\pi B}$$

Dividing (2) by (1), we get

$$\frac{e^{-\frac{\pi}{2}B}\sin\left(\frac{\pi}{2}A\right)}{e^{-\frac{\pi}{2}B}\cos\left(\frac{\pi}{2}A\right)}=\frac{B}{A} \quad\Rightarrow\quad \tan\left(\frac{\pi}{2}A\right)=\frac{B}{A}$$

Example 6.29*a*: Prove that $\sin\left(\log i^{i}\right)=-1$.

Solution: LHS $= \sin\left(\log i^{i}\right)=\sin\left(i\log i\right)$ $\quad\left[\because\ i=\cos\left(\frac{\pi}{2}\right)+i\sin\left(\frac{\pi}{2}\right)=e^{\frac{\pi}{2}i}\right]$

$$=\sin\left(i\log e^{\frac{\pi}{2}i}\right)$$

$$=\sin\left(i\times\frac{\pi}{2}i\log e\right)=\sin\left(i^{2}\times\frac{\pi}{2}\times 1\right)$$

$$=\sin\left(-\frac{\pi}{2}\right)=-1=\text{ RHS}$$

Hence, the result is proved.

Example 6.29*b*: Prove that $\text{Log}i^{i}=\frac{4m+1}{4n+1}$, m,n are integers..

Solution: LHS $= \text{Log}_{i}i=\frac{\text{Log}i}{\text{Log}i}=\frac{2m\pi i+\log i}{2n\pi i+\log i}=\frac{2m\pi i+\log e^{\frac{\pi}{2}i}}{2n\pi i+\log e^{\frac{\pi}{2}i}}$ $\quad\left[\because\ \text{Log}_{a}b=\frac{\text{Log}b}{\text{Log}b}\right]$

$$=\frac{2m\pi i+\frac{\pi}{2}i}{2n\pi i+\frac{\pi}{2}i}=\frac{(4m+1)\frac{\pi}{2}i}{(4n+1)\frac{\pi}{2}i}$$

$$=\frac{(4m+1)}{(4n+1)},\ m,\ n\notin Z=\text{RHS}$$

Hence, the result is proved.

EXERCISE 6.3

1. Split up into real and imaginary parts:

   (*a*) $e^{5+\frac{1}{2}i\pi}$ (*b*) $e^{(5+3i)^2}$

2. Prove that $\exp(z+\pi i)=-\exp(z)$.

3. Prove that $|\exp(-2z)|<1$ iff $\text{Re}(z)>0$.

4. Show that $\left|\exp(z^2)\right|\le\left|\exp(|z|^2)\right|$.

5. Show that

   (*a*) $\text{Log}(i)=(4n+1)\frac{\pi}{2}i$ (*b*) $\text{Log}(1-i)=\frac{1}{2}\log z-\frac{\pi}{4}i$

6. Verify that

   (*a*) $\text{Log}(e)=1+2n\pi i$ (*b*) $\text{Log}(i)=\left(2n+\frac{1}{2}\right)\pi i$

   (c) $\text{Log}\left(-1+\sqrt{3}i\right)=\log 2+2\left(n+\frac{1}{3}\right)\pi i$

7. Show that

   (*a*) $\text{Log}(1+i)^2=2\log(1+i)$ (*b*) $\text{Log}(-1+i)^2\ne 2\log(-1+i)$

8. Show that the principle value of $(-i)^i$ is $\exp\left(\frac{\pi}{2}\right)$.

9. Prove that

   (*a*) $(-i)^{-i}=e^{(4n-1)\frac{\pi}{2}}$ (*b*) $i^{\sqrt{2}}=\text{cis}\frac{\sqrt{2}\pi}{2}$

10. Prove that

    (*a*) $\left((i)^i\right)^i=\cos\theta-i\sin\theta$, where $\theta=(4n+1)\frac{\pi}{2}$, $n\in I$

    (*b*) $(i)^{i^2}=\cos\theta+i\sin\theta$, where $\theta=\frac{1}{2}(4m+1)\pi e^{-(4n+1)\frac{\pi}{2}}$, $m,n\in I$

11. Find the general value of $\log(-3)$.

12. Prove that $\tan\left(i\log\frac{a-ib}{a+ib}\right)=\frac{2ab}{a^2-b^2}$.

13. If $\tan\left(\log\left(x+iy\right)\right)=a+ib$ and $a^2+b^2\neq 1$, prove that $\tan\left(\log\left(x^2+y^2\right)\right)=\dfrac{2a}{1-a^2-b^2}$.

ANSWERS

1. (a) Real part = 0; Imaginary part = e^5

   (b) Real part = $\dfrac{\sqrt{3}}{2}e^{16}$; Imaginary part = $\dfrac{e^{16}}{2}$

11. $\log 3+i(2n+1)\pi$; Imaginary part = 0

6.6 CIRCULAR FUNCTIONS

For a real number x, we have proved that $\cos x=\dfrac{e^{ix}+e^{-ix}}{2},\ \sin x=\dfrac{e^{ix}-e^{-ix}}{2i}$.

Similarly, we can prove for any complex number z.

$$\therefore\quad \cos z=\frac{e^{iz}+e^{-iz}}{2},\quad \sin z=\frac{e^{iz}-e^{-iz}}{2i}$$

Now, $$\tan z=\frac{\sin z}{\cos z}=\frac{e^{iz}-e^{-iz}}{i(e^{iz}+e^{-iz})},\quad \cot z=\frac{1}{\tan z}=\frac{i(e^{iz}+e^{-iz})}{e^{iz}-e^{-iz}}$$

$$\sec z=\frac{1}{\cos z}=\frac{2}{e^{iz}+e^{-iz}},\quad \operatorname{cosec} z=\frac{1}{\sin z}=\frac{2i}{e^{iz}-e^{-iz}}$$

Again, $$e^z=1+z+\frac{z^2}{2!}+\frac{z^3}{3!}+\ldots$$

$$e^{iz}=1+iz+\frac{(iz)^2}{2!}+\frac{(iz)^3}{3!}+\ldots$$

$$e^{-iz}=1-iz+\frac{(-iz)^2}{2!}+\frac{(-iz)^3}{3!}+\ldots$$

$$\therefore\quad \cos z=\frac{e^{iz}+e^{-iz}}{2}=\frac{1}{2}\left[2-\frac{2.z^2}{2!}+\frac{2.z^4}{4!}\mp\ldots\right]=1-\frac{z^2}{2!}+\frac{z^4}{4!}\mp\ldots$$

$$\therefore\quad \sin z=\frac{e^{iz}-e^{-iz}}{2i}=\frac{1}{2i}\left[2iz-\frac{iz^3}{3!}+\frac{iz^5}{5!}\mp\ldots\right]=z-\frac{z^3}{3!}+\frac{z^5}{5!}\mp\ldots$$

Properties

1. Euler's Theorem

By definition $\cos z + i\sin z = \dfrac{e^{iz}+e^{-iz}}{2} + i\dfrac{e^{iz}-e^{-iz}}{2i} = e^{iz}$, where $z = x+iy$.

Also we know that $e^{iy} = \cos y + i\sin y$, where y is real.

Hence $e^{i\theta} = \cos\theta + i\sin\theta$, where θ is real or complex. This is called Euler's theorem.

2. De-Moivre's theorem for complex number.

For θ real or complex,

$$(\cos\theta + i\sin\theta)^n = \left(e^{i\theta}\right)^n = e^{ni\theta} = \cos n\theta + i\sin n\theta$$

Hence De-Moivre's theorem is true for all θ.

3. Periods of circular functions, whether real or complex, we know that

$$\sin z = \frac{e^{iz}-e^{-iz}}{2i}$$

$$\therefore\ \sin(z+2\pi n) = \frac{e^{i(z+2\pi n)} - e^{-i(z+2\pi n)}}{2i} = \frac{e^{iz}e^{2\pi n} - e^{-iz}e^{-2\pi n}}{2i}$$

$$= \frac{e^{iz}\times 1 - e^{-iz}\times 1}{2i} = \frac{e^{iz}-e^{-iz}}{2i} = \sin z$$

$$\left[\begin{aligned} \because\ e^{2n\pi i} &= \text{cis}(2\pi n) \\ &= \cos 2n\pi + i\sin 2n\pi \\ &= (-1)^2 + i.0 = 1 \end{aligned}\right]$$

Thus, $\sin z$ remains unchanged when z is increased by any multiple of 2π. Hence $\sin z$ is periodic with period 2π. Similarly, $\cos z$, $\sec z$, $\operatorname{cosec} z$ are all periodic functions of 2π.

$i \qquad i$

Again by definition, $\tan z = \dfrac{e^{z} - e^{-z}}{i\left(e^{iz} + e^{-iz}\right)}$

$$\therefore\ \tan(z+n\pi) = \frac{e^{i(z+n\pi)} - e^{-i(z+n\pi)}}{i\left[e^{i(z+n\pi)} + e^{-i(z+n\pi)}\right]} = \frac{e^{iz}e^{in\pi} - e^{-iz}e^{-in\pi}}{i\left[e^{iz}e^{in\pi} + e^{-iz}e^{-in\pi}\right]} = \frac{e^{iz}e^{2n\pi i} - e^{-iz}}{i\left[e^{iz}e^{2n\pi i} + e^{-iz}\right]}$$

Multiply the numerator and denominator by $e^{in\pi}$, we get

$$\tan(z+n\pi) = \frac{e^{iz}\times 1 - e^{-iz}}{i\left[e^{iz}\times 1 + e^{-iz}\right]} = \frac{e^{iz} - e^{-iz}}{i\left[e^{iz} + e^{-iz}\right]} = \tan z$$

Thus, $\tan z$ remains unchanged when z is increased by any multiple of π.

$\therefore$ $\tan z$ is periodic with period π.

Similarly, $\cot z$ is periodic with period π.

Hence, the complex circular functions are all periodic.

6.7 HYPERBOLIC FUNCTION

We defined $\sinh x = \dfrac{e^{x} - e^{-x}}{2}$; $\cosh x = \dfrac{e^{x} + e^{-x}}{2}$

Also $\tanh x = \dfrac{\sinh x}{\cosh x} = \dfrac{e^{x} - e^{-x}}{e^{x} + e^{-x}}$; $\coth x = \dfrac{1}{\tanh x} = \dfrac{\cosh x}{\sinh x} = \dfrac{e^{x} + e^{-x}}{e^{x} - e^{-x}}$

$\operatorname{sech} x = \dfrac{1}{\cosh x} = \dfrac{2}{e^{x} + e^{-x}}$; $\operatorname{cosech} x = \dfrac{1}{\sinh x} = \dfrac{2}{e^{x} - e^{-x}}$

Remarks:

1. $\sinh 0 = \dfrac{e^{0} - e^{-0}}{2} = \dfrac{1-1}{2} = 0$

2. $\cosh 0 = \dfrac{e^{0} + e^{-0}}{2} = \dfrac{1+1}{2} = 1$

3. $\cosh x + \sinh x = \dfrac{e^{x} + e^{-x}}{2} + \dfrac{e^{x} - e^{-x}}{2} = \dfrac{2e^{x}}{2} = e^{x}$

4. $\cosh x - \sinh x = \dfrac{e^{x} + e^{-x}}{2} - \dfrac{e^{x} - e^{-x}}{2} = \dfrac{2e^{-x}}{2} = e^{-x}$

6.7.1 Relations between Hyperbolic and Circular Functions

1. $\cos(i\theta) = \cosh\theta$
2. $\sin(i\theta) = i\sinh\theta$
3. $\tan(i\theta) = \dfrac{\sin(i\theta)}{\cos(i\theta)} = \dfrac{i\sinh\theta}{\cosh\theta} = i\tanh\theta$
4. $\cot(i\theta) = \dfrac{\cos(i\theta)}{\sin(i\theta)} = \dfrac{\cosh\theta}{i\sinh\theta} = -i\coth\theta$
5. $\sec(i\theta) = \dfrac{1}{\cos(i\theta)} = \dfrac{1}{\cosh\theta} = \text{sech}\,\theta$
6. $\text{cosec}(i\theta) = \dfrac{1}{\sin(i\theta)} = \dfrac{1}{i\sinh} = -i\text{cosech}\,\theta$

$$\left[\because \frac{1}{i} = -i\right]$$

6.7.2 Important Formulae

Following are some important formulae. The proof of these formulae is beyond the scope of the present book. We demonstrate their use in solving variety of problems.

1. (*a*) $\cosh^2 x - \sinh^2 x = 1$
   (*b*) $\text{sech}^2 x + \tanh^2 x = 1$
   (*c*) $\coth^2 x - \text{cosec}\,\text{h}^2 x = 1$
2. (*a*) $\sinh 2x = 2\sinh x\cosh x = \dfrac{2\tanh x}{1-\tanh^2 x}$
   (*b*) $\cosh 2x = \cosh^2 x + \sinh^2 x = 2\cosh^2 x - 1 = 1 + \sinh^2 x = \dfrac{1+\tanh^2 x}{1-\tanh^2 x}$
3. (*a*) $\sinh 3x = 3\sinh x - 4\sinh^3 x$
   (*b*) $\cosh 3x = 4\cosh^3 x - 3\cosh x$
   (*c*) $\tanh 3x = \dfrac{3\tanh x + \tanh^3 x}{1 - 3\tanh^2 x}$

ILLUSTRATIVE EXAMPLES 6.5

Example 6.30: Prove that (*i*) $\sin 3z = 3\sin z - 4\sin^3 z$ (*ii*) $\cos(-z) = \cos z$

Solution: (*i*) RHS $= 3\sin z - 4\sin^3 z$

$$= 3\left(\frac{e^{iz} - e^{-iz}}{2i}\right) - 4\left(\frac{e^{iz} - e^{-iz}}{2i}\right)^3 = \frac{e^{iz} - e^{-iz}}{2i}\left[3 - 4\left(\frac{e^{iz} - e^{-iz}}{2i}\right)^2\right]$$

$$= \frac{e^{iz}-e^{-iz}}{2i}\left[3-4\left(\frac{e^{2iz}+e^{-2iz}-2\left(e^{2iz}\right)\left(e^{-2iz}\right)}{-4}\right)\right]$$

$$= \frac{e^{iz}-e^{-iz}}{2i}\left[3+e^{2iz}+e^{-2iz}-2\right] = \frac{e^{iz}-e^{-iz}}{2i}\left[e^{2iz}+e^{-2iz}+1\right]$$

$$= \frac{e^{iz}-e^{-iz}}{2i}\left[e^{2iz}+e^{-2iz}+1\right]$$

$$= \frac{\left[e^{iz}.e^{2iz}+e^{iz}.e^{-2iz}+e^{iz}-e^{-iz}.e^{2iz}-e^{-iz}.e^{-2iz}-e^{-iz}\right]}{2i}$$

$$= \frac{\left[e^{3iz}+e^{-iz}+e^{iz}-e^{iz}-e^{-3iz}-e^{-iz}\right]}{2i} = \frac{\left[e^{3iz}-e^{-3iz}\right]}{2i}$$

$= \sin 3z = \text{LHS}$ Hence proved.

(*ii*) $\text{LHS} = \cos(-z) = \frac{e^{i(-z)}+e^{-i(-z)}}{2} = \frac{e^{-iz}+e^{iz}}{2} = \frac{e^{iz}+e^{-iz}}{2} = \cos z = \text{RHS}$

Hence proved.

Example 6.31: Separate the following into real and imaginary parts:

(*i*) $\sin(x+iy)$ (*ii*) $\cosh(x+iy)$

Solution: (*i*) $\sin(x+iy) = \sin x\cos(iy)+\cos x\sin(iy)$

$$= \sin x\cosh y+\cos x(i\sinh y)$$

$$= \sin x\cosh y+i\cos x\sinh y$$

$\therefore$ $\text{Re}\left[\sin(x+iy)\right] = \sin x\cos y$, $\text{Im}\left[\sin(x+iy)\right] = \cos x\sinh y$

(*ii*) $\cosh(x+iy) = \cos\left[i(x+iy)\right]$ $\left[\because \cosh(\theta)=\cos(i\theta)\right]$

$$= \cos(ix-y)$$

$$= \cos(ix)\cos y+\sin(ix)\sin y = \cosh x\cos y+i\sinh x\sin y$$

$\therefore$ $\text{Re}\left[\cosh(x+iy)\right] = \cosh x\cos y$, $\text{Im}\left[\cosh(x+iy)\right] = \sinh x\sin y$

Example 6.32: If $\cosh x = \sec\theta$, prove that $\tanh^2 \frac{x}{2} = \tan^2 \frac{\theta}{2}$.

Solution: Given $\cosh x = \sec\theta = \frac{1}{\cos\theta}$

$$\Rightarrow \quad \frac{1+\tanh^2 \frac{x}{2}}{1-\tanh^2 \frac{x}{2}} = \frac{1+\tan^2 \frac{\theta}{2}}{1-\tan^2 \frac{\theta}{2}}$$

Applying componendo and dividendo, we get

$$\frac{\left(1+\tanh^2 \frac{x}{2}\right)+\left(1-\tanh^2 \frac{x}{2}\right)}{\left(1-\tan^2 \frac{x}{2}\right)-\left(1-\tanh^2 \frac{x}{2}\right)} = \frac{\left(1+\tan^2 \frac{\theta}{2}\right)+\left(1+\tan^2 \frac{\theta}{2}\right)}{\left(1+\tan^2 \frac{\theta}{2}\right)-\left(1-\tan^2 \frac{\theta}{2}\right)}$$

$$\Rightarrow \quad \frac{2}{2\tanh^2 \frac{x}{2}} = \frac{2}{2\tan^2 \frac{\theta}{2}}$$

$$\Rightarrow \quad \tanh^2 \frac{x}{2} = \tan^2 \frac{\theta}{2}$$

Hence proved.

Example 6.33: If $\tan(A+iB) = x+iy;\ A, B, x, y \in E$. Prove that $x^2 + y^2 + 2x\cot(2A) = 1$.

Solution: Given $\tan(A+iB) = x+iy$ . . . (1)

Changing i to $-i$, we get

$$\tan(A-iB) = x-iy \qquad \text{. . . (2)}$$

Now we can write

$$2A = (A+iB)+(A-iB)$$

$$\Rightarrow \quad \tan(2A) = \tan\left[(A+iB)+(A-iB)\right]$$

$$= \frac{\tan(A+iB)+\tan(A-iB)}{1-\tan(A+iB)\tan(A-iB)} = \frac{(x+iy)+(x-iy)}{1-(x+iy)(x-iy)}$$

$$= \frac{(x+iy)+(x-iy)}{1-(x+iy)(x-iy)} = \frac{2x}{1-\left[x^2-(iy)^2\right]} = \frac{2x}{1-\left[x^2-i^2y^2\right]}$$

$$= \frac{2x}{1-x^2-y^2}$$

$$\therefore \cot(2A) = \frac{1}{\tan(2A)} = \frac{1-x^2-y^2}{2x} \Rightarrow 2x\cot(2A) = 1-x^2-y^2$$

$$\Rightarrow x^2+y^2+2x\cot(2A)=1$$

Hence the proof.

EXERCISE 6.4

1. Separate into real and imaginary parts

   (*a*) $\exp(z^2)$, where $z = x+iy$ (*b*) $\exp\left(\sin\frac{\pi}{2}\right)$

2. If z is any complex number, show that

   (*a*) $\cos(2z) = \cos^2 z - \sin^2 z$ (*b*) $\tan(2z) = \frac{2\tan z}{1-\tan^2 z}$

3. If z_1 and z_2 are any complex number, show that

   (*a*) $\sin(z_1 \pm z_2) = \sin z_1 \cos z_2 \pm \cos z_1 \sin z_2$

   (*b*) $\cos(z_1+z_2) = \cos z_1 \cos z_2 - \sin z_1 \sin z_2$

   (*c*) $\cos(z_1-z_2) = \cos z_1 \cos z_2 + \sin z_1 \sin z_2$

   (*d*) $\tan(z_1+z_2) = \frac{\tan z_1 + \tan z_2}{1-\tan z_1 \tan z_2}$

   (*e*) $\tan(z_1-z_2) = \frac{\tan z_1 - \tan z_2}{1+\tan z_1 \tan z_2}$

4. If $\cosh(u+vi) = x+iy$, prove that

   (*a*) $\frac{x^2}{\cosh^2 u} + \frac{y^2}{\sinh^2 u} = 1$ (*b*) $\frac{x^2}{\cos^2 v} - \frac{y^2}{\sin^2 v} = 1$

5. If $\tanh(x+iy) = \theta + i\phi$; $x, y, \theta, \phi \in R$, prove that

   $$\theta^2+\phi^2 = \frac{\cosh^2 y - \cos^2 x}{\cosh^2 y - \sin^2 x}$$

6. If $\tanh(u+iv) = x+iy$; $x, y, u, v \in R$, show that

   $$\tan(2u) = \frac{2x}{1-x^2-y^2} \text{ and } \tan(2v) = \frac{2y}{1-x^2-y^2}$$

7. If $\tan\frac{x}{2} = \tanh\frac{x}{2}$, show that $\cos x \cosh x = 1$.

8. If $\sin\phi = i\tan\theta$, prove that $\cos\theta + i\sin\theta = \tan\left(\frac{\pi}{4}+\frac{\phi}{2}\right)$.

9. Prove that $(\cosh x+\sinh x)^n=\cosh(nx)+\sinh(nx),\ n\in Z^+$.

10. Show that $\left(\dfrac{1+\tanh x}{1-\tanh x}\right)^3=\cosh(6x)+\sinh(6x)$.

11. If $\tan(x+iy)=\sin(u+iv)$, prove that

$$\frac{\sin(2x)}{\sinh(2y)}=\frac{\tan u}{\tanh u}$$

12. Show that $\sinh u=\tan\theta,\ \tanh u=\ \sin(n\theta)$.

13. If $\sin^{-1}(u+iv)=\alpha+i\beta$; then show that $\sin^2\alpha$ and $\cosh^2\beta$ are the roots of the equation $x^2-(1+u^2+v^2)x+u^2=0$.

14. If $u=\log\left[\tan\left(\dfrac{\pi}{4}+\dfrac{\theta}{2}\right)\right]$, then prove that

(*i*) $\tanh u=\sin\theta$
(*ii*) $\sinh u=\tan\theta$

6.8 INVERSE TRIGNOMETRIC FUNCTIONS

As for a real variable *x*, we can define inverse *sine* function as

$$y=\sin^{-1}x,\ \text{when } x=\sin y$$

Similarly, we can define inverse *sine* function for a complex variable *z* as

$$w=\sin^{-1}z,\ \text{when } z=\sin w$$

Now $\quad z=\dfrac{e^{iw}-e^{-iw}}{2i}$

$\Rightarrow \quad 2iz=e^{iw}-e^{-iw}$

Solving for e^{iw}

$$2iz=e^{iw}-\frac{1}{e^{-iw}}$$

$\Rightarrow \quad 2ize^{iw}=e^{i2w}-1$

$\Rightarrow \quad e^{i2w}-2ize^{iw}-1=0$

which is a quadratic equation in e^{iw} with $a=1, b=-2iz$ and $c=-1$, solving which gives

$$e^{iw}=\frac{2iz\pm\sqrt{4i^2z^2+4}}{2}=\frac{2iz\pm\sqrt{4\left(i^2z^2+1\right)}}{2}=\frac{2iz\pm2\sqrt{\left(-z^2+1\right)}}{2}$$

$$=\frac{2\left[iz\pm\sqrt{\left(1-z^2\right)}\right]}{2}=iz\pm\sqrt{1-z^2}$$

Now taking logarithm on both sides, we get

$$\log\left(e^{iw}\right)=\log\left(iz\pm\sqrt{1-z^2}\right)\quad\Rightarrow\quad iw=\log\left(iz\pm\sqrt{1-z^2}\right)$$

$$\Rightarrow\quad w=\frac{1}{i}\log\left(iz\pm\sqrt{1-z^2}\right)$$

$$\therefore\quad w=\sin^{-1}z=-i\log\left(iz\pm\sqrt{1-z^2}\right)$$

Since $\sin^{-1}z$ is defined for all values of z except

$$iz+\sqrt{1-z^2}=0$$

$$\Rightarrow\quad iz=-\sqrt{1-z^2}$$

Squaring both sides, we get

$$(iz)^2=\left(-\sqrt{1-z^2}\right)^2\quad\Rightarrow\quad i^2z^2=1-z^2\quad\Rightarrow\quad -z^2=1-z^2$$

$\Rightarrow\ 0=1$ which is impossible.

$$\therefore\quad \sin^{-1}z=-i\log\left(iz+\sqrt{1-z^2}\right)$$

Similarly, other complex functions are defined by the following:

$$\cos^{-1}z=-i\log\left(z+\sqrt{1-z^2}\right)$$

$$\tan^{-1}z=\frac{-i}{2}\log\left(\frac{1+iz}{1-iz}\right)=\frac{i}{2}\log\left(\frac{i+z}{i-z}\right);\ z\neq\pm i$$

$$\operatorname{cosec}^{-1} z = \sin^{-1}\frac{1}{z} = -i\log\left(i\frac{1}{z} + \sqrt{1-\frac{1}{z^2}}\right)$$

$$= -i\log\left(i\frac{1}{z} + \sqrt{1-\frac{1}{z^2}}\right) = -i\log\left(\frac{i+\sqrt{z^2-1}}{z}\right);\ z \neq 0$$

$$\sec^{-1} z = \cos^{-1}\frac{1}{z} = -i\log\left(\frac{1}{z} + \sqrt{1-\frac{1}{z^2}}\right) = -i\log\left(\frac{1+\sqrt{z^2-1}}{z}\right);\ z \neq 0$$

$$\cot^{-1} z = \tan^{-1}\frac{1}{z} = \frac{-i}{2}\log\left(\frac{1+i\frac{1}{z}}{1-i\frac{1}{z}}\right) = \frac{-i}{2}\log\left(\frac{z+i}{z-i}\right);\ z \neq \pm i$$

6.9 INVERSE HYPERBOLIC FUNCTIONS

As for a real variable x, we can define inverse *sine* function as

Let $\sinh^{-1} z = w \Rightarrow z = \sinh w$

Now $\quad z = \dfrac{e^w - e^{-w}}{2} \Rightarrow 2z = e^w - e^{-w} \Rightarrow 2z = e^w - \dfrac{1}{e^w}$

$$\Rightarrow 2ze^w = e^{2w} - 1 \Rightarrow e^{2w} - 2ze^w - 1 = 0$$

which is a quadratic equation in e^w with $a = 1,\ b = -2z,\ c = -1$, solving which gives

$$e^w = \frac{2z \pm \sqrt{4z^2+4}}{2} = \frac{2z \pm \sqrt{4(z^2+1)}}{2} = \frac{2z \pm 2\sqrt{(z^2+1)}}{2}$$

$$= \frac{2\left[z \pm \sqrt{(z^2+1)}\right]}{2} = z \pm \sqrt{(z^2+1)}$$

which gives $\boxed{\sinh^{-1} z = \log\left(z + \sqrt{(z^2+1)}\right)}$

Let $\cosh^{-1} z = w \Rightarrow z = \cosh w$

Now $\quad z = \dfrac{e^w + e^{-w}}{2} \Rightarrow 2z = e^w - e^{-w} \Rightarrow 2z = e^w + \dfrac{1}{e^w}$

$$\Rightarrow \quad 2ze^{w} = e^{2w} + 1 \quad \Rightarrow \quad e^{2w} - 2ze^{w} + 1 = 0$$

which is a quadratic equation in e^{w} with $a = 1,\ b = -2z,\ c = 1$, solving which gives

$$e^{w} = \frac{2z \pm \sqrt{4z^2 - 4}}{2} = \frac{2z \pm \sqrt{4(z^2 - 1)}}{2} = \frac{2z \pm 2\sqrt{(z^2 - 1)}}{2}$$

$$= \frac{2\left[z \pm \sqrt{(z^2 - 1)}\right]}{2} = z \pm \sqrt{(z^2 - 1)}$$

which gives $\boxed{\cosh^{-1} z = \log\left(z + \sqrt{(z^2 - 1)}\right)}$

Let $\tanh^{-1} z = w \;\Rightarrow\; z = \tanh w$

Now $\Rightarrow \quad z = \dfrac{e^{w} - e^{-w}}{e^{w} - e^{-w}} \quad \Rightarrow \quad \dfrac{z}{1} = \dfrac{e^{w} - e^{-w}}{e^{w} - e^{-w}}$

Applying componendo and dividendo, we get

$$\frac{1 + z}{1 - z} = \frac{(e^{w} - e^{-w}) + (e^{w} + e^{-w})}{(e^{w} - e^{-w}) - (e^{w} + e^{-w})} = \frac{2e^{w}}{2e^{-w}} = e^{2w}$$

$$e^{2w} = \frac{1 + z}{1 - z} \quad \Rightarrow \quad 2w = \log\frac{1 + z}{1 - z};\ z \neq 1$$

$$\Rightarrow \quad w = \frac{1}{2}\log\frac{1 + z}{1 - z};\ z \neq 1$$

which gives $\boxed{\tanh^{-1} z = \dfrac{1}{2}\log\left(\dfrac{1 + z}{1 - z}\right)}$

Now $\operatorname{cosech}^{-1} z = \sinh^{-1}\dfrac{1}{z} = \log\left(\dfrac{1 + \sqrt{1 + z^2}}{z}\right);\ z \neq 0$

$$\operatorname{sech}^{-1} z = \cosh^{-1}\frac{1}{z} = \log\left(\frac{1 + \sqrt{1 - z^2}}{z}\right);\ z \neq 0$$

$$\coth^{-1} z = \tanh^{-1}\frac{1}{z} = \frac{1}{2}\log\left(\frac{z + 1}{z - 1}\right);\ z \neq \pm 1$$

ILLUSTRATIVE EXAMPLES 6.6

Example 6.34: If $\tan(A+iB)=x+iy$, find real and imaginary parts.

Solution: Let $\tan^{-1}(x+iy)=u+iv$ ... (1)

$$\tan^{-1}(x-iy)=u-iv \qquad \text{... (2)}$$

Adding (1) and (2), we get

$$2u=\tan^{-1}(x+iy)+\tan^{-1}(x-iy)=\tan^{-1}\frac{(x+iy)+(x-iy)}{1-(x+iy)(x-iy)}=\tan^{-1}\frac{2x}{1-x^2-y^2}$$

$$\Rightarrow \quad u=\frac{1}{2}\tan^{-1}\frac{2x}{1-x^2-y^2}$$

Subtracting (2) from (1), we get

$$2iv=\tan^{-1}(x+iy)-\tan^{-1}(x-iy)=\tan^{-1}\frac{(x+iy)-(x-iy)}{1+(x+iy)(x-iy)}=\tan^{-1}\frac{2iy}{1+x^2+y^2}$$

$$\Rightarrow \quad \tan(2iv)=\frac{2iy}{1+x^2+y^2} \quad \Rightarrow \quad i\tanh(2v)=\frac{2iy}{1+x^2+y^2}$$

$$\Rightarrow \quad \tanh(2v)=\frac{2y}{1+x^2+y^2} \quad \Rightarrow \quad 2v=\tanh^{-1}\frac{2y}{1+x^2+y^2} \Rightarrow v=\frac{1}{2}\tanh^{-1}\frac{2y}{1+x^2+y^2}$$

$$\therefore \quad \text{Re}\left[\tan^{-1}(x+iy)\right]=\frac{1}{2}\tan^{-1}\frac{2x}{1-x^2-y^2}, \text{ and}$$

$$\text{Im}\left[\tan^{-1}(x+iy)\right]=\frac{1}{2}\tanh^{-1}\frac{2y}{1+x^2+y^2}$$

Example 6.35: If $\sin^{-1}(u+iv)=\alpha+i\beta$, prove that $\sin^2\alpha$ and $cash^2\beta$ are the roots of the equation $x^2-(1+u^2+v^2)x+u^2=0$.

Solution: Since $\sin^{-1}(u+iv)=\alpha+i\beta$

$$\therefore \quad u+iv=\sin(\alpha+i\beta)=\sin\alpha\cos(i\beta)+\sin(i\beta)\cos\alpha$$

$$=\sin\alpha\cosh\beta+i\sinh\beta\cos\alpha$$

Comparing real and imaginary parts on both sides of the equations, we get

$$\sin\alpha\cosh\beta = u \qquad \ldots(1)$$

$$\sinh\beta\cos\alpha = v \qquad \ldots(2)$$

$$1+u^2+v^2 = 1+\sin^2\alpha\cosh^2\beta+\cos^2\alpha\sinh^2\beta$$

$$=1+\sin^2\alpha\cosh^2\beta+\left(1-\sin^2\alpha\right)\left(\cosh^2\beta-1\right)$$

$$=1+\sin^2\alpha\cosh^2\beta+\cosh^2\beta-1-\sin^2\alpha\cosh^2\beta+\sin^2\alpha$$

$$=\cosh^2\beta+\sin^2\alpha \qquad \ldots(3)$$

Now equation whose roots are $\sin^2\alpha$ and $\cosh^2\beta$ is

$$x^2-\left(\sin^2\alpha+\cosh^2\beta\right)x+\sin^2\alpha\cosh^2\beta=0 \quad \left[\because\ x^2-Sx+P=0\right]$$

Using Eqs. (2) and (3), we get

$$x^2-\left(1+u^2+v^2\right)x+u^2=0$$

Hence, $\sin^2\alpha$ and $cash^2\beta$ are roots of equation $x^2-\left(1+u^2+v^2\right)x+u^2=0$.

Example 6.36: Separate $\cos^{-1}\left(\frac{3i}{4}\right)$ into real and imaginary parts.

Solution: Let $\cos^{-1}\left(\frac{3i}{4}\right)=x+iy$

$$\Rightarrow \quad \frac{3i}{4}=\cos(x+iy) \quad \Rightarrow \quad \frac{3i}{4}=\cos x\cosh y-i\sin x\sinh y$$

$$\therefore \quad \cos x\cosh y=0 \quad \Rightarrow \quad \cos x=0 \quad \Rightarrow \quad x=\frac{\pi}{2}$$

$$\sin x\sinh y=\frac{3}{4} \quad \Rightarrow \quad \sin x=1 \quad \text{and} \quad -\sinh y=\frac{3}{4}$$

$$\Rightarrow \quad \sin x=1 \quad \text{and} \quad \sinh y=-\frac{3}{4}$$

$$\Rightarrow \quad \sin x=\sin\frac{\pi}{2} \quad \text{and} \quad y=\log\left[-\frac{3}{4}+\sqrt{\frac{9}{16}+1}\right] \quad \left[\because\ \sinh^{-1}z=\log\left(z+\sqrt{z^2+1}\right)\right]$$

$$\Rightarrow \quad x=\frac{\pi}{2} \quad \text{and} \quad y=\log\left[-\frac{3}{4}+\frac{5}{4}\right]=\log\frac{2}{4}=\log\frac{1}{2}=\log 1-\log 2=-\log 2$$

EXERCISE 6.5

1. If $u = \log\left[\tan\left(\frac{\pi}{4}+\frac{\theta}{2}\right)\right]$, prove that

   (*a*) $\tanh\left(\frac{u}{2}\right) = \tan\left(\frac{\theta}{2}\right)$

   (*b*) $\theta = -i\log\left[\tan\left(\frac{\pi}{4}+\frac{iu}{2}\right)\right]$

2. Show that $\tanh^{-1}(\cos\theta) = \cosh^{-1}(\operatorname{cosec}\theta)$.

3. Find $\tanh x$ if $5\sinh x - \cosh x = 5$.

4. $\sinh^{-1} x = \cosh^{-1}\left(\sqrt{1+x^2}\right) = \tanh^{-1}\left(\frac{x}{\sqrt{1-x^2}}\right) = \frac{1}{2}\operatorname{cosech}^{-1}\left(\frac{1}{2x\sqrt{1+x^2}}\right)$, prove it.

5. Prove that $\tanh^{-1} x = \sinh^{-1}\left(\frac{x}{\sqrt{1-x^2}}\right)$.

6. Find all values of $\sin^{-1} 2$ treating 2 as a complex number.

7. Separate $\sin^{-1}(\cos\theta + i\sin\theta)$, $0 < \theta < \frac{\pi}{2}$.

8. Separate $\cos^{-1}\left(e^{i\theta}\right)$, θ is an acute angle.

9. If $\tan(x+iy) = A+iB$, show that $\frac{A}{B} = \frac{\sin 2x}{\sinh 2y}$.

10. If $\tan\left(\frac{x}{2}\right) = \tan\left(\frac{y}{2}\right)$, prove that $\cos x \cosh x = 1$.

ANSWERS

3. $\tanh x = \frac{4}{5}$ or $\frac{-3}{5}$

6. $\sin^{-1} 2 = \left(\frac{4n+1}{2}\right)\pi - i\cosh^{-1} 2$

7. $y = \log\left(\sqrt{\sin\theta} + \sqrt{1+\sin\theta}\right)$, $x = \cos^{-1}\left(\sqrt{\sin\theta}\right)$

8. $x = \sin^{-1}\left(\sqrt{\sin\theta}\right)$, $y = \log\left[\sqrt{1+\sin\theta} - \sqrt{\sin\theta}\right]$

6.10 SUMMATION OF TRIGONOMETRIC FUNCTIONS – C+iS METHOD

This is the most general method and is applied to find the sum of the series of the form

$$a_0 \sin\alpha + a_1 \sin(\alpha+\beta) + a_2 \sin(\alpha+2\beta) + \ldots$$

$$a_0 \cos\alpha + a_1 \cos(\alpha+\beta) + a_2 \cos(\alpha+2\beta) + \ldots$$

Procedure:

1. Put the given Series = S(or C) according as it is a series of Sines or Cosines. Then write C(or S) = a similar series of Cosines or Sines.

   For example, if $S = a_0 \sin\alpha + a_1 \sin(\alpha+\beta) + a_2 \sin(\alpha+2\beta) + \ldots$, and

   $$C = a_0 \cos\alpha + a_1 \cos(\alpha+\beta) + a_2 \cos(\alpha+2\beta) + \ldots$$

2. Multiply the series of Sines by i and add to the series of Cosines, so that

   $$C + iS = a_0[\cos\alpha + i\sin\alpha] + a_1[\cos(\alpha+\beta) + i\sin(\alpha+\beta)]$$
   $$+ a_2[\cos(\alpha+2\beta) + i\sin(\alpha+2\beta)] + \ldots$$
   $$= a_0 e^{i\alpha} + a_1 e^{i(\alpha+\beta)} + a_2 a_1 e^{i(\alpha+2\beta)} + \ldots$$

3. Sum up the above series using any of the following series:

   (a) *Exponential Series:*

   $$1 + x + \frac{x^2}{2!} + \frac{x3}{3!} + \ldots\infty = e^x$$

   (b) *Sine, Cosine, Sinh or Cosh Series:*

   $$x - \frac{x^3}{3!} + \frac{x^5}{5!} - \ldots\infty = \sin x \qquad 1 - \frac{x^2}{2!} + \frac{x^4}{4!} - \ldots\infty = \cos x$$

   $$x + \frac{x^3}{3!} + \frac{x^5}{5!} + \ldots\infty = \sinh x \qquad 1 + \frac{x^2}{2!} + \frac{x^4}{4!} + \ldots\infty = \cosh x$$

   (c) *Logarithmic Series:*

   $$x - \frac{x^2}{2} + \frac{x^3}{3} - \ldots\infty = \log(1+x)$$

   $$-\left(x + \frac{x^2}{2} + \frac{x^3}{3} + \ldots\infty\right) = \log(1-x)$$

(d) *Gregory's Series:*

$$x-\frac{x^3}{3}+\frac{x^5}{5}-\ldots\infty=\tan^{-1}x \qquad x+\frac{x^3}{3}+\frac{x^5}{5}+\ldots\infty=\tanh^{-1}x$$

(v) *Binomial Series:*

$$1+nx+\frac{n(n+1)}{2!}x^2+\frac{n(n+1)(n+2)}{3!}x^3+\ldots\infty=(1+x)^n$$

$$1-nx+\frac{n(n+1)}{2!}x^2-\frac{n(n+1)(n+2)}{3!}x^3+\ldots\infty=(1+x)^{-n}$$

$$1+nx+\frac{n(n+1)}{2!}x^2+\frac{n(n+1)(n+2)}{3!}x^3+\ldots\infty=(1-x)^{-n}$$

(v) *Geometric Series:*

$$a+ar+ar^2+ar^3+\ldots\text{to } n \text{ terms}=\frac{a(1-r^n)}{1-r}$$

$$a+ar+ar^2+ar^3+\ldots\infty=\frac{a}{1-r} \qquad |r|<1$$

4. Finally, express the sum thus obtained in the form $A+iB$ so that by equating the real and imaginary parts, we get $C=A$ and $S=B$.

Note: While forming *C* and *S* equation, *Sines* or *Cosines* of multiple angle (i.e., $\sin n\theta$, $\cos n\theta$) should be replaced by *Cosines* and *Sines*, respectively, whereas *Sines* as *Cosines* with powers, if any will remain same.

ILLUSTRATIVE EXAMPLES 6.7

Example 6.37: Sum to infinity: $1+x\cos\theta+\frac{x^2\cos 2\theta}{2!}+\frac{x^3\cos 3\theta}{3!}+\ldots\infty$

Solution: Let $C=1+x\cos\theta+\frac{x^2\cos 2\theta}{2!}+\frac{x^3\cos 3\theta}{3!}+\ldots\infty$

$$S=x\sin\theta+\frac{x^2\sin 2\theta}{2!}+\frac{x^3\sin 3\theta}{3!}+\ldots\infty$$

$$C+iS=\left[1+x\cos\theta+\frac{x^2\cos 2\theta}{2!}+\frac{x^3\cos 3\theta}{3!}+\ldots\infty\right]$$

$$+i\left[x\sin\theta+\frac{x^2\sin 2\theta}{2!}+\frac{x^3\sin 3\theta}{3!}+\ldots\infty\right]$$

$$=1+x(\cos\theta+i\sin\theta)+\frac{x^2}{2!}(\cos 2\theta+i\sin 2\theta)+\frac{x^3}{3!}(\cos 3\theta+i\sin 3\theta)+\ldots\infty$$

$$=1+xe^{i\theta}+\frac{x^2}{2!}e^{i2\theta}+\frac{x^3}{3!}e^{i3\theta}+\ldots\infty$$

$$=1+t+\frac{t^2}{2!}+\frac{t^3}{3!}+\ldots\infty \qquad \text{where } xe^{i\theta}=t$$

$$=e^t=e^{xe^{i\theta}}$$

$$=e^{x(\cos\theta+i\sin\theta)}=e^{x\cos\theta+ix\sin\theta}$$

$$=e^{x\cos\theta+ix\sin\theta}$$

$$=e^{x\cos\theta}\,e^{ix\sin\theta}$$

$$=e^{x\cos\theta}\left[\cos(x\sin\theta)+i\sin(x\sin\theta)\right]$$

Equating real, we get

$$C=e^{x\cos\theta}\cos(x\sin\theta)$$

Example 6.38: Sum the series: $\sin\alpha\cos\beta-\frac{1}{2}\sin^2\alpha\cos 2\beta+\frac{1}{3}\sin^3\alpha\cos 3\beta-\ldots\infty$

Solution: Let $C=\sin\alpha\cos\beta-\frac{1}{2}\sin^2\alpha\cos 2\beta+\frac{1}{3}\sin^3\alpha\cos 3\beta-\ldots\infty$

$$S=\sin\alpha\sin\beta-\frac{1}{2}\sin^2\alpha\sin 2\beta+\frac{1}{3}\sin^3\alpha\sin 3\beta-\ldots\infty$$

$$C+iS=\left[\sin\alpha\cos\beta-\frac{1}{2}\sin^2\alpha\cos 2\beta+\frac{1}{3}\sin^3\alpha\cos 3\beta-\ldots\infty\right]$$

$$+i\left[\sin\alpha\sin\beta-\frac{1}{2}\sin^2\alpha\sin 2\beta+\frac{1}{3}\sin^3\alpha\sin 3\beta-\ldots\infty\right]$$

$$=\sin\alpha(\cos\beta+i\sin\beta)-\frac{1}{2}\sin^2\alpha(\cos 2\beta+i\sin 2\beta)$$

$$+\frac{1}{3}\sin^3\alpha(\cos 3\beta+i\sin 3\beta)-\ldots\infty$$

$$=\sin\alpha e^{i\beta}-\frac{1}{2}\sin^2\alpha e^{i2\beta}+\frac{1}{3}\sin^3\alpha e^{i3\beta}-\ldots\infty$$

$$=x-\frac{x^2}{2}+\frac{x^3}{3}-\ldots\infty \qquad \text{where } x=\sin\alpha e^{i\beta}$$

$$=\log(1+x)=\log\left(1+\sin\alpha e^{i\beta}\right)=\log\left[1+\sin\alpha(\cos\beta+i\sin\beta)\right]$$

$$=\log[1+\sin\alpha\cos\beta+i\sin\alpha\sin\beta] \qquad \because \log(\alpha+i\beta)=\frac{1}{2}\log(\alpha^2+\beta^2)+i\tan^{-1}\left(\frac{\beta}{\alpha}\right)$$

$$=\frac{1}{2}\log\left[(1+\sin\alpha\cos\beta)^2+\sin^2\alpha\sin^2\beta\right]+i\tan^{-1}\left(\frac{\sin\alpha\sin\beta}{1+\sin\alpha\cos\beta}\right)$$

$$=\frac{1}{2}\log\left[1+2\sin\alpha\cos\beta+\sin^2\alpha\left(\cos^2\beta+\sin^2\beta\right)\right]$$

$$+i\tan^{-1}\left(\frac{\sin\alpha\sin\beta}{1+\sin\alpha\cos\beta}\right)$$

$$=\frac{1}{2}\log\left[1+2\sin\alpha\cos\beta+\sin^2\alpha\right]+i\tan^{-1}\left(\frac{\sin\alpha\sin\beta}{1+\sin\alpha\cos\beta}\right)$$

Equating real, we get

$$C=\frac{1}{2}\log\left[1+2\sin\alpha\cos\beta+\sin^2\alpha\right]$$

Example 6.39: Sum the series: $e^{\alpha}\cos\beta-\frac{e^{3\alpha}}{3}\cos 3\beta+\frac{e^{5\alpha}}{5}\cos 5\beta-\ldots\infty$

Solution: Let $C=e^{\alpha}\cos\beta-\frac{e^{3\alpha}}{3}\cos 3\beta+\frac{e^{5\alpha}}{5}\cos 5\beta-\ldots\infty$

$$S = e^{\alpha}\sin\beta - \frac{e^{3\alpha}}{3}\sin 3\beta + \frac{e^{5\alpha}}{5}\sin 5\beta - \dots\infty$$

$$C + iS = e^{\alpha}\left(\cos\beta + i\cos\beta\right) - \frac{e^{3\alpha}}{3}\left(\cos 3\beta + i\cos 3\beta\right) + \frac{e^{5\alpha}}{5}\left(\cos 5\beta + i\cos 5\beta\right) - \dots\infty$$

$$= e^{\alpha}.e^{i\beta} - \frac{e^{3\alpha}}{3}.e^{i3\beta} + \frac{e^{5\alpha}}{5}.e^{i5\beta} - \dots\infty = e^{(\alpha+i\beta)} - \frac{e^{3(\alpha+i\beta)}}{3} + \frac{e^{5(\alpha+i\beta)}}{5} - \dots\infty$$

$$= x - \frac{x^3}{3} + \frac{x^5}{5} - \dots\infty \qquad \text{where } x = e^{(\alpha+i\beta)}$$

$$= \tan^{-1} x = \tan^{-1}(e^{\alpha+i\beta}) = \tan^{-1}\left[e^{\alpha}\left(\cos\beta + i\sin\beta\right)\right]$$

$$= \tan^{-1}\left(e^{\alpha}\cos\beta + ie^{\alpha}\sin\beta\right)$$

We know (from Example 6.34) that

$$\tan^{-1}\left(x+iy\right) = \frac{1}{2}\tan^{-1}\left(\frac{2x}{1-x^2-y^2}\right) + i\frac{1}{2}\tanh^{-1}\left(\frac{2y}{1+x^2+y^2}\right)$$

$$= \frac{1}{2}\tan^{-1}\left(\frac{2e^{\alpha}\cos\beta}{1-e^{2\alpha}\cos^2\beta - e^{2\alpha}\sin^2\beta}\right) + i\frac{1}{2}\tanh^{-1}\left(\frac{2e^{\alpha}\sin\beta}{1+e^{2\alpha}\cos^2\beta - e^{2\alpha}\sin^2\beta}\right)$$

$$= \frac{1}{2}\tan^{-1}\left(\frac{2e^{\alpha}\cos\beta}{1-e^{2\alpha}}\right) + i\frac{1}{2}\tanh^{-1}\left(\frac{2e^{\alpha}\sin\beta}{1-e^{2\alpha}}\right)$$

Comparing real part, we get

$$C = \frac{1}{2}\tan^{-1}\left(\frac{2e^{\alpha}\cos\beta}{1-e^{2\alpha}}\right) = \frac{1}{2}\tan^{-1}\left(\frac{2\cos\beta}{e^{-\alpha}-e^{\alpha}}\right) = -\frac{1}{2}\tan^{-1}\left(\frac{2\cos\beta}{e^{\alpha}-e^{-\alpha}}\right)$$

$$= -\frac{1}{2}\tan^{-1}\left(\frac{\cos\beta}{\frac{e^{\alpha}-e^{-\alpha}}{2}}\right) = -\frac{1}{2}\tan^{-1}\left(\frac{\cos\beta}{\sinh\alpha}\right)$$

$$= -\frac{1}{2}\tan^{-1}\left(\text{cosech}\,\alpha\cos\beta\right) \qquad \text{which is required answer.}$$

Example 6.40: Sum the following series

$$\sin\alpha+\frac{n(n+1)}{1.2}\sin 2\alpha+\frac{n(n+1)(n+2)}{1.2.3}\sin 3\alpha+\ldots\infty$$

Solution: Let $S=\sin\alpha+\frac{n(n+1)}{1.2}\sin 2\alpha+\frac{n(n+1)(n+2)}{1.2.3}\sin 3\alpha+\ldots\infty$

$$C\ =1+n\cos\alpha+\frac{n(n+1)}{1.2}\cos 2\alpha+\frac{n(n+1)(n+2)}{1.2.3}\cos 3\alpha+\ldots\infty$$

$$C+iS=1+ne^{i\alpha}+\frac{n(n+1)}{1.2}e^{i2\alpha}+\frac{n(n+1)(n+2)}{1.2.3}e^{i3\alpha}+\ldots\infty$$

$$=\left(1-e^{i\alpha}\right)^{-n}=\frac{1}{\left(1-e^{i\alpha}\right)^{n}}=\frac{1}{\left(1-e^{i\alpha}\right)^{n}}\times\frac{\left(1-e^{-i\alpha}\right)^{n}}{\left(1-e^{-i\alpha}\right)^{n}}=\frac{\left(1-e^{-i\alpha}\right)^{n}}{\left[\left(1-e^{i\alpha}\right)\left(1-e^{-i\alpha}\right)\right]^{n}}$$

$$=\frac{[1-(\cos\alpha-i\sin\alpha)]^{n}}{\left(1-e^{i\alpha}-e^{-i\alpha}+1\right)^{n}}=\frac{[1-\cos\alpha+i\sin\alpha]^{n}}{(2-2\cos\alpha)^{n}}$$

$$=\frac{\left[2\sin^{2}\frac{\alpha}{2}+2i\sin\frac{\alpha}{2}.\cos\frac{\alpha}{2}\right]^{n}}{2^{n}\left(2\sin^{2}\frac{\alpha}{2}\right)^{n}}=\frac{2^{n}\sin^{n}\frac{\alpha}{2}\left[\sin\frac{\alpha}{2}+i.\cos\frac{\alpha}{2}\right]^{n}}{2^{n}\left(\sin^{n}\frac{\alpha}{2}\right)\left(2\sin\frac{\alpha}{2}\right)^{n}}$$

$$=\frac{\left[\cos\left(\frac{\pi}{2}-\frac{\alpha}{2}\right)+i.\sin\left(\frac{\pi}{2}-\frac{\alpha}{2}\right)\right]^{n}}{2^{n}\left(\sin^{n}\frac{\alpha}{2}\right)}=\frac{\left[\cos n\left(\frac{\pi}{2}-\frac{\alpha}{2}\right)+i.\sin n\left(\frac{\pi}{2}-\frac{\alpha}{2}\right)\right]}{2^{n}\left(\sin^{n}\frac{\alpha}{2}\right)}$$

Comparing imaginary parts, we get

$$S=\frac{\left[\sin n\left(\frac{\pi}{2}-\frac{\alpha}{2}\right)\right]}{2^{n}\left(\sin^{n}\frac{\alpha}{2}\right)}$$

Example 6.41: Sum the following series

$$\frac{c}{1}\cos\theta+\frac{c^2}{2}\cos 2\theta+\frac{c^3}{3}\cos 3\theta+\ldots\infty$$

Solution:

Let $C = c\cos\theta+\frac{c^2}{2}\cos 2\theta+\frac{c^3}{3}\cos 3\theta+\ldots\infty$

$$S = c\sin\theta+\frac{c^2}{2}\sin 2\theta+\frac{c^3}{3}\sin 3\theta+\ldots\infty$$

$$C+iS = c(\cos\theta+i\sin\theta)+\frac{c^2}{2}(\cos 2\theta+i\sin 2\theta)+\frac{c^3}{3}(\cos 3\theta+i\sin 3\theta)+\ldots\infty$$

$$= ce^{i\theta}+\frac{c^2}{2}e^{i2\theta}+\frac{c^3}{3}e^{i3\theta}+\ldots\infty = z+\frac{z^2}{2}+\frac{z^3}{3}+\ldots\infty,\ \text{where } z = ce^{i\theta}$$

$$= -\log(1-z) = -\log(1-ce^{i\theta})$$

$$= -\log[1-c(\cos\theta+i\sin\theta)] = -\log[1-c\cos\theta-ic\sin\theta]$$

$$= -\left[\frac{1}{2}\log\left[(1-\cos\theta)^2+c^2\sin^2\theta\right]-i\tan^{-1}\left(\frac{c\sin\theta}{1-c\cos\theta}\right)\right]$$

$$\left[\because \log(\alpha-i\beta) = \frac{1}{2}\log\left(\alpha^2+\beta^2\right)-i\tan^{-1}\left(\frac{\beta}{\alpha}\right)\right]$$

$$= -\left[\frac{1}{2}\log\left[1+c^2\cos^2\theta-2c\cos\theta+c^2\sin^2\theta\right]-i\tan^{-1}\left(\frac{c\sin\theta}{1-c\cos\theta}\right)\right]$$

$$= -\left[\frac{1}{2}\log\left[1-2c\cos\theta+c^2\left(\cos^2\theta+\sin^2\theta\right)\right]-i\tan^{-1}\left(\frac{c\sin\theta}{1-c\cos\theta}\right)\right]$$

$$= \left[-\frac{1}{2}\log\left(1-2c\cos\theta+c^2\right)+i\tan^{-1}\left(\frac{c\sin\theta}{1-c\cos\theta}\right)\right]$$

Comparing real parts, we get

$$C = -\frac{1}{2}\log\left(1-2c\cos\theta+c^2\right)$$

EXERCISE 6.6

1. Sum the series $\cos\theta+\sin\theta\cos 2\theta+\frac{\sin^2\theta}{2!}\cos 3\theta+\frac{\sin^3\theta}{3!}\cos 4\theta+...\infty$

2. Sum to the n terms of the series:

   (*a*) $\sin\alpha\sec\alpha+\sin 2\alpha\sec^2\alpha+\sin 3\alpha\sec^3\alpha+...$

   (*b*) $\sin^2\theta+\sin^2\alpha\sin 2\theta+\sin^3\alpha\sin 3\theta+...$

3. Sum the series: $1+x\cos\alpha+x^2\cos 2\alpha+x^3\cos 3\alpha+...$. Also find the sum to infinity, $|x|<1$.

4. Find the sum to the n terms of the series:

   (*a*) $\sin\theta+x\sin 2\theta+x^2\sin 3\theta+...$

   (*b*) $\cos\theta+x\cos 2\theta+x^2\cos 3\theta+...$

5. Find the sum to the n terms of the series

   (*a*) $\cos\theta+\frac{1}{3}\cos 2\theta+\frac{1}{3^2}\cos 3\theta+...$

   (*b*) $\sin\theta+\frac{1}{3}\sin 2\theta+\frac{1}{3^2}\sin 3\theta+...$

6. Sum to the n terms of the series:

   (*a*) $1-\frac{1}{2}\cos\theta+\frac{1\cdot 3}{2\cdot 4}\cos 2\theta-\frac{1\cdot 3\cdot 5}{2\cdot 4\cdot 6}\cos 3\theta+...$ [PTU May 2012]

   (*b*) $-\frac{1}{2}\sin\theta+\frac{1\cdot 3}{2\cdot 4}\sin 2\theta-\frac{1\cdot 3\cdot 5}{2\cdot 4\cdot 6}\sin 3\theta+...$

7. Sum the series: $\cos^n\alpha-n\cos^{n-1}\alpha\cos\alpha+\frac{n(n-1)}{2!}\cos^{n-2}\alpha\cos 2\alpha-...(n+1)$ terms.

8. Sum to the series:

   (*a*) $\sin\alpha-\frac{\sin(\alpha+2\beta)}{2!}+\frac{\sin(\alpha+4\beta)}{4!}-...\infty$

(*b*) $\cos\alpha-\dfrac{\cos(\alpha+2\beta)}{2!}+\dfrac{\cos(\alpha+4\beta)}{4!}-\ldots\infty$

9. Sum to the *n* terms of the series

(*a*) $\cos\theta-2\cos 2\theta+3\cos 3\theta-4\cos 4\theta+\ldots$

(*b*) $\sin\theta-2\sin 2\theta+3\sin 3\theta-4\sin 4\theta+\ldots$

10. Sum to the *n* terms of the series

(*a*) $1+{}^nc_1\cos\alpha+{}^nc_2\cos 2\alpha+{}^nc_3\cos 3\alpha+\ldots$

(*b*) $\sin\alpha+n\sin(\alpha+\beta)+\dfrac{n(n-1)}{2!}\sin(\alpha+2\beta)+\ldots$

11. Sum the series: $1+\dfrac{1}{2!}\cos 2\theta+\dfrac{1}{4!}\cos 4\theta+\ldots\infty$

12. Sum the series: $\sin\theta-\dfrac{1}{3!}\sin 3\theta+\dfrac{1}{5!}\sin 5\theta+\ldots\infty$

13. Sum the series to *n* terms: $\cos\theta+\sin\theta\cos 2\theta+\dfrac{\sin^2\theta}{2!}\cos 3\theta+\ldots$

ANSWERS

2. (*a*) $\cot\alpha\left[1-\sec^n\alpha\cos n\alpha\right]+\sec^n\alpha\sin n\alpha$

(*b*) $\dfrac{\sin^2\theta-\sin(n+1)\theta\sin^{n+1}\theta+\sin n\theta\sin^{n+2}\theta}{1-\sin 2\theta+\sin^2\theta}$

3. $\dfrac{1-x\cos\alpha-x^n\cos n\alpha+x^{n+1}\cos(n-1)\alpha}{1-2x\cos\alpha+x^2}$

4. (*a*) $\dfrac{\sin\theta-x^n\sin(n+1)\theta+x^{n+1}\sin n\theta}{1-2x\cos\theta+x^2}$, (*b*) $\dfrac{\cos\theta-x-x^n\cos(n+1)\theta+x^{n+1}\cos n\theta}{1-2x\cos\theta+x^2}$

5. (*a*) $\dfrac{\cos\theta-\frac{1}{3}-\frac{1}{3^n}\cos(n+1)\theta+\frac{1}{3^{n+1}}\cos n\theta}{\frac{10}{9}-\frac{2}{3}\cos\theta}$, (*b*) $\dfrac{\sin\theta-\frac{1}{3^n}\sin(n+1)\theta+\frac{1}{3^{n+1}}\sin n\theta}{\frac{10}{9}-\frac{2}{3}\cos\theta}$

6. (*a*) $\left[2\cos\left(\frac{\theta}{2}\right)\right]^{\frac{-1}{2}}\cos\left(\frac{\theta}{4}\right)$, (*b*) $\left[2\cos\left(\frac{\theta}{2}\right)\right]^{\frac{-1}{2}}\left[-\sin\left(\frac{\theta}{4}\right)\right]$

7. 0 if n is odd integer, $(-1)^{\frac{n}{2}}\sin^n\alpha$ if n is even integer.

8. (*a*) $\sin\alpha\cos(\cos\beta)\text{cash}(\sin\beta)-\cos\alpha\sin(\cos\beta)\sinh(\sin\beta)$

   (*b*) $\cos\alpha\cos(\cos\beta)\text{cash}(\sin\beta)+\sin\alpha\sin(\cos\beta)\sinh(\sin\beta)$

9. (*a*) $\dfrac{1+(-1)^{n-1}(n+1)\cos n\theta+(-1)^{n-1}n\cos(n+1)\theta}{4\cos^2\frac{\theta}{2}}$

   (*b*) $\dfrac{(n+1)(-1)^{n-1}\sin n\theta+n(-1)^{n-1}\sin(n+1)\theta}{4\cos^2\frac{\theta}{2}}$

10. (*a*) $\left[2\cos\left(\frac{\alpha}{2}\right)\right]^{n}\cos\left(\frac{n\alpha}{2}\right)$, (*b*) $\left[2\cos\left(\frac{\beta}{2}\right)\right]^{n}\left[\sin\left(\frac{n\beta}{2}\right)\right]$

11. $\cos(\sin\theta)\cosh(\cos\theta)$

12. $\cos(\cos\theta)\sinh(\sin\theta)$

❑ ❑ ❑

Appendix A

Solutions of Selected Exercises

Chapter 1: Ordinary Differential Equations of First Order

EXERCISE 1.2

Exercise 8: Solve $\left(xy\, e^{\frac{x}{y}} + y^2\right)dx - x^2\, e^{\frac{x}{y}} dy = 0$.

Solution: Given differential equation is

$$\left(xy\, e^{\frac{x}{y}} + y^2\right)dx - x^2\, e^{\frac{x}{y}} dy = 0 \qquad \ldots (1)$$

Comparing (1) with $Mdx + Ndy = 0$, we have

$$M = xy\, e^{\frac{x}{y}} + y^2, \quad N = -x^2\, e^{\frac{x}{y}}$$

$$\therefore \qquad \frac{\partial M}{\partial y} = x\left[y.e^{\frac{x}{y}}\left(-\frac{x}{y^2}\right) + e^{\frac{x}{y}}.1\right] + 2y, \quad \frac{\partial N}{\partial x} = -\left[x^2.e^{\frac{x}{y}}.\frac{1}{y} + e^{\frac{x}{y}}.2x\right]$$

Since $\dfrac{\partial M}{\partial y} \neq \dfrac{\partial N}{\partial x}$

$\therefore$ Differential Eq. (1) is not exact.

Dividing Eq. (1) by x^2, we get

$$\left(\frac{y}{x}e^{\frac{x}{y}}+\frac{y^2}{x^2}\right)dx-e^{\frac{x}{y}}dy=0 \qquad \ldots (2)$$

Here $M=\dfrac{y}{x}e^{\frac{x}{y}}+\dfrac{y^2}{x^2}$, $N=-e^{\frac{x}{y}}$ are homogeneous function of x and y.

$$\therefore\ I.F.=\frac{1}{Mx+Ny}=\frac{1}{\left(\frac{y}{x}\,e^{\frac{x}{y}}+\frac{y^2}{x}\right)+\left(\left(-e^{\frac{x}{y}}\right)y\right)}=\frac{1}{\frac{y}{x}\,e^{\frac{x}{y}}+\frac{y^2}{x}+-ye^{\frac{x}{y}}}=\frac{1}{\frac{y^2}{x}}=\frac{x}{y^2}$$

Multiplying Eq. (2) by $\dfrac{x}{y^2}$, we get

$$\frac{x}{y^2}\left(\frac{y}{x}e^{\frac{x}{y}}+\frac{y^2}{x^2}\right)dx-\frac{x}{y^2}e^{\frac{x}{y}}dy=0 \quad\Rightarrow\quad \left(\frac{e^{\frac{x}{y}}}{y}+\frac{1}{x}\right)dx-\frac{x}{y^2}e^{\frac{x}{y}}dy=0 \qquad \ldots (3)$$

Now $M=\dfrac{e^{\frac{x}{y}}}{y}+\dfrac{1}{x}$, $N=-\dfrac{x}{y^2}e^{\frac{x}{y}}$

$$\therefore\quad \frac{\partial M}{\partial y}=\frac{1}{y}.\,e^{\frac{x}{y}}.\left(-\frac{x}{y^2}\right)+e^{\frac{x}{y}}.\left(\frac{-1}{y^2}\right)=e^{\frac{x}{y}}\left(-\frac{x}{y^3}-\frac{1}{y^2}\right)$$

$$\frac{\partial N}{\partial x}=-\frac{1}{y^2}\left[x.e^{\frac{x}{y}}.\frac{1}{y}+e^{\frac{x}{y}}.1\right]=e^{\frac{x}{y}}\left(-\frac{x}{y^3}-\frac{1}{y^2}\right)$$

Since $\dfrac{\partial M}{\partial y}=\dfrac{\partial N}{\partial x}$

$\therefore$ Differential Eq. (3) is exact and its solution is given by

$$\int_y M dx + \int (\text{terms of } N \text{ not containing } x)\, dy = c$$

$$\Rightarrow \quad \int_y \left(\frac{e^{\frac{x}{y}}}{y} + \frac{1}{x} \right) dx + \int 0 dy = c \qquad \Rightarrow \quad \frac{1}{y} \int_y e^{\frac{x}{y}} dx + \int \frac{1}{x} dx = c$$

$$\Rightarrow \quad \frac{1}{y} \frac{e^{\frac{x}{y}}}{\frac{1}{y}} + \log x = c \qquad \text{or} \qquad e^{\frac{x}{y}} + \log x = c$$

which is the required solution.

Exercise 18: Solve $\left(y + \frac{y^3}{3} + \frac{x^2}{2} \right) dx + \frac{1}{4}\left(x + xy^2 \right) dy = 0$.

Solution: Given differential equation is

$$\left(y + \frac{y^3}{3} + \frac{x^2}{2} \right) dx + \frac{1}{4}\left(x + xy^2 \right) dy = 0 \qquad \ldots (1)$$

Comparing Eq. (1) with $Mdx + Ndy = 0$, we have

$$M = y + \frac{y^3}{3} + \frac{x^2}{2}, \quad N = \frac{1}{4}\left(x + xy^2 \right)$$

$$\therefore \quad \frac{\partial M}{\partial y} = 1 + y^2, \quad \frac{\partial N}{\partial x} = \frac{1 + y^2}{4}$$

Since $\frac{\partial M}{\partial y} \neq \frac{\partial N}{\partial x}$, therefore, Eq. (1) is not exact. We apply Rule III.

Consider $\dfrac{\frac{\partial M}{\partial y} - \frac{\partial N}{\partial x}}{N} = \dfrac{1 + y^2 - \frac{1}{4}\left(1 + y^2\right)}{\frac{1}{4}\left(x + xy^2 \right)} = \dfrac{3}{4} \cdot \dfrac{\left(1 + y^2\right)}{\frac{1}{4} x\left(1 + y^2\right)} = \dfrac{3}{x} = f(x)$, say

$$\therefore \quad \text{I.F.} = e^{\int f(x) dx} = e^{\int \frac{3}{x} dx} = e^{3 \log x} = e^{\log x^3} = x^3$$

Multiply Eq. by x^3, we get

$$x^3\left(y+\frac{y^3}{3}+\frac{x^2}{2}\right)dx+\frac{x^3}{4}\left(x+xy^2\right)dy=0$$

$$\Rightarrow \quad \left(x^3y+\frac{x^3y^3}{3}+\frac{x^5}{2}\right)dx+\left(\frac{x^4}{4}+\frac{x^4y^2}{4}\right)dy=0$$

Here $\quad M=x^3y+\dfrac{x^3y^3}{3}+\dfrac{x^5}{2},\ N=\dfrac{x^4}{4}+\dfrac{x^4y^2}{4}$

$$\therefore \quad \frac{\partial M}{\partial y}=x^3+x^3y^2,\ \frac{\partial N}{\partial x}=x^3+x^3y^2$$

Since $\dfrac{\partial M}{\partial y}=\dfrac{\partial N}{\partial x}$, therefore, Eq. (2) is exact and its required solution is

$$\int_y Mdx+\int(\text{terms of } N \text{ not containing } x)\,dy=c$$

$$\Rightarrow \quad \int_y\left(x^3y+\frac{x^3y^3}{3}+\frac{x^5}{2}\right)dx+\int 0dy=c$$

$$\Rightarrow \quad y\int x^3dx+\frac{y^3}{3}\int x^3dx+\frac{1}{2}\int x^5dx=c$$

$$\Rightarrow \quad \frac{1}{4}x^4y+\frac{x^4y^3}{12}+\frac{x^6}{12}=c$$ which is the required solution.

EXERCISE 1.3

Exercise 2: Solve $xy\left(\dfrac{dy}{dx}\right)^2-\left(x^2+y^2\right)\dfrac{dy}{dx}+xy=0$

Solution: Given differential equation

$$xy\left(\frac{dy}{dx}\right)^2-\left(x^2+y^2\right)\frac{dy}{dx}+xy=0 \qquad \text{. . . (1)}$$

Taking $\dfrac{dy}{dx}=p$, the given differential equation reduces to

$$xy\,p^2-\left(x^2+y^2\right)p+xy=0.$$

which is quadratic in p, and on solving gives

$$p = \frac{(x^2+y^2) \pm \sqrt{(x^2+y^2)^2 - 4xy.xy}}{2xy} = \frac{(x^2+y^2) \pm \sqrt{x^4+y^4+2x^2y^2-4x^2y^2}}{2xy}$$

$$= \frac{(x^2+y^2) \pm \sqrt{x^4+y^4-2x^2y^2}}{2xy} = \frac{(x^2+y^2) \pm \sqrt{(x^2-y^2)^2}}{2xy}$$

$$= \frac{(x^2+y^2) \pm (x^2-y^2)}{2xy} = \frac{x^2+y^2+x^2-y^2}{2xy}, \frac{x^2+y^2-x^2+y^2}{2xy}$$

$$= \frac{2x^2}{2xy}, \frac{2y^2}{2xy} = \frac{x}{y}, \frac{y}{x}$$

When $p = \frac{x}{y}$,

$$\frac{dy}{dx} = \frac{x}{y}$$

Separating the variables, we get

$$y\, dy = x\, dx$$

Now integrating, we get

$$\int y\, dy = \int x\, dx + c$$

$$\Rightarrow \quad \frac{y^2}{2} = \frac{x^2}{2} + c$$

$$\Rightarrow \quad y^2 = x^2 + 2c$$

$$\Rightarrow \quad y^2 = x^2 + c_1 \qquad [\text{Take } c_1 = 2c] \qquad \ldots (2)$$

When $p = \frac{y}{x}$

$$\frac{dy}{dx} = \frac{y}{x}$$

Separating the variables, we get

$$\frac{1}{y} dy = \frac{dx}{x}$$

Integrating, we get

$$\int \frac{1}{y}\,dy = \int \frac{1}{x}\,dx + c$$

$\Rightarrow \quad \log y = \log x + c$

$\Rightarrow \quad \log y - \log x = c$

$\Rightarrow \quad \log \frac{y}{x} = \log c_1 \qquad [\text{Take } c = \log c_1]$

$\Rightarrow \quad \frac{y}{x} = c_1$

$\Rightarrow \quad y = c_1 x \qquad \ldots (3)$

Exercise 5: Solve $\frac{dy}{dx} - \frac{dx}{dy} = \frac{x}{y} - \frac{y}{x}$.

Solution: Given differential equation

$$\frac{dy}{dx} - \frac{dx}{dy} = \frac{x}{y} - \frac{y}{x} \qquad \ldots (1)$$

Taking $\frac{dy}{dx} = p$, the given differential equation reduces to

$$p - \frac{1}{p} = \frac{x}{y} - \frac{y}{x}$$

$\Rightarrow \quad \frac{p^2 - 1}{p} = \frac{x^2 - y^2}{xy}$

$\Rightarrow \quad (p^2 - 1)xy - p(x^2 - y^2) = 0$

$\Rightarrow \quad p^2 xy - p(x^2 - y^2) - xy = 0 \qquad \ldots (2)$

which is quadratic in *p*, and on solving gives

$$p = \frac{x^2 - y^2 \pm \sqrt{(x^2 - y^2)^2 + 4xy.xy}}{2xy} = \frac{x^2 - y^2 \pm \sqrt{x^4 + y^4 - 2x^2y^2 + 4x^2y^2}}{2xy}$$

$$= \frac{x^2 - y^2 \pm \sqrt{x^4 + y^4 + 2x^2y^2}}{2xy} = \frac{x^2 - y^2 \pm \sqrt{(x^2+y^2)^2}}{2xy} \frac{x^2 - y^2 \pm (x^2 + y^2)}{2xy}$$

$$= \frac{x^2 - y^2 + x^2 + y^2}{2xy}, \frac{x^2 - y^2 - x^2 - y^2}{2xy} = \frac{x}{y}, \frac{-y}{x}$$

When $p = \frac{x}{y}$

$$\frac{dy}{dx} = \frac{x}{y}$$

Separating the variables, we get

$$y\,dy = x\,dx$$

Integrating, we get

$$\Rightarrow \quad \frac{y^2}{2} = \frac{x^2}{2} + \frac{c}{2}$$

$$\Rightarrow \quad y^2 = x^2 + c \qquad \ldots (3)$$

When $p = \frac{-y}{x}$

$$\frac{dy}{dx} = \frac{-y}{x}$$

Separating the variables, we get

$$\frac{1}{y}dy + \frac{1}{x}dx = 0,$$

Integrating, we get

$$\log y + \log x = \log c$$

$$\Rightarrow \quad \log xy = \log c$$

$$\Rightarrow \quad xy = c \qquad \ldots (4)$$

EXERCISE 1.4

Exercise 2: Solve $y-2px=\tan^{-1}(xp^2)$

Solution: Given differential equation is

$$y-2px=\tan^{-1}(xp^2) \qquad \ldots (1)$$

Solving the given equation for *y*, we have

$$y=2px+\tan^{-1}(xp^2) \qquad \ldots (2)$$

Differentiating Eq. (2) w.r.t. *x*, we get

$$\frac{dy}{dx}=2p+2x\frac{dp}{dx}+\frac{1}{1+x^2p^4}\cdot\frac{d}{dx}(xp^2)$$

$$\Rightarrow \quad p=2p+2x\frac{dp}{dx}+\frac{1}{1+x^2p^4}\left(x.2p\frac{dp}{dx}+p^2.1\right) \qquad \left[\because \frac{dy}{dx}=p\right]$$

$$\Rightarrow \quad \left(p+2x\frac{dp}{dx}\right)+\frac{p}{1+x^2p^4}\left(2x\frac{dp}{dx}+p\right)=0$$

$$\Rightarrow \quad \left(p+2x\frac{dp}{dx}\right)\left(1+\frac{p}{1+x^2p^4}\right)=0$$

Dividing by $\left(1+\frac{p}{1+x^2p^4}\right)$, we get

$$p+2x\frac{dp}{dx}=0 \quad \Rightarrow \quad p=-2x\frac{dp}{dx}$$

Separating the variables, we get

$$2\frac{dp}{p}=\frac{-dx}{x}$$

Integrating, we get

$$2\log p=-\log x+\log c \quad \Rightarrow \quad \log p^2=\log\frac{c}{x}$$

$$\Rightarrow \quad p^2=\frac{c}{x} \quad \Rightarrow p=\sqrt{\frac{c}{x}}$$

Substituting value of in Eq. (2), we get

$$y = 2x.\sqrt{\frac{c}{x}} + \tan^{-1}\left(x.\frac{c}{x}\right) = 2\sqrt{c}\sqrt{x} + \tan^{-1} c$$

$$y = 2\sqrt{cx} + \tan^{-1} c$$ is the required solution.

Exercise 6: Solve $x - yp = ap^2$

Solution: Given differential equation is

$$x - yp = ap^2 \qquad \ldots (1)$$

Solving for y, we have

$$x = yp + ap^2 \text{ or } y = \frac{x}{p} - ap \qquad \ldots (2)$$

Differentiating Eq. (1) w.r.t. x, we get

$$\frac{dy}{dx} = \frac{p.1 - x.\frac{dp}{dx}}{p^2} - a\frac{dp}{dx}$$

$$\Rightarrow \quad p = \frac{1}{p} - \frac{x}{p^2}\frac{dp}{dx} - a\frac{dp}{dx} \qquad \left[\because \frac{dy}{dx} = p\right]$$

$$\Rightarrow \quad p^3 = p - x\frac{dp}{dx} - ap^2\frac{dp}{dx} = p - \left(x + ap^2\right)\frac{dp}{dx}$$

$$\Rightarrow \quad \left(x + ap^2\right)\frac{dp}{dx} = p - p^3$$

$$\Rightarrow \quad \frac{dx}{dp} = \frac{x + ap^2}{p - p^3} = \frac{x}{p - p^3} + \frac{ap^2}{p - p^3}$$

$$\Rightarrow \quad \frac{dx}{dp} + \frac{1}{p^3 - p}x = \frac{ap^2}{p - p^3}$$

which is of the form $\frac{dx}{dp} + Px = Q$, where P, Q are function of p only, where, $P=\dfrac{1}{p^3-p}$, $Q=\dfrac{ap^2}{p-p^3}$

Here I.F. = $\int e^{\left(\frac{1}{p^3-p}\right)dp} = e^{\int\frac{1}{p(p-1)(p+1)}dp}$

Consider $\dfrac{1}{p(p-1)(1+p)}=\dfrac{A}{p}+\dfrac{B}{p-1}+\dfrac{C}{1+p}$ . . .(3)

Multiplying (3) by $p(p-1)(1+p)$, we get

$$1=A(p-1)(p+1)+Bp(p+1)+C(p-1)p \quad \text{. . . (4)}$$

Putting p = 1 in Eq. (4), we get $1=2B \Rightarrow B=\dfrac{1}{2}$

Putting p = −1 in Eq. (4), we get $1=2C \Rightarrow C=\dfrac{1}{2}$

Putting p = 0 in Eq. (4), we get $1=-A \Rightarrow A=-1$

$$\therefore \quad \frac{1}{p(p-1)(p+1)}=\frac{-1}{p}+\frac{1}{2}.\frac{1}{p-1}+\frac{1}{2}.\frac{1}{p+1}$$

$$\therefore \quad e^{\int\left(-\frac{1}{p}+\frac{1}{2}.\frac{1}{p-1}+\frac{1}{2}.\frac{1}{p+1}\right)dp} = e^{-\log p+\frac{1}{2}\log(p-1)+\frac{1}{2}\log(p+1)}$$

$$e^{\frac{1}{2}\log\frac{(p-1)(p+1)}{p}} = e^{\log\frac{\sqrt{(p-1)(p+1)}}{p}} = e^{\log\frac{\sqrt{p^2-1}}{p}} = \frac{\sqrt{p^2-1}}{p}$$

∴ Solution of Eq. (2) is given by

$$\frac{x.\sqrt{p^2-1}}{p} = \int\frac{ap^2}{p-p^3}.\frac{\sqrt{p^2-1}}{p}dp+c$$

$$=-\int\frac{a\sqrt{p^2-1}}{p^2-1}dp+c=-a\int\frac{1}{\sqrt{p^2-1}}dp+c$$

$$=-a\log\left|p+\sqrt{p^2-1}\right|+c$$

$$\Rightarrow \quad x = \frac{p}{\sqrt{p^2-1}}\left(c - a\log\left(p+\sqrt{p^2-1}\right)\right)$$

Also from Eq. (1), we hace

$$y = \frac{x}{p} - ap = \frac{1}{p}.\frac{p}{\sqrt{p^2-1}}\left(c - a\log\left(p+\sqrt{p^2-1}\right)\right) - ap$$

$$\Rightarrow \quad y = \frac{1}{\sqrt{p^2-1}}\left(c - a\log\left(p+\sqrt{p^2-1}\right)\right) - ap$$ is the required solution.

EXERCISE 1.5

Exercise 2: Solve $y = 3px + 6p^2y^2$.

Solution: Given differential equation is

$$y = 3px + 6p^2y^2 \qquad \ldots (1)$$

Solving for *x*, we have

$$3px = y - 6p^2y^2$$

$$3x = \frac{y}{p} - 6y^2p \qquad \ldots (2)$$

Differentiating Eq. (2) w.r.t. *y*, we have

$$3\frac{dx}{dy} = \frac{p.1 - y\frac{dp}{dy}}{p^2} - 6\left[p.2y + y^2\frac{dp}{dy}\right]$$

$$\Rightarrow \quad \frac{3}{p} = \frac{1}{p} - \frac{y}{p^2}.\frac{dp}{dy} - 12py - 6y^2\frac{dp}{dy} \qquad \left[\therefore \frac{dy}{dx} = p\right]$$

$$\Rightarrow \quad 3p = p - y\frac{dp}{dy} - 12p^3y - 6y^2p^2\frac{dp}{dy}$$

$$\Rightarrow \quad 2p + 12p^3y = -y\frac{dp}{dy}\left(1+6p^2y\right)$$

$$\Rightarrow \quad 2p\left(1+6p^2y\right) + y\frac{dp}{dy}\left(1+6p^2y\right) = 0$$

$$\Rightarrow \quad (1+6p^2y)\left(2p+y\frac{dp}{dy}\right)=0$$

Dividing by $(1+6p^2y)$, we get

$$\Rightarrow \quad 2p+y\frac{dp}{dy}=0$$

$$\Rightarrow \quad 2p=-y\frac{dp}{dy}$$

Separating variables, we get

$$\frac{dp}{p}=-2\frac{dy}{y}$$

Integrating, we get

$$\log p=-2\log y+\log c$$

$$\Rightarrow \quad \log p=\log\frac{c}{y^2}$$

$$\Rightarrow \quad p=\frac{c}{y^2} \qquad \ldots (3)$$

Substituting p in Eq. (2), we get

$$3x=\frac{y^3}{c}-6c$$

$$\Rightarrow \quad 3cx=y^3-6c^2$$

$$\Rightarrow \quad y^3=3cx+6c^2$$ is the required solution.

EXERCISE 1.6

Exercise 10: Solve $(y+px)^2=x^2p$.

Solution: Given differential equation is

$$(y+px)^2=x^2p \qquad \ldots (1)$$

Put $xy=Y$, $x=X$ so that

$$x\frac{dy}{dx}+y=\frac{dY}{dx}=p \text{ , say}$$

$$xp+y=p, \quad \text{where } \frac{dy}{dx}=p$$

Also $\quad dx=dX$

$\therefore$ Given equation reduces to

$$P^2=x^2\left(\frac{P-y}{x}\right)=x\left(P-\frac{Y}{x}\right)=xP-Y$$

$$Y=Px-P^2$$

which is a Clairaut's form and hence its solution is given by

$$Y=cx-c^2 \qquad \text{(Replacing } P \text{ by } c)$$

or $\quad xy=cx-c^2 \quad$ is the required solution.

Chapter 2: Linear Differential Equations

EXERCISE 2.2

Exercise 8: Solve $(D^2+2D+2)y=e^{-x}\sec x$.

Solution: A.E. of the given differential equation is

$$D^2+2D+2=0$$

which is quadratic in D, and on solving gives

$$D=\frac{-2\pm\sqrt{4-8}}{2}=\frac{-2\pm 2i}{2}=-1\pm i$$

$$\therefore \quad C.F.=e^{-x}(c_1\cos x+c_2\sin x)$$

Now, $\quad P.I.=\dfrac{1}{D^2+2D+2}e^{-x}\sec x=e^{-x}\dfrac{1}{(D-1)^2+2(D-1)+2}\sec x$

$$= e^{-x}\frac{1}{D^2+1}\sec x = e^{-x}\frac{1}{(D+i)^2+2(D-1)+2}\sec x$$

$$= e^{-x}\left(\frac{\frac{1}{2i}}{D-i}-\frac{\frac{1}{2i}}{D+i}\right)\sec x$$

By partial fractions

$$= \frac{e^{-x}}{2i}\left(\frac{1}{D-i}-\frac{1}{D+i}\right)\sec x$$

Now, $\frac{1}{D-i}\sec x = e^{ix}\int e^{-ix}\sec x\,dx = e^{ix}\int(\cos x - i\sin x)\frac{1}{\cos x}dx$

$$= e^{ix}\int(1-i\tan x)dx = e^{ix}(x+i\log\cos x)$$

Similarly, $\frac{1}{D+i}\sec x = e^{-ix}(x-i\log\cos x)$

$\therefore \quad P.I. = \frac{e^{-x}}{2i}\left[e^{ix}(x+i\log\cos x)-e^{-ix}(x-i\log\cos x)\right]$

$$= \frac{e^{-x}}{2i}\left[x\left(e^{ix}-e^{-ix}\right)+i\log\cos x\left(e^{ix}+e^{-ix}\right)\right]$$

$$= \frac{e^{-x}}{2i}(x.2i\sin x + i\log\cos x.2\cos x) = e^{-x}(x\sin x+\cos x\log\cos x)$$

$\therefore \quad C.S. = C.F. + P.I.$

$\Rightarrow \quad y = e^{-x}(c_1\cos x + c_2\sin x) + e^{-x}(x\sin x + \cos x\log\cos x)$

$\Rightarrow y = e^{-x}(c_1\cos x + c_2\sin x + x\sin x + \cos x\log\cos x)$.

EXERCISE 2.3

Exercise 2: Solve, by the method of variation of parameter,

$$\frac{d^2y}{dx^2}+4y=\tan 2x \qquad \ldots (1)$$

Solution: Given differential equation in symbolical form is

$$\left(D^2+4\right)y=\tan 2x$$

It's A.E. is

$$D^2+4=0$$

on solving it , we get

$$D=\pm 2i=0\pm 2i$$

$\therefore$ C.F $=e^{0x}\left(c_1\cos 2x+c_2\sin 2x\right)=c_1\cos 2x+c_2\sin 2x$ . . . (2)

Here $y_1=\cos 2x,\ y_2=\sin 2x, X=\tan 2x$

Now $W=\begin{vmatrix} y_1 & y_2 \\ y_1' & y_2' \end{vmatrix}=\begin{vmatrix} \cos 2x & \sin 2x \\ -2\sin 2x & 2\cos 2x \end{vmatrix}$

$$=2\left(\cos^2 2x+\sin^2 2x\right)=2(1)=2\neq 0$$

Let P.I. $=y_p=uy_1+vy_2$

where $u=-\int\frac{y_2X}{W}dx=-\int\frac{\sin 2x.\tan 2x}{2}dx=-\frac{1}{2}\int\frac{\sin^2 2x}{\cos 2x}dx$

$$=-\frac{1}{2}\int\left(\frac{1-\cos^2 2x}{\cos 2x}\right)dx=-\frac{1}{2}\int(\sec 2x-\cos 2x)dx$$

$$=-\frac{1}{2}\frac{\log(\sec 2x+\tan 2x)}{2}+\frac{1}{2}\frac{\sin 2x}{2}$$

$$=-\frac{1}{4}\log(\sec 2x+\tan 2x)+\frac{1}{4}\sin 2x$$

And $v=\int\frac{y_1X}{W}dx=\int\frac{\cos 2x.\tan 2x}{2}dx=\frac{1}{2}\int\sin 2x dx=-\frac{1}{4}\cos 2x$

Hence P.I. $=uy_1+vy_2$

$$=\left(-\frac{1}{4}\log(\sec 2x+\tan 2x)+\frac{1}{4}\sin 2x\right)\cos 2x-\frac{1}{4}\cos 2x.\sin 2x$$

$$=-\frac{\cos 2x}{4}\log(\sec 2x+\tan 2x)$$ . . . (3)

From Eq. (2) and Eq. (3), the required solution is given by

$$y = \text{C.F.} + \text{P.I.} = y_c + y_p = c_1 \cos 2x + c_2 \sin 2x - \frac{\cos 2x}{4} \log(\sec 2x + \tan 2x).$$

Exercise 5: Solve, by the method of variation of parameter.

$$y'' - 2y' + 2y = e^x \tan x.$$

Solution: Given differential equation, in symbolical form, is

$$(D^2 - 2D + 2)y = e^x \tan x.$$

It's A.E. is

$$D^2 - 2D + 2 = 0$$

Which is a quadratic equation in D, and on solving it gives

$$D = \frac{2 \pm \sqrt{4-8}}{2} = \frac{2 \pm 2i}{2} = 1 \pm i$$

$\therefore$ $$\text{C.F} = y_c = e^x (c_1 \cos x + c_2 \sin x)$$

$$= c_1 e^x \cos x + c_2 e^x \sin x \qquad \ldots (2)$$

Here $y_1 = e^x \cos x,\ y_2 = e^x \sin x,\ X = e^x \tan x$

Now $$W = \begin{vmatrix} y_1 & y_2 \\ y_1' & y_2' \end{vmatrix} = \begin{vmatrix} e^x \cos x & e^x \sin x \\ -e^x \sin x + \cos x . e^x & e^x \cos x + \sin x e^x \end{vmatrix}$$

$$= e^x \cos x (e^x \cos x + e^x \sin x) - e^x \sin x (-e^x \sin x + \cos x\, e^x)$$

$$= e^{2x} \cos^2 x + e^{2x} \sin x \cos x + e^{2x} \sin^2 x - e^{2x} \sin x \cos x$$

$$= e^{2x} (\cos^2 x + \sin^2 x) = e^{2x}(1) = e^{2x} \neq 0$$

Let $\text{P.I.} = y_p = u y_1 + v y_2$

where $$u = -\int \frac{y_2 X}{W} dx = -\int \frac{e^x \sin x . e^x \tan x}{e^{2x}} dx = -\int \frac{\sin^2 x}{\cos x} dx$$

$$= -\int \frac{1 - \cos^2 x}{\cos x} dx = -\int (\sec x - \cos x) dx$$

$$= -\log(\sec x + \tan x) + \sin x$$

$$v = \int \frac{y_1 X}{W} dx = \int \frac{e^x \cos x . e^x \tan x}{e^{2x}} dx = \int \sin x \, dx = -\cos x$$

$\therefore$ $$\text{P.I.} = y_p = uy_1 + vy_2 = \left[-\log(\sec x + \tan x) + \sin x\right] e^x \cos x + (-\cos x) e^x \sin x$$

$$= -e^x \cos x . \log(\sec x + \tan x) \qquad \ldots (3)$$

Hence from Eq. (2) and Eq. (3), the required solution is given by

$$y = \text{C.F.} + \text{P.I.} = y_c + y_p$$

$$= e^x (c_1 \cos x + c_2 \sin x) - e^x \cos x . \log(\sec x + \tan x).$$

EXERCISE 2.4

Exercise 6: Solve, $x^2 \frac{d^2 y}{dx^2} - 3x \frac{dy}{dx} + 5y = x^2 \sin(\log x)$.

Solution: Given differential equation is

$$x^2 \frac{d^2 y}{dx^2} - 3x \frac{dy}{dx} + 5y = x^2 \sin(\log x) \qquad \ldots (1)$$

which is a Cauchy's homogeneous linear differential equation.

So put $x = e^z \quad \Rightarrow \quad z = \log x$

$$\Rightarrow \quad \frac{dz}{dx} = \frac{1}{x}$$

Now $$\frac{dy}{dx} = \frac{dy}{dz} . \frac{dz}{dx} = \frac{1}{x} \frac{dy}{dz}$$

$$\Rightarrow \quad x \frac{dy}{dx} = \frac{dy}{dz} = Dy, D = \frac{d}{dz}$$

Similarly, $x^2 \frac{d^2 y}{dx^2} = D(D-1) y, D = \frac{d}{dz}$

Putting all these values in the given differential equation, we get

$$(D(D-1) - 3D + 5) y = e^{2z} \sin z, D = \frac{d}{dz}$$

$$\Rightarrow \quad \left(D^2-4D+5\right)y = e^{2z}\sin z, \qquad \ldots (2)$$

which is a linear differential equation, with constant coefficients.

It's A.E. is

$$D^2-4D+5=0$$

which is a quadratic equation, and on solving it gives

$$D=\frac{4\pm\sqrt{16-20}}{2}=\frac{4\pm 2i}{2}=2\pm i$$

$$\therefore \quad \text{C.F.} = e^{2z}\left(c_1\cos z + c_2\sin z\right) = x^2\left(c_1\cos(\log x)+c_2\sin(\log x)\right) \qquad \ldots (3)$$

Also $\text{P.I.} = \dfrac{1}{D^2-4D+5}e^{2z}\sin z$

$$= e^{2z}\frac{1}{(D+2)^2-4(D+2)+5}e^{2z}\sin z \qquad \left[\because \frac{1}{f(D)}e^{az}.V = e^{az}\frac{1}{f(D+a)}V\right]$$

$$= e^{2z}\frac{1}{D^2+4+4D-4D-8+5}\sin z$$

$$= e^{2z}\frac{1}{D^2+1}\sin z = e^{2z}.z.\frac{1}{2D}\sin z = e^{2z}.z.\frac{1}{2}(-\cos z) = -\frac{1}{2}z\cos ze^{2z}$$

$$= -\frac{1}{2}x^2(\log x)\cos(\log x) \qquad \ldots (4)$$

Therefore, Eq. (3) and Eq. (3), the required solution is given by

$$y = \text{C.F.+P.I} = x^2\left(c_1\cos(\log x)+c_2\sin(\log x)\right)-\frac{1}{2}x^2\log x.\cos(\log x).$$

Exercise 7: Solve $x^2\dfrac{d^2y}{dx^2}-x\dfrac{dy}{dx}+2y = x\log x.$

Solution: Given differential equation is

$$x^2\frac{d^2y}{dx^2}-x\frac{dy}{dx}+2y = x\log x \qquad \ldots (1)$$

which is a Cauchy's homogeneous linear differential equation.

So put $x = e^z \quad \Rightarrow \quad z = \log x$

$$\Rightarrow \quad \frac{dz}{dx} = \frac{1}{x}$$

Now $\frac{dy}{dx} = \frac{dy}{dz} \cdot \frac{dz}{dx} = \frac{1}{x}\frac{dy}{dz}$

$$\Rightarrow \quad x\frac{dy}{dx} = \frac{dy}{dz} = Dy, D = \frac{d}{dz}$$

Similarly, $x^2 \frac{d^2y}{dx^2} = D(D-1)y, D = \frac{d}{dz}$

Putting all these values in the given differential equation, we get

$$\left(D(D-1) - D + 2\right) y = e^z z \qquad \text{where } D = \frac{d}{dz}$$

$$\Rightarrow \quad \left(D^2 - 2D + 2\right) y = ze^z \qquad \ldots (2)$$

which is a linear differential equation with constant coefficients.

It's A.E. is

$$D^2 - 2D + 2 = 0$$

which is a quadratic equation on D, and on solving gives

$$D = \frac{2 \pm \sqrt{4-8}}{2} = \frac{2 \pm 2i}{2} = 1 \pm i$$

$$\therefore \quad \text{C.F.} = e^z\left(c_1 \cos z + c_2 \sin z\right) = x\left(c_1 \cos(\log x) + c_2 \sin \log x\right) \qquad \ldots (2)$$

Also $\text{P.I.} = \frac{1}{D^2 - 2D + 2} e^z . z$

$$= e^z \frac{1}{(D+1)^2 - 2(D+1) + 2} z$$

$$= e^z . \frac{1}{D^2 + 1 + 2D - 2D - 2 + 2} z \qquad \left[\because \frac{1}{f(D)} e^{az} . V = e^{az} \frac{1}{f(D+a)} V\right]$$

$$= e^z . \frac{1}{D^2+1} z = e^z \left(1+D^2\right)^{-1} . z$$

$$= e^z \left(1 - D^2 + ...\right) z \quad \text{|Retaining up to } D \text{ as the highest power of } z \text{ is one}$$

$$= e^z (z) = (\log x) x \qquad \text{. . . (3)}$$

Therefore, from Eq. (2) and Eq. (3), the required complete solution is given by

$$y = \text{C.F.+P.I.} = x\left(c_1 \cos(\log x) + c_2 \sin(\log x)\right) + x \log x$$

EXERCISE 2.6

Exercise 2: Solve, the following simultaneous linear differential equation

$$\frac{dx}{dt} + y = \sin t, \quad \frac{dy}{dt} + x = \cos t$$

given that $x = 2$ and $y = 0$ when $t = 0$.

Solution: Given simultaneous equations, in symbolical form, are

$$Dx + y = \sin t \qquad \text{. . . (1)}$$

$$Dy + x = \cos t \qquad \text{. . . (2)}$$

To eliminate y from Eq. (1) and Eq. (2), multiply Eq. (1) by D and subtracting from (2), we get

$$\left(D^2 - 1\right) x = D(\sin t) - \cos t = \cos t - \cos t = 0$$

$$\Rightarrow \quad \left(D^2 - 1\right) x = 0$$

It's A.E. is

$$D^2 - 1 = 0$$

which on solving gives

$$D = \pm 1$$

$$\therefore \quad \text{C.F.} = c_1 e^t + c_2 e^{-t}$$

Hence $x = \text{C.F.+P.I.} = c_1 e^t + c_2 e^{-t}$ . . . (3) $[\because \text{P.I.} = 0]$

To find y, differentiate Eq. (3) w.r.t. t, we get

$$\frac{dx}{dt} = c_1 e^t - c_2 e^{-t} \quad \text{. . . (4)}$$

Using the Eqs. (3) and (4) in Eq. (1), we get

$$c_1 e^t - c_2 e^{-t} + y = \sin t$$

$$\Rightarrow \quad y = -c_1 e^t + c_2 e^{-t} + \sin t \quad \text{. . . (5)}$$

Now given $x = 2$ when $t = 0$

$\therefore$ Eq. (3) gives $2 = c_1 + c_2$ . . . (6)

Also $y = 0$ when $t = 0$

$\therefore$ Eq. (5) gives

$$0 = -c_1 + c_2 \quad \text{. . . (7)}$$

Solving Eqs. (6) and (7), we get $c_2 = 1$, $c_1 = 1$

Hence, from Eq. (3) and Eq. (5), the required solutions of the given simultaneous equations are

$$x = e^t + e^{-t}, \qquad y = e^{-t} - e^t + \sin t.$$

Exercise 6: Solve, the following simultaneous equations,

$$(D-1)x + Dy = 2t + 1 \quad \text{. . . (1)}$$

$$(2D+1)x + 2Dy = t \quad \text{. . . (2)}$$

Solution: To find x, multiplying Eq. (1) by 2 and subtracting from Eq. (2), we get

$$(2D - 2 - 2D - 1)x = 4t + 2 - t = 3t + 2 \quad \text{. . (3)}$$

$$\Rightarrow \quad -3x = 3t + 2$$

$$\Rightarrow \quad x = -t - \frac{2}{3} \quad \text{. . . (4)}$$

To find y, differentiate Eq. (4) w.r.t. t, we get

$$\frac{dx}{dt} = -1$$

Putting this value in Eq. (1), we get

$$-1 - \left(-t - \frac{2}{3}\right) + \frac{dy}{dt} = 2t + 1$$

$$-1 + t + \frac{2}{3} + \frac{dy}{dt} = 2t + 1$$

$$\frac{dy}{dt} = 2t + 1 + 1 - t - \frac{2}{3} = t + \frac{4}{3}$$

Integrating, we get

$$y = \frac{t^2}{2} + \frac{4}{3}t + c_1 \qquad \ldots (5)$$

Hence from Eqs. (5) and (6), the required solution to the given simultaneous equations is

$$x = -t - \frac{2}{3}, \quad y = \frac{t^2}{2} + \frac{4}{3}t + c_1$$

Chapter 4: Linear Algebra

EXERCISE 4.3

Exercise 1: Examine the consistency of the following equation and if consistent, find the complete solution.

$$x + y + z = 6$$
$$x + 2y + 3z = 14$$
$$x + 4y + 7z = 30$$

Solution: The augmented matrix is $[A:B] = \begin{pmatrix} 1 & 1 & 1 & : & 6 \\ 1 & 2 & 3 & : & 14 \\ 1 & 4 & 7 & : & 30 \end{pmatrix}$ $\quad R_2 \to R_2 - R_1,$ $R_3 \to R_3 - R_1$

$$\begin{pmatrix} 1 & 1 & 1 & : & 6 \\ 0 & 1 & 2 & : & 8 \\ 0 & 3 & 6 & : & 24 \end{pmatrix} \qquad R_3 \to \frac{1}{3} R_3$$

$$\begin{pmatrix} 1 & 1 & 1 & : & 6 \\ 0 & 1 & 2 & : & 8 \\ 0 & 1 & 2 & : & 8 \end{pmatrix} \qquad R_3 \to R_3 - R_2$$

$$\begin{pmatrix} 1 & 1 & 1 & : & 6 \\ 0 & 1 & 2 & : & 8 \\ 0 & 0 & 0 & : & 0 \end{pmatrix}$$

Clearly $\rho(A) = \rho([A:B]) = 2$

Hence, the system is consistent.

Since $\rho(A) < 3$ (number of unknowns)

$\therefore$ the system has an infinite number of solutions

Now the reduced system of linear equation is

$$x + y + z = 6$$

$$y + 2z = 8$$

$$\Rightarrow \quad \begin{aligned} y &= 8 - 2z \\ x &= 6 - y - z = 6 - (8 - 2z) - z = -2 + z \end{aligned}$$

Taking $z = k$, we have

$$x = k - 2$$

$$y = 8 - 2k$$

$$z = k$$

where k is arbitrary, is the general solution.

Exercise 2: Test for consistency the following system of equations and if found consistent solve it.

$$5x+3y+7z=4$$
$$3x+26y+2z=9$$
$$7x+2y+10z=5$$

Solution: The given equation can be written in the matrix form as

$$\begin{pmatrix} 5 & 3 & 7 \\ 3 & 26 & 2 \\ 7 & 2 & 10 \end{pmatrix}\begin{pmatrix} x \\ y \\ z \end{pmatrix}=\begin{pmatrix} 4 \\ 9 \\ 5 \end{pmatrix}$$

Now the augmented matrix $[A:B]=\begin{pmatrix} 5 & 3 & 7 & : & 4 \\ 3 & 26 & 2 & : & 9 \\ 7 & 2 & 10 & : & 5 \end{pmatrix}$

$$\begin{pmatrix} 1 & 3/5 & 7/5 & : & 4/5 \\ 3 & 26 & 2 & : & 9 \\ 7 & 2 & 10 & : & 5 \end{pmatrix} \qquad R_1 \to \frac{1}{5}R_1$$

$$\begin{pmatrix} 1 & 3/5 & 7/5 & : & 4/5 \\ 0 & 121/5 & -11/5 & : & 33/5 \\ 0 & -11/5 & 1/5 & : & -3/5 \end{pmatrix} \qquad R_2 \to R_2-3R_1, R_3 \to R_3-7R_1$$

$$\begin{pmatrix} 1 & 3/5 & 7/5 & : & 4/5 \\ 0 & 121/5 & -11/5 & : & 33/5 \\ 0 & 0 & 0 & : & 0 \end{pmatrix} \qquad R_3 \to 11R_3+R_2$$

Rank A = Rank $[A: B]$ =2 < 3, the number of unknown.

Thus, there are infinite numbers of solutions and we have

$$\begin{pmatrix} 1 & 3/5 & 7/5 \\ 0 & 121/5 & -11/5 \\ 0 & 0 & 0 \end{pmatrix}\begin{pmatrix} x \\ y \\ z \end{pmatrix}=\begin{pmatrix} 4/5 \\ 33/5 \\ 0 \end{pmatrix}$$

$\therefore \quad x+\frac{3}{5}y+\frac{7}{5}z=\frac{4}{5}$

$$\frac{121}{5}y-\frac{11}{5}z=\frac{33}{5} \quad \Rightarrow \quad 5x+3y+7z=4$$

$$11y-z=3$$

Let $z=k$

$\therefore \quad 11y=k+3 \quad \Rightarrow \quad y=\frac{k+3}{11}$

$$5x=4-3y-7z=4-\frac{3k+9}{11}-7k$$

$$=\frac{44-3k-9-77k}{11}=\frac{-80k+35}{11}$$

$\therefore \quad x=\frac{-16k+7}{11}$

Hence $x=\frac{-16}{11}k+\frac{7}{11},\ y=\frac{k}{11}+\frac{3}{11}, z=k,$ where k is arbitrary

Exercise 3: Solve the system of equations:

$$2x_1+x_2+2x_3+x_4=6$$

$$6x_1-6x_2+6x_3+12x_4=36$$

$$4x_1+3x_2+3x_3-3x_4=-1$$

$$2x_1+2x_2-x_3+x_4=10$$

Solution: In matrix notation, the given system of equations can be written as

$$AX=B$$

where $A=\begin{pmatrix} 2 & 1 & 2 & 1 \\ 6 & -6 & 6 & 12 \\ 4 & 3 & 3 & -3 \\ 2 & 2 & -1 & 1 \end{pmatrix},\ X=\begin{pmatrix} x_1 \\ x_2 \\ x_3 \\ x_4 \end{pmatrix},\ B=\begin{pmatrix} 6 \\ 36 \\ -1 \\ 10 \end{pmatrix}$

Augmented matrix $(A:B)=\begin{pmatrix} 2 & 1 & 2 & 1 & : & 6 \\ 6 & -6 & 6 & 12 & : & 36 \\ 4 & 3 & 3 & -3 & : & -1 \\ 2 & 2 & -1 & 1 & : & 10 \end{pmatrix}$ $\quad R_2 \to R_2 - 3R_1,\ R_3 \to R_3 - 2R_1,\ R_4 \to R_4 - R_1$

$$\sim \begin{pmatrix} 2 & 1 & 2 & 1 & : & 6 \\ 0 & -9 & 0 & 9 & : & 18 \\ 0 & 1 & -1 & -5 & : & -13 \\ 0 & 1 & -3 & 0 & : & 4 \end{pmatrix} \quad R_2 - \frac{1}{9}R_2$$

$$\sim \begin{pmatrix} 2 & 1 & 2 & 1 & : & 6 \\ 0 & 1 & 0 & -1 & : & -2 \\ 0 & 1 & -1 & -5 & : & -13 \\ 0 & 1 & -3 & 0 & : & 4 \end{pmatrix} \quad R_1 \to R_1 - R_2, R_3 \to R_3 - R_2, R_4 \to R_4 - R_2$$

$$\sim \begin{pmatrix} 2 & 0 & 2 & 2 & : & 8 \\ 0 & 1 & 0 & -1 & : & -2 \\ 0 & 0 & -1 & -4 & : & -11 \\ 0 & 0 & -3 & 1 & : & 6 \end{pmatrix} \quad R_4 \to R_4 - 3R_3, R_1 \to \frac{1}{2}R_1$$

$$\sim \begin{pmatrix} 1 & 0 & 1 & 1 & : & 4 \\ 0 & 1 & 0 & -1 & : & -2 \\ 0 & 0 & -1 & -4 & : & -11 \\ 0 & 0 & 0 & 13 & : & 39 \end{pmatrix} \quad R_1 \to R_1 + R_3, R_4 \to \frac{1}{13}R_4$$

$$\sim \begin{pmatrix} 1 & 0 & 0 & -3 & : & -7 \\ 0 & 1 & 0 & -1 & : & -2 \\ 0 & 0 & -1 & -4 & : & -11 \\ 0 & 0 & 0 & 1 & : & 3 \end{pmatrix} \quad R_1 \to R_1 + 3R_4, R_2 \to R_2 + R_4, R_3 \to R_3 + 4R_4$$

$$\sim \begin{pmatrix} 1 & 0 & 0 & 0 & : & -2 \\ 0 & 1 & 0 & 0 & : & 1 \\ 0 & 0 & -1 & 0 & : & 1 \\ 0 & 0 & 0 & 1 & : & 3 \end{pmatrix} \quad R_3 \to -R_3$$

$$\sim \begin{pmatrix} 1 & 0 & 0 & 0 & : & -2 \\ 0 & 1 & 0 & 0 & : & 1 \\ 0 & 0 & 1 & 0 & : & -1 \\ 0 & 0 & 0 & 1 & : & 3 \end{pmatrix}$$

Hence, $x_1 = -2,\ x_2 = 1,\ x_3 = -1,\ x_4 = 3.$

Exercise 4: Using matrix method, show that the equations:

$$3x+3y+2z=1$$

$$x+2y=4$$

$$10y+3z=-2$$

$$2x-3y-z=5$$

are consistent and hence obtain the solution for x, y and z.

Solution: In matrix notation, the given system of equations can be written as $AX=B$

where $A=\begin{pmatrix} 3 & 3 & 2 \\ 1 & 2 & 0 \\ 0 & 10 & 3 \\ 2 & -3 & -1 \end{pmatrix}$, $X=\begin{pmatrix} x \\ y \\ z \end{pmatrix}$, $B=\begin{pmatrix} 1 \\ 4 \\ -2 \\ 5 \end{pmatrix}$

Augmented matrix $(A \vdots B) = \begin{pmatrix} 3 & 3 & 2 & \vdots & 1 \\ 1 & 2 & 0 & \vdots & 4 \\ 0 & 10 & 3 & \vdots & -2 \\ 2 & -3 & -1 & \vdots & 5 \end{pmatrix}$ $\quad R_1 \leftrightarrow R_2$

$$\sim \begin{pmatrix} 1 & 2 & 0 & \vdots & 4 \\ 3 & 3 & 2 & \vdots & 1 \\ 0 & 10 & 3 & \vdots & -2 \\ 2 & -3 & -1 & \vdots & 5 \end{pmatrix} \quad R_2 \to R_2 - 3R_1, R_4 \to R_4 - 2R_1$$

$$\sim \begin{pmatrix} 1 & 2 & 0 & \vdots & 4 \\ 0 & -3 & 2 & \vdots & -11 \\ 0 & 10 & 3 & \vdots & -2 \\ 0 & -7 & -1 & \vdots & -3 \end{pmatrix} \quad R_3 \to R_3 + 3R_2, R_4 \to R_4 - 2R_2$$

$$\sim\begin{pmatrix} 1 & 2 & 0 & \vdots & 4 \\ 0 & -3 & 2 & \vdots & -11 \\ 0 & 1 & 9 & \vdots & -35 \\ 0 & -1 & -5 & \vdots & -19 \end{pmatrix} \quad R_1 \to R_1 - 2R_3, R_2 \to R_2 + 3R_3, R_4 \to R_4 + R_3$$

$$\sim\begin{pmatrix} 1 & 0 & 18 & \vdots & 74 \\ 0 & 0 & 29 & \vdots & -116 \\ 0 & 1 & 9 & \vdots & -35 \\ 0 & 0 & 4 & \vdots & -16 \end{pmatrix} \quad R_2 \leftrightarrow R_3, R_4 \to \frac{1}{4} R_4$$

$$\sim\begin{pmatrix} 1 & 0 & -18 & \vdots & 74 \\ 0 & 1 & 9 & \vdots & -35 \\ 0 & 0 & 29 & \vdots & -116 \\ 0 & 0 & 1 & \vdots & -4 \end{pmatrix} \quad R_1 \to R_1 + 18R_4, R_2 \to R_2 - 9R_4, R_3 \to R_3 - 29R_4$$

$$\sim\begin{pmatrix} 1 & 0 & 0 & \vdots & 2 \\ 0 & 1 & 0 & \vdots & 1 \\ 0 & 0 & 0 & \vdots & 0 \\ 0 & 0 & 1 & \vdots & -4 \end{pmatrix} \quad R_3 \leftrightarrow R_4$$

$$\sim\begin{pmatrix} 1 & 0 & 0 & \vdots & 2 \\ 0 & 1 & 0 & \vdots & 1 \\ 0 & 0 & 1 & \vdots & -4 \\ 0 & 0 & 0 & \vdots & 0 \end{pmatrix}$$

$\rho(A) = \rho(A \vdots B) = 3 =$ number of unknowns

$\Rightarrow$ The given system of equation is consistent and the solution is

$x = 2,\ y = 1,\ z = -4.$

Chapter 5: Infinite Series

EXERCISE 5.1

Exercise 6 (*iii*): $\frac{1}{\sqrt{1\cdot 2}}+\frac{1}{\sqrt{2\cdot 3}}+\frac{1}{\sqrt{3\cdot 4}}+...\text{to }\infty$

Solution: Here $u_n=\frac{1}{\sqrt{n(n+1)}}=\frac{1}{n\sqrt{1+\frac{1}{n}}}$

Using $T_n=a+(n-1)d$,
Here, $T_n=n$ th term of 1,3,5, ...
$=1+(n-1).2$
$=2n-1$

Take $v_n=\frac{1}{n}$

Consider $\underset{n\to\infty}{\text{Lt}}\frac{u_n}{v_n}=\underset{n\to\infty}{\text{Lt}}\frac{1}{\sqrt{\left(1+\frac{1}{n}\right)}}\frac{1}{\sqrt{1+0}}=1,$ which is finite and non-zero.

Therefore, by comparison test, $\sum_{n=1}^{\infty}u_n$ and $\sum_{n=1}^{\infty}v_n$ converges or diverges together.

But $\sum_{n=1}^{\infty}v_n=\sum_{n=1}^{\infty}\frac{1}{n}$ is a p-series with $p=1$, and hence divergent.

$\therefore$ The given series $\sum_{n=1}^{\infty}u_n$ is also divergent.

Exercise 11 (*k*): $\sum_{n=1}^{\infty}\left(\sqrt{n^3+1}-\sqrt{n^3}\right).$

Solution. Here $u_n=\sqrt{n^3+1}-\sqrt{n^3}=\sqrt{n^3+1}-\sqrt{n^3}\times\frac{\sqrt{n^3+1}+\sqrt{n^3}}{\sqrt{n^3+1}+\sqrt{n^3}}$

$$=\frac{n^3+1-n^3}{\sqrt{n^3+1}+\sqrt{n^3}}=\frac{1}{\sqrt{n^3+1}+\sqrt{n^3}}=\frac{1}{n^{\frac{3}{2}}\left(\sqrt{1+\frac{1}{n^3}}+1\right)}$$

Take $v_n=\frac{1}{n^{\frac{3}{2}}}$

Consider $\underset{n\to\infty}{\text{Lt}}\dfrac{u_n}{v_n}=\underset{n\to\infty}{\text{Lt}}\dfrac{1}{\sqrt{1+\dfrac{1}{n^3}}+1}=\dfrac{1}{\sqrt{1+0}+1}=\dfrac{1}{2}$, which is finite and non-zero.

$\therefore$ By comparison test, $\sum\limits_{n=1}^{\infty}u_n$ and $\sum\limits_{n=1}^{\infty}v_n$ converges or diverges together.

But $\sum\limits_{n=1}^{\infty}v_n=\sum\limits_{n=1}^{\infty}\dfrac{1}{n^{\frac{3}{2}}}$ is $p-$series with $p=\dfrac{3}{2}>1$ and hence convergent.

Therefore, the given series $\sum\limits_{n=1}^{\infty}u_n$ is also convergent.

EXERCISE 5.3

Exercise 11 (*c*): $\sum\limits_{n=1}^{\infty}\dfrac{n}{n^2+1}x^n, x>0$.

Solution: Here $u_n=\dfrac{n}{n^2+1}x^n$

$$u_{n+1}=\frac{n+1}{(n+1)^2+1}x^{n+1}$$

$x>0$ Suggests that the given series is of positive terms. Hence we can apply D' Alembert's Ratio test.

$$\frac{u_n}{u_{n+1}}=\frac{n}{n^2+1}x^n.\frac{(n+1)^2+1}{n+1}.\frac{1}{x^{n+1}}$$

$$=\frac{n}{n+1}.\frac{(n+1)^2+1}{n^2+1}.\frac{1}{x}$$

$$\Rightarrow \quad \underset{n\to\infty}{\text{Lt}}\frac{u_n}{u_{n+1}}=\underset{n\to\infty}{\text{Lt}}\frac{n}{n+1}.\frac{(n+1)^2+1}{n^2+1}.\frac{1}{x}$$

$$=\underset{n\to\infty}{\text{Lt}}\frac{n}{n\left(1+\frac{1}{n}\right)}.\frac{n^2\left[\left(1+\frac{1}{n}\right)^2+\frac{1}{n^2}\right]}{n^2\left(1+\frac{1}{n^2}\right)}.\frac{1}{x}$$

$$= \underset{n\to\infty}{\text{Lt}} \frac{1}{1+\frac{1}{n}} \cdot \frac{\left(1+\frac{1}{n}\right)^2 + \frac{1}{n^2}}{1+\frac{1}{n^2}} \cdot \frac{1}{x} = \frac{1}{1+0} \cdot \frac{(1+0)^2+0}{1+0} \cdot \frac{1}{x} = \frac{1}{x}$$

$\therefore$ By D' Alembert's Ratio test, the given series $\sum_{n=1}^{\infty} u_n$ is convergent for $\frac{1}{x} > 1$ and divergent for $\frac{1}{x} < 1$. When $x = 1$, Ratio test fails.

Now, when $x = 1,\ u_n = \frac{n}{n^2+1} = \frac{n}{n^2\left(1+\frac{1}{n^2}\right)} = \frac{1}{n\left(1+\frac{1}{n^2}\right)}$

Take $v_n = \frac{1}{n}$

$$\therefore \quad \frac{u_n}{v_n} = \frac{1}{1+\frac{1}{n^2}}$$

$$\Rightarrow \quad \underset{n\to\infty}{\text{Lt}} \frac{u_n}{v_n} = \underset{n\to\infty}{\text{Lt}} \frac{1}{1+\frac{1}{n^2}} = \frac{1}{1+0} = 1 \quad \text{which is finite and non-zero.}$$

$\therefore$ By Comparison test, $\sum_{n=1}^{\infty} u_n$ and $\sum_{n=1}^{\infty} v_n$ converges or diverges together.

But $\sum_{n=1}^{\infty} v_n = \sum_{n=1}^{\infty} \frac{1}{n}$ is *p*-series with *p=1* and hence divergent.

Therefore, the series $\sum_{n=1}^{\infty} u_n$ is divergent when $x = 1$, i.e., we can say that the given series is convergent for $x < 1$ and divergent for $x \geq 1$.

Exercise 13 (*a*): Discuss the convergence of the following series;

$$1 + \frac{x}{2} + \frac{x^2}{5} + \frac{x^3}{10} + \ldots + \frac{x^n}{n^2+1} + \ldots \text{to } \infty,\ x > 0.$$

Solution. Neglecting first term,

$$u_n = \frac{x^n}{n^2+1},\ x > 0$$

Here $x > 0$, suggests that the given series is of positive terms.

$$u_{n+1}=\frac{x^{n+1}}{(n+1)^2+1}$$

$$\therefore \quad \frac{u_n}{u_{n+1}}=\frac{x^n}{n^2+1}\cdot\frac{(n+1)^2+1}{x^{n+1}}=\frac{(n+1)^2+1}{n^2+1}\cdot\frac{1}{x}$$

$$\Rightarrow \quad \underset{n\to\infty}{\text{Lt}}\frac{u_n}{u_{n+1}}=\underset{n\to\infty}{\text{Lt}}\frac{n^2\left[\left(1+\frac{1}{n}\right)^2+\frac{1}{n^2}\right]}{n^2\left(1+\frac{1}{n^2}\right)}\cdot\frac{1}{x}=\underset{n\to\infty}{\text{Lt}}\frac{\left(1+\frac{1}{n}\right)^2+\frac{1}{n^2}}{\left(1+\frac{1}{n^2}\right)}\cdot\frac{1}{x}$$

$$=\frac{(1+0)^2+0}{(1+0)}\cdot\frac{1}{x}=\frac{1}{x}$$

$\therefore$ By D' Alembert's Ratio test, the given series $\sum_{n=1}^{\infty}u_n$ is convergent for $\frac{1}{x}>1$ and divergent for $\frac{1}{x}<1$. When $\frac{1}{x}=1$, i.e., $x=1$, Ratio test fails.

For $x=1$, $u_n=\frac{1}{n^2+1}=\frac{1}{n^2\left(1+\frac{1}{n^2}\right)}$

Take $v_n=\frac{1}{n^2}$

$$\therefore \quad \frac{u_n}{v_n}=\frac{1}{1+\frac{1}{n^2}}$$

$$\Rightarrow \quad \underset{n\to\infty}{\text{Lt}}\frac{u_n}{v_n}=\underset{n\to\infty}{\text{Lt}}\frac{1}{1+\frac{1}{n^2}}=\frac{1}{1+0}=1,$$ which is finite and non-zero.

$\therefore$ By Comparison test, $\sum_{n=1}^{\infty}u_n$ and $\sum_{n=1}^{\infty}v_n$ converges or diverges together. But $\sum_{n=1}^{\infty}v_n=\sum_{n=1}^{\infty}\frac{1}{n^2}$ is $p-$series with $p=2$ and hence divergent.

Therefore, the given series $\sum_{n=1}^{\infty}u_n$ is also convergent when $x=1$.

EXERCISE 5.4

Exercise 13 (*a*): Discuss the convergence of the following series.

$$1+\frac{3}{7}x+\frac{3\cdot 6}{7\cdot 10}x^2+\frac{3\cdot 6\cdot 9}{7\cdot 10\cdot 13}x^3+\ldots\text{to }\infty,\ x>0$$

Solution: Neglecting the first term,

Using $T_n = a+(n-1)d$,
$T_n = n$th term of 3, 6, ,...
$= 3+(n-1).3 = 3n$
Also nth term of 7,10,13,...
$= 7+(n-1).3$
$= 3n+4$

$$u_n = \frac{3\cdot 6\cdot 9\ldots 3n}{7\cdot 10\cdot 13\ldots(3n+4)}\cdot x^n$$

$$u_{n+1} = \frac{3\cdot 6\cdot 9\ldots 3n(3n+3)}{7\cdot 10\cdot 13\ldots(3n+4)(3n+7)}\cdot x^{n+1}$$

$$\therefore \quad \frac{u_n}{u_{n+1}} = \frac{3n+7}{3n+3}.\frac{1}{x} \qquad \ldots(1)$$

$$\Rightarrow \quad \underset{n\to\infty}{\text{Lt}}\frac{u_n}{u_{n+1}} = \underset{n\to\infty}{\text{Lt}}\frac{3n+7}{3n+3}.\frac{1}{x} = \underset{n\to\infty}{\text{Lt}}\frac{1+\frac{7}{3n}}{1+\frac{1}{n}}.\frac{1}{x} = \frac{1+0}{1+0}.\frac{1}{x} = \frac{1}{x}$$

$\therefore$ By D' Alembert's Ratio test, the given series $\sum_{n=1}^{\infty} u_n$ is convergent for $\frac{1}{x}>1$ and divergent for $\frac{1}{x}<1$. For $\frac{1}{x}=1$, i.e., $x=1$. Ratio test fails.

Therefore, we apply Raabe's test, we have, from (1),

$$n\left(\frac{u_n}{u_{n+1}}-1\right) = n\left(\frac{3n+7}{3n+3}-1\right) = \frac{n(3n+7-3n-3)}{3n+3} = \frac{4n}{3n+3}$$

$$\Rightarrow \quad \underset{n\to\infty}{\text{Lt}}\, n\left(\frac{u_n}{u_{n+1}}-1\right) = \underset{n\to\infty}{\text{Lt}}\frac{4n}{3n+3} = \underset{n\to\infty}{\text{Lt}}\frac{4}{3+\frac{3}{n}} = \frac{4}{3+0} = \frac{4}{3} > 1$$

$\therefore$ By Raabe's test, the given series $\sum_{n=1}^{\infty} u_n$ is convergent for $x=1$.

Aliter. We can also apply Gauss test, after the failure of D'Alembert Ratio test.

From(1), when $x=1$, we have

$$\frac{u_n}{u_{n+1}} = \frac{3n+7}{3n+3} = \frac{1+\frac{7}{3n}}{1+\frac{1}{n}} = \left(1+\frac{7}{3}n\right)\left(1+\frac{1}{n}\right)^{-1}$$

$$= \left(1+\frac{7}{3}n\right)\left(1-\frac{1}{n}+\frac{1}{n^2}-\ldots\right) \qquad (1+x)^{-1} = 1-x+x^2-\ldots$$

$$= 1-\frac{1}{n}+\frac{7}{3n}+O\left(\frac{1}{n^2}\right)$$

$$= 1+\frac{\frac{4}{3}}{n}+O\left(\frac{1}{n^2}\right)$$

Here $\lambda = \frac{4}{3} > 1$

$\therefore$ By Gauss test, the given series $\sum_{n=1}^{\infty} u_n$ is convergent for $x = 1$.

Hence, we can say, that the given series is convergent for $x \le 1$ and divergent.

Chapter 6: Complex Numbers

EXERCISE 6.2

Exercise 2: Given $2\cos\theta = x+\frac{1}{x}$ and $2\cos\phi = y+\frac{1}{y}$, prove that $2\cos(\theta+\phi) = xy+\frac{1}{xy}$.

Solution: Given that

$$2\cos\theta = x+\frac{1}{x} \quad \Rightarrow \quad x^2 - 2x\cos\theta + 1 = 0$$

$$\Rightarrow \quad x = \frac{2\cos\theta \pm \sqrt{4\cos^2\theta - 4}}{2} = \frac{2\cos\theta \pm 2\sqrt{\cos^2\theta - 1}}{2} = \cos\theta \pm i\sin\theta$$

Take $a = \cos\theta + i\sin\theta$

Similarly, $y = \cos\phi + i\sin\phi$

Now $xy+\frac{1}{xy} = (\cos\theta + i\sin\theta)(\cos\phi + i\sin\phi) + \frac{1}{(\cos\theta + i\sin\theta)(\cos\phi + i\sin\phi)}$

$$= \cos(\theta+\phi) + i\sin(\theta+\phi) + \frac{1}{\cos(\theta+\phi)+i\sin(\theta+\phi)}$$

$$\left[\because \text{cis}\,\theta_1 \text{cis}\,\theta_2 = \text{cis}(\theta_1+\theta_2)\right]$$

$$= \cos(\theta+\phi) + i\sin(\theta+\phi) + \left(\cos(\theta+\phi)+i\sin(\theta+\phi)\right)^{-1}$$

$$= \cos(\theta+\phi) + i\sin(\theta+\phi) + \cos(\theta+\phi) - i\sin(\theta+\phi)$$

$$= 2\cos(\theta+\phi)$$

Hence $\quad 2\cos(\theta+\phi) = xy + \dfrac{1}{xy}.$

Exercise 3: Prove that $(1+i)^n + (1-i)^n = 2^{\frac{n}{2}+1}\cos\dfrac{n\pi}{4}$

Solution: Let $1+i = r(\cos\theta + i\sin\theta)$

Equating real and imaginary parts, we get

$$1 = r\cos\theta \qquad \ldots (1)$$

$$1 = r\sin\theta \qquad \ldots (2)$$

Squaring and adding Eq. (1) and Eq. (2) , we get

$$1^2 + 1^2 = r^2 \quad \Rightarrow \quad r = \sqrt{2}$$

Also from Eq. (1) $\cos\theta = \dfrac{1}{r} = \dfrac{1}{\sqrt{2}}$, and from Eq. (2) $\sin\theta = \dfrac{1}{r} = \dfrac{1}{\sqrt{2}}$

Both these equations are satisfied when $\theta = \dfrac{\pi}{4}$

$$\therefore \quad 1+i = \sqrt{2}\left(\cos\frac{\pi}{4} + i\sin\frac{\pi}{4}\right)$$

Similarly

$$1-i = \sqrt{2}\left(\cos\frac{\pi}{4} - i\sin\frac{\pi}{4}\right)$$

$$\therefore \qquad (1+i)^n + (1-i)^n = \left(\sqrt{2}\left(\cos\frac{\pi}{4} + i\sin\frac{\pi}{4}\right)\right)^n + \left(\sqrt{2}\left(\cos\frac{\pi}{4} - i\sin\frac{\pi}{4}\right)\right)^n$$

$$= 2^{\frac{n}{2}}\left(\cos\frac{n\pi}{4} + i\sin\frac{n\pi}{4}\right) + 2^{\frac{n}{2}}\left(\cos\frac{n\pi}{4} - i\sin\frac{n\pi}{4}\right) \quad \left[\begin{array}{l}\because [r(\cos\theta + i\sin\theta)] \\ = r^n \cos n\theta + i\sin n\theta\end{array}\right]$$

$$= 2^{\frac{n}{2}}\left(\cos\frac{n\pi}{4} + i\sin\frac{n\pi}{4} + \cos\frac{n\pi}{4} - i\sin\frac{n\pi}{4}\right)$$

$$= 2^{\frac{n}{2}}\left(2\cos\frac{n\pi}{4}\right) = 2^{\frac{n}{2}+1}\cos\frac{n\pi}{4}$$ Hence the proof.

EXERCISE 6.3

Exercise 2: Prove that $\exp(z+\pi i) = -\exp(z)$.

Solution: $LHS = \exp(z+\pi i) = e^{z+i\pi} = e^z e^{i\pi}$

$$= e^z(\cos\pi + i\sin\pi) = e^z(-1+i.0) = -e^z = -\exp(z)$$

Exercise 5: Show that (*a*) $\text{Log}(i) = (4n+1)\frac{\pi}{2}i$, (*b*) $\text{Log}(1-i) = \frac{1}{2}\log 2 - \frac{\pi}{4}i + 2n\pi i$

Solution: (*a*) $\text{Log}(i) = \log i + 2n\pi i = \log(0+i) + 2n\pi i = \frac{1}{2}\log(0+1) + i\tan^{-1}\frac{1}{0} + 2n\pi i$

$$= i\frac{\pi}{2} + 2n\pi i = (4n+1)\frac{\pi}{2}i$$

(*b*) $\text{Log}(1-i) = 2n\pi i + \log(1-i) = 2n\pi i + \frac{1}{2}\log(1+1) + i\tan^{-1}\left(\frac{-1}{1}\right)$

$$= 2n\pi i + \frac{1}{2}\log 2 + i\tan^{-1}(-1) = 2n\pi i + \frac{1}{2}\log 2 + i\left(-\frac{\pi}{4}\right) = 2n\pi i + \frac{1}{2}\log 2 - \frac{\pi}{4}i$$

$$= \frac{1}{2}\log 2 - \frac{\pi}{4}i + 2n\pi i$$

Exercise 9: Prove that (*a*) $(-i)^{-i}=e^{(4n-1)\frac{\pi}{2}}$, (*b*) $i^{\sqrt{2}}=\text{cis}\dfrac{\sqrt{2}\pi}{2}$

Solution: (*a*) $(-i)^{-i}=e^{\log(-i)^{-i}}=e^{-i[\text{Log}(-i)]}=e^{-i\left[2n\pi i+i\left(-\frac{\pi}{2}\right)\right]}$

$$=e^{-i^2(4n-1)\frac{\pi}{2}}=e^{(4n-1)\frac{\pi}{2}}$$

(*b*) $i^{\sqrt{2}}=e^{\log i^{\sqrt{2}}}=e^{\sqrt{2}\log i}=e^{\sqrt{2}\frac{\pi}{2}i}=e^{\frac{\sqrt{2}\pi}{2}i}=\text{cis}\dfrac{\sqrt{2}\pi}{2}$

EXERCISE 6.6

Exercise 6: Find the sum to infinity of the series

$$1-\frac{1}{2}\cos\theta+\frac{1\cdot3}{2\cdot4}\cos 2\theta-\frac{1\cdot3\cdot3}{2\cdot4\cdot6}\cos 3\theta+\ldots,\ (-\pi<\theta<\pi).$$

Solution: $C=1-\dfrac{1}{2}\cos\theta+\dfrac{1\cdot3}{2\cdot4}\cos 2\theta-\dfrac{1\cdot3\cdot5}{2\cdot4\cdot6}\cos 3\theta+\cdots\infty$

$$S=-\frac{1}{2}\sin\theta+\frac{1\cdot3}{2\cdot4}\sin 2\theta-\frac{1\cdot3\cdot5}{2\cdot4\cdot6}\sin 3\theta+\cdots\infty$$

$$\therefore\ C+iS=1-\frac{1}{2}e^{i\theta}+\frac{1\cdot3}{2\cdot4}e^{2i\theta}-\frac{1\cdot3\cdot5}{2\cdot4\cdot6}e^{3i\theta}+\cdots\infty=(1+e^{i\theta})^{-\frac{1}{2}}$$

$$=(1+\cos\theta+i\sin\theta)^{-\frac{1}{2}}$$

$$=\left[2\cos^2\frac{\theta}{2}+2i\sin\frac{\theta}{2}\cos\frac{\theta}{2}\right]^{-\frac{1}{2}}=\left(2\cos\frac{\theta}{2}\right)^{-\frac{1}{2}}\left[\cos\frac{\theta}{2}+i\sin\frac{\theta}{2}\right]^{-\frac{1}{2}}$$

$$=\left(2\cos\frac{\theta}{2}\right)^{-\frac{1}{2}}\left(\text{cis}\frac{\theta}{2}\right)^{-\frac{1}{2}}=\left(2\cos\frac{\theta}{2}\right)^{-\frac{1}{2}}\left(\cos\frac{\theta}{4}-i\sin\frac{\theta}{4}\right)$$

Comparing real part on both sides, we get

$$C=\left(2\cos\frac{\theta}{2}\right)^{-\frac{1}{2}}\cos\frac{\theta}{4}$$

Exercise 8: Sum the series $\sin\alpha-\dfrac{\sin(\alpha+2\beta)}{2!}+\dfrac{\sin(\alpha+4\beta)}{4!}-\ldots\infty$

Solution: Let $S=\sin\alpha-\dfrac{\sin(\alpha+2\beta)}{2!}+\dfrac{\sin(\alpha+4\beta)}{4!}-\ldots\infty$

$$C = \cos\alpha - \frac{\cos(\alpha+2\beta)}{2!} + \frac{\cos(\alpha+4\beta)}{4!} - \ldots\infty$$

$$\therefore \quad C + iS = [\cos\alpha + i\sin\alpha] - \left[\frac{\cos(\alpha+2\beta)+i\sin(\alpha+2\beta)}{2!}\right]$$

$$+\left[\frac{\cos(\alpha+4\beta)+i\sin(\alpha+4\beta)}{4!}\right] - \ldots$$

$$= e^{i\alpha} - \frac{1}{2!}e^{i(\alpha+2\beta)} + \frac{1}{4!}e^{i(\alpha+4\beta)} - \ldots$$

$$= e^{i\alpha}\left[1 - \frac{1}{2!}e^{2i\beta} + \frac{1}{4!}e^{4i\beta} - \ldots\right] = e^{i\alpha}\left[1 - \frac{x^2}{2!} + \frac{x^4}{4!} - \ldots\right] \quad \text{where } x = e^{i\beta}$$

$$= e^{i\alpha}.\cos x = (\cos\alpha + i\sin\alpha).\cos(e^{i\beta})$$

$$= (\cos\alpha + i\sin\alpha).\cos(\cos\beta + i\sin\beta)$$

$$= (\cos\alpha + i\sin\alpha).[\cos(\cos\beta)\cos(i\sin\beta) - \sin(\cos\beta)\sin(i\sin\beta)]$$

$$= (\cos\alpha + i\sin\alpha).[\cos(\cos\beta)\cosh(\sin\beta) - \sin(\cos\beta).i\sinh(\sin\beta)]$$

$$= [\cos\alpha\cos(\cos\beta)\cosh(\sin\beta) + \sin\alpha\sin(\cos\beta).\sinh(\sin\beta)]$$

$$+ i[\sin\alpha\cos(\cos\beta)\cosh(\sin\beta) - \cos\alpha\sin(\cos\beta).\sinh(\sin\beta)]$$

Equating imaginary parts, we get

$$S = \sin\alpha\cos(\cos\beta)\cosh(\sin\beta) - \cos\alpha\sin(\cos\beta).\sinh(\sin\beta).$$

❑ ❑ ❑

Appendix B

Latest PTU Question Papers and Their Solutions

ENGINEERING MATHEMATICS-II (BTAM-102)
APRIL/MAY, 2012

Time: 3 Hrs. Max. Marks: 60

INSTRUCTION TO CANDIDATES:

1. SECTION–A is COMPULSORY.
2. Attempt any FIVE questions from SECTION - B & C.
3. Select atleast TWO questions from SECTION – B & C.

SECTION–A

(10 × 2 = 20 Marks)

1. Solve the following sums:

(*a*) Find the general value of $\log\left(-1+\sqrt{3}i\right)$.

(*b*) Solve the differential equation $\dfrac{d^3y}{dx^3}+y=0$.

(*c*) Under what conditions on '*a*, *b*, *c* and d', the differential equation

$$(a\sinh x\cos y+b\cosh x\sin y)dx+\left(c\sinh x\cos y+d\cosh x\sin y\right)dy=0$$

is exact?

(*d*) Find all the roots of $(-8i)^{1/3}$.

(*e*) Test for what values of '*k*' the set of vectors $\{(k,1,1),\ (0,1,1),(k,0,k)\}$ is linearly independent.

(*f*) Examine the convergence/divergence of the series $\sum_{n=1}^{\infty}\dfrac{x^n}{(2n)!}$.

(*g*) Test the absolute convergence of the series $\sum_{n=2}^{\infty}\dfrac{(-1)^n}{n(\log n)^2}$.

(*h*) Test whether the matrix $\begin{pmatrix} 3 & 1 & -1 \\ -2 & 1 & 2 \\ 0 & 1 & 2 \end{pmatrix}$ is diagonalizable or not?

(*i*) For what values of '*k*', the system of equations

$$x+y+z=2;\ x+2y+z=-2;\ x+y+(k-5)z=k$$

has no solution.

(*j*) Express $\cos 6\theta$ in terms of powers of $\cos\theta$.

SECTION–B **(4 × 5 = 20 Marks)**

2. (*a*) Find the general solution of the differential equation
$(3x^2y^3e^y + y^3 + y^2)dx + (x^2y^3e^y - xy)dy = 0$.

(*b*) Obtain the general solution of the differential equation $y'' + 3y' + 2y = \sin(e^x)$, by using method of variation of parameters.

3. (*a*) Solve the following simultaneous differential equations
$\frac{dx}{dt} + y = \sin t,\ \frac{dy}{dt} + x = \cos t,\ y(0) = 0,\ x(0) = 2.$

(*b*) Find the complete solution of the differential equation
$(x+1)^2 y'' + (x+1)y' + y = \sin(2\log(1+x))$, by using operator method.

4. (*a*) Solve the differential equation $x\left(\frac{dy}{dx} + y\right) = 1 - y$.

(*b*) Find the particular solution of the differential equation $y'' + a^2 y = \sec ax$.

5. An L-C-R circuit with battery e.m.f. '$E \sin pt$' is tuned to resonance so that $p^2 = \frac{1}{LC}$. If initially the current i and the charge q be zero, then show that for small value of $\frac{R}{L}$, the current in the circuit at time t is given by $\frac{E}{2L} t \sin pt$.

SECTION–C **(10 × 2 = 20 Marks)**

6. (*a*) Find the eigen values and the corresponding eigen vectors of the matrix
$$\begin{pmatrix} -2 & 2 & -3 \\ 2 & 1 & -6 \\ -1 & -2 & 0 \end{pmatrix}.$$

(*b*) Test the absolute convergence of the series $\sum_{n=1}^{\infty} \frac{\cos n\pi}{n^2+1}$.

7. (*a*) Test the consistency of the system of equations
$x + 2y - z = 3;\ 3x - y + 2z = 1;\ 2x - 2y + 3z = 2;\ x - y + z = -1,$
and if consistent, then solve it completely.

(*b*) Reduce the matrix $\begin{pmatrix} 0 & 1 & -3 & -1 \\ 1 & 0 & 1 & 1 \\ 3 & 1 & 0 & 2 \\ 1 & 1 & -2 & 0 \end{pmatrix}$ to normal form and hence find its rank.

8. (*a*) Discuss for what values of '*x*' does the series

$$\frac{1}{2}x + x^2 + \frac{9}{8}x^3 + x^4 + \frac{25}{32}x^5 + \ldots\infty,$$ converge/diverge.

(*b*) Examine the convergence/diverge of the series $\sum_{n=1}^{\infty}\left[(n^3+1)^{1/3} - n\right]$.

9. (*a*) Use De Moivre's theorem to find all the roots of the equation $z^4 - (1-z)^4 = 0$.

(*b*) Find the sum to infinity of the series

$$1 - \frac{1}{2}\cos\theta + \frac{1\cdot 3}{2\cdot 4}\cos 2\theta - \frac{1\cdot 3\cdot 3}{2\cdot 4\cdot 6}\cos 3\theta + \ldots, \ (-\pi < \theta < \pi).$$

❑ ❑ ❑

ENGINEERING MATHEMATICS-II (BTAM-102)
November/December, 2012

Time: 3 Hrs. Max. Marks: 60

INSTRUCTION TO CANDIDATES:

4. SECTION–A is COMPULSORY consisting of TEN questions carrying TWO marks each.
5. Section–B & C have FOUR questions each.
6. Attempt any FIVE questions from SECTION B & C carrying EIGHT marks each.
7. Select atleast TWO questions from SECTION – B & C.

SECTION–A

1. Solve the following sums:

   (*a*) Test whether the set of vectors $\{(1,1,1),(1,-1,1),(3,-1,3)\}$ are LI or LD by giving suitable reasons.

   (*b*) Find the rank of the matrix $\begin{pmatrix} 0 & 1 & -3 & -1 \\ 1 & 0 & 1 & 1 \\ 3 & 1 & 0 & 2 \\ 1 & 1 & -2 & 0 \end{pmatrix}$

   (*c*) Reduce the matrix $\begin{pmatrix} 1 & 0 & -1 \\ 1 & 2 & 1 \\ 2 & 2 & 3 \end{pmatrix}$ to diagonal form.

   (*d*) If $x = \cos\theta + i\sin\theta$, and $y = \cos\phi + i\sin\phi$, then show that $\dfrac{x-y}{x+y} = i\tan\dfrac{\theta-\phi}{2}$.

   (*e*) Find all the values of $\left(\dfrac{1}{2}+\dfrac{\sqrt{3}}{2}i\right)^{3/4}$.

   (*f*) Examine the conditional convergence of the series $\sum_{n=1}^{\infty}\dfrac{(-1)^{n-1}}{2n-1}$.

   (*g*) Test the convergence of the series $\sum_{n=1}^{\infty}\left(1+\dfrac{1}{n}\right)^{-n^2}$.

   (*h*) Show that the necessary condition for the differential equation $Mdx + Ndy = 0$ to be exact is $\dfrac{\partial M}{\partial y} = \dfrac{\partial N}{\partial x}$.

(*i*) Find the particular solution of equation $\frac{d^3y}{dx^3}+4\frac{dy}{dx}=\sin 2x$.

(*j*) Solve the equation $e^{2z-1}=1+i$.

SECTION–B

2. (*a*) Obtain the general solution of the differential equation $y''-6y'+9y=e^{3x}/x^2$, by using method of variation of parameters.

   (*b*) Find the solution of the differential equation $y''-2y'+y=xe^x \sin x$.

3. (*a*) Solve the following simultaneous differential equations

   $\frac{dx}{dt}-2y+5x=t,\ \frac{dy}{dt}+2x+y=0.$ Given that $x(0)=0,\ y(0)=0.$

   (*b*) Find the complete solution of the differential equation

   $(1+x^2)y''+(1+x)y'+y=2\sin\log(1+x)$ by using operator method.

4. (*a*) Solve the differential equation $\left(xy^2-e^{1/x^3}\right)dx-x^2 ydy=0.$

   (*b*) Solve the equation $y=2px+yp^2$ where p has its usual meaning.

5. An e.m.f. $E\sin pt$ is applied at $t=0$ to a circuit containing a capacitance C and inductance L. The current i satisfies the equation $L\frac{di}{dt}+\frac{1}{C}\int idt=E\sin pt$. If $p^2=\frac{1}{LC}$ and initially the current and the charge are zero, then find the current at any time t.

SECTION–C

6. (*a*) Use the rank method to test the consistency of the system of equations

   $4x-y=12;\ -x+5y-2z=0;\ -2y+4z=-8;$

   if consistent then solve it completely.

   (*b*) State Cayley-Hamilton theorem. Use it to find the inverse of the matrix

   $$\begin{pmatrix} 4 & 3 & 1 \\ 2 & 1 & -2 \\ 1 & 2 & 1 \end{pmatrix}$$

7. (*a*) Find the eigen values and the corresponding eigen vectors of the matrix

   $$\begin{pmatrix} 1 & 1 & 3 \\ 1 & 5 & 1 \\ 3 & 1 & 1 \end{pmatrix}$$

   (*b*) Prove that eigen values of a skew hermetian matrix are either zero or purely imaginary.

8. (*a*) Test for values of x the series

$$\sum_{n=1}^{\infty} \frac{\sqrt{n}}{\sqrt{n^2+1}} x^n, \; x > 0 \text{ converges/diverges.}$$

(*b*) Test convergence/diverge of the series

$$\sum_{n=1}^{\infty} \left[\sqrt{(n^4+1)} - \sqrt{(n^4-1)} \right].$$

9. (*a*) Use De Moivre's theorem to solve the equation $(z-1)^5 + z^5 = 0$.

(*b*) Separate $\sin^{-1}\left(e^{i\theta}\right)$ into real and imaginary parts, where θ is a positive acute angle.

SOLUTIONS

ENGINEERING MATHEMATICS-II (BTAM-102)
April/May, 2012

SECTION–A

1. Solve the following sums:

(a) Find the general value of $\log\left(-1+\sqrt{3}i\right)$.

Solution: $\log\left(-1+\sqrt{3}i\right)=\log\left|-1+\sqrt{3}i\right|+i\tan^{-1}(-\sqrt{3})+2n\pi i$

$$=\log\sqrt{1+3}+i\left(\frac{-\pi}{3}\right)+2n\pi i$$

$$=\log 2+i\left(\frac{-\pi}{3}\right)+2n\pi i$$

(b) Solve the differential equation $\frac{d^3y}{dx^3}+y=0$.

Solution: $\frac{d^3y}{dx^3}+y=0$

It's A.E. is

$D^3+1=0$

$$\Rightarrow\ (D+1)(D^2-D+1)=0\ \Rightarrow\quad D=-1,\frac{1+\sqrt{3}i}{2},\frac{1-\sqrt{3}i}{2}$$

$$\therefore\ y=c_1e^{-x}+e^{\frac{1}{2}x}\left(c_2\cos\frac{\sqrt{3}}{2}x+c_3\sin\frac{\sqrt{3}}{2}x\right)$$

(c) Under what conditions on '*a*, *b*, *c* and *d*', the differential equation

$$(a\sinh x\cos y+b\cosh x\sin y)dx+\left(c\sinh x\cos y+d\cosh x\sin y\right)dy=0$$

is exact?

Solution: Given differential equation

$$(a\sinh x\cos y+b\cosh x\sin y)dx+(c\sinh x\cos y+d\cosh x\sin y)dy=0$$

$M = a\sinh x\cos y + b\cosh x\sin y \qquad N = c\sinh x\cos y + d\cosh x\sin y$

$$\frac{\partial M}{\partial y} = -a\sinh x\sin y + b\cosh x\cos y \qquad \frac{\partial N}{\partial x} = c\cosh hx\cos y + d\sinh x\sin y$$

As given differential equation is exact, therefore

$$\frac{\partial M}{\partial y} = \frac{\partial N}{\partial x}$$

$$\Rightarrow -a\sinh x\sin y + b\cosh x\cos y = c\cosh hx\cos y + d\sinh x\sin y$$

$$-a = d, b = c \qquad \Rightarrow \qquad a = -d, b = c$$

(*d*) Find all the roots of $(-8i)^{1/3}$.

Solution: Let $z = (-8i)^{\frac{1}{3}}$

Cubing both the Sides

$$z^3 = -8i$$

Let $z = r\text{cis}\,\theta$

$$r^3\text{cis}3\theta = 8\left(\text{cis}\frac{-\pi}{2}\right)$$

Comparing LHS and RHS, we get

$$r^3 = 8 \qquad 3\theta = \frac{-\pi}{2} + 2k\pi,\ k = 0, 1, 2$$

$$r = 2 \qquad \theta = (4k-1)\frac{\pi}{12},\ k = 0, 1, 2$$

$$\therefore\ z = 2\text{cis}(4k-1)\frac{\pi}{12},\ k = 0,1,2$$

(*e*) Test for what values of '*k*' the set of vectors $\{(k,1,1),\ (0,1,1),(k,0,k)\}$ is linearly independent.

Solution: Given $\{(k,1,1),(0,1,1),(k,0,k)\}$ is L.I.

$$\therefore \quad \begin{vmatrix} k & 1 & 1 \\ 0 & 1 & 1 \\ k & 0 & k \end{vmatrix} = 0$$

Expanding by first row, we get

$$k(k-0)-1(0-k)+1(0-k)=0$$

$$\Rightarrow \quad k^2+k-k=0 \quad \Rightarrow \quad k^2=0$$

$$\Rightarrow \quad k=0$$

(*f*) Examine the convergence/divergence of the series $\sum_{n=1}^{\infty} \frac{x^n}{(2n)!}$.

Solution: Given series $\sum_{n=1}^{\infty} \frac{x^n}{2n!}$

Here $u_n = \frac{1}{\underline{|2n}}$, $u_{n+1} = \frac{1}{\underline{|2(n+1)}} = \frac{1}{\underline{|2n+2}}$

$$\operatorname*{Lt}_{n\to\infty} \frac{u_n}{u_{n+1}} = \operatorname*{Lt}_{n\to\infty} \frac{\underline{|2n+2}}{\underline{|2n}} = \operatorname*{Lt}_{n\to\infty} \frac{(2n+2)(2n+1)\underline{|2n}}{\underline{|2n}} = \operatorname*{Lt}_{n\to\infty} (2n+2)(2n+1) = \infty > 1$$

Therefore, $\operatorname*{Lt}_{n\to\infty} \frac{u_n}{u_{n+1}} = \infty > 1$

Therefore, by Ratio Test the given series is convergent.

(g) Test the absolute convergence of the series $\sum_{n=2}^{\infty} \frac{(-1)^n}{n(\log n)^2}$.

Solution: Given series is $\sum_{n=1}^{\infty} \frac{(-1)^n}{n(\log n)^2}$

Here $|u_n| = \left| \frac{(-1)^n}{n(\log n)^2} \right| = \frac{1}{n(\log n)^2}$

$$\operatorname*{Lt}_{n\to\infty} |u_n| = \operatorname*{Lt}_{n\to\infty} \left| \frac{1}{n(\log n)^2} \right| = 0$$

It is Convergent; therefore the given series is absolutely convergent.

(*h*) Test whether the matrix $\begin{pmatrix} 3 & 1 & -1 \\ -2 & 1 & 2 \\ 0 & 1 & 2 \end{pmatrix}$ is diagonalizable or not?

Solution: Let $A = \begin{pmatrix} 3 & 1 & -1 \\ -2 & 1 & 2 \\ 0 & 1 & 2 \end{pmatrix}$

Taking $|A - \lambda I| = 0$

$$\begin{vmatrix} 3-\lambda & 1 & -1 \\ -2 & 1-\lambda & 2 \\ 0 & 1 & 2-\lambda \end{vmatrix} = 0 \qquad \text{Expanding by first row, we get}$$

$$(3-\lambda)(2+\lambda^2-3\lambda-2)-1(4+2\lambda)-1(-2)=0$$

$$\Rightarrow (3-\lambda)(-3\lambda+\lambda^2)+4-2\lambda+2=0$$

$$\Rightarrow -9\lambda+3\lambda^2+3\lambda^2-\lambda^3+6-2\lambda=0$$

$$\Rightarrow -\lambda^3+6\lambda^2-11\lambda+6=0$$

$$\Rightarrow \lambda^3-6\lambda^2+11\lambda-6=0$$

$$\Rightarrow \quad (\lambda-1)(\lambda-2)(\lambda-3)=0$$

$$\Rightarrow \quad \lambda=1,2,3$$

When $\lambda = 1$, $(A-I)X = O$

$$\begin{pmatrix} 2 & 1 & -1 \\ -2 & 0 & 2 \\ 0 & 1 & 1 \end{pmatrix}\begin{pmatrix} x \\ y \\ z \end{pmatrix} = \begin{pmatrix} 0 \\ 0 \\ 0 \end{pmatrix} \qquad R_2 \to R_2 + R_1$$

$$\begin{pmatrix} 2 & 1 & -1 \\ 0 & 1 & 1 \\ 0 & 1 & 1 \end{pmatrix}\begin{pmatrix} x \\ y \\ z \end{pmatrix} = \begin{pmatrix} 0 \\ 0 \\ 0 \end{pmatrix} \qquad R_3 \to R_3 - R_2$$

$$\begin{pmatrix} 2 & 1 & -1 \\ 0 & 1 & 1 \\ 0 & 0 & 0 \end{pmatrix}\begin{pmatrix} x \\ y \\ z \end{pmatrix} = \begin{pmatrix} 0 \\ 0 \\ 0 \end{pmatrix}$$

$$2x + y - z = 0 \quad \Rightarrow \quad x = z$$

$$y + z = 0 \quad \Rightarrow \quad y = -z \qquad z \neq 0$$

$$\therefore \quad \begin{pmatrix} x \\ y \\ z \end{pmatrix} = \begin{pmatrix} 1 \\ -1 \\ 1 \end{pmatrix} z \quad \Rightarrow \quad X_1 = \begin{pmatrix} 1 \\ -1 \\ 1 \end{pmatrix}$$

When $\lambda = 2$, $(A - 2I)X = O$

$$\begin{pmatrix} 1 & 1 & -1 \\ -2 & -1 & 2 \\ 0 & 1 & 0 \end{pmatrix} \begin{pmatrix} x \\ y \\ z \end{pmatrix} = \begin{pmatrix} 0 \\ 0 \\ 0 \end{pmatrix} \qquad R_2 \to R_2 + 2R_1$$

$$\begin{pmatrix} 1 & 1 & -1 \\ 0 & 1 & 0 \\ 0 & 1 & 0 \end{pmatrix} \begin{pmatrix} x \\ y \\ z \end{pmatrix} = \begin{pmatrix} 0 \\ 0 \\ 0 \end{pmatrix} \qquad R_3 \to R_3 - R_2$$

$$\begin{pmatrix} 1 & 1 & -1 \\ 0 & 1 & 0 \\ 0 & 0 & 0 \end{pmatrix} \begin{pmatrix} x \\ y \\ z \end{pmatrix} = \begin{pmatrix} 0 \\ 0 \\ 0 \end{pmatrix}$$

$$x + y - z = 0 => x = z$$

$$y = 0$$

$$z \neq 0$$

$$\therefore \quad \begin{pmatrix} x \\ y \\ z \end{pmatrix} = \begin{pmatrix} 1 \\ 0 \\ 1 \end{pmatrix} z \quad \Rightarrow \quad X_2 = \begin{pmatrix} 1 \\ 0 \\ 1 \end{pmatrix}$$

When $\lambda = 3$, $\qquad (A - 3I)X = O$

$$\begin{pmatrix} 0 & 1 & -1 \\ -2 & -2 & 2 \\ 0 & 1 & -1 \end{pmatrix} \begin{pmatrix} x \\ y \\ z \end{pmatrix} = \begin{pmatrix} 0 \\ 0 \\ 0 \end{pmatrix} \qquad R_3 \to R_3 - R_1$$

$$\begin{pmatrix} 0 & 1 & -1 \\ -2 & -2 & 2 \\ 0 & 0 & 0 \end{pmatrix} \begin{pmatrix} x \\ y \\ z \end{pmatrix} = \begin{pmatrix} 0 \\ 0 \\ 0 \end{pmatrix} \qquad R_2 \to \frac{1}{2} R_2$$

$$\begin{pmatrix} 0 & 1 & -1 \\ -1 & -1 & 1 \\ 0 & 0 & 0 \end{pmatrix}\begin{pmatrix} x \\ y \\ z \end{pmatrix} = \begin{pmatrix} 0 \\ 0 \\ 0 \end{pmatrix} \qquad R_2 \to \frac{1}{2} R_2$$

$y - z = 0 \Rightarrow y = z$

$-x - y + z = 0$

$x = 0$

$$\therefore \quad \begin{pmatrix} x \\ y \\ z \end{pmatrix} = \begin{pmatrix} 0 \\ 1 \\ 1 \end{pmatrix} z \quad \Rightarrow \quad X_3 = \begin{pmatrix} 0 \\ 1 \\ 1 \end{pmatrix}$$

$$\therefore P = \begin{pmatrix} 1 & 1 & 0 \\ -1 & 0 & 1 \\ 1 & 1 & 1 \end{pmatrix}$$

$|P| = 1(0-1) - 1(-1-1) = -1 + 2 = 1 \neq 0$ [Expanding by first row]

$\therefore \quad P^{-1}$ exists

Now $P^{-1} = \dfrac{\text{adj}P}{|P|}$

adj $P \Rightarrow c_{11} = -1 \quad c_{21} = -1 \quad c_{31} = 1$

$c_{12} = 2 \quad c_{22} = 1 \quad c_{32} = -1$

$c_{13} = -1 \quad c_{23} = 0 \quad c_{33} = 1$

$$\text{adj } P = \begin{pmatrix} -1 & 2 & -1 \\ -1 & 1 & 0 \\ 1 & -1 & 1 \end{pmatrix}^T = \begin{pmatrix} -1 & -1 & 1 \\ 2 & 1 & -1 \\ -1 & 0 & 1 \end{pmatrix}$$

$$P^{-1}AP = \begin{pmatrix} -1 & -1 & 1 \\ 2 & 1 & -1 \\ -1 & 0 & 1 \end{pmatrix}\begin{pmatrix} 3 & 1 & -1 \\ -2 & 1 & 2 \\ 0 & 1 & 2 \end{pmatrix}\begin{pmatrix} 1 & 1 & 0 \\ -1 & 0 & 1 \\ 1 & 1 & 1 \end{pmatrix}$$

$$= \begin{pmatrix} -1 & -1 & 1 \\ 4 & 2 & -2 \\ -3 & 0 & 3 \end{pmatrix}\begin{pmatrix} 1 & 1 & 0 \\ -1 & 0 & 1 \\ 1 & 1 & 1 \end{pmatrix} = \begin{pmatrix} 1 & 0 & 0 \\ 0 & 2 & 0 \\ 0 & 0 & 3 \end{pmatrix}$$ which is the diagonal matrix.

(i) For what values of 'k', the system of equations

$x + y + z = 2; \; x + 2y + z = -2; \; x + y + (k-5)z = k$

has no solution.

Solution: Given system of equations is

$x+y+z=2$

$x+2y+z=-2$

$x+y+(k-5)z=k$

Let $A=\begin{pmatrix}1 & 1 & 1\\ 1 & 2 & 1\\ 1 & 1 & k-5\end{pmatrix}$, $X=\begin{pmatrix}x\\ y\\ z\end{pmatrix}$, $B=\begin{pmatrix}2\\ -2\\ k\end{pmatrix}$

Now $[A:B]=\begin{pmatrix}1 & 1 & 1 & : & 2\\ 1 & 2 & 1 & : & -2\\ 1 & 1 & k-5 & : & k\end{pmatrix}$ $\quad R_2 \to R_2 - R_1,\ R_3 \to R_3 - R_1$

$=\begin{pmatrix}1 & 1 & 1 & : & 2\\ 0 & 1 & 0 & : & -4\\ 0 & 0 & k-6 & : & k-2\end{pmatrix}$

For no solution $\rho(A)<\rho(A:B)$

$\Rightarrow\ k-6=0$ and $k-2\neq 0$

$\Rightarrow\ k=6$

(*j*) Express $\cos 6\theta$ in terms of powers of $\cos\theta$.

Solution: We know that $\cos n\theta + i\sin n\theta = (\cos\theta + i\sin\theta)^n$

Taking $n=6$

$$\cos 6\theta + i\sin 6\theta = (\cos\theta + i\sin\theta)^6$$

$$= {}^6C_0\cos^6\theta + {}^6C_1\cos^5\theta(i\sin\theta) + {}^6C_2\cos^4\theta(i\sin\theta)^2 + {}^6C_3\cos^3\theta(i\sin\theta)^3 + {}^6C_4\cos^2\theta(i\sin\theta)^4 + {}^6C_5\cos^5\theta(i\sin\theta)^5 + {}^6C_6\cos\theta(i\sin\theta)^6$$

$$=\cos^6\theta + 6i\cos^5\theta\sin\theta - 15\cos^4\theta\sin^2\theta + 20i\cos^3\theta\sin^3\theta + 15\cos^2\theta\sin^4\theta + 6i\cos\theta\sin^5\theta - \sin^6\theta$$

Comparing real part on both the sides

$$\cos 6\theta = \cos^6\theta - 15\cos^4\theta(1-\cos^2\theta) + 15\cos^2\theta(1-\cos^2\theta)^2 - (1-\cos^2\theta)^3$$

$$\cos 6\theta = \cos^6\theta - 15\cos^4\theta + 15\cos^6\theta + 15\cos^2\theta - 30\cos^4\theta + 15c os^6\theta - 1 + 15\cos^6\theta + 3\cos^2\theta - 3\cos^4\theta$$

$$\cos 6\theta = 32\cos^6\theta - 48\cos^4\theta + 18\cos^2\theta - 1$$

SECTION B

2. (*a*) Find the general solution of the differential equation
$(3x^2y^3e^y + y^3 + y^2)dx + (x^3y^3e^y - xy)dy = 0$.

Solution: Comparing it with $Mdx + Ndy = 0$, we get

$$M = 3x^2y^3e^y + y^3 + y^2,\; N = x^3y^3e^y - xy$$

$$\frac{\partial M}{\partial y} = 9x^2y^2e^y + 3x^2y^3e^y + 3y^2 + 2y,\; \frac{\partial N}{\partial x} = 3x^2y^3e^y - y$$

Since $\frac{\partial M}{\partial y} \neq \frac{\partial N}{\partial x}$, therefore given equation is not exact. To make it exact we will multiply it with Integrating Factor, which is found as shown below:

$$\frac{\frac{\partial N}{\partial x} - \frac{\partial M}{\partial y}}{M} = \frac{3x^2y^3e^y - y - 9x^2y^2e^y - 3x^2y^3e^y - 3y^2 - 2y}{3x^2y^3e^y + y^3 + y}$$

$$= \frac{-3(3x^2y^2e^y + y^2 + y)}{y(3x^2y^2e^y + y^2 + y)} = \frac{-3}{y}$$

Therefore, $I.F. = e^{\int \frac{-3}{y}dy} = e^{-3\log y} = e^{\log y^{-3}} = y^{-3} = \frac{1}{y^3}$

On multiplying given equation with $I.F. = \frac{1}{y^3}$, we get

$\left(3x^2e^y + 1 + \frac{1}{y}\right)dx + \left(x^3e^y - \frac{x}{y^2}\right)dy = 0$, which is exact differential equation.

Its solution is

$$\int_{y\text{ constant}} Mdx + \int(\text{terms of } N \text{ not containg } x)dy = c$$

$$\int\left(3x^2e^y + 1 + \frac{1}{y}\right)dx = c$$

$\Rightarrow\; e^yx^3 + x + \frac{x}{y} = c$ which is the required solution.

(*b*) Obtain the general solution of the differential equation $y'' + 3y' + 2y = \sin(e^x)$, by using method of variation of parameters.

Solution: Given differential equation is $y''+3y'+2y=\sin(e^x)$

It's A.E. is

$$D^2+3D+2=0 \Rightarrow (D+1)(D+2)=0 \Rightarrow D=-1, -2$$

$$\therefore \; C.F. = c_1e^{-x}+c_2e^{-2x}$$

Let its solution be

$$y = uy_1+vy_2 \qquad \ldots (1)$$

Let $y_1=e^{-x}$, $y_2=e^{-2x}$

$$w=\begin{vmatrix} y_1 & y_2 \\ y_1' & y_2' \end{vmatrix} = \begin{vmatrix} e^{-x} & e^{-2x} \\ -e^{-x} & -2e^{-2x} \end{vmatrix} = -2e^{-3x}+e^{-3x} = -e^{-3x}$$

Now Let

$$u=-\int \frac{y_2X}{w}dx = -\int \frac{e^{-2x}\sin(e^x)}{e^{-3x}}dx$$

$$=-\int e^x \sin e^x dx = -\int \sin t dt \qquad \left[\text{Put } e^x=t,\ e^x dx = dt\right]$$

$$=\cos t + c_1 = \cos e^x + c_1$$

$$v=\int \frac{y_1X}{w}dx = \int \frac{e^{-x}\sin(e^x)}{e^{-3x}}dx$$

$$=\int e^{2x}\sin(e^x)dx = \int t \sin t dt \qquad \left[\text{Put } e^x=t,\ e^x dx = dt\right]$$

$$=-t\cos t - \int 1.(-\cos t)dt \qquad [\text{Integrating by parts}]$$

$$=-t\cos t+\sin t + c_2$$

$$=-e^x\cos e^x + \sin e^x + c_2$$

$\therefore$ (1) becomes

$$y=(\cos e^x + c_1)e^{-x} + (-e^x\cos e^x + \sin e^x + c_2)e^{-2x}$$

3.(*a*) Solve the following simultaneous differential equations

$$\frac{dx}{dt}+y=\sin t, \quad \frac{dy}{dt}+x=\cos t, \quad y(0)=0,\ x(0)=2.$$

Solution: Given differential equations are

$$\frac{dx}{dt}+y=\sin t \qquad \frac{dy}{dt}+x=\cos t$$

Writing $D=\frac{d}{dt}$, we have

$$Dx+y=\sin t \qquad \ldots (1)$$

$$Dy + x = \cos t \qquad \ldots (2)$$

Now $(1)D-(2)$ gives

$$(D^2-1)x = \cos t - \cos t$$

$$\Rightarrow (D^2-1)x = 0$$

Its A.E. is

$$D^2 - 1 = 0 \quad \Rightarrow \quad D = \pm 1$$

$$\therefore \; x = c_1 e^t + c_2 e^{-t} \qquad \ldots (3)$$

Putting in (1), we get

$$c_1 e^t - c_2 e^{-t} + y = \sin t$$

$$\Rightarrow y = -c_1 e^t + c_2 e^{-t} + \sin t \qquad \ldots (4)$$

Given $x(0) = 2$, $y(0) = 0$, putting in (3) and (4), we get

$$2 = c_1 + c_2, \quad 0 = -c_1 + c_2$$

$$\Rightarrow c_1 + c_2 = 2, \quad -c_1 + c_2 = 0$$

Solving, these we get

$$c_1 = 1, c_2 = 1$$

$$\therefore \quad x = e^t + e^{-t}$$

$$y = -e^t + e^{-t} + \sin t$$

is the required solution.

(b) Find the complete solution of the differential equation $(x+1)^2 y'' + (x+1)y' + y = \sin(2\log(1+x))$, by using operator method.

Solution: Given differential equation is

$$(x+1)^2 y'' + (x+1)y' + y = \sin(2(\log(1+x)) \qquad \ldots (1)$$

Put $(x+1) = e^z \quad \Rightarrow \quad \log(1+x) = z$

Further put $(x+1)D = Q$

$$(x+1)^2 D^2 = Q(Q-1) \qquad \text{where } Q = \frac{d}{dz}$$

$\because$ (1) becomes

$$[Q(Q-1) + Q + 1]z = \sin(2z)$$

$$\Rightarrow \quad (Q^2+1)z = \sin 2z$$

Its A.E. is

$$Q^2 + 1 = 0 \quad \Rightarrow \quad Q = \pm i$$

$\therefore \quad C.F. = c_1 \cos z + c_2 \sin zy = c_1 \cos(\log(1+x)) + c_2 \sin(\log(1+x))$

Now, $P.I. = \dfrac{1}{Q^2+1}\sin 2z$

Put $Q^2 = -4$

$P.I. = \dfrac{\sin 2z}{-3} = -\dfrac{1}{3}\sin(2\log(1+x))$

$\therefore$ Complete solution = $C.F.$ + $P.I.$

$$\Rightarrow \quad y = c_1 \cos(\log(1+x)) + c_2 \sin(\log(1+x)) - \frac{1}{3}\sin 2(\log(1+x))$$

4.(*a*) Solve the differential equation $x\left(\dfrac{dy}{dx}+y\right) = 1-y.$

Solution: Given differential equation is $x\left(\dfrac{dy}{dx}+y\right) = 1-y$

$$\Rightarrow \quad \frac{dy}{dx}+y = \frac{1-y}{x}$$

$\Rightarrow \quad \dfrac{dy}{dx}+y\left(1+\dfrac{1}{x}\right) = \dfrac{1}{x}$ which is linear equation with $P = 1+\dfrac{1}{x}$, $Q = \dfrac{1}{x}$

$$I.F. = e^{\int P dx} = e^{\int 1+\frac{1}{x}dx} = e^{x+\log x} = e^x.e^{\log x} = xe^x$$

$$\therefore \quad y \times (I.F.) = \int Q(I.F.)dx + c$$

$$y \times xe^x = e^x + C \quad \Rightarrow \quad y = \frac{1}{x} + \frac{c}{xe^x}$$

(*b*) Find the particular solution of the differential equation $y'' + a^2 y = \sec ax.$

Solution: Given differential equation is

$$y'' + a^2 = \sec ax$$

Rewriting in S. F. we have

$$(D^2 + a^2)y = \sec ax$$

$$P.I. = \frac{\sec ax}{D^2+a^2} = \frac{\sec ax}{(D+ai)(D-ai)} = \frac{1}{2ai}\left(\frac{1}{D-ai} - \frac{1}{D+ai}\right)\sec ax$$

$$= \frac{1}{2ai}\left(\frac{1}{D-ai}\sec ax - \frac{1}{D+ai}\sec ax\right) \quad \ldots (A)$$

Here, $\frac{1}{D-ai}\sec ax = e^{iax}\int \sec ax \cdot e^{-iax}dx$ $\left[\because \frac{1}{D-a}X = e^{ax}\int Xe^{-ax}dx\right]$

$$= e^{iax}\int \frac{\cos ax - i\sin ax}{\cos ax}dx = e^{iax}\int (1 - i\tan ax)dx$$

$$= e^{aix}\left(x + \frac{i}{a}\log\sec ax\right)$$

Changing i to $-i$, we get

$$\frac{1}{D-ai}\sec ax = e^{-aix}\left(x - \frac{i}{a}\log\sec ax\right)$$

$\therefore$ from (A), we have

$$P.I. = \frac{1}{2ai}\left[e^{aix}\left(x + \frac{i}{a}\log\sec ax\right) - e^{-aix}\left(x - \frac{i}{a}\log\sec ax\right)\right]$$

$$= \frac{1}{2ai}\left(x(e^{aix} - e^{-aix}) + \frac{i}{a}\log\sec ax(e^{aix} + e^{-aix})\right)$$

$$= \frac{1}{a}\left(x\frac{(e^{aix} - e^{-aix})}{2i} + \frac{1}{a}\log\sec ax\frac{(e^{aix} + e^{-aix})}{2}\right)$$

$$= \frac{1}{a}\left(x\sin ax + \frac{\log(\sec ax)}{a}\cos ax\right) \qquad \left[\cos ax = \frac{e^{aix} + e^{-aix}}{2},\ \sin ax = \frac{e^{aix} - e^{-aix}}{2i}\right]$$

5. An L-C-R circuit with battery e.m.f. '$E\sin pt$' is tuned to resonance so that $p^2 = \frac{1}{LC}$. If initially the current i and the charge q be zero, then show that for small value of $\frac{R}{L}$, the current in the circuit at time t is given by $\frac{E}{2L}t\sin pt$.

Solution: Given Differential Equation is

$$L\frac{d^2q}{dt^2} + R\frac{dq}{dt} + \frac{q}{C} = E\sin pt$$

In S.F., we have

$$\left(LD^2 + RD + \frac{1}{C}\right)q = E\sin pt$$

Its A.E. is

$$LD^2 + RD + \frac{1}{C} = 0 \quad \Rightarrow \quad D = \frac{-R}{2L} \pm \sqrt{\frac{R^2}{4L^2} - \frac{1}{LC}}$$

$$\Rightarrow D = \frac{-R}{2L} \pm i\sqrt{\frac{1}{LC}} \qquad \because \frac{R}{L} \text{ is small, } \therefore \frac{R^2}{L^2} \text{ is neglected.}$$

$$D = \frac{-R}{2L} \pm ip \quad \left(\because p^2 = \frac{1}{LC}\right)$$

$$C.F. = e^{-\frac{R}{2L}t}(c_1 \cos pt + c_2 \sin pt)$$

$$P.I. = \frac{E \sin pt}{LD^2 + RD + \frac{1}{C}} \qquad \text{Put } D^2 = -P^2 = -\frac{1}{LC}$$

$$= \frac{E \sin pt}{-\frac{1}{C} + RD + \frac{1}{C}} = \frac{E \sin pt}{RD} = \frac{E}{R}\int \sin pt\, dt = \frac{-E}{R}\frac{\cos pt}{p}$$

$$C.S. = C.F. + P.I.$$

$$q = e^{-\frac{R}{2L}t}(c_1 \cos pt + c_2 \sin pt) - \frac{E}{Rp}\cos pt \qquad \ldots (1)$$

$$\text{Now } e^{-\frac{R}{2L}t} = \left[1 + \frac{-R}{2L}t + \frac{1}{2!}\left(\frac{-R}{2L}t\right)^2 + \ldots\right]$$

$$= 1 - \frac{R}{2L}t \qquad \left[\because \frac{R}{L} \text{ is small}\right]$$

$\therefore$ (1) becomes

$$q = \left(1 - \frac{R}{2L}t\right)(c_1 \cos pt + c_2 \sin pt) - \frac{E}{Rp}\cos pt \qquad \ldots (2)$$

When $t = 0, q = 0 \qquad \therefore \qquad 0 = c_1 - \frac{E}{pR} \Rightarrow c_1 = \frac{E}{pR}$

$\therefore$ (2) becomes

$$q = \left(1 - \frac{R}{2L}t\right)\left(\frac{E}{pR}\cos pt + c_2 \sin pt\right) - \frac{E}{Rp}\cos pt$$

$$i = \frac{dq}{dt} = \left(1 - \frac{R}{2L}t\right)\left[\frac{-E}{pR}p \sin pt + c_2 p \cos pt\right]$$

$$-\frac{R}{2L}\left(\frac{E}{pR}\cos pt + c_2 \sin pt\right) + \frac{E}{R}\sin pt$$

Now $t = 0, i = 0$

$$\therefore \quad 0 = c_2 p - \frac{E}{2Lp} \quad \Rightarrow \quad c_2 = \frac{E}{2Lp^2}$$

$$\therefore \quad i = \left(1 - \frac{R}{2L}t\right)\left(-\frac{E}{R}\sin pt + \frac{E}{2Lp}\cos pt\right)$$

$$-\frac{R}{2L}\left[\frac{E}{pR}\cos pt + \frac{E}{2Lp^2}\sin pt\right] + \frac{E}{R}\sin pt$$

$$= \frac{Et}{2L}\sin pt - \frac{ERt}{4L^2 p}\cos pt - \frac{ER}{4L^2 p^2}\sin pt$$

$$= \frac{Et}{2L}\sin pt \qquad \left[\because \frac{R}{L} \text{ is very small}\right]$$

$$\therefore \quad i = \frac{Et}{2L}\sin pt \quad \text{which is the required answer.}$$

SECTION C

6. (*a*) Find the eigen values and the corresponding eigen vectors of the matrix

$$\begin{pmatrix} -2 & 2 & -3 \\ 2 & 1 & -6 \\ -1 & -2 & 0 \end{pmatrix}$$

Solution: Let $A = \begin{pmatrix} -2 & 2 & -3 \\ 2 & 1 & -6 \\ -1 & -2 & 0 \end{pmatrix}$

Now, $|A - \lambda I| = \begin{vmatrix} -2-\lambda & 2 & -3 \\ 2 & 1-\lambda & -6 \\ -1 & -2 & -\lambda \end{vmatrix} = 0$

$$\Rightarrow \quad (-2-\lambda)[-\lambda + \lambda^2 - 12] - 2(-2\lambda - 6) - 3(-3-\lambda) = 0$$

$$\Rightarrow \quad 2\lambda - 2\lambda^2 + 24 + \lambda^2 - \lambda^3 + 12\lambda + 4\lambda + 12 + 9 + 3\lambda = 0$$

$$\Rightarrow \quad -\lambda^3 - \lambda^2 + 21\lambda + 45 = 0$$

$$\Rightarrow \quad \lambda^3 + \lambda^2 - 21\lambda - 45 = 0$$

$$\Rightarrow \quad (\lambda + 3)^2(\lambda - 5) = 0$$

$$\Rightarrow \quad \lambda = -3, -3, 5$$

When $\lambda = 3$, $(A + 3I)X = O$

$$\begin{pmatrix} 1 & 2 & -3 \\ 2 & 4 & -6 \\ -1 & -2 & 3 \end{pmatrix}\begin{pmatrix} x \\ y \\ z \end{pmatrix} = \begin{pmatrix} 0 \\ 0 \\ 0 \end{pmatrix} \quad R_2 \to R_2 - 2R_1,\ R_3 \to R_3 + R_1$$

$$\begin{pmatrix} 1 & 2 & -3 \\ 0 & 0 & 0 \\ 0 & 0 & 0 \end{pmatrix}\begin{pmatrix} x \\ y \\ z \end{pmatrix} = \begin{pmatrix} 0 \\ 0 \\ 0 \end{pmatrix}$$

$$x + 2y - 3z = 0$$

$$x = -2y + 3z$$

$$y \neq 0$$

$$z \neq 0$$

$$\begin{pmatrix} x \\ y \\ z \end{pmatrix} = \begin{pmatrix} -2y+3z \\ y \\ z \end{pmatrix} = \begin{pmatrix} -2 \\ 1 \\ 0 \end{pmatrix} y + \begin{pmatrix} 3 \\ 0 \\ 1 \end{pmatrix} z$$

Eigen vector are $(-2,1,0)$ and $(3,0,1)$

When $\lambda = 5$, $(A - 5I)X = O$

$$\begin{pmatrix} -7 & 2 & -3 \\ 2 & -4 & -6 \\ -1 & -2 & -5 \end{pmatrix}\begin{pmatrix} x \\ y \\ z \end{pmatrix} = \begin{pmatrix} 0 \\ 0 \\ 0 \end{pmatrix} \quad R_1 \to R_1 - 7R_3,\ R_2 \to R_2 + 2R_3$$

$$\begin{pmatrix} 0 & 16 & 32 \\ 0 & -8 & -16 \\ -1 & -2 & -5 \end{pmatrix}\begin{pmatrix} x \\ y \\ z \end{pmatrix} = \begin{pmatrix} 0 \\ 0 \\ 0 \end{pmatrix} \quad R_1 \to R_1 + 2R_2$$

$$\begin{pmatrix} 0 & 0 & 0 \\ 0 & -8 & -16 \\ -1 & -2 & -5 \end{pmatrix}\begin{pmatrix} x \\ y \\ z \end{pmatrix} = \begin{pmatrix} 0 \\ 0 \\ 0 \end{pmatrix}$$

$$8y + 16 = 0 \quad \Rightarrow \quad y = -2z$$

$$x + 2y + 5z = 0 \quad \Rightarrow \quad x = -z \quad \therefore \quad z = z$$

$$\begin{pmatrix} x \\ y \\ z \end{pmatrix} = \begin{pmatrix} -z \\ -2z \\ z \end{pmatrix} = \begin{pmatrix} -1 \\ -2 \\ 1 \end{pmatrix} z$$

Eigen vector is $(1, 2, -1)$.

(b) Test the absolute convergence of the series $\sum_{n=1}^{\infty} \frac{\cos n\pi}{n^2+1}$.

Solution: Given series is $\sum_{n=1}^{\infty} \frac{\cos n\pi}{n^2+1} = \sum_{n=1}^{\infty} \frac{(-1)^n}{n^2+1}$

Here $a_n = \frac{1}{n^2+1}$ It is an alternative series.

$\therefore$ (*i*) Here $a_n > 0 \quad \forall \quad n \in N$

(*ii*) To prove $a_{n+1} < a_n \quad \forall n \in N$

$\therefore \; a_{n+1} = \frac{1}{(n+1)^2+1}$

We have $n+1 > n \quad \forall n \in N$

$$(n+1)^2 > n^2 \quad \forall n \in N$$

$$(n+1)^2 + 1 > n^2 + 1 \quad \forall n \in N$$

$$\frac{1}{(n+1)^2+1} < \frac{1}{n^2+1} \quad \forall n \in N$$

$$a_{n+1} < a_n \quad \forall n \in N$$

(*iii*) $\underset{n\to\infty}{\text{Lt}}\, a_n = \underset{n\to\infty}{\text{Lt}} \frac{1}{n^2+1} = 0$

$\therefore$ By Leibnitz's test, given series is convergent.

$\therefore$ It satisfy all the condition of Leibnitz's test.

7. (a) Test the consistency of the system of equations
$x+2y-z=3;\ 3x-y+2z=1;\ 2x-2y+3z=2;\ x-y+z=-1,$
and if consistent, then solve it completely.

Solution: Given system of equations is

$$x+2y-z=3$$
$$3x-y+2z=1$$
$$2x-2y+3z=2$$
$$x-y+z=-1$$

Here $[A:B] = \begin{pmatrix} 1 & 2 & -1 & : & 3 \\ 3 & -1 & 2 & : & 1 \\ 2 & -2 & 3 & : & 2 \\ 1 & -1 & 1 & : & -1 \end{pmatrix}$ $R_2 \to R_2 - 3R_1,\ R_3 \to R_3 - 2R_1,\ R_4 \to R_4 - R_1$

$$\begin{pmatrix} 1 & 2 & -1 & : & 3 \\ 0 & -7 & 5 & : & -8 \\ 0 & -6 & 5 & : & -4 \\ 0 & -3 & 2 & : & -4 \end{pmatrix} \quad R_3 \to R_3 - \frac{6}{7}R_2,\ R_4 \to R_4 - \frac{3}{7}R_2$$

$$\begin{pmatrix} 1 & 2 & -1 & : & 3 \\ 0 & -7 & 5 & : & -8 \\ 0 & 0 & \frac{5}{7} & : & \frac{20}{7} \\ 0 & 0 & -\frac{1}{7} & : & -\frac{4}{7} \end{pmatrix} \quad R_4 \to R_4 + \frac{1}{5}R_3$$

$$\begin{pmatrix} 1 & 2 & -1 & : & 3 \\ 0 & -7 & 5 & : & -8 \\ 0 & 0 & \frac{5}{7} & : & \frac{20}{7} \\ 0 & 0 & 0 & : & 0 \end{pmatrix}$$

$\because$ $\rho(A) = \rho(A:B) = 3 =$ No. of Unknown

$\therefore$ it is consistent.

Now the equivalent system of equation can be written as

$$x + 2y - z = 3 \qquad \ldots (1)$$

$$-7y + 5z = -8 \qquad \ldots (2)$$

$$\frac{5}{7}z = \frac{20}{7} \qquad \ldots (3)$$

From (3), we have $z = 4$

From (2), we have $y = 4$

From (1), we have $x = -1$

$\therefore$ $x = -1, y = 4, z = 4$

(*b*) Reduce the matrix $\begin{pmatrix} 0 & 1 & -3 & -1 \\ 1 & 0 & 1 & 1 \\ 3 & 1 & 0 & 2 \\ 1 & 1 & -2 & 0 \end{pmatrix}$ to normal form and hence find its rank.

Solution: $A=\begin{pmatrix} 0 & 1 & -3 & -1 \\ 1 & 0 & 1 & 1 \\ 3 & 1 & 0 & 2 \\ 1 & 1 & -2 & 0 \end{pmatrix}$ $\quad R_1 \leftrightarrow R_2$

$$\begin{pmatrix} 1 & 0 & 1 & 1 \\ 0 & 1 & -3 & -1 \\ 3 & 1 & 0 & 2 \\ 1 & 1 & -2 & 0 \end{pmatrix} \quad R_1 \to R_3 - 3R_1, R_4 \to R_4 - R_1$$

$$\begin{pmatrix} 1 & 0 & 1 & 1 \\ 0 & 1 & -3 & -1 \\ 0 & 1 & -3 & -1 \\ 0 & 1 & -3 & -1 \end{pmatrix} \quad R_3 \to R_3 - R_2, R_4 \to R_4 - R_2$$

$$\begin{pmatrix} 1 & 0 & 1 & 1 \\ 0 & 1 & -3 & -1 \\ 0 & 0 & 0 & 0 \\ 0 & 0 & 0 & 0 \end{pmatrix} \quad C_3 \to C_3 - C_1 + 3C_2, C_4 \to C_4 - C_1 + C_2$$

$$\begin{pmatrix} 1 & 0 & 0 & 0 \\ 0 & 1 & 0 & 0 \\ 0 & 0 & 0 & 0 \\ 0 & 0 & 0 & 0 \end{pmatrix}$$ which is required normal form.

$\therefore \quad \rho(A) = 2 =$ Order of identity matrix

8. (*a*) Discuss for what values of '*x*' does the series

$$\frac{1}{2}x + x^2 + \frac{9}{8}x^3 + x^4 + \frac{25}{32}x^5 + \ldots\infty, \text{ converge/diverge.}$$

Solution: $\frac{1}{2}x + x^2 + \frac{9}{8}x^3 + x^4 + \frac{25}{32}x^5 + \ldots\infty,$

$$\Rightarrow \quad \frac{1^2}{2^1}x + \frac{2^2}{2^2}x^2 + \frac{3^2}{2^3}x^3 + \frac{4^2}{2^4}x^4 + \frac{5^2}{2^5}x^5 + \ldots\infty$$

Here $a_n = \frac{n^2}{2^n}x^n, \; a_{n+1} = \frac{(n+1)^2}{2^{n+1}}x^{n+1}$

$$\underset{n\to\infty}{\text{Lt}} \frac{a_n}{a_{n+1}} = \frac{n^2}{2^n} x^n \times \frac{2^{n+1}}{(n+1)^2 x^{n+1}} = \frac{2}{x}$$

By Ratio test, it is converges if $\frac{2}{x} > 1 \Rightarrow x < 2$ and divergent if $\frac{2}{x} < 1 \Rightarrow x > 2$,

when $\frac{2}{x} = 1$, test fails.

Putting $x = 2$, we have

$$a_n = \frac{n^2}{2^n} \quad \Rightarrow \quad \underset{n\to\infty}{\text{Lt}}\, a_n = \underset{n\to\infty}{\text{Lt}} \frac{n^2}{2^n} = \frac{\infty}{\infty} \quad \text{(in-determine form)}$$

Apply L'Hospital rule

$$\underset{n\to\infty}{\text{Lt}}\, a_n = \underset{n\to\infty}{\text{Lt}} \frac{2n}{2^n \log 2} = \frac{\infty}{\infty} \quad \text{(again in-determine form)}$$

Again apply L'Hospital rule

$$\underset{n\to\infty}{\text{Lt}}\, a_n = \underset{n\to\infty}{\text{Lt}} \frac{2}{2^n (\log 2)^2} = \frac{2}{\infty} = 0$$

$\therefore \underset{n\to\infty}{\text{Lt}}\, a_n \to 0$, i.e., it is convergent

Therefore, given series is convergent when $x \le 2$ and divergent when $x > 2$.

(*b*) Examine the convergence/diverge of the series $\sum_{n=1}^{\infty} \left[(n^3+1)^{1/3} - n\right]$.

Solution: Given series is $\sum_{n=1}^{\infty} \left[(n^3+1)^{\frac{1}{3}} - n\right]$

Here $a_n = (n^3+1)^{\frac{1}{3}} - n = n\left[1 + \frac{1}{n^3}\right]^{\frac{1}{3}} - n = n\left[1 + \frac{1}{3n^3} - \frac{1}{9n^6} + \cdots\right] - n$

$$\Rightarrow \quad a_n = \frac{1}{3n^2} - \frac{1}{9n^5} + \cdots \infty$$

Let $b_n = \frac{1}{n^2}$, therefore, $\underset{n\to\infty}{\text{Lt}} \frac{a_n}{b_n} = \frac{1}{3}$ which is finite.

$\therefore$ By Comparison test $\sum a_n$ and $\sum b_n$ both behave alike.

Also $\sum b_n = \sum \frac{1}{n^2}$, were $p = 2 > 1$,

$\therefore$ By *P*-test $\sum b_n$ is convergent as well as $\sum a_n$ is also convergent.

9. (*a*) Use De Moivre's theorem to find all the roots of the equation $z^4 - (1-z)^4 = 0$.

Solution: Given $z^4-(1-z)^4=0 \Rightarrow \dfrac{(1-z)^4}{z^4}=1$

$$\Rightarrow \frac{1-z}{z}=(1)^{\frac{1}{4}}=\text{cis}\frac{2k\pi}{4},\ k=0,1,2,3=\text{cis}\frac{k\pi}{2},\ k=0,1,2,3$$

Let $\dfrac{k\pi}{2}=\theta$

$$\therefore \quad \frac{1-z}{z}=\text{cis}\,\theta \quad \Rightarrow \quad \frac{1}{z}-1=\text{cis}\,\theta$$

$$\frac{1}{z}=1+\text{cis}\,\theta=1+\cos\theta+i\sin\theta$$

$$\frac{1}{z}=2\cos^2\frac{\theta}{2}+2i\sin\frac{\theta}{2}\cos\frac{\theta}{2}=2\cos\frac{\theta}{2}\left(\cos\frac{\theta}{2}+i\sin\frac{\theta}{2}\right)=2\cos\frac{\theta}{2}\left(\text{cis}\frac{\theta}{2}\right)$$

$$z=\frac{1}{2\cos\frac{\theta}{2}\,\text{cis}\left(\frac{\theta}{2}\right)}=\frac{1}{2}\sec\frac{\theta}{2}\,\text{cis}\left(-\frac{\theta}{2}\right)$$

$$=\frac{1}{2}\sec\left(\frac{k\pi}{4}\right)\text{cis}\left(\frac{-k\pi}{4}\right),\ k=0,1,2,3$$

$$=\frac{1}{2}\frac{1}{\cos\left(\frac{k\pi}{4}\right)}\left[\cos\left(\frac{k\pi}{4}\right)-i\sin\left(\frac{k\pi}{4}\right)\right],\ k=0,1,2,3$$

$$=\frac{1}{2}\left[1-i\tan\left(\frac{k\pi}{4}\right)\right],\ k=0,1,2,3 \quad \text{which is required answer.}$$

(*b*) Find the sum to infinity of the series

$$1-\frac{1}{2}\cos\theta+\frac{1\cdot 3}{2\cdot 4}\cos 2\theta-\frac{1\cdot 3\cdot 3}{2\cdot 4\cdot 6}\cos 3\theta+\ldots,\ (-\pi<\theta<\pi).$$

Solution: $C=1-\dfrac{1}{2}\cos\theta+\dfrac{1.3}{2.4}\cos 2\theta-\dfrac{1.3.5}{2.4.6}\cos 3\theta+\cdots\infty$

$$S=-\frac{1}{2}\sin\theta+\frac{1.3}{2.4}\sin 2\theta-\frac{1.3.5}{2.4.6}\sin 3\theta+\cdots\infty$$

$$C+iS=1-\frac{1}{2}e^{i\theta}+\frac{1.3}{2.4}e^{2i\theta}-\frac{1.3.5}{2.4.6}e^{3i\theta}+\cdots\infty=(1+e^{i\theta})^{-\frac{1}{2}}$$

$$=(1+\cos\theta+i\sin\theta)^{-\frac{1}{2}}$$

$$= \left[2\cos^2\frac{\theta}{2} + 2i\sin\frac{\theta}{2}\cos\frac{\theta}{2}\right]^{-\frac{1}{2}} = \left(2\cos\frac{\theta}{2}\right)^{-\frac{1}{2}}\left[\cos\frac{\theta}{2} + i\sin\frac{\theta}{2}\right]^{-\frac{1}{2}}$$

$$= \left(2\cos\frac{\theta}{2}\right)^{-\frac{1}{2}}\left(\text{cis}\frac{\theta}{2}\right)^{-\frac{1}{2}} = \left(2\cos\frac{\theta}{2}\right)^{-\frac{1}{2}}\left(\cos\frac{\theta}{4} - i\sin\frac{\theta}{4}\right)$$

Comparing real part on both sides, we get

$$C = \left(2\cos\frac{\theta}{2}\right)^{-\frac{1}{2}}\cos\frac{\theta}{4}$$

❑ ❑ ❑

SOLUTIONS

ENGINEERING MATHEMATICS-II (BTAM-102)
November/December, 2012

SECTION–A

1. Solve the following sums:

(*a*) Test whether the set of vectors $\{(1,1,1),(1,-1,1),(3,-1,3)\}$ are LI or LD by giving suitable reasons.

Solution: Let X_1, X_2 and X_3 be the four vectors such that $X_1 = (1,1,1)$, $X_2 = (1,-1,1)$, and $X_3 = (3,-1,3)$. Then for any three scalars k_1, k_2 and k_3, we have

$$k_1X_1 + k_2X_2 + k_3X_3 = 0 \qquad \ldots (1)$$

$$\Rightarrow \qquad k_1 + k_2 + 3k_3 = 0$$

$$k_1 - k_2 - k_3 = 0$$

$$k_1 + k_2 + 3k_3 = 0$$

Consider $A = \begin{pmatrix} 1 & 1 & 3 \\ 1 & -1 & -1 \\ 1 & 1 & 3 \end{pmatrix}$ $\qquad R_3 \to R_3 - R_1$

$$A = \begin{pmatrix} 1 & 1 & 3 \\ 1 & -1 & -1 \\ 0 & 0 & 0 \end{pmatrix} \qquad R_2 \to R_2 - R_1$$

$$A = \begin{pmatrix} 1 & 1 & 3 \\ 0 & -2 & -4 \\ 0 & 0 & 0 \end{pmatrix}$$ which is in echelon form.

Since the echelon form contains a zero row, hence the vectors X_1, X_2 and X_3, i.e., $\{(1,1,1),(1,-1,1),(3,-1,3)\}$ are LD.

(*b*) Find the rank of the matrix $\begin{pmatrix} 0 & 1 & -3 & -1 \\ 1 & 0 & 1 & 1 \\ 3 & 1 & 0 & 2 \\ 1 & 1 & -2 & 0 \end{pmatrix}$

Solution: To find the rank of the matrix, we reduce the matrix to echelon form. The number of non-zero rows in echelon form gives the rank of the matrix.

Given $A = \begin{pmatrix} 0 & 1 & -3 & -1 \\ 1 & 0 & 1 & 1 \\ 3 & 1 & 0 & 2 \\ 1 & 1 & -2 & 0 \end{pmatrix}$ $\quad R_1 \leftrightarrow R_2$

$$A = \begin{pmatrix} 1 & 0 & 1 & 1 \\ 0 & 1 & -3 & -1 \\ 3 & 1 & 0 & 2 \\ 1 & 1 & -2 & 0 \end{pmatrix} \quad R_3 \to R_3 - 3R_1,\ R_4 \to R_4 - R_1$$

$$A = \begin{pmatrix} 1 & 0 & 1 & 1 \\ 0 & 1 & -3 & -1 \\ 0 & 1 & -3 & -1 \\ 0 & 1 & -3 & -1 \end{pmatrix} \quad R_3 \to R_3 - R_2,\ R_4 \to R_3 - R_2$$

$$A = \begin{pmatrix} 1 & 0 & 1 & 1 \\ 0 & 1 & -3 & -1 \\ 0 & 0 & 0 & 0 \\ 0 & 0 & 0 & 0 \end{pmatrix} \quad \text{which is in echlon form.}$$

Since number of rows of non-zero rows is 2, therefore, $\rho(A) = 2$.

(*c*) Reduce the matrix $\begin{pmatrix} 1 & 0 & -1 \\ 1 & 2 & 1 \\ 2 & 2 & 3 \end{pmatrix}$ to diagonal form.

Solution: Given $A = \begin{pmatrix} 1 & 0 & -1 \\ 1 & 2 & 1 \\ 2 & 2 & 3 \end{pmatrix}$

$$|A - \lambda I| = \begin{vmatrix} 1-\lambda & 0 & -1 \\ 1 & 2-\lambda & 1 \\ 2 & 2 & 3-\lambda \end{vmatrix}$$

Expanding by first row, we get

$$|A - \lambda I| = (1-\lambda)[(2-\lambda)(3-\lambda)-2]-0-1[2-2(2-\lambda)]$$

$$= (1-\lambda)\left[\lambda^2 - 5\lambda + 6\right] - 1[-2+2\lambda]$$

$$= -\lambda^3 + 6\lambda^2 - 11\lambda + 6$$

For eigen values, $|A - \lambda I| = 0$

$$\Rightarrow \quad -\lambda^3 + 6\lambda^2 - 11\lambda + 6 = 0$$

$$\Rightarrow \quad \lambda^3 - 6\lambda^2 + 11\lambda - 6 = 0$$

Put $\lambda = 1$, $1 - 6 + 11 - 6 = 12 - 12 = 0$, therefore, $\lambda = 1$ is a root.

$$\lambda^3 - 6\lambda^2 + 11\lambda - 6 = (\lambda - 1)(\lambda^2 - 5\lambda + 6) = 0$$

$$\Rightarrow \quad (\lambda-1)(\lambda-2)(\lambda-3) = 0$$

Hence, eigen values are 1, 2, 3

$\therefore$ The diagonal matrix D is given as $\begin{pmatrix} 1 & 0 & 0 \\ 0 & 2 & 0 \\ 0 & 0 & 3 \end{pmatrix}$

(*d*) If $x = \cos\theta + i\sin\theta$, and $y = \cos\phi + i\sin\phi$, then show that $\dfrac{x-y}{x+y} = i\tan\dfrac{\theta-\phi}{2}$.

Solution:

$$\frac{x-y}{x+y} = \frac{(\cos\theta + i\sin\theta) - (\cos\phi + i\sin\phi)}{(\cos\theta + i\sin\theta) + (\cos\phi + i\sin\phi)} = \frac{\cos\theta + i\sin\theta - \cos\phi - i\sin\phi}{\cos\theta + i\sin\theta + \cos\phi + i\sin\phi}$$

$$= \frac{(\cos\theta - \cos\phi) + i(\sin\theta - \sin\phi)}{(\cos\theta + \cos\phi) + i(\sin\theta + \sin\phi)}$$

$$= \frac{\left(-2\sin\frac{\theta+\phi}{2}\sin\frac{\theta-\phi}{2}\right) + i\left(2\cos\frac{\theta+\phi}{2}\sin\frac{\theta-\phi}{2}\right)}{\left(2\cos\frac{\theta+\phi}{2}\cos\frac{\theta-\phi}{2}\right) + i\left(2\sin\frac{\theta+\phi}{2}\cos\frac{\theta-\phi}{2}\right)}$$

$$= \frac{2\sin\frac{\theta-\phi}{2}\left(-\sin\frac{\theta+\phi}{2} + i\cos\frac{\theta+\phi}{2}\right)}{2\cos\frac{\theta-\phi}{2}\left(\cos\frac{\theta+\phi}{2} + i\sin\frac{\theta+\phi}{2}\right)} = \frac{\sin\frac{\vartheta-\phi}{2}\left(i^2\sin\frac{\theta-\phi}{2} + i\cos\frac{\theta+\phi}{2}\right)}{\cos\frac{\theta-\phi}{2}\left(\cos\frac{\theta-\phi}{2} + i\sin\frac{\theta+\phi}{2}\right)}$$

$$\left[\because\ i^2 = -1\right]$$

$$= \frac{i\sin\frac{\theta-\phi}{2}\left(\cos\frac{\theta+\phi}{2} + i\sin\frac{\theta+\phi}{2}\right)}{\cos\frac{\theta-\phi}{2}\left(\cos\frac{\theta+\phi}{2} + i\sin\frac{\theta+\phi}{2}\right)} = i\tan\frac{\theta-\phi}{2}$$

(*e*) Find all the values of $\left(\frac{1}{2} + \frac{\sqrt{3}}{2}i\right)^{3/4}$.

Solution: $\frac{1}{2} + i\frac{\sqrt{3}}{2} = r(\cos\theta + i\sin\theta) = r\cos\theta + ir\sin\theta$

Equating real and imaginary parts, we get

$$r\cos\theta = \frac{1}{2}, \quad r\sin\theta = \frac{\sqrt{3}}{2}$$

Squaring and adding, we get

$$r^2(\cos^2\theta + \sin^2\theta) = \frac{1}{4} + \frac{3}{4} = 1 \qquad \Rightarrow \qquad r = 1$$

Dividing imaginary part by real part, we get

$$\frac{r\sin\theta}{r\cos\theta}=\frac{\sqrt{3}/2}{1/2}=\sqrt{3} \Rightarrow \quad \tan\theta=\sqrt{3} \quad \Rightarrow \quad \theta=\frac{\pi}{3}$$

$$\therefore \quad \frac{1}{2}+i\frac{\sqrt{3}}{2}=\cos\frac{\pi}{3}+i\sin\frac{\pi}{3}$$

$$\Rightarrow \quad \left(\frac{1}{2}+i\frac{\sqrt{3}}{2}\right)^{3/4}=\left(\cos\frac{\pi}{3}+i\sin\frac{\pi}{3}\right)^{3/4}=\left[\cos\left(2k\pi+\frac{\pi}{3}\right)+i\sin\left(2k\pi+\frac{\pi}{3}\right)\right]^{3/4}$$

$$k=0,1,2,3$$

Apply De-Moiver's theorem, we get

$$=\left[\cos\frac{3}{4}\left(2k\pi+\frac{\pi}{3}\right)+i\sin\frac{3}{4}\left(2k\pi+\frac{\pi}{3}\right)\right] \quad k=0,1,2,3$$

$$=\left[\cos\left(\frac{3}{2}k\pi+\frac{\pi}{4}\right)+i\sin\left(\frac{3}{2}k\pi+\frac{\pi}{4}\right)\right] \quad k=0,1,2,3$$

The values are given below:

For $k=0$, value is $\cos\frac{\pi}{4}+i\sin\frac{\pi}{4}=\frac{1}{\sqrt{2}}+i\frac{1}{\sqrt{2}}$

For $k=1$, value is $\cos\frac{7\pi}{4}+i\sin\frac{7\pi}{4}=\cos\left(2\pi-\frac{\pi}{4}\right)+i\sin\left(2\pi-\frac{\pi}{4}\right)$

$$=\cos\frac{\pi}{4}-i\sin\frac{\pi}{4}=\frac{1}{\sqrt{2}}-i\frac{1}{\sqrt{2}}$$

For $k=2$, value is $\cos\frac{13\pi}{4}+i\sin\frac{13\pi}{4}=\cos\left(3\pi+\frac{\pi}{4}\right)+i\sin\left(3\pi+\frac{\pi}{4}\right)$

$$=-\cos\frac{\pi}{4}-i\sin\frac{\pi}{4}=-\frac{1}{\sqrt{2}}-i\frac{1}{\sqrt{2}}$$

For $k=3$, value is $\cos\frac{19\pi}{4}+i\sin\frac{19\pi}{4}=\cos\left(5\pi-\frac{\pi}{4}\right)+i\sin\left(5\pi-\frac{\pi}{4}\right)$

$$=-\cos\frac{\pi}{4}+i\sin\frac{\pi}{4}=-\frac{1}{\sqrt{2}}+i\frac{1}{\sqrt{2}}$$

$\therefore$ The values are $\pm\frac{1}{\sqrt{2}}\pm i\frac{1}{\sqrt{2}}$

(*f*) Examine the conditional convergence of the series $\sum_{n=1}^{\infty}\frac{(-1)^{n-1}}{2n-1}$.

Solution: We know that an alternating series is said to be conditionally convergent if it is convergent but does not converge when all its terms are made positive.

Take $u_n = \frac{1}{2n-1}$ $\qquad u_{n+1} = \frac{1}{2(n+1)-1} = \frac{1}{2n+1}$

(*i*) $u_n - u_{n+1} = \frac{1}{2n-1} - \frac{1}{2n+1} = \frac{2n+1-(2n-1)}{(2n-1)(2n+1)} = \frac{2}{4n^2-1} > 0$

$$\Rightarrow \quad u_n - u_{n+1} > 0 \quad \Rightarrow \quad u_n > u_{n+1}$$

(*ii*) $\underset{n\to\infty}{\text{Lt}}\, u_n = \underset{n\to\infty}{\text{Lt}}\frac{1}{2n-1} = 0$

Hence, all the conditions of Leibnitz's test are satisfied. Therefore, the given series is convergent.

When all the terms of the given series are made positive, we have

$$\sum_{n=1}^{\infty}\left|(-1)^{n-1}\cdot\frac{1}{2n-1}\right| = \sum_{n=1}^{\infty}\frac{1}{2n-1}$$

Let $u_n = \frac{1}{2n-1}$, take $v_n = \frac{1}{n}$, then $\frac{u_n}{v_n} = \frac{n}{2n-1} = \frac{n}{n\left(2-\frac{1}{n}\right)} = \frac{1}{2-\frac{1}{n}}$

$\underset{n\to\infty}{\text{Lt}}\frac{u_n}{v_n} = \underset{n\to\infty}{\text{Lt}}\frac{1}{2-\frac{1}{n}} = \frac{1}{2-0} = \frac{1}{2}$ which is finite

$\therefore$ By comparison test, $\sum_{n=1}^{\infty}u_n$ and $\sum_{n=1}^{\infty}v_n$ converges or diverges together,

But $\sum_{n=1}^{\infty}v_n = \sum_{n=1}^{\infty}\frac{1}{n}$ is *p*-series with $p = 1$, and hence diverges.

$\therefore \quad \sum_{n=1}^{\infty} \frac{(-1)^{n-1}}{2n-1}$ is conditionally convergent.

(g) Test the convergence of the series $\sum_{n=1}^{\infty} \left(1+\frac{1}{n}\right)^{-n^2}$.

Solution: Here $u_n = \left(1+\frac{1}{n}\right)^{-n^2} \Rightarrow (u_n)^{\frac{1}{n}} = \left[\left(1+\frac{1}{n}\right)^{-n^2}\right]^{\frac{1}{n}} = \left(1+\frac{1}{n}\right)^{-n} = \frac{1}{\left(1+\frac{1}{n}\right)^n}$

$$\Rightarrow \quad \underset{n\to\infty}{\text{Lt}} (u_n)^{\frac{1}{n}} = \underset{n\to\infty}{\text{Lt}} \frac{1}{\left(1+\frac{1}{n}\right)^n} = \frac{1}{e} < 1 \qquad \left[\because \underset{x\to\infty}{\text{Lt}} \left(1+\frac{1}{x}\right)^x = e\right]$$

$\therefore$ By Cauchy's Root test, the given series is convergent.

(h) Show that the necessary condition for the differential equation $Mdx + Ndy = 0$ to be exact is $\frac{\partial M}{\partial y} = \frac{\partial N}{\partial x}$.

Solution: The equation $Mdx + Ndy = 0$ will be exact, if $du = Mdx + Ndy$

But $du = \frac{\partial u}{\partial x} dx + \frac{\partial u}{\partial y} dy$

$$\therefore \quad Mdx + Ndy = \frac{\partial u}{\partial x} dx + \frac{\partial u}{\partial y} dy$$

Equating coefficients of dx and dy, we get

$$M = \frac{\partial u}{\partial x} \quad \text{and} \quad N = \frac{\partial u}{\partial y}$$

$$\therefore \quad \frac{\partial M}{\partial y} = \frac{\partial^2 u}{\partial y \partial x} \quad \text{and} \quad \frac{\partial N}{\partial x} = \frac{\partial^2 u}{\partial x \partial y}$$

Also $\dfrac{\partial^2 u}{\partial y \partial x} = \dfrac{\partial^2 u}{\partial x \partial y}$ $\quad\therefore\quad$ $\dfrac{\partial M}{\partial y} = \dfrac{\partial N}{\partial x}$

which is the necessary condition of exactness.

(*i*) Find the particular solution of equation $\dfrac{d^3 y}{dx^3} + 4\dfrac{dy}{dx} = \sin 2x$.

Solution: The characteristic equation is $(D^3 + 4D)y = \sin 2x$

$$P.I. = \frac{\sin 2x}{D^3 + 4D} = \frac{\sin 2x}{D^2 \cdot D + 4D} = \frac{\sin 2x}{-4D + 4D} \qquad \left[\text{Put } D^2 = -(2)^2\right]$$

$$\Rightarrow \quad P.I. = \frac{\sin 2x}{0} \qquad \text{which is a case of failure}$$

Multiply numerator by x and differentiate denominator w.r.t. D, we get

$$P.I. = x\frac{1}{3D^2 + 4}\cos 2x = \frac{x\cos 2x}{-12 + 4} = \frac{x\cos 2x}{-8} \qquad \left[\text{Put } D^2 = -(2)^2\right]$$

Thus, $P.I. = -\dfrac{1}{4}\cos 2x$

(*j*) Solve the equation $e^{2z-1} = 1 + i$.

Solution: Given $e^{2z-1} = 1 + i$

Taking log on both sides, we get

$$2z - 1 = \log(1+i) \quad \Rightarrow \quad 2z = 1 + \log(1+i) \quad \Rightarrow \quad z = \frac{1}{2}[1 + \log(1+i)]$$

$$\Rightarrow z = \frac{1}{2}\left[1 + \frac{1}{2}\log(1+1) + i\tan^{-1}\frac{1}{1}\right] \qquad \left[\because \log(x+iy) = \frac{1}{2}\log\left(x^2+y^2\right) + i\tan^{-1}\frac{y}{x}\right]$$

$$= \frac{1}{2}\left[1 + \frac{1}{2}\log 2 + i\tan^{-1}1\right] = \frac{1}{2}\left[1 + \frac{1}{2}\log 2 + i\tan^{-1}\left(\tan\frac{\pi}{4}\right)\right] \qquad \left[\because \tan\frac{\pi}{4} = 1\right]$$

$$\Rightarrow z = x + iy = \frac{1}{2}\left[1 + \frac{1}{2}\log 2 + i\frac{\pi}{4}\right] = \frac{1}{2} + \frac{1}{4}\log 2 + i\frac{\pi}{8}$$

Equating real and imaginary parts, we get $x = \dfrac{1}{2} + \dfrac{1}{4}\log 2$, $y = \dfrac{\pi}{8}$

SECTION–B

2.(*a*) Obtain the general solution of the differential equation $y''-6y'+9y=e^{3x}/x^2$, by using method of variation of parameters.

Solution: Given differential equation is $y''-6y'+9y=e^{3x}/x^2$

It's A.E. is

$$D^2-6D+9=0 \;\Rightarrow\; (D-3)^2=0 \;\Rightarrow\; D=3,\ 3$$

$$\therefore\ C.F.=(c_1+c_2x)e^{3x}=c_1e^{3x}+c_2xe^{3x}$$

Take $y_1=e^{3x}$ and $y_2=xe^{3x}$; $X=\dfrac{e^{3x}}{x^2}$

$$\therefore\ w=\begin{vmatrix} y_1 & y_2 \\ y_1' & y_2' \end{vmatrix}=\begin{vmatrix} e^{3x} & xe^{3x} \\ 3e^{3x} & (3x+1)e^{3x} \end{vmatrix}$$

$$=(3x+1)e^{6x}-3xe^{6x}=3xe^{6x}+e^{6x}-3xe^{6x}=e^{6x}$$

Let general solution $y=uy_1+vy_2$ where $u=-\int\dfrac{y_2X}{W}dx,\ v=\int\dfrac{y_1X}{W}dx$

$$\text{Now } u=-\int\frac{y_2X}{W}dx=-\int\frac{xe^{3x}\cdot(e^{3x}/x^2)}{e^{6x}}dx=-\int\frac{1}{x}dx=-\log x+c_1$$

$$v=\int\frac{y_1X}{W}dx=\int\frac{e^{3x}\cdot(e^{3x}/x^2)}{e^{6x}}dx=\int\frac{1}{x^2}dx=-\frac{1}{x}+c_2$$

$$\therefore\ y=(-\log x+c_1)e^{3x}+\left(-\frac{1}{x}+c_2\right)xe^{3x}=(-\log x)e^{3x}-e^{3x}+c_1e^{3x}+xc_2e^{3x}$$

(*b*) Find the solution of the differential equation $y''-2y'+y=xe^x\sin x$.

Solution: Given differential equation is $y''-2y'+y=xe^x\sin x$.

It's A.E. is

$$D^2-2D+1=0 \;\Rightarrow\; (D-1)^2=0 \;\Rightarrow\; D=1,\ 1$$

$$\therefore\ C.F.=(c_1+c_2x)e^x$$

Now

$$P.I. = \frac{xe^x \sin x}{D^2 - 2D + 1} = \frac{(x \sin x) \cdot e^x}{(D+1)^2 - 2(D+1) + 1} \quad \left[\begin{array}{l}\text{If } X = e^{ax} \cdot V, \\ \text{Replace } D \text{ by } (D+a)\end{array}\right]$$

$$= e^x \cdot \left(\frac{x \sin x}{D^2 + 2D + 1 - 2D - 2 + 1} \right) = e^x \cdot \left(\frac{x \sin x}{D^2} \right)$$

$$= \frac{e^x}{D} \cdot \frac{1}{D} (x \sin x) = \frac{e^x}{D} \int x \sin x \, dx$$

$$= \frac{e^x}{D} \left[x(-\cos x) - \int 1 \cdot (-\cos x) \, dx \right]$$

$$= \frac{e^x}{D} \left[-x \cos x + \int \cos x \, dx \right] = \frac{e^x}{D} [-x \cos x + \sin x]$$

$$= e^x \int (-x \cos x + \sin x) dx$$

$$= e^x \left[-\left\{ x \cos x - \int 1 \cdot (\sin x) \, dx \right\} + \cos x \right]$$

$$= e^x \left[-\left\{ x \sin x + (-\cos x) \right\} + \cos x \right] = e^x [-x \sin x - \cos x + \cos x]$$

$$= -xe^x \sin x$$

$\therefore \quad C.S. = C.F. + P.I. = (c_1 + c_2 x) e^x - xe^x \sin x$

3.(*a*) Solve the following simultaneous differential equations

$$\frac{dx}{dt} - 2y + 5x = t, \quad \frac{dy}{dt} + 2x + y = 0. \text{ Given that } x(0) = 0, \ y(0) = 0.$$

Solution: Given simultaneous differential equations

$$\frac{dx}{dt} - 2y + 5x = t \qquad \ldots (1)$$

$$\frac{dy}{dt} + 2x + y = 0 \qquad \ldots (2)$$

Characteristic equations

$$Dx - 2y + 5x = t, \quad Dy + 2x + y = 0$$

$$(D+5)x-2y=t \qquad \ldots(3)$$

$$Dy+2x+y=0 \qquad \ldots(4)$$

Multiply (1) by 2 and (2) by $(D+5)$, and then subıract (2) from (1), we get

$$-4y-(D+1)(D+5)y=2t$$

$$\Rightarrow -\left[4-D^2+6D+5\right]y=2t$$

$$\Rightarrow \left(D^2+6D+9\right)y=-2t$$

It's A.E. is

$$D^2+6D+9=0 \quad \Rightarrow \quad (D+3)^2=0 \quad \Rightarrow \quad D=-3,-3$$

$$\therefore \quad C.F.=(c_1+c_2t)e^{-3t}$$

$$P.I.=\frac{-2t}{D^2+6D+9}=\frac{-2t}{(D+3)^2}=-2t(D+3)^{-2}$$

$$=(-2t)\cdot\left(3^{-2}\right)\left(\frac{D}{3}+1\right)^{-2}=-\frac{2}{9}t\left(1+\frac{D}{3}\right)^{-2}$$

$$=-\frac{2}{9}t\left(1-\frac{2D}{3}\right)$$

$$\left[\begin{array}{l}\text{Apply Bionomial theorem}\\ (1+x)^n=1+nx+\ldots\end{array}\right]$$

$$=-\frac{2}{9}t+\frac{4D(t)}{27}=-\frac{2}{9}t+\frac{4}{27}$$

$$\therefore \quad C.S.=C.F.+P.I.=(c_1+c_2t)e^{-3t}-\frac{2}{9}t+\frac{4}{27}$$

$$\therefore \quad y=(c_1+c_2t)e^{-3t}-\frac{2}{9}t+\frac{4}{27}$$

Since $y(0)=0$, putting $t=0$ and $y=0$, we get

$$0=c_1e^0-0+\frac{4}{27} \quad \Rightarrow \quad c_1+\frac{4}{27}=0 \quad \Rightarrow \quad c_1=-\frac{4}{27}$$

$$\therefore \quad y=\left(-\frac{4}{27}+c_2t\right)e^{-3t}-\frac{2}{9}t+\frac{4}{27}$$

$$\frac{dy}{dt}=-\frac{4}{27}e^{-3t}(-3)+c_2\{te^{-3t}(-3)+e^{-3t}.1\}-\frac{2}{9}$$

$$=\frac{4}{9}e^{-3t}+c_2(-3te^{-3t}+e^{-3t})-\frac{2}{9}=\left[\frac{4}{9}+c_2(-3t+1)\right]e^{-3t}-\frac{2}{9}$$

Substituting value of y and $\frac{dy}{dt}$ in (2), we get

$$\left[\frac{4}{9}+c_2(-3t+1)\right]e^{-3t}-\frac{2}{9}+2x+\left(-\frac{4}{27}+c_2t\right)e^{-3t}-\frac{2}{9}t+\frac{4}{27}=0$$

When $t=0$, $x=0$, therefore

$$\frac{4}{9}+c_2-\frac{2}{9}+0+-\frac{4}{27}-0+\frac{4}{27}=0 \quad\Rightarrow\quad c_2=-\frac{2}{9}$$

$$\therefore \quad \boxed{y=-\frac{2}{27}(2+3t)e^{-3t}+\frac{2}{27}(2-3t)}$$

Again from (2), we have

Now $y=-\dfrac{2}{27}(2+3t)e^{-3t}+\dfrac{2}{27}(2-3t)$

$$\frac{dy}{dt}=-\frac{2}{27}\left[(2+3t)e^{-3t}(-3)+e^{-3t}(3)\right]+\frac{2}{27}(-3)=-\frac{2}{27}\left[-3(2+3t)e^{-3t}+3e^{-3t}\right]-\frac{2}{9}$$

$$2x=-\frac{dy}{dt}-y$$

$$=-\left[-\frac{2}{27}\left[-3(2+3t)e^{-3t}+3e^{-3t}\right]-\frac{2}{9}\right]-\left[-\frac{2}{27}(2+3t)e^{-3t}+\frac{2}{27}(2-3t)\right]$$

$$\Rightarrow \quad \boxed{x=-\frac{1}{27}(1+6t)e^{-3t}+\frac{1}{27}(1+3t)}$$

(*b*) Find the complete solution of the differential equation

$$(1+x^2)y''+(1+x)y'+y=2\sin\log(1+x)$$

by using operator method.

Solution: Put $1+x=e^z \Rightarrow z=\log(1+x) \Rightarrow \frac{dz}{dx}=\frac{1}{1+x}$

$$\frac{dy}{dx}=\frac{dy}{dz}\cdot\frac{dz}{dx}=\frac{1}{1+x}\frac{dy}{dz} \Rightarrow (1+x)\frac{dy}{dx}=\frac{dy}{dz}=Dy$$

$$\frac{d^2y}{dx^2}=\frac{d}{dx}\left(\frac{dy}{dx}\right)=\frac{d}{dx}\left(\frac{1}{1+x}\frac{dy}{dz}\right)=\frac{1}{1+x}\frac{d^2y}{dz^2}\cdot\frac{dz}{dx}-\frac{1}{(1+x)^2}\frac{dy}{dz}$$

$$=\frac{1}{1+x}\frac{d^2y}{dz^2}\cdot\frac{1}{1+x}-\frac{1}{(1+x)^2}\frac{dy}{dz}=\frac{1}{(1+x)^2}\frac{d^2y}{dz^2}-\frac{1}{(1+x)^2}\frac{dy}{dz}$$

$$(1+x)^2\frac{d^2y}{dx^2}=\frac{d^2y}{dz^2}-\frac{dy}{dz}=(D^2-D)y=D(D-1)y \qquad \left[D\equiv\frac{d}{dz}\right]$$

Putting the value of $\frac{dy}{dx}, \frac{d^2y}{dx^2}$ in given equation, we get

$$[D(D-1)+D+1]y=2\sin z \quad \Rightarrow \quad (D^2+1)y=2\sin z$$

It's A.E. is $D^2+1=0 \Rightarrow D^2=-1 \Rightarrow D=0\pm i$

$$\therefore \quad C.F.=(c_1\cos z+c_2\sin z)e^{0z}=c_1\cos z+c_2\sin z$$

Now $P.I.=\frac{2\sin z}{D^2+1}=\frac{2\sin z}{-1+1}=\frac{2\sin z}{0} \qquad [D^2=-1]$

It is a case of failure.

$$\therefore \quad P.I.=\frac{2z\sin z}{2D}=\frac{z\sin z}{D}=\int z\sin z\,dz=z\cdot(-\cos z)-\int 1\cdot(-\cos z)dz$$

$$=-z\cos z+\int\cos z\,dz=-z\cos z+\sin z$$

$\therefore$ Complete solution

$$y=C.F.+P.I.=c_1\cos z+c_2\sin z+\sin z-z\cos z$$

Putting $z=\log(1+x)$, we get

$$y=c_1\cos\log(1+x)+c_2\sin\log(1+x)+\sin\log(1+x)-\log(1+x)\cos\log(1+x)$$

4.(*a*) Solve the differential equation $\left(xy^2 - e^{1/x^3}\right)dx - x^2 y dy = 0.$

Solution: Given differential equation $\left(xy^2 - e^{1/x^3}\right)dx - x^2 y dy = 0.$

Comparing it with $Mdx + Ndy = 0,$ we get

$$M = xy^2 - e^{1/x^3}, \quad N = -x^2 y$$

$$\frac{\partial M}{\partial y} = 2xy, \quad \frac{\partial N}{\partial x} = -2xy$$

Since $\frac{\partial M}{\partial y} \neq \frac{\partial N}{\partial x}$, therefore, the given equation is not exact.

We will find integrating factor (*I.F.*) to make it exact.

Now $\frac{\partial N/\partial x - \partial M/\partial y}{N} = \frac{-2xy - 2xy}{-x^2 y} = \frac{-4xy}{-x^2 y} = -\frac{4}{x} = f(x)$ only

$$\therefore \quad I.F. = e^{\int -\frac{4}{x}dx} = e^{-4\log x} = e^{\log x^{-4}} = x^{-4} = \frac{1}{x^4}$$

After multiplying with *I.F.*, given equation becomes

$$\left(\frac{xy^2 - e^{1/x^3}}{x^4}\right)dx - \frac{x^2 y}{x^4}dy = 0 \Rightarrow \left(\frac{y^2}{x^3} - \frac{e^{1/x^3}}{x^4}\right)dx - \frac{y}{x^2}dy = 0 \quad \text{which is exact.}$$

Its solution is

$$\int_{y \text{ constant}} Mdx + \int (\text{terms of } N \text{ not containg } x) dy = c$$

$$\Rightarrow \quad \int \left(\frac{y^2}{x^3} - \frac{e^{1/x^3}}{x^4}\right)dx = c$$

$$\Rightarrow \quad y^2 \int x^{-3}dx - \int \frac{e^{1/x^3}}{x^4}dx = c$$

Put $\frac{1}{x^3} = t \Rightarrow -3x^{-4} = dt$
$\frac{1}{x^4} = -\frac{1}{3}dt$

$$\Rightarrow \quad y^2\left(\frac{x^{-2}}{-2}\right)+\frac{1}{3}\int e^{-t}dt=c$$

$$\Rightarrow \quad y^2\left(\frac{x^{-2}}{-2}\right)+\frac{1}{3}e^{-t}=c$$

$$\Rightarrow \quad -\frac{1}{2}\frac{y^2}{x^2}+\frac{1}{3}e^{\frac{1}{x^3}}=c$$

(*b*) Solve the equation $y=2px+yp^2$ where p has its usual meaning.

Solution: Given differential equation $y=2px+yp^2$

$$\Rightarrow 2px=y-yp^2 \quad \Rightarrow \quad 2x=\frac{y-yp^2}{p} \quad \Rightarrow \quad 2x=\frac{y}{p}-yp$$

Differentiating w. r. t. y, we get

$$2\frac{dx}{dy}=\frac{1}{p}-\frac{y}{p^2}\frac{dp}{dy}-y\frac{dp}{dy}-p \quad \Rightarrow \quad \frac{2}{p}=\frac{1}{p}-\frac{y}{p^2}\frac{dp}{dy}-y\frac{dp}{dy}-p$$

$$\Rightarrow \quad \frac{1}{p}+\frac{y}{p^2}\frac{dp}{dy}+y\frac{dp}{dy}+p=0$$

$$\Rightarrow \quad y\frac{dp}{dy}\left(\frac{1}{p^2}+1\right)+p\left(\frac{1}{p^2}+1\right)=0$$

$$\Rightarrow \quad \left(\frac{1}{p^2}+1\right)\left(y\frac{dp}{dy}+p\right)=0$$

$$\Rightarrow \quad \text{Either } \frac{1}{p^2}+1=0 \quad \text{or} \quad y\frac{dp}{dy}+p=0$$

Now, $\frac{1}{p^2}+1\neq 0$ Rejected

Therefore, $y\frac{dp}{dy}+p=0 \Rightarrow \quad ydp+pdy=0 \quad \Rightarrow \quad d(py)=0$

Integrating both sides, we get

$$py = c \quad \Rightarrow \quad p = \frac{c}{y}$$

Put the value of p in the given equation to get the required solution as

$$y = \frac{2cx}{y} + \frac{c}{y^2}$$

5. An e.m.f. $E\sin pt$ is applied at $t = 0$ to a circuit containing a capacitance C and inductance L. The current i satisfies the equation $L\frac{di}{dt} + \frac{1}{C}\int i\,dt = E\sin pt$. If $p^2 = \frac{1}{LC}$ and initially the current and the charge are zero, then find the current at any time t.

Solution: Given Differential Equation is

$$L\frac{di}{dt} + \frac{1}{C}\int i\,dt = E\sin pt$$

$$\Rightarrow L\frac{d^2q}{dt^2} + \frac{1}{C}q = E\sin pt \qquad \left[\because q = \int i\,dt,\ i = \frac{dq}{dt}\right]$$

In S.F., we have

$$\left(LD^2 + \frac{1}{C}\right)q = E\sin pt$$

Its A.E. is

$$LD^2 + \frac{1}{C} = 0 \quad \Rightarrow \quad D = \pm\frac{1}{\sqrt{LC}} = \pm p, \text{ where } p = \frac{1}{\sqrt{LC}}$$

$$C.F. = c_1\cos pt + c_2\sin pt$$

$$P.I. = \frac{1}{LD^2 + \frac{1}{C}}E\sin pt \qquad \text{Put } D^2 = -p^2 = -\frac{1}{LC}$$

$$= \frac{1}{-Lp^2 + \frac{1}{C}}E\sin pt = \frac{EC\sin pt}{1 - LCp^2} \qquad \text{But } p^2 = \frac{1}{LC}$$

∴ Case of failure.

$$\therefore \quad P.I. = \frac{t}{2LD} E \sin pt \qquad \text{[Multiply by t and differentiate the denominator]}$$

$$= \frac{Et}{2L} \frac{-\cos pt}{p} = -\frac{Et \cos pt}{2Lp}$$

C.S. = C.F. + P.I.

$$q = c_1 \cos pt + c_2 \sin pt - \frac{Et}{2Lp} \cos pt \qquad \ldots (1)$$

When $t = 0,\ q = 0 \quad \therefore \quad c_1 = 0$

$\therefore$ (2) becomes

$$q = c_2 \sin pt - \frac{Et}{2Lp} \cos pt$$

$$i = \frac{dq}{dt} = c_2 p \cos pt - \frac{E}{2Lp}[-tp \sin pt + \cos pt]$$

Now $t = 0,\ i = 0$

$$\therefore \quad 0 = c_2 p - \frac{E}{2Lp} \quad \Rightarrow \quad c_2 = \frac{E}{2Lp^2}$$

$$\therefore \quad i = \frac{E}{2Lp^2} p \cos pt + \frac{Et}{2p} \sin pt - \frac{E}{2Lp} \cos pt$$

$$= \frac{E}{2Lp} \cos pt + \frac{Et}{2p} \sin pt - \frac{E}{2Lp} \cos pt = \frac{Et}{2p} \sin pt$$

$\therefore \quad i = \frac{Et}{2p} \sin pt$ which is the required answer.

SECTION–C

6.(*a*) Use the rank method to test the consistency of the system of equations

$$4x - y = 12; \ -x + 5y - 2z = 0; \ -2y + 4z = -8;$$

if consistent then solve it completely.

Solution: Given system of equation can be written in the matrix form as

$$\begin{pmatrix} 4 & -1 & 0 \\ -1 & 5 & -2 \\ 0 & -2 & 4 \end{pmatrix} \begin{pmatrix} x \\ y \\ z \end{pmatrix} = \begin{pmatrix} 12 \\ 0 \\ -8 \end{pmatrix}, \text{ i.e., } AX = B$$

The augmented matrix may be written as

$$\begin{pmatrix} 4 & -1 & 0 & \vdots 12 \\ -1 & 5 & -2 & \vdots 0 \\ 0 & -2 & 4 & \vdots -8 \end{pmatrix} \qquad R_2 \leftrightarrow R_3$$

$$\begin{pmatrix} 4 & -1 & 0 & \vdots 12 \\ 0 & -2 & 4 & \vdots -8 \\ -1 & 5 & -2 & \vdots 0 \end{pmatrix} \qquad R_1 \to \frac{1}{4} R_1$$

$$\begin{pmatrix} 1 & -1/4 & 0 & \vdots 3 \\ 0 & -2 & 4 & \vdots -8 \\ -1 & 5 & -2 & \vdots 0 \end{pmatrix} \qquad R_3 \to R_3 + R_1$$

$$\begin{pmatrix} 1 & -1/4 & 0 & \vdots 3 \\ 0 & -2 & 4 & \vdots -8 \\ 0 & 19/4 & -2 & \vdots 3 \end{pmatrix} \qquad R_3 \to R_3 + \frac{19}{8} R_2$$

$$\begin{pmatrix} 1 & -1/4 & 0 & \vdots 3 \\ 0 & -2 & 4 & \vdots -8 \\ 0 & 0 & 15/2 & \vdots -16 \end{pmatrix}$$

Since $\rho(A:B) = \rho(A) =$ No. of unknowns, therefore, the system of equations has unique solution.

Now the equivalent system of equation can be written as

$$x - \frac{1}{4} y = 3 \qquad \ldots (1)$$

$$-2y + 4z = -8 \quad \Rightarrow \quad y - 2z = 8 \qquad \ldots (2)$$

$$\frac{15}{2} z = -16 \quad \Rightarrow \quad 15z = -32 \qquad \ldots (3)$$

From (3), we have $z = -\dfrac{32}{15}$

From (2), we have

$$y = 2z + 4 = 2\times\left(-\frac{32}{15}\right) + 4 = -\frac{64}{15} + 4 = \frac{-64+60}{15} = -\frac{4}{15}$$

From (1), we have

$$x = \frac{1}{4}y + 3 = \frac{1}{4}\left(-\frac{4}{15}\right) + 3 = -\frac{1}{15} + 3 = \frac{-1+45}{15} = \frac{44}{15}$$

(*b*) State Cayley-Hamilton theorem. Use it to find the inverse of the matrix

$$\begin{pmatrix} 4 & 3 & 1 \\ 2 & 1 & -2 \\ 1 & 2 & 1 \end{pmatrix}$$

Solution: ***Statement:*** Every square matrix satisfies its own characteristic equation.

$$\text{Let } A = \begin{pmatrix} 4 & 3 & 1 \\ 2 & 1 & -2 \\ 1 & 2 & 1 \end{pmatrix}$$

Characteristic equation is given by $|A - \lambda I| = 0$

$$|A - \lambda I| = \begin{vmatrix} 4-\lambda & 3 & 1 \\ 2 & 1-\lambda & -2 \\ 1 & 2 & 1-\lambda \end{vmatrix} = 0$$

$$\Rightarrow \quad (4-\lambda)\left[(1-\lambda)(1-\lambda)+4\right] - 3\left[2(1-\lambda)+2\right] + 1\left[4-(1-\lambda)\right] = 0$$

$$\Rightarrow \quad (4-\lambda)\left[1-2\lambda+\lambda^2+4\right] - 3[2-2\lambda+2] + 1[4-1+\lambda] = 0$$

$$\Rightarrow \quad (4-\lambda)(\lambda^2-2\lambda+5) - 3(4-2\lambda) + (3+\lambda) = 0$$

$$\Rightarrow \quad 4\lambda^2 - 8\lambda + 20 - \lambda^3 + 2\lambda^2 - 5\lambda - 12 + 6\lambda + 3 + \lambda = 0$$

$$\Rightarrow \quad -\lambda^3 + 6\lambda^2 - 6\lambda + 11 = 0$$

$$\Rightarrow \quad \lambda^3 - 6\lambda^2 + 6\lambda - 11 = 0$$

By Cayley-Hamilton theorem, we have

$$A^3 - 6A^2 + 6A - 11 = 0$$

Multiplying both sides by A^{-1}, we get

$$A^2 - 6A + 6I - 11A^{-1} = 0$$

$$11A^{-1} = A - 6A + 6I$$

$$A^{-1} = \frac{1}{11}\left(A^2 - 6A + 6I\right)$$

New, $A^2 = \begin{pmatrix} 4 & 3 & 1 \\ 2 & 1 & -2 \\ 1 & 2 & 1 \end{pmatrix}\begin{pmatrix} 4 & 3 & 1 \\ 2 & 1 & -2 \\ 1 & 2 & 1 \end{pmatrix} = \begin{pmatrix} 23 & 17 & -1 \\ 8 & 3 & -2 \\ 9 & 7 & -2 \end{pmatrix}$

$$A^2 - 6A - 6I = \begin{pmatrix} 23 & 17 & -1 \\ 8 & 3 & -2 \\ 9 & 7 & -2 \end{pmatrix} - 6\begin{pmatrix} 4 & 3 & 1 \\ 2 & 1 & -2 \\ 1 & 2 & 1 \end{pmatrix} + 6\begin{pmatrix} 1 & 0 & 0 \\ 0 & 1 & 0 \\ 0 & 0 & 1 \end{pmatrix}$$

$$= \begin{pmatrix} 23-24+6 & 17-18+0 & -1-6+0 \\ 8-12+0 & 3-6+6 & -2+12+0 \\ 9-6+0 & 7-12+0 & -2-6+6 \end{pmatrix} = \begin{pmatrix} 5 & -1 & -7 \\ -4 & 3 & 10 \\ 3 & -5 & -2 \end{pmatrix}$$

$$\therefore \quad A^{-1} = \frac{1}{11}\left(A^2 - 6A + 6I\right) = \frac{1}{11}\begin{pmatrix} 5 & -1 & -7 \\ -4 & 3 & 10 \\ 3 & -5 & -2 \end{pmatrix}$$

7.(*a*) Find the eigen values and the corresponding eigen vectors of the matrix $\begin{pmatrix} 1 & 1 & 3 \\ 1 & 5 & 1 \\ 3 & 1 & 1 \end{pmatrix}$.

Solution: Let $A = \begin{pmatrix} 1 & 1 & 3 \\ 1 & 5 & 1 \\ 3 & 1 & 1 \end{pmatrix}$

Characteristic equation is $|A - \lambda I| = 0$

$$\Rightarrow \begin{vmatrix} 1-\lambda & 1 & 3 \\ 1 & 5-\lambda & 1 \\ 3 & 1 & 1-\lambda \end{vmatrix} = 0$$

$$\Rightarrow (1-\lambda)\left[(5-\lambda)(1-\lambda)-1\right]-1[1-\lambda-3]+3\left[1-3(5-\lambda)\right]=0$$

$$\Rightarrow (1-\lambda)\left(5-6\lambda+\lambda^2-1\right)-1(-2-\lambda)+3(1-15+3\lambda)=0$$

$$\Rightarrow (1-\lambda)\left(\lambda^2-6\lambda+4\right)+2+\lambda+3(3\lambda-14)=0$$

$$\Rightarrow \lambda^2-6\lambda+4-\lambda^3+6\lambda^2-4\lambda+2+\lambda+9\lambda-42=0$$

$$\Rightarrow -\lambda^3+7\lambda^2-36=0$$

$$\Rightarrow \lambda^3-7\lambda^2+36=0$$

$$\Rightarrow (\lambda+2)\left(\lambda^2-9\lambda+18\right)=0$$

$$\Rightarrow (\lambda+2)(\lambda-3)(\lambda-6)=0$$

$\therefore \quad \lambda = 3, 6, -2$ are the eigen values.

When $\lambda=-2$, corresponding eigin vector is given by

$$\begin{pmatrix} 3 & 1 & 3 \\ 1 & 7 & 1 \\ 3 & 1 & 3 \end{pmatrix}\begin{pmatrix} x_1 \\ y_1 \\ z_1 \end{pmatrix} = \begin{pmatrix} 0 \\ 0 \\ 0 \end{pmatrix} \quad \Rightarrow \quad \begin{aligned} 3x_1 + y_1 + 3z_1 &= 0 \\ x_1 + 7y_1 + z_1 &= 0 \\ 3x_1 + y_1 + 3z_1 &= 0 \end{aligned}$$

Solving first and second equations, we get

$$\frac{x_1}{-20} = \frac{y_1}{0} = \frac{z_1}{20} = K(say) \qquad \Rightarrow \qquad X_1 = K(-1,0,1)$$

When $\lambda = 3$, corresponding eigen vectors is given by

$$\begin{pmatrix} 2 & 1 & 3 \\ 1 & 2 & 1 \\ 3 & 1 & -2 \end{pmatrix}\begin{pmatrix} x_2 \\ y_2 \\ z_2 \end{pmatrix} = \begin{pmatrix} 0 \\ 0 \\ 0 \end{pmatrix} \quad \Rightarrow \quad \begin{aligned} 2x_2 + y_2 + 3z_2 &= 0 \\ x_2 + 2y_2 + z_2 &= 0 \\ 3x_2 + y_2 - 2z_2 &= 0 \end{aligned}$$

Solving second and third equations, we get

$$\frac{x_2}{-5} = \frac{y_2}{+5} = \frac{z_2}{-5} = K\,(say) \quad \Rightarrow \quad X_2 = K(-1,1,-1)$$

When $\lambda = 6$, corresponding eigen vector is given by

$$\begin{pmatrix} -5 & 1 & 3 \\ 1 & -1 & 1 \\ 3 & 1 & -5 \end{pmatrix}\begin{pmatrix} x_3 \\ y_3 \\ z_3 \end{pmatrix} = \begin{pmatrix} 0 \\ 0 \\ 0 \end{pmatrix} \quad \Rightarrow \quad \begin{matrix} -5x_3 + y_3 + 3z_3 = 0 \\ x_3 - y_3 + z_3 = 0 \\ 3x_3 + y_3 - 5z_3 = 0 \end{matrix}$$

Solving first and second equation, we get

$$\frac{x_3}{4} = \frac{y_3}{8} = \frac{z_3}{4} \quad \Rightarrow \quad \frac{x_3}{1} = \frac{y_3}{2} = \frac{z_3}{1} = K\,(say) \quad \Rightarrow \quad X_3 = K(1,2,1)$$

Thus, eigen vectors are given by

$$X_1 = K(-1,0,1),\; X_2 = K(-1,1,-1),\; X_3 = k(1,2,1)$$

7. (*b*) Prove that eigen values of a skew hermitian matrix are either zero or purely imaginary.

Solution: Denote the matrix by A.

Since A is skew hermitian, therefore, iA is hermitian

Let λ be the eigen value of A.

$\therefore$ there exists a non-zero vector X such that

$$AX = \lambda X \quad \Rightarrow \quad (iA)X = (i\lambda)X$$

$\Rightarrow$ $i\lambda$ is an eigen value of iA, which is hermitian. Therefore, $i\lambda$ is real.

Hence, λ must be either zero or purely imaginary.

8.(*a*) Test for values of x the series $\sum_{n=1}^{\infty} \frac{\sqrt{n}}{\sqrt{n^2+1}} x^n$, $x > 0$ converges/diverges.

Solution: $u_n = \frac{\sqrt{n}}{\sqrt{n^2+1}}.x^n, \quad u_{n+1} \frac{\sqrt{n+1}}{\sqrt{(n+1)^2+1}} x^{n+1}$

$$\frac{u_n}{u_{n+1}} = \frac{\sqrt{n}}{\sqrt{n^2+1}}.x^n \times \frac{\sqrt{(n+1)^2+1}}{\sqrt{n+1}} \cdot \frac{1}{x^{n+1}}$$

$$= \frac{\sqrt{n}}{n\sqrt{1+\frac{1}{n^2}}} \cdot \frac{n\sqrt{\left(1+\frac{1}{n}\right)^{2+\frac{1}{n}}}}{\sqrt{n}\sqrt{1+\frac{1}{n}}} \cdot \frac{1}{x} = \frac{\sqrt{\left(1+\frac{1}{n}\right)^2+\frac{1}{n}}}{\sqrt{1+\frac{1}{n^2}}\sqrt{1+\frac{1}{n}}} \cdot \frac{1}{x}$$

$$\underset{n\to\infty}{\text{Lt}} \frac{u_n}{u_{n+1}} = \underset{n\to\infty}{\text{Lt}} \frac{\sqrt{\left(1+\frac{1}{n}\right)^2+\frac{1}{n}}}{\sqrt{1+\frac{1}{n^2}}\sqrt{1+\frac{1}{n}}} \cdot \frac{1}{x} = \frac{\sqrt{(1+0)^2+0}}{\sqrt{1+0}\sqrt{1+0}} \cdot \frac{1}{x} = \frac{1}{x}$$

$\therefore$ By D' Alembert's Ratio test, the given series $\sum_{n=1}^{\infty} u_n$ is convergent for $\frac{1}{x} > 1$, i.e., $x < 1$; and divergent for $\frac{1}{x} < 1$, i.e., for $x > 1$. For $\frac{1}{x} = 1$, i.e., $x = 1$, this test fails.

Applying comparison test for $x = 1$, u_n becomes

$$u_n = \frac{\sqrt{n}}{\sqrt{n^2+1}} = \frac{\sqrt{n}}{n\sqrt{1+\frac{1}{n^2}}} \frac{1}{\sqrt{n}\sqrt{1+\frac{1}{n^2}}}$$

Let $v_n = \frac{1}{\sqrt{n}}$ such that $\frac{u_n}{v_n} = \frac{\frac{1}{\sqrt{n}\sqrt{1+\frac{1}{n^2}}}}{\frac{1}{\sqrt{n}}} = \frac{1}{\sqrt{1+\frac{1}{n^2}}}$

$\underset{n\to\infty}{\text{Lt}} \frac{u_n}{v_n} = \frac{1}{\sqrt{1+0}} = 1$, which is finite and non-zero.

Therefore, $\sum u_n$ and $\sum v_n$ converge or diverge together. Now $\sum v_n = \frac{1}{\sqrt{n}} = \frac{1}{n^{\frac{1}{2}}}$, which is p-series with $p = \frac{1}{2} < 1$ and hence divergent.

Therefore, the given series convergent for $x < 1$ and diverges for $x \geq 1$.

(b) Test convergence/diverge of the series $\sum_{n=1}^{\infty}\left[\sqrt{(n^4+1)}-\sqrt{(n^4-1)}\right]$.

Solution: Let $u_n=\sqrt{n^4+1}-\sqrt{n^4-1}$

Rationalizing u_n, we get

$$u_n=\frac{\sqrt{n^4+1}-\sqrt{n^4-1}}{\sqrt{n^4+1}+\sqrt{n^4-1}}\times\sqrt{n^4+1}+\sqrt{n^4-1}$$

$$=\frac{\left(\sqrt{n^4+1}\right)^2-\left(\sqrt{n^4-1}\right)^2}{\sqrt{n^4+1}+\sqrt{n^4-1}}=\frac{n^4+1-n^4+1}{\sqrt{n^4+1}+\sqrt{n^4-1}}$$

$$=\frac{2}{\sqrt{n^4+1}+\sqrt{n^4-1}}=\frac{2}{n^2\left(\frac{1}{\sqrt{1+\frac{1}{n^4}}}+\frac{1}{\sqrt{1-\frac{1}{n^4}}}\right)}$$

Let $v_n=\frac{1}{n^2}$ such that $\frac{u_n}{v_n}=\frac{\dfrac{2}{n^2\left(\frac{1}{\sqrt{1+\frac{1}{n^u}}}+\frac{1}{\sqrt{1-\frac{1}{n^v}}}\right)}}{\dfrac{1}{n^2}}=\frac{2}{\left(\frac{1}{\sqrt{1+\frac{1}{n^u}}}+\frac{1}{\sqrt{1-\frac{1}{n^v}}}\right)}$

$$\Rightarrow\ \underset{n\to\infty}{\text{Lt}}\frac{u_n}{v_n}=\underset{n\to\infty}{\text{Lt}}\frac{2}{\sqrt{1+\frac{1}{n^4}}+\sqrt{1-\frac{1}{n^4}}}=\frac{2}{\sqrt{1+0}+\sqrt{1-0}}=\frac{2}{2}=1$$

Thus, $\underset{n\to\infty}{\text{Lt}}\frac{u_n}{v_n}=1$, finite and non-zero.

Therefore, by comparison test $\sum u_n$ and $\sum v_n$ converge or divege together.

Now, $\sum v_n=\sum\frac{1}{n^2}$ is a *p*-series with $p=2>1$ and hence converges.

9.(*a*) Use De Moiver's theorem to solve the equation $(z-1)^5+z^5=0$.

Solution: The given equation is $(z-1)^5+z^5=0$

$$\Rightarrow (z-1)^5=-z^5 \quad \Rightarrow \quad \frac{(z-1)^5}{z^5}=-1=\cos\pi+i\sin\pi$$

$$\Rightarrow \left(\frac{z-1}{z}\right)^5=\cos\pi+i\sin\pi \quad \Rightarrow \frac{z-1}{z}=(\cos\pi+i\sin\pi)^{\frac{1}{5}}$$

Adding $2k\pi$ in the angles of both cosine and $\sin e$, we get

$$1-\frac{1}{z}=\left[\cos(2k\pi+\pi)+i\sin(2k\pi+\pi)\right]^{\frac{1}{5}}, \quad k=0,1,2,3,4$$

Apply De Moiver's theorem, we get

$$1-\frac{1}{z}=\cos\frac{(2k+1)}{5}\pi+i\sin\frac{(2k+1)}{5}\pi, \quad k=0,1,2,3,4.$$

Putting $\left(\frac{2k+1}{5}\right)\pi=\theta$, we get

$$1-\frac{1}{z}=\cos\theta+i\sin\theta \quad \Rightarrow \quad 1-\cos\theta+i\sin\theta=\frac{1}{z}$$

$$\Rightarrow z=\frac{1}{1-\cos\theta-i\sin\theta}=\frac{1}{2\sin^2\frac{\theta}{2}-i2\sin\frac{\theta}{2}\cos\frac{\theta}{2}}$$

$$\Rightarrow z=\frac{1}{2\sin\frac{\theta}{2}\left(\sin\frac{\theta}{2}-i\cos\frac{\theta}{2}\right)}=\frac{i}{2\sin\frac{\theta}{2}\left(\cos\frac{\theta}{2}+i\sin\frac{\theta}{2}\right)}$$

$$\Rightarrow z=\frac{i}{2\sin\frac{\theta}{2}}\left(\cos\frac{\theta}{2}+i\sin\frac{\theta}{2}\right)^{-1}=\frac{i}{2\sin\frac{\theta}{2}}\left(\cos\frac{\theta}{2}-i\sin\frac{\theta}{2}\right)^{-1}$$

$$\Rightarrow z=\frac{1}{2}\left(i\cot\frac{\theta}{2}+1\right)=\frac{1}{2}\left[i\cot(2K+1)\frac{\pi}{10}+1\right], K=0,1,2,3,4$$

9. (*b*) Separate $\sin^{-1}\left(e^{i\theta}\right)$ into real and imaginary parts, where θ is a +ve acute angle.

Solution: Let $z = \sin^{-1}\left(e^{i\theta}\right)$ where $z = x + iy$

$$\Rightarrow \quad \sin z = e^{i\theta} = \cos\theta + i\sin\theta \qquad \Rightarrow \qquad \sin(x+iy) = \cos\theta + i\sin\theta$$

$$\Rightarrow \quad \sin x\cos(iy) + \cos x\sin(iy) = \cos\theta + i\sin\theta$$

$$\Rightarrow \quad \sin x\cosh y + i\cos x\sinh y = \cos\theta + i\sin\theta$$

Equating real and imaginary parts, we get

$$\sin x\cosh y = \cos\theta \quad \text{and} \quad \cos x\sinh y = \sin\theta$$

$$\cosh y = \frac{\cos\theta}{\sin x} \quad \text{and} \quad \sinh y = \frac{\sin\theta}{\cos x}$$

Now, $\cosh^2 y - \sinh^2 y = 1 \quad \Rightarrow \quad \dfrac{\cos^2\theta}{\sin^2 x} - \dfrac{\sin^2\theta}{\cos^2 x} = 1$

$$\Rightarrow \quad \frac{\cos^2\theta\cos^2 x - \sin^2\theta\sin^2 x}{\sin^2 x\cos^2 x} = 1 \quad \Rightarrow \quad \cos^2\theta\cos^2 x - \sin^2\theta\sin^2 x = \sin^2 x\cos^2 x$$

$$\Rightarrow \quad \left(1 - \sin^2\theta\right)\cos^2 x - \sin^2\theta\left(1 - \cos^2 x\right) = \sin^2 x\cos^2 x$$

$$\cos^2 x - \sin^2\theta\cos^2 x - \sin^2\theta + \sin^2\theta\cos^2 x = (1 - \cos^2 x)\cos^2 x$$

$$\cos^2 x - \sin^2\theta = \cos^2 x - \cos^4 x \qquad \Rightarrow \qquad \sin^2\theta = \cos^4 x$$

$$\Rightarrow \quad \cos^2 x = \sin\theta \qquad \Rightarrow \quad \cos x = \sqrt{\sin\theta} \qquad \Rightarrow \quad \boxed{x = \cos^{-1}\sqrt{\sin\theta}}$$

$$\sinh y = \frac{\sin\theta}{\cos x} = \frac{\sin\theta}{\sqrt{\sin\theta}} = \sqrt{\sin\theta} \quad \Rightarrow \quad \boxed{y = \sinh^{-1}\sqrt{\sin\theta}}$$

❑ ❑ ❑

Index

E, F

G

H

I

J, K

L

M

N

O

P

Q, R

S

T

U

V, W, Z

❑ ❑ ❑